Springer Series in **Materials Science** 14

Edited by Manuel Cardona

Springer Series in Materials Science

Volume 1 **Chemical Processing with Lasers**
By D. Bäuerle

Volume 2 **Laser-Beam Interactions with Materials**
Physical Principles and Applications
By M. von Allmen

Volume 3 **Laser Processing of Thin Films and Microstructures**
Oxidation, Deposition and Etching of Insulators
By I. W. Boyd

Volume 4 **Microclusters**
Editors: S. Sugano, Y. Nishina, and S. Ohnishi

Volume 5 **Graphite Fibers and Filaments**
By M. S. Dresselhaus, G. Dresselhaus,
K. Sugihara, I. L. Spain, and H. A. Goldberg

Volume 6 **Elemental and Molecular Clusters**
Editors: G. Benedek, T. P. Martin, and G. Pacchioni

Volume 7 **Molecular Beam Epitaxy** Fundamentals and Current Status
By M. A. Herman and H. Sitter

Volume 8 **Physical Chemistry of, in and on Silicon**
By G. F. Cerofolini and L. Meda

Volume 9 **Tritium and Helium-3 in Metals**
By R. Lässer

Volume 10 **Computer Simulation of Ion–Solid Interactions**
By W. Eckstein

Volume 11 **Mechanisms of High Temperature Superconductivity**
Editors: H. Kamimura and A. Oshiyama

Volume 12 **Laser Technology in Microelectronics**
Editors: S. Metev and V. P. Veiko

Volume 13 **Semiconductor Silicon**
Materials Science and Technology
Editors: G. C. Harbeke and M. J. Schulz

Volume 14 **Graphite Intercalation Compounds I**
Structure and Dynamics Editors: H. Zabel and S. A. Solin

Volume 15 **Crystal Chemistry of High T_c Superconducting Copper Oxides**
By B. Raveau, C. Michel, M. Hervieu and D. Groult

H. Zabel S.A. Solin (Eds.)

Graphite Intercalation Compounds I

Structure and Dynamics

With Contributions by
D.M. Hwang G. Kirczenow P. Lagrange
A. Magerl R. Moret S.C. Moss
R. Setton S.A. Solin H. Zabel

With 180 Figures

Springer-Verlag Berlin Heidelberg New York
London Paris Tokyo Hong Kong Barcelona

Professor Dr. Hartmut Zabel

Loomis Laboratory of Physics, University of Illinois at Urbana-Champaign,
1110 West Green Street, Urbana, IL 61801, USA

Current address: Ruhr-Universität Bochum, Fakultät für Physik und Astronomie,
Universitätsstrasse 150, D-4630 Bochum 1, Fed. Rep. of Germany

Stuart Solin, Ph. D.

NEC Research Institute, Inc.
4 Independence Way, Princeton, NJ 08540, USA

Guest Editor: Professor Dr. Manuel Cardona

Max-Planck-Institut für Festkörperforschung, Heisenbergstrasse 1
D-7000 Stuttgart 80, Fed. Rep. of Germany

ISBN-13:978-3-642-75272-8 e-ISBN-13:978-3-642-75270-4
DOI: 10.1007/978-3-642-75270-4

Library of Congress Cataloging-in-Publication Data. Graphite intercalation compounds I : structure and dynamics / H. Zabel, S. A. Solin [editors]. p. cm. – (Springer series in materials science ; v. 14) Includes bibliographical references. 1. Clathrate compounds. 2. Graphite. I. Zabel, H. (Hartmut), 1946– . II. Solin, S. A. (Stuart A.), 1942– . III. Series. QD474.G74 1990 541.2'2–dc20 89-26292

2154/3150(3011)-543210 – Printed on acid-free paper

Preface

The progress of materials science depends on the development of novel materials and the development of novel experimental techniques. The research on graphite intercalation compounds combines both aspects: new compounds with strikingly new and anisotropic properties have been synthesized and analyzed during the past couple of years by means of state-of-the-art experimental methods. At the same time, the preparation of the compounds already known has improved considerably, giving increased reliability and reproducibility of the experimental results. The high quality experimental data now available have stimulated theoretical work. Moreover, the theoretical work has had a great impact on further experimental studies, with the effect of a much improved understanding of this class of materials.

This volume is dedicated to a thorough description of all relevant experimental and theoretical aspects of the structural and dynamical properties of graphite intercalation compounds. Because of the large number of topics, a second volume, which is now in preparation, will follow and will treat the electronic, transport, magnetic, and optical properties. The second volume will also contain a chapter on applications of graphite intercalation compounds. There have been a number of reviews written on selected aspects of these compounds in various journals and conference proceedings during the last couple of years, but this is the first comprehensive review since the thorough overview provided by M.S. Dresselhaus and G. Dresselhaus appeared ten years ago.

There is another aspect to graphite intercalation compounds which has appeared in recent years. Because of the considerably improved understanding of these materials, graphite intercalation compounds have been used more and more as a testing ground for physical phenomena in reduced dimensions, including two-dimensional commensurate and incommensurate phase transitions, cross-over effects, layer lattice dynamics, melting in two dimensions on a periodic substrate, the kinetics of domain formation, etc., to mention just a few of them. These new developments are fully recognized in this volume.

Because of the many different facets of graphite intercalation compounds, many scientists from various disciplines, such as physics, materials science, physical chemistry, and electrical engineering, are contributing to this area of research. It is the intention of the editors to provide the active researchers in those fields with an expert and up-to-date overview, to discuss open questions and to indicate new research opportunities. To this end, we hope that newcomers to this fascinating field of layered materials will find this volume a valuable introduction and source book for further studies.

The editors are grateful to the contributors to this volume for their efforts to provide up-to-date reviews and their cooperation in all details. The editors are also grateful to the publishers and their staff, in particular to the managing editor Dr. H.K.V. Lotsch, for their encouragement and patience.

Urbana, Princeton
February 1990

H. Zabel
S.A. Solin

Contents

Contributors

Hwang, Dah-Min
Bell Communications Research, 331 Newman Springs Road, NVC 3X-171, Red Bank, NJ 07701-7040, USA

Kirczenow, George
Department of Physics, Simon Fraser University, Burnaby, British Columbia, Canada V5A 1S6

Lagrange, Philippe
Laboratoire de Chimie Minérale Appliquée, Université de Nancy I, BP 239, F-54506 Vandœuvre les Nancy Cedex, France

Magerl, Andreas
Institut Laue-Langevin, Avenue des Martyrs, 156X, F-38042 Grenoble Cedex, France

Moret, Roger
Department of Physics, University of Illinois, 1110 West Green Street, Urbana, IL 61801, USA

Moss, Simon C.
Department of Physics, University of Houston, University Park, Houston, TX 77004, USA

Setton, Ralph
Centre de Recherche sur les Solides à Organisation Cristalline Imparfaite, 1b, Rue de la Férollerie, F-45071 Orléans Cedex 2, France

Solin, Stuart A.
NEC Research Institute, Inc., 4 Independence Way, Princeton, NJ 08540, USA

Zabel, Hartmut
Loomis Laboratory of Physics, University of Illinois at Urbana-Champaign, 1110 West Green Street, Urbana, IL 61801, USA
Current address:
Ruhr-Universität Bochum, Fakultät für Physik und Astronomie, Universitätsstr. 150, D-4630 Bochum 1, Fed. Rep. of Germany

1. Introduction

By Hartmut Zabel and Stuart A. Solin

The field of graphite intercalation compounds (GICs) has been witnessing a tremendous worldwide growth in research activity in recent years. Many of the open questions which were of pressing urgency ten years ago have meanwhile been dealt with and in many cases satisfactory answers have been found. These were questions concerning the origin of staging, the kinetics of stage formation and transitions, the complex structure of modulated intercalate layers in higher stage compounds, the vibrational excitations and the diffusivity of the intercalates, just to name a few. In all of these areas, major theoretical and experimental advances have been made and documented in part by a sequence of conference proceedings over the last couple of years. A number of larger or smaller review articles have also appeared during this period. The first international symposium on GICs was held in La Napoule, France (1977) [1.1], followed by symposia at Cape Code, USA (1981) [1.2], Pont-à-Mousson, France (1983) [1.3], Tsukuba, Japan (1985) [1.4], Jerusalem, Israel (1987) [1.6], and Berlin, Germany (1989) [1.6]. Further conferences on the properties of layered solids including GICs were held at Nijmegen, The Netherlands (1979) [1.7], Trieste, Italy (1982) [1.8] and Ericc, Italy (1986) [1.9]. Meetings on the topic of graphite intercalation compounds have been held at the biannual Carbon Conferences and the Materials Research Society Fall Meetings in Boston since 1982.

The time is, therefore, ripe for a compilation of in-depth expert reviews of the various topics on GICs. Such a compilation will clarify the immense progress made in the GIC field to date and provide a signpost for future research in the area. Accordingly, we have prepared these two volumes of comprehensive reviews of GICs. The present volume addresses the structures and dynamics of graphite intercalation compounds, while the other volume concentrates on electronic properties and applications of graphite intercalation compounds.

Some of the terminology of graphite intercalation compounds has developed into an expert jargon which is used throughout these two volumes. We will, therefore, provide a few definitions here, which will hopefully make it easier for newcomers to the field to focus in the intriguing properties of these compounds.

For experimental studies on GICs, two main types of host graphite material are currently used: naturally occurring graphite and synthetically produced graphite. Small single crystals for X-ray structural work are readily available from coal mines in New York and Madagascar. A similar form of single crystal is found in blast furnaces for steel production. These single crystal graphite flakes, which are deposited on the walls of the furnaces, are called Kish graphite, from the German word *Kiesel*, which means gravel. Kish graphite was first noticed by *Schafhäutl* in 1841 [1.10], who also accidentally prepared the first GIC while analyzing the crystal flakes in a solution of sulfuric acid. Natural crystals as well as the Kish graphite usually contain a number of single crystalline domains, which need to be separated out before starting single crystal structural work.

The most commonly used host material in GIC research is a synthetic form of graphite, called highly oriented pyrolitic graphite (HOPG). This is a textured polycrystal with a well-defined c-axis and random a- and b-axis orientation. The best material available has a c-axis mosaic distribution of about 0.2° and structural coherency lengths of 5 000 to 10 000 Å parallel and perpendicular, respectively, to the c-axis. Another form of synthetic graphite is graphite fiber material, which comes in various structural forms and which is renowned for its mechanical strength. The intercalation compounds derived from graphite fiber materials are reviewed by M.S. Dresselhaus in the other volume of this compilation.

The GICs are categorized according to their charge transfer, their stage, and the number of different atomic or molecular species within the host galleries. GICs in which the intercalates donate electrons to the host graphite layers are called donor compounds, while those which extract electrons from the host are labeled as acceptor compounds. In the chemical nomenclature, the anion is usually listed first. Fot instance, KC_{24} is a donor compound, but $C_{4.8}CoCl_2$ is an acceptor compound. By the stage of a compound, we understand the number of graphite planes between any successive intercalate layers. Therefore, a stage n compound consists of a periodic sequence of an intercalate layer followed by n graphite layers. An alternative definition of staging is based on the ratio of the total number of graphite gallery spacings to those which are filled (intercalated) per repeat unit. This definition has the advantage that fractional stages can also be taken into account. For instance, a periodic sequence of two filled galleries and one empty corresponds to a stage $3/2$. Finally, we distinguish between binary and ternary compounds. Binary compounds are those which contain only one type of atomic or molecular species within the intercalate layer. For instance, KC_{24} and $C_{4.8}CoCl_2$ are both stage 2 binary compounds. The mixture of two donor, two acceptor, or donor and acceptor species leads to ternary compounds. For instance, $K_{1-x}Rb_xC_8$ is a ternary donor compound. Clearly, this definition can be extended to quarternary and subsequent compounds.

The graphite planes in graphite intercalation compounds are often referred to as graphite basal planes. Recently, "graphene layer" has been recommended for designating the graphite basal planes, and this is gradually becoming the accepted wording.

This volume contains nine chapters, including this introduction. The second chapter by *Moss* and *Moret* introduces the main structural concepts of graphite intercalation compounds, including stacking and staging arrangements, and in particular describes the structural phase transitions that occur in binary donor and acceptor compounds. The third chapter, by *Kirczenow*, reviews theory and experiments that answer questions about the origin of staging, the stage stability range, and staging kinetics during the intercalation process. Chapters 4 and 5 deal with the lattice dynamics of graphite intercalation compounds, subdivided into studies mainly carried out via inelastic neutron scattering, reviewed by *Zabel*, and lattice dynamical information derived from optical probes such as infrared and Raman scattering, which is reviewed by *Solin*. Chapter 6 by *Magerl* describes high resolution neutron scattering studies of the translational diffusion of intercalated atoms and rotational diffusion of molecules in graphite. Chapter 7 by *Hwang* summarizes structural investigations via microprobes on a microscopic scale, such as domain structures and stacking arrangements. Chapter 8 by *Lagrange* and *Setton* tackles new horizons, in the form of newly developed donor intercalation compounds, and Chapter 9, by *Setton*, provides an overview of the chemical and synthesis properties of ternary compounds. The chapter focuses mainly on the synthesis and chemistry of ternary compounds and is thus complementary to a recent review paper on the physics of ternary graphite intercalation compounds [1.11].

References

1.1 Proceedings of the Franco-American Conference of La Napoule, France, 1977, Mater. Sci. Eng. **31** (1977)

1.2 F.L. Vogel, A. Hérold (eds.): Proc. of the Second Int. Conf. on Intercalation Compounds of Graphite, Cape Code, 1980, Synth. Met. **2** (1980)

1.3 A. Hérold, D. Guérard, P. Lagrange (eds.): Proc. of the Third Int. Conf. on Graphite Intercalation Compounds, Pont-à-Mousson, France, 1983, Synth. Met. **7** (1983)

1.4 K. Nakao, S.A. Solin (eds.): Proc. of the Int. Symp. on Graphite Intercalation Compounds, Tsukuba, Japan, 1985, Synth. Met. **12** (1985)

1.5 D. Davidov, H. Selig (eds.): Proc. of the Fourth Int. Symp. on Graphite Intercalation Compounds, Jerusalem, Israel, 1987, Synth. Met. **23** (1988)

1.6 K. Lüders, R. Schöllhorn (eds.): Proc. of the Fifth Int. Symp. on Graphite Intercalation Compounds, Berlin, Germany, 1989, to be published

1.7 C.F. van Bruggen, C. Haas, H.W. Myron (eds.): Proc. of the Int. Conf. on Layered Materials and Intercalates, University of Nijmegen, The Netherlands, 1979, Physica **99B** (1980)

1.8 L. Pietronero, E. Tosatti (eds.): *Physics of Intercalation Compounds*, Springer Ser. Solid-State Sci., Vol. 38 (Springer, Berlin, Heidelberg 1981)

1.9 M.S. Dresselhaus (ed.): *Intercalation in Layered Materials*, Summer School at Erice, Italy, 1986, NATO ASI Series, Ser. B: Physics, Vol. 148 (Plenum, New York 1986)

1.10 C. Schafhäutl: J. Prakt. Chem. **21**, 129 (1841)

1.11 S.A. Solin, H. Zabel: Adv. Phys. **37**, 87–254 (1988)

2. Structural Properties and Phase Transitions

Simon C. Moss and Roger Moret

With 35 Figures

In this chapter we review the subject of structure and phase transitions in the graphite intercalation compounds (GICs). Results on both donor and acceptor compounds are covered, for both liquid and solid states, and a range of structural phenomena are treated. This discussion of structure serves two purposes. Firstly, we are interested in structure per se and the GICs offer an extraordinary abundance of interesting phenomena. This abundance comes about through the interplay between intercalant–intercalant forces and graphite–intercalant forces, which ultimately are responsible for the intercalation process. Depending in detail upon these competing interactions, which are naturally sensitive to the choice of intercalant, a great variety of structural effects appear, ranging from altered host stackings to Daumas–Herold domains and staging and including host-modulated two-dimensional (2d) incommensurate liquids which freeze to incommensurate solids for which domain-discommensuration states have been derived. In fact, it may be said that the GICs, by providing a sequential stacking of intercalant layers, have substantially enhanced our understanding of structure and phase transitions in two dimensions.

The second purpose served by a detailed discussion of structure is in the observation that properties depend ultimately upon structure; that, if we wish to understand the detailed electronic, magnetic and optical properties of GICs, we must know the structure. To this truism we should add that many properties depend only upon grosser structural aspects such as stage index, in-plane density and charge transfer, while others are more sensitive to fine details. This is just as well, because it is only recently that we have been able to formulate a more complete picture of the structure of GICs.

Although X-ray and neutron scattering have both been crucial to the elucidation of the structure, we will confine our attention mainly to X-ray results. Neutron scattering has been of prime importance for understanding the dynamics of GICs and for studying both intra- and interplanar forces. It has also been used to reveal magnetic structures and phase transitions in acceptor compounds. Neither of these topics, however, is covered in this review although we do occasionally make reference to some of the dynamical results. Electron microscopy and diffraction are not treated here either, except in passing, as they are dealt with elsewhere in this volume.

There are several reviews [2.1–4] and conference proceedings (published by the MRS or in *Synthetic Metals* and referenced here through individual papers) that provide a good deal of structural background for our discussion of GICs.

Recently, one of us reviewed structure and phase transitions in some detail [2.5] and *Zabel* and *Chow* [2.6] have discussed the test character of GICs in various structural, dynamical and magnetic studies. Our purpose here, then, is to bring the subject up to date and to present what we believe are some reasonably sound aspects of the structures and phase transitions in these materials, aspects that have evolved through the combination of X-ray scattering, theory and modeling. Because structure includes deviation from perfection, we will also be concerned with the disorder induced in the graphite upon intercalation, along with that found in the starting material. The phenomenon of staging, whereby the graphite host is filled by a regular sequence of intercalant layers each separated by n graphite layers for stage n, is reviewed in Chap. 3 of this volume, by *Kirczenow*. Stage disorder, as elucidated by X-ray scattering, is also discussed in Chap. 3 but will be briefly treated here as well.

It is clear that the acceptor GICs are more varied in the structures they assume than are the donor alkali GICs. We have therefore grouped the alkalis together in presenting their structures and transitions. Aside, however, from some general remarks about incommensurability in the liquid and solid states and the relevance of the pristine acceptor crystal structures, our discussion of the molecular acceptor species treats them more or less individually.

2.1 Effects of Intercalation on the Graphite Host

2.1.1 Pure Graphite

The graphite structure shown in Fig. 2.1 consists of hexagonal net planes of carbon stacked along the c-axis in a staggered array usually denoted as $ABAB\ldots$, where the lateral shift, on going from layer A to layer B, is $\boldsymbol{\Delta}_{AB} = (1/3)\boldsymbol{a}_1 + (2/3)\boldsymbol{a}_2, |\boldsymbol{a}_1| = |\boldsymbol{a}_2| = a_G$. While this is the normal graphite stacking, $ABCA\ldots$ or $ABAC\ldots$, etc. are possible sequences (the main requirement is that a layer be followed by one shifted by $\pm\boldsymbol{\Delta}_{AB}$) and the commonly observed stacking faults in graphite are a reflection of these discrete stacking alternatives.

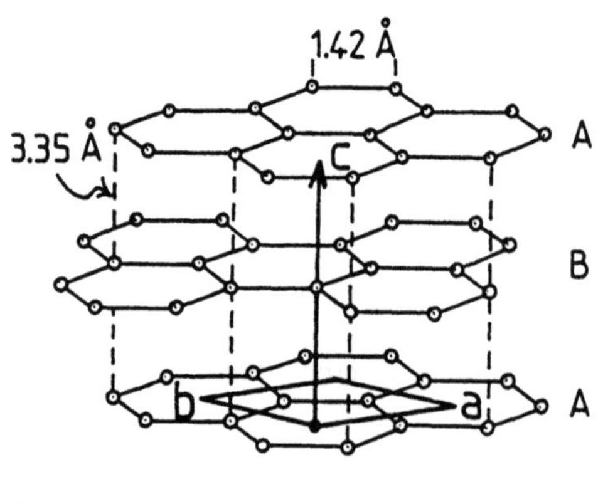

Fig. 2.1. The structure of hexagonal graphite showing the $ABAB$ stacking of honeycomb carbon layers (space group $P6_3/mmc$, a_G = 2.46 Å= b_G, c_G = 6.70 Å)

The graphite samples discussed here are of two varieties. For single crystals, both natural samples, found in coal or lead mines, and synthetic (Kish) crystals of various degrees of perfection are available. Choosing a good crystal from a batch generally involves many Laue photographs followed by diffractometer rocking (omega) scans on the best ones, which are often rather small ($\sim 1-2\,\mathrm{mm}^2 \times 0.2\,\mathrm{mm}$). As a common alternative one may also use highly oriented pyrolytic graphite (HOPG) which for many years has been available thanks to the kindness of Dr. A.W. Moore of Union Carbide. This material consists of a polycrystalline array of graphite grains (crystallites) randomly oriented about a common c-axis. The stationary X-ray pattern of such a sample, with a monochromatic beam normal to the c-axis, is identical to a c-axis rotation photograph with integer (L) layer lines. A measure of the perfection of HOPG is usually the width of an $00.L$ rocking curve which can be as small as $\sim 0.1°$. The starting graphite may also be characterized by the extension along L of $HK.L$ reflections of the class $H-K = 3n \pm 1$. [We shall here use H, K, L to denote Miller indices or reciprocal lattice points: $HK.L$ signifies the formal 4-index hexagonal description $HK(-H-K)L$.] These reflections are sensitive to stacking faults and may be used to determine the fault density in the pristine (and intercalated) material [2.7–11]. Figure 2.2 presents a schematic reciprocal lattice diagram for a HOPG stage 2 alkali compound (Rb) at room temperature showing nominally sharp $00.L$ and $HK.L$ ($H-K=3n$) reflections and streaked $HK.L$ reflections for $H-K=3n\pm 1$. The continuous extension of alkali diffuse scattering along L is characteristic of the 2d liquid state and is discussed below.

While the best available HOPG appears to be quite well oriented, a recent study by *Kan* et al. [2.12] revealed, in contrast to single crystals, a high density of twist or T-defects as discussed earlier by *Maire* and *Mering* [2.13] for considerably more defective graphites. These defects arise during the pyrolysis and growth and represent small rotational misorientations within the mosaic blocks of a single grain. The spacing between the T-defects is estimated by *Kan* et al. [2.12] to be only about 110 Å and the change in interlayer spacing across the

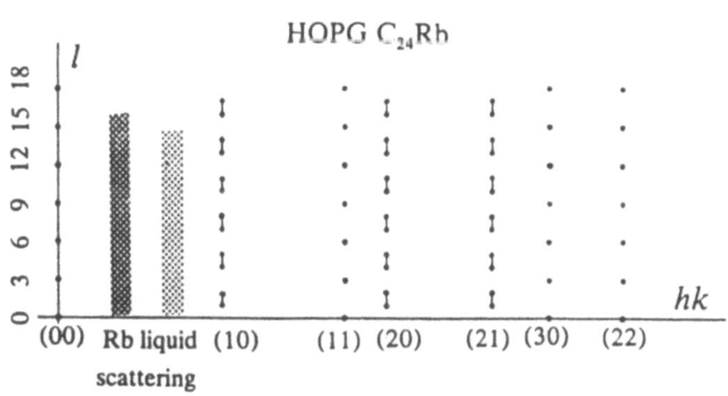

Fig. 2.2. Schematic of the reciprocal space for stage 2 Rb in HOPG. The reflections along l for $10.l$, $20.l$ and $21.l$ are smeared by stacking faults in the graphite. The extension along l of the in-plane (2d) Rb liquid scattering is indicated; $HK.L$ of the text are here denoted $hk.l$. (From [2.11])

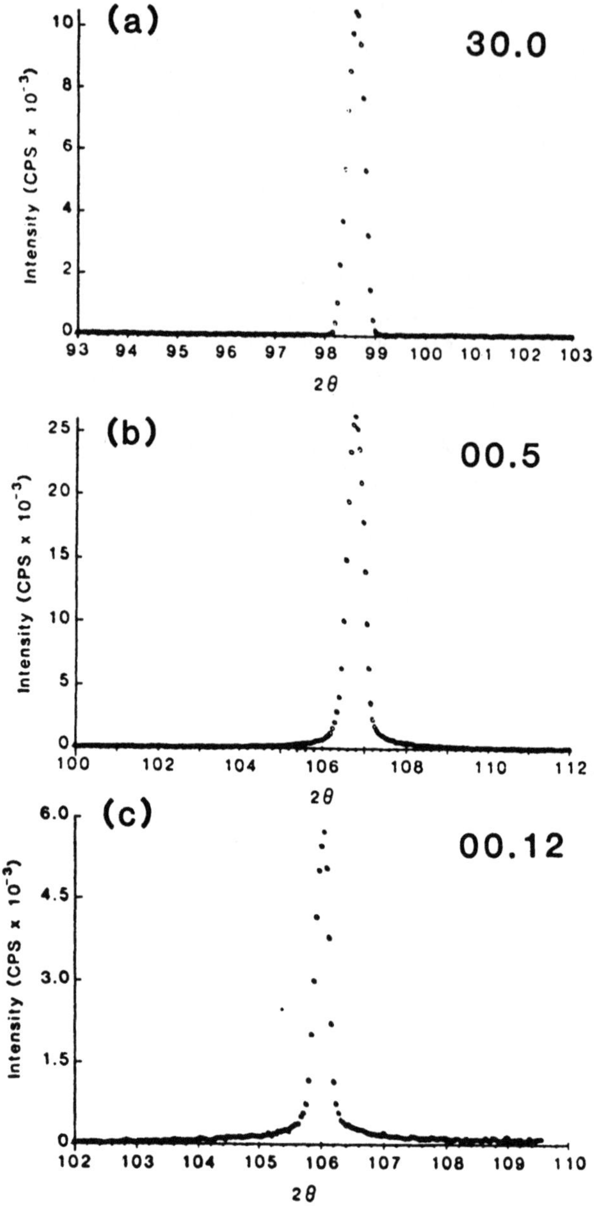

Fig. 2.3. Bragg reflection profiles taken at the Cornell High Energy Synchrotron Source (CHESS) on HOPG in (a) and (b) and on HOPG intercalated with K to stage 2 ($C_{24}K$) in (c). Note extensive wings on the $00.L$ reflections in (b) and (c) that are absent at 30.0 in (a). The traces were selected to be at nearly the same 2θ in order to compare width and wings directly. (From [2.12])

planar defect is $\sim$ 0.008 Å. The effect of these twist defects on the diffraction pattern is shown in Fig. 2.3 where it is clear that the extended wings on the $00.L$ reflections do not appear on the $HK.0$ scans. The L scans at $H, K = 1, 1$, however, showed mainly broadened peaks, indicating that across the planar T-defects, lattice coherence is lost. A single mosaic block of HOPG thus consists of stacks of $\sim$ 110 Å-thick units whose lateral coherence is several thousand angstroms.

2.1.2 Intercalation Effects

The stacking correlation in pure HOPG can otherwise be quite long range even though there will exist, in both HOPG and single crystal samples, a small density of stacking faults which are manifested in Bragg streaking of the type depicted in Fig. 2.2. Because it is so disruptive, however, the intercalation process results in a large increase in stacking fault density. *Hastings* et al. [2.7], for example, from an anlysis of the profiles of the graphite $10.L$ reflections in $C_{24}K$, determined that one fault occurred for about every 2.5 graphite–metal–graphite sandwiches. This short stacking correlation distance, which is less than the c-axis repeat distance for the stage 2 compound, was also recorded by *Rousseaux* et al. [2.9]. In $C_{24}Rb$, *Suzuki* et al. [2.8] studied both in-plane ordering and graphite stacking using neutron scattering and found a c-axis graphite stacking correlation of about 3–4 sandwich layers. In a more recent study of a similar stage 2 Rb-GIC, *Ohshima* et al. [2.10] reported a fault for approximately every 7 graphite–alkali packages, while *Thompson* et al. [2.11] found a higher fault density of 1 in 4. In a $C_{24}K$ crystal, which will be discussed later, the faulting across any two unintercalated graphite planes occurred completely at random.

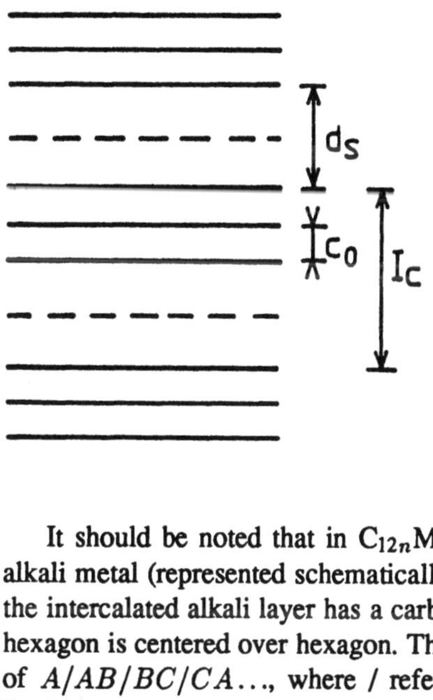

Fig. 2.4. Schematic model for a stage 3 GIC. Solid and dashed lines represent carbon and intercalant layers respectively. d_s: intercalate sandwich or gallery spacing; c_0: carbon layer separation; I_c: c-axis length or repeat distance. The parameter C_1 in (2.1) is given by $C_1 = d_s/I_c$

It should be noted that in $C_{12n}M$, where n is the stage number and M the alkali metal (represented schematically for $n = 3$ in Fig. 2.4), the stacking across the intercalated alkali layer has a carbon layer always positioned over itself, i.e., hexagon is centered over hexagon. This leads to a sequence in intercalated $C_{24}M$ of $A/AB/BC/CA\ldots$, where / refers to the alkali metal in its disordered or liquid state at room temperature [2.14]. Perhaps not surprisingly this is also true for registered stage 1 alkali compounds; it also occurs in stage 2 graphite–HNO_3 ($A/AB/BA$) [2.15, 16] and graphite–Br_2 ($A/AB/BC/CA$) [2.17]. In the donor alkali metal compounds there seems to be no inclination to other than A/A (or B/B, C/C), even in the higher stages. The faulting is thus associated with mistakes in graphite sequencing, i.e., $A/AB/BC/C\ldots \rightarrow A/AB/BA/A$, which is quite similar to a deformation fault in a fcc metal [2.18]. In the acceptor compounds, the separation of the gallery carbon planes, d_s, is larger, which tends to diminish the direct carbon–carbon interaction. In these cases, the graphite correlation across the gallery is mainly determined by more complex interactions through the intercalant molecules and can in some cases lead to stacking sequences other than A/A or A/B. This was first reported for stage 5 HNO_3 [2.16] and has been recently documented for stage 4 $FeCl_3$ [2.19]. Figure 2.5 presents a set of scans for stage 4 $FeCl_3$ in HOPG intercalated in situ on the X-ray diffractometer and held in a thin glass tube at 350°C. All L scans show very diffuse profiles indicating a loss of graphite correlation across the intercalated gallery. The profiles are well fit by assuming packages or sets of four graphite planes of stacking $ABAB$, separated by $FeCl_3$; 40% of these are ran-

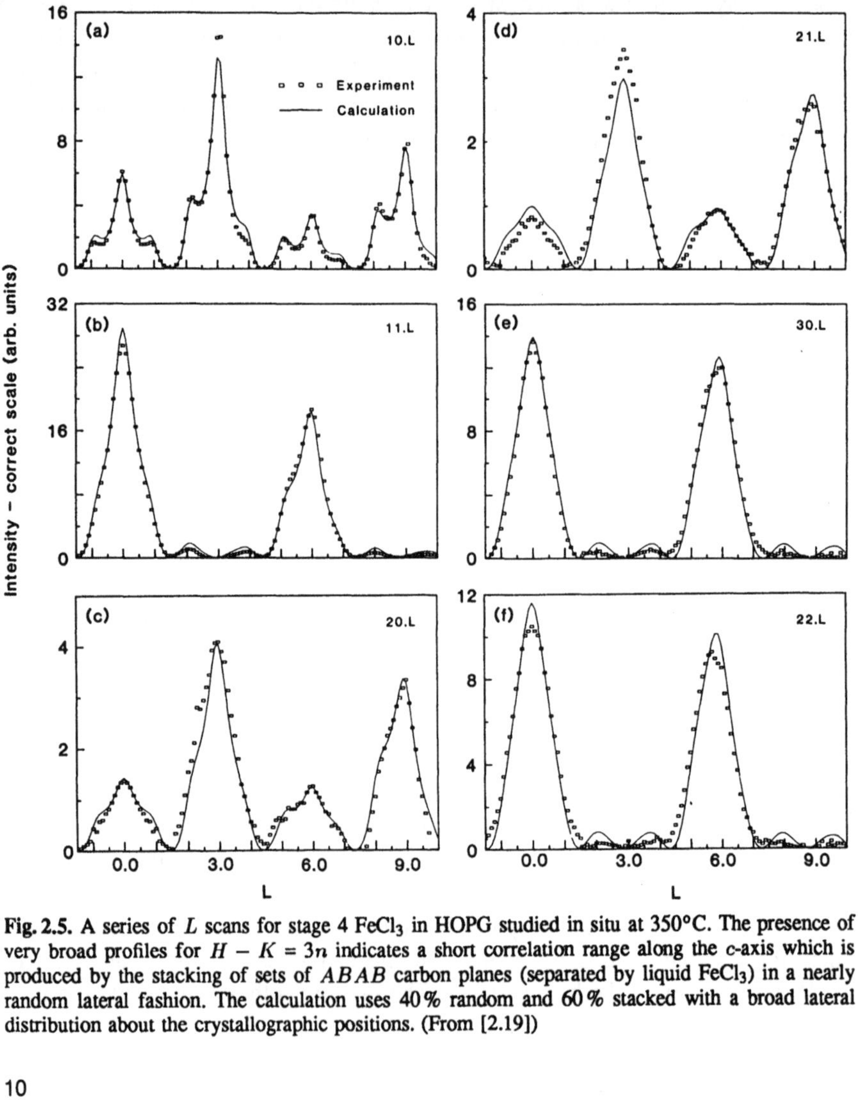

Fig. 2.5. A series of L scans for stage 4 $FeCl_3$ in HOPG studied in situ at 350°C. The presence of very broad profiles for $H - K = 3n$ indicates a short correlation range along the c-axis which is produced by the stacking of sets of $ABAB$ carbon planes (separated by liquid $FeCl_3$) in a nearly random lateral fashion. The calculation uses 40% random and 60% stacked with a broad lateral distribution about the crystallographic positions. (From [2.19])

domly positioned over each other while 60 % show a preference for an $ABAB$ set to be followed by $ABAB$ (B followed by A across the gallery). The large lateral distribution about this discrete position, however, propagates along L to destroy the stacking correlation after a few sets.

The effects of intercalation on the bond lengths of the carbon atoms in the bounding layers also depends on whether one is dealing with donors or acceptors. As reviewed by *Moret* [2.5], with the alkalis there is a small expansion of d_{C-C} over the pristine value of 1.420 Å that is roughly proportional to the valence and inversely proportional to the stage index and ionic radius of the metal. For the acceptors, d_{C-C} shrinks.

A more speculative aspect of the intercalation process is in the deformation or rumpling of the carbon layer by the intercalant. Clearly, for staggered Daumas-Herold domains [2.20 and see also Chap. 3], there will be an attractive interdomain interaction through the deformation of the graphite planes at the domain boundary (Chap. 3) where the "healing" of the deformation is thought to occur over several lattice spacings. In addition, it has been suggested [2.21] that the intercalated alkali species in stage $n \geq 2$ induces a local deformation of the graphite host in analogy with balls on a rubber sheet. As *Mori* et al. [2.21] discuss, the graphite is flexible but not very stretchable, the ratio of the two elastic stiffnesses, C_{44}/C_{11}, being $\sim 1/300$. This anisotropy leads to a possible contribution of the elastic deformation of the graphite to the intralayer ordering, although it is not the dominant interaction. An indication of the tendency for the carbon planes to deform in the vicinity of an intercalated alkali atom came initially from the EXAFS measurements by *Caswell* et al. [2.22]. More recently, *Fischer* and *Kim* [2.23] presented very convincing evidence for the corrugation of graphite about the intercalated Li ions in stage 1 Li_xC_6, where $0 < x \leq 0.7$; and for the ternary GICs $K_{1-x}Rb_xC_8$, *Chow* and *Zabel* [2.24] have found a d_s spacing vs x which increases continuously from $x = 0$ to $x = 1$ and which would also seem to require a local "buckling" of the carbon layers.

2.2 Intercalate Structures: General Features

2.2.1 Stage Disorder

In both donor and acceptor compounds, staged intercalation is achieved through a careful choice of the temperature of the graphite host and the intercalant chemical potential – usually through the control of vapor pressure [2.1–4]. A well-intercalated sample may then be evaluated by routine $00.L$ diffraction scans to determine the nominal stage index n and repeat distance I_c. Imperfect staging is increasingly found with increasing stage index [2.25–27 and Chap. 3]. It manifests itself both as a stage mixture, in which discrete $00.L$ peaks of the separate stages may be detected, and as stage disorder, in which there is a more uniform distribution of the layer spacings associated with each stage. The 1d diffraction

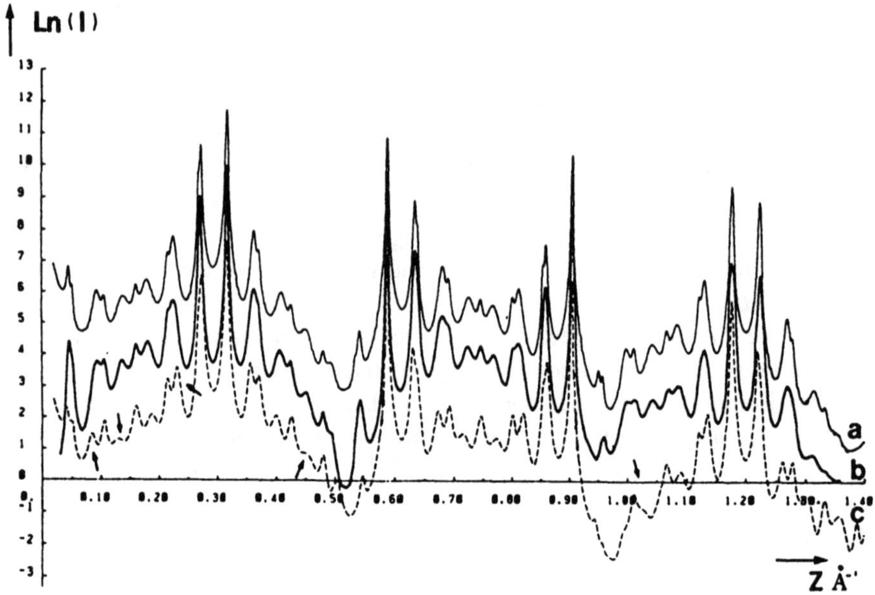

Fig. 2.6. Comparison of the experimental X-ray profile along 00.L, plotted in curve (**b**) as $\ln[I(Z)]$ vs Z [Å^{-1}], with calculated intensity distributions for stage disorder in a nominally stage 5 K-GIC. Curve (**a**) is based on a model with 2-layer combinations (5 + C) where C indicates carbon, and curve (**c**) illustrates a 3-layer model (5 + 6 + C). The experimental curve is corrected for Lorentz and polarization effects and the origins of the calculated curves are shifted. The agreement with curve (**a**) is seen to be quite good (From [2.31])

model developed for layer disorder by *Hendricks* and *Teller* [2.28], and refined by *Kakinoki* and *Komura* [2.29], has been quite successfully employed to characterize both the degree of uniformity and the distribution of stages, particularly in the higher stage alkali GICs. These studies of stage disorder have been done both in situ [2.25, 30], to directly observe the staging transitions, and on quenching, to characterize the nominally single-stage compounds [2.26, 27, 31–33]. Figure 2.6 presents an excellent fit to a nominally stage 5 K-GIC in which the 1d modeling has incorporated packages of stage 5 and graphite. Figures 2.7 and 2.8 demonstrate the fit that can be achieved for two stage 2 K-GIC ($\sim C_{24}K$) samples of slightly different in-plane densities. Both consisted of better than 98 % stage 2 but they showed a somewhat different distribution of the stage 3 contaminant. The small difference in in-plane density may nevertheless have an influence on the low temperature in-plane order. Here the less dense sample shows multiphase behavior at 45 K while the denser sample, at 90 K, is single phase. Because 90 K is close to the ordering temperature of 95 K, one may not directly compare the two low-temperature patterns. It remains interesting, however, to consider the ordering behavior of two essentially pure stage samples of different in-plane density.

The existence of even a small degree of stage disorder introduces into an effectively 1d system of stacked layers a random sequence of spacing mistakes.

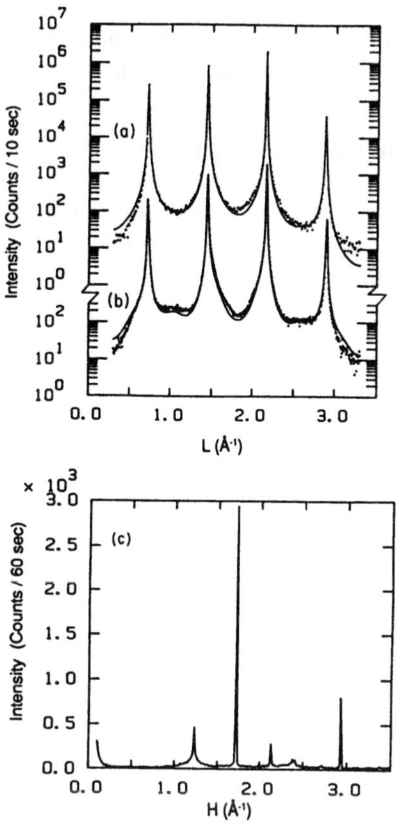

Fig. 2.7. With Fig. 2.8, a presentation of a Hendricks–Teller fit to 00.L profiles at both 300 K (**a**) and 90 K (**b**) for a nominally stage 2 K-GIC in HOPG. In this figure, the results for the densest sample are shown, for which the stage 2 purity is ~ 99 %. In (**c**) the in-plane scan shows the "usual" ordered diffraction peaks for stage 2. (From [2.33])

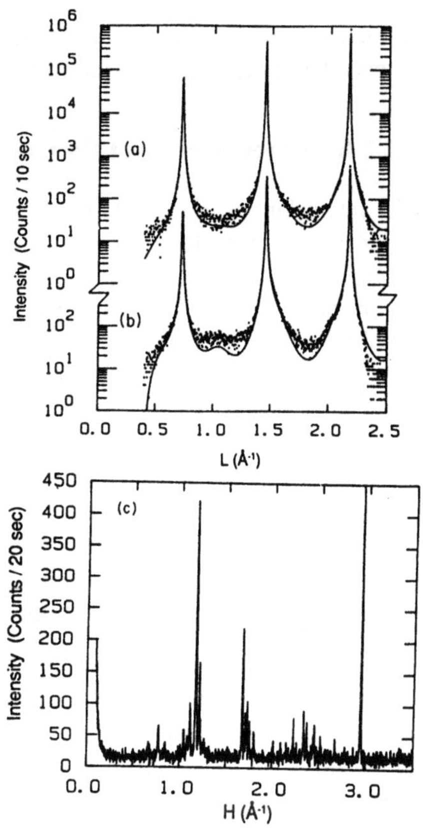

Fig. 2.8. As in Fig. 2.7 but at a lower in-plane density. (**a**) is at 300 K; (**b**) and (**c**) are at 45 K. While the stage purity in (**a**) and (**b**) is still ~ 98 %, the ordering in (**c**) is clearly complex and has been given conflicting interpretations (Sect. 2.4.2b). (From [2.33])

As with a defective diffraction grating, or a vibrating 1d crystal [2.34, 35], such disorder reduces the correlation range and broadens the diffraction spectra [2.34–36] (in 3d, vibrations and point defects do not broaden the Bragg peaks). The T-defects noted above, whose diffraction effects are shown in Fig. 2.3, are also good examples of this 1d disorder; on intercalation, the wings of Fig. 2.3b are enhanced (Fig. 2.3c) due to the increased spacing disorder. The log scale in Figs. 2.7 and 2.8 emphasizes the disorder-induced scattering for which a careful analysis of the continuous 00.L line profile may yield the distribution of spacings [2.25, 28, 31-33]. Because, strictly speaking, these 00.L reflections are no longer Bragg peaks, it becomes increasingly difficult with increasing stage disorder to measure their integrated intensities. Where the 00.L intensities are used to evaluate both Debye-Waller factors and in-plane densities [2.37, 38], caution is therefore urged, especially in conjunction with the large mosaic spreads found in these materials [2.39]. On the other hand, the mixed stage compounds serve as excellent testing grounds for 1d disorder and phase transitions [2.6].

2.2.2 Interlayer Intercalant Correlations in the Liquid State

The intercalant usually enters the graphite from the vapor phase and forms a 2d liquid which crystallizes (freezes) on cooling. For the higher stage GICs, these 2d liquid layers are spatially well-separated along the c-axis, which accounts for the mistakes in the regular staging that would otherwise follow from the screened repulsive interlayer potential (Chap. 3). The spatial separation of the 2d liquids also means that they will, in general, be uncorrelated from layer to layer – that an atom at position r_i in layer i will not prejudice the position r_j of an atom in layer j. For the liquid state in stages $n \geq 2$ this seems invariably to be the case. The absence of intercalant correlations along c^* (reciprocal to the c-axis) was first demonstrated by *Parry* and *Nixon* [2.40, 41] whose X-ray photographs revealed the continuous streaking of the in-plane liquid diffuse peak along L, as shown schematically in Fig. 2.2 for a stage 2 Rb-GIC in HOPG. In order to explore these c-axis correlations in HOPG (or single crystal samples), a transmission scattering geometry is employed in the X-ray experiment. The symmetrical situation where the specimen surface bisects the incident and scattered waves permits scans at $L = 0$ in the H, K plane. Decoupling the θ and 2θ motions of a standard two-axis $(\theta - 2\theta)$ diffractometer then permits, in the asymmetric transmission scattering geometry, a simple exploration of the entire reciprocal space covered in Fig. 2.2. One must, however, take care with the absorption corrections [2.11], and account properly for the mosaic spread of the sample [2.11, 12, 39]. The L-independent extension of the liquid scattering, first seen photographically by *Parry* and *Nixon* [2.40] for $C_{24}K$ has since been quantitatively confirmed by *Zabel* et al. [2.42] and corroborated in $C_{24}Cs$ by *Clarke* et al. [2.43] and in $C_{24}Rb$ by *Ohshima* et al. [2.10].

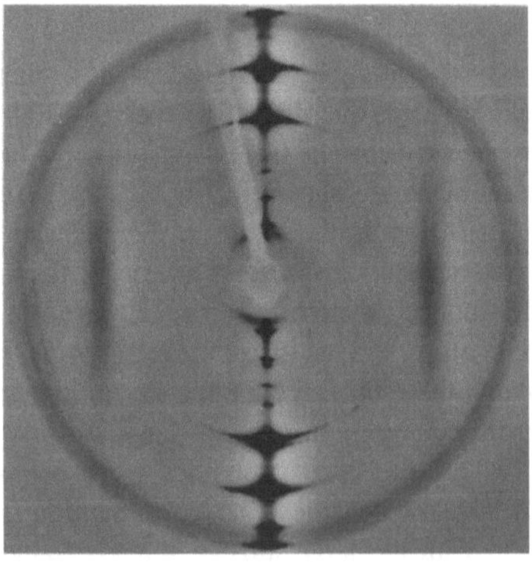

Fig. 2.9. $HH.L$ single crystal precession photograph of a stage 2 HNO_3 GIC with the L directoin vertical. The photograph is overexposed to reveal the extended diffuse scattering along L characteristic of a 2d liquid. The mosaic spread in the $00.L$ reflections is also clear. (From *Samuelsen* in [2.44])

For the acceptor GICs, the evidence for 2d liquid scattering is similar to the situation in the donors. Figure 2.9 obtained by *Samuelsen* [2.44], is an $(HH.L)$ precession photo of a single crystal of stage 2 HNO_3 in graphite. It shows clear evidence of the streaking along L (vertical) of the liquid scattering at the first diffuse peak in the in-plane structure factor. For stage 4 Br_2 in graphite, *Erbil* et al. [2.45] discuss the melting of a 2d incommensurate solid into its liquid phase. The stage 4 $FeCl_3$ GIC studied in situ at 350°C by *Hashimoto* et al. [2.19] was also shown to consist of an uncorrelated stack of 2d liquids.

Because both donor and acceptor liquids are modulated by their graphite host, there will be a component of the liquid structure, i.e., the induced density wave, which is, in fact, correlated along the c-axis. This component, as we shall see, contributes only to the δ-function-like graphite Bragg peaks; the liquid-like diffuse scattering remains 2d.

2.2.3 Two-Dimensional Melting

The issue of two-dimensionality is so germane to our discussion of these layered intercalants that it is important to note briefly those aspects of structure that are peculiar to two dimensions. Two recent reviews of this subject are fortunately available [2.46, 47] and the reader is directed to them for details. As mentioned earlier, a vibrating or defective 1d crystal can formally be taken to be a liquid in that the "Bragg peaks" are of finite width at finite temperature, T, with this width increasing with increasing reflection order [2.34–36]. Three-dimensional crystals are, of course, crystals at all temperatures with δ-function-like Bragg peaks whose intensity, but not width, decreases with increasing T, as described by the Debye-Waller factor. In two dimensions it has been well-established [2.48–51] that the Fourier transform of an infinite vibrating crystal is characterized by a set of power-law singularities as in a system at its critical point [2.46, 47]. In real space, the pair correlations for this 2d periodic array decay as $r^{-\eta}$, which is referred to as marginal long-range order. The exponent η is a function of the reciprocal lattice vector, $\boldsymbol{q}_{HK} = 2\pi(H\boldsymbol{b}_1 + K\boldsymbol{b}_2)$, so that the breadth of the power-law "Bragg peak" also increases with increasing distance in reciprocal space. The detailed predictions of $\eta(T)$ have been borne out in the highest resolution X-ray experiments on the incommensurate 2d adsorbate, Xe on graphite [2.52]. In these experiments, both the power-law line shapes and the theoretical value for η at the melting point have been determined. In the GICs, the study of a stage 4 Br_2-intercalated crystal [2.45] has provided perhaps the most elegant demonstration of the evolution of power-law peaks into Lorentzian line shapes on passing through the melting transition.

It has also been predicted that 2d crystals will melt via a dislocation-mediated process [2.53] that may proceed in two steps. The first is a change from power-law positional correlations to an exponentially decaying correlation function (i.e. power-law Bragg peak to a Lorentzian "liquid" shape for the translational order) but with power-law bond-orientational order preserved. This intermediate *hexatic* (6-fold oriented) phase, which has been the subject of a great deal of discussion [2.46, 47], was elegantly displayed in a model experiment which varied the areal density of a monodispersion of plastic spheres between glass plates [2.54] to induce hexatic order with increasing density. More recently, both electron diffraction [2.55] and X-ray scattering [2.56] have succeeded in showing a reversible hexatic regime in the melting of thin freely suspended liquid crystal films. All experiments, however, on real adsorbed gases or intercalates suffer, to a greater or lesser extent, from the influence of the periodic substrate crystal potential, which for graphite will naturally induce a 6-fold modulation having little or nothing to do with the inherent tendency of the adsorbate to undergo the two-stage melting: crystal $\rightarrow$ hexatic liquid $\rightarrow$ isotropic liquid. In particular, for the intercalate systems, we know of no examples where the substrate influence has been demonstrated to be small enough to ignore. As we shall show, the alkali metal donor liquids and their corresponding solids are examples of strong substrate-mediated systems; in the acceptor liquids the modulation effects can be weaker and there is at least one example of $GaCl_3$ [2.57] that appears to be unmodulated in the liquid state.

2.3 The Two-Dimensional Liquid State

2.3.1 Alkali Liquids – Experimental Results

Following initial observations on the disordered state in the stage $n \geq 2$ alkali GICs [2.14, 15], *Parry* [2.58] published stationary transmission single crystal X-ray photographs of a stage 2 Cs-GIC for both disordered and low-temperature ordered states. These photographs showed all of the basic features of both liquid and solid phases that we shall be discussing. In particular, the liquid pattern revealed strong 6-fold anisotropic scattering about the origin, as well as the reproduction (in the form of halos about the six graphite $\{10.0\}$ reciprocal lattice positions) of the first sharp liquid diffraction peak, or FSDP [2.59], which comes at a diffraction vector incommensurate with the reciprocal graphite spacing. Both the anisotropy about the origin and the halos – from which all of the complex modulation features of the 2d alkali ordered states develop on cooling – are unmistakable signs of a modulated liquid [2.60].

A third aspect of the modulated liquid was also first discussed by *Nixon* and *Parry* [2.41] in their analysis of the intensities of the graphite Bragg peaks from the disordered state in the stage 2 alkali GICs. Their intensity fits were substantially improved by including contributions from between 50 % and 75 % of the intercalated alkali atoms to the graphite peaks. Arguing, thereby, for a registry or lattice-gas model of the liquid state, these results have informed all of the work that followed, including the aforementioned studies by *Rousseaux* et al. [2.9] on faulting which required an alkali contribution to the graphite profiles.

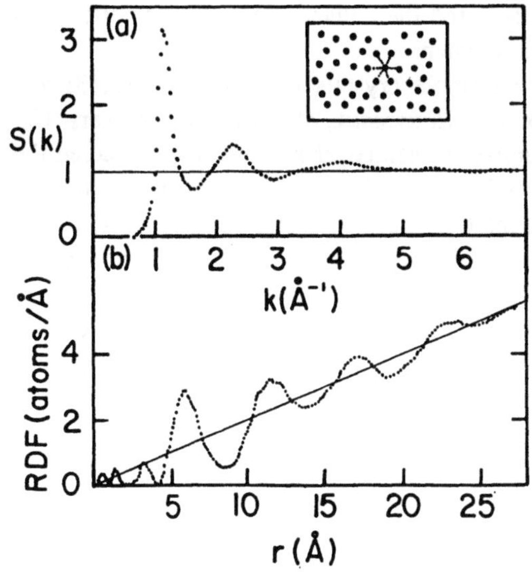

Fig. 2.10. (a) Structure factor $S(k)$ for the 2d Cs liquid layer in a stage 2 Cs HOPG sample at 300 K. (b) The (Bessel) transform of $S(k)$ in (a) to give the area radial distribution function (RDF). The inset shows a schematic of the liquid-like local order. (From [2.43])

The issue of the liquid state as a registered (hopping) lattice gas vs an incommensurate diffusing liquid is an interesting one and, as we shall see, depending upon the time scales involved, both descriptions have merit. *Clarke* et al. [2.43] were the first to characterize carefully the 2d liquid, *qua* liquid, for a stage 2 Cs-GIC in HOPG. Figure 2.10 shows their X-ray result for the circularly averaged structure factor, $S(k)$ [our notation will eventually be $S(q)$], where $k = (4\pi \sin\theta)/\lambda$, 2θ is the angle between incident and scattered waves and λ is the wavelength. The X-ray data were reduced in the standard way [2.18] and the 2d Fourier-Bessel transform was taken to obtain an area radial density function (RDF). The position, r_1, of the first peak in the RDF was 5.95 ± 0.1 Å with a coordination number of 6 ± 0.25 atoms. This number of nearest neighbors is consistent with a close-packed array whose average area density, at the concentration of $C_{12}Cs$ per gallery, would give $r_1 = 6.02$ Å. It is also consistent with a simple estimate of nearest neighbor distance obtained from the Ehrenfest equation for diatomic molecules [2.61] that is often applied to the first sharp diffraction peak (FSDP) of a liquid [2.59]. For 2d, $k_1 r_1 \simeq 6.97$ [2.62], which gives $r_1 = 6.01$ Å for $C_{24}Cs$ where $k_1 = 1.16$ Å^{-1} [2.43]. This relationship is in fact quite useful and enables us to estimate nearest neighbor distances and trends in these liquids without recourse to the transform of $S(k)$.

While the coordination number of six is consistent with some form of hexagonal registry, the value of r_1 for Cs is *not*, and *Clarke* et al. [2.43] concluded that the disordered state in their Cs-GIC was an unregistered liquid whose short-range order may be given schematically by the insert in Fig. 2.10. The question of the 6-fold modulation of this liquid was also addressed using azimuthal single crystal scans, at a fixed $k_1 = 1.16$ Å^{-1}, around the first strong diffraction circle. *Clarke* et al. [2.43] attributed the observed variation with azimuthal angle to the influence of the substrate potential which gives rise to static density waves

(SDWs) in the liquid. They also discussed the enhancement of these effects on cooling, which we postpone for the moment.

Extensive photographic work on single crystals of both $C_{24}K$ [2.63] and $C_{24}Rb$ [2.64] has also been reported by *Rousseaux*, *Moret* and co-workers. Using a fixed crystal–fixed film transmission method, these researchers studied the liquid state and its temperature dependence into the ordered state. The anisotropic modulation of the liquid scattering and the halos about graphite {10.0} positions, first observed by *Parry* [2.58], were clearly seen and will be treated in more detail in the next section.

For K-intercalated GICs, *Zabel* et al. [2.62] presented in-plane diffractometer data for the liquid state of (nominally) $C_{24}K$, $C_{36}K$ and $C_{48}K$. For the stage 2 sample, $k_1 = 1.23\,Å^{-1}$ while for stages 3 and 4, $k_1 = 1.16\,Å^{-1}$. This variation in k_1, and thus r_1 and in-plane density, is found not only on changing intercalant or stage index. It is also a sensitive function of intercalation conditions, as shown in Fig. 2.11. This figure was first published by *Nishitani* et al. [2.65], who studied the evolution of the in-plane X-ray intensity along $q_0(k)$ for a K-GIC in HOPG liquid as a function of alkali vapor pressure. These experiments were performed at the KEK synchrotron (Photon Factory) with in situ K-intercalation on an X-ray spectrometer using white radiation and an energy dispersive solid-state detector. The profiles in Fig. 2.11 show a continuous shift of the in-plane density of K

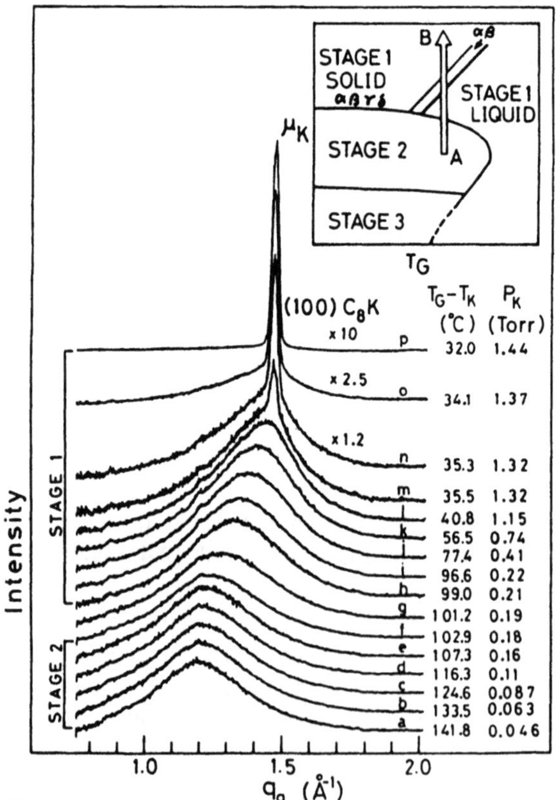

Fig. 2.11. Evolution of the X-ray diffraction pattern for a K-GIC as recorded with a solid-state detector using white radiation at the KEK synchrotron. q_0 is an in-plane wave vector and the scans are recorded as the potassium vapor pressure is increased from $a - p$ to follow the stage 2–stage 1 transition while the graphite is held at 390°C. In the schematic diagram in the inset, the scan is along AB. (From [2.65])

with increasing vapor pressure at a constant graphite temperature of 390°C. While stage 2 clearly may not be characterized by a single in-plane density or nearest neighbor spacing, the transition from stage 2 liquid to stage 1 liquid does seem to be accompanied by a jump in nearest neighbor distance (not shown) [2.65].

There appears, then, to be no fixed atom size associated with a particular intercalated alkali liquid. Rather, this size depends upon vapor pressure during intercalation; i.e., these donor compounds permit some extended range of stoichiometry at each stage. The range of in-plane atom or ion sizes measured for the K, Rb and Cs liquids is about 5.67–6.10 Å with most of the data for all of them lying between 5.85 Å and 6.10 Å. These sizes are all much larger than those measured for the 3d alkali liquids in which $r_1(\mathrm{K}) < r_1(\mathrm{Rb}) < r_1(\mathrm{Cs})$ [2.66], and they are substantially greater than the ion core radii. They represent effective interionic separations of the incompletely screened donor species and are therefore sensitive less to atomic number than to in-plane density and layer separation or stage index.

Zabel et al. [2.62] also attempted to fit the K liquid X-ray data with a model of lattice-gas registry. They created a random structure of registered K atoms with nearest- and next-nearest-neighbor hexagon exclusion rules and calculated the circularly averaged X-ray intensity from several such assemblies. The agreement with all of the K-GIC experiments was sufficiently poor to rule out this model. *Winokur* et al. [2.67] subsequently explored the lattice gas decoration in great detail and came to the same conclusion for stage $n \geq 2$.

The above results, while compelling, had to be reconciled with the Bragg contributions of the alkali atoms in the liquid state [2.41,9]. *Ohshima* et al. [2.10] developed this theme in more detail, measuring both the liquid structure factor and a set of 85 Bragg peaks from a stage 2 Rb GIC. The crystallographic analysis was somewhat compromised by the problems of mosaic spread and stacking faults discussed previously. The authors were quite fortunate, however, in having a well-staged sample with a modest mosaic spread and relatively low stacking fault density. This permitted a "reasonable" estimate of the integrated Bragg intensities over background and thus, via a nonlinear least squares fitting procedure, of the gallery spacing, four Debye-Waller factors (in- and out-of-plane for Rb and C) and the (so-called) registered fraction, X, of Rb atoms in hexagon centers. With a crystallographic R-factor of ~ 0.09, X was determined to be about 0.71, for an assumed in-plane density of $C_{12}Rb$. All other parameters were deemed reasonable except for the in-plane Rb Debye-Waller factor (DWF), which yielded a rms planar displacement of 0.32 Å! This is both unphysical and well outside the regime of validity of the harmonic Debye-Waller theory. What it reflects is a rapid damping in reciprocal space of the alkali contribution. Formally, the graphite unit cell structure factor can be written

$$F(HK.L) = [4f_{\mathrm{C}} \cos 2\pi(H+2K)/3 \cos \pi L C_1 + X/6 f_{\mathrm{Rb}}] \times [1 + \cos 2\pi(H+2K \pm L)/3] , \quad (2.1)$$

where f_{C} and f_{Rb} both have incorporated DWFs and $F(HK.L)$ does not depend upon the details of the local liquid order. If $X = 0$, however, there is, of course, no Rb contribution; if the Rb DFW falls off rapidly in reciprocal space, the contribution of alkali will be only to the first few low-angle Bragg peaks. Because the fit, while reasonably successful, yielded an unphysical DWF, another way of combining the idea of registry with that of a (modulated) unregistered and incommensurate liquid was required.

2.3.2 Modulated 2d Liquids – Theory and Application to Alkalis

Motivated by the above results, *Reiter* and *Moss* [2.60] developed a theory for the X-ray scattering from a 2d liquid modulated by its periodic host. The theory is quite general for all fluids within or on periodic substrates, but it concentrates on the alkalis in graphite to clarify the origin of (a) the alkali contribution to the graphite Bragg peaks; (b) the replication of the alkali–metal liquid scattering about the graphite reciprocal lattice points (halos); and (c) the anisotropic 6-fold angular modulation of the 2d liquid scattering. The theory followed earlier work by *Bunde* and *Dietrich* [2.68], *Plischke* [2.69] and *Plischke* and *Leckie* [2.70]. The treatment was also motivated by the dynamical results of *Kamitakahara* and *Zabel* [2.71], who used incoherent inelastic neutron scattering to measure a two-peak vibrational density of states for alkali GICs; it was also (ultimately) influenced by diffusion data, obtained from quasi-elastic neutron scattering [2.72, 73], which show both a solid-like and liquid-like response in the liquid state.

The basic idea of the theory is that the 2d liquid in its periodic host will be subjected to a periodic host–intercalant modulation potential of the form

$$V_{\mathrm{m}}(\boldsymbol{r}) = \sum_{H,K} V_{HK}(z) \mathrm{e}^{2\pi \mathrm{i} \boldsymbol{q}_{HK} \cdot \boldsymbol{r}} , \tag{2.2}$$

where z is an out-of-plane coordinate along the c-axis and $\boldsymbol{q}_{HK} = 2\pi(H\boldsymbol{b}_1 + K\boldsymbol{b}_2)$ is, as before, a reciprocal lattice vector of the graphite. Separating the in-plane ($\|$) and z ($\perp$) parts of this potential, we have

$$V_{\mathrm{m}}(\boldsymbol{r}_{\|}, z) = V_0(z) + \sum_{H,K} V_{HK} \mathrm{e}^{\mathrm{i} \boldsymbol{q}_{HK} \cdot \boldsymbol{r}_{\|}} , \tag{2.3}$$

where it is assumed that the entire z dependence, which gives rise to a c-axis DWF, is contained in $V_0(z)$ and that V_{HK} represents an average over the (thermal) z motion. We henceforth drop the notation $\|$ and write $\boldsymbol{r}_{\|} = \boldsymbol{r}$ implying only in-plane dependence. The effect of this potential will be to introduce static density waves (SDWs) into the alkali, as noted by *Clarke* et al. [2.43]. These SDWs will both influence the liquid correlations and give rise to a static Bragg-like component from the modulated alkali at the reciprocal lattice points of graphite. The job of the theory, then, is to calculate both $\langle \varrho_{q_{HK}} \rangle$ for this Bragg amplitude [contribution to $F(HK.L)$ in (2.1)] and $\langle \varrho_q \varrho_{-q} \rangle$ for the liquid scattering. It can be seen that $\langle \varrho_{q_{HK}} \rangle$, which is the ensemble-averaged transform of the Rb number density, is nearly linear in the relevant Fourier coefficients V_{HK}, and $\langle \varrho_q \varrho_{-q} \rangle$ involves products of $V_{HK} V_{H'K'} \ldots$ etc., summed over all contributions. *Reiter* and *Moss* [2.60] were able to demonstrate the origin of all of the liquid-like and Bragg effects but the analytics are rather complicated for $\langle \varrho_q \varrho_{-q} \rangle$ in other than trivial approximations. For the Bragg contribution, a simple expression results, for $H + 2K \pm L = 3n$,

$$F(HK.L) = 4 f_C \cos(2\pi/3)(H + 2K) \cos \pi L C_1 + (X'/6) f_{Rb} e^{-M_{Rb}^{\perp}} \langle \varrho'_{q_{HK}} \rangle \,, \quad (2.4)$$

where $\langle \varrho'_{q_{HK}} \rangle$ is the normalized value of $\langle \varrho_{q_{HK}} \rangle$ at a graphite Bragg peak; f_C still contains a DWF while $\exp(-M_{Rb}^{\perp})$ is the DWF of Rb normal to the layer; X' now refers to an adjusted in-plane density.

If the modulation potential is weak, i.e. βV_{HK} ($\beta = 1/k_B T$) ≤ -0.10, $\langle \varrho'_{q_{HK}} \rangle = -\beta V_{HK}$, as suggested above; otherwise the measurement of a set of $\langle \varrho'_{q_{HK}} \rangle$'s, the alkali contributions to the graphite structure factors, has to be analyzed more carefully to extract a set of βV_{HK}'s. This procedure, outlined by *Reiter* and *Moss* [2.60], was applied to two separate sets of data on the liquid state at ~ 300 K of a stage 2 Rb-GIC in HOPG [2.74]; the Bragg data collected by *Ohshima* et al. [2.10] were re-analyzed and data on a second sample collected by *Thompson* [2.38] were treated both by profile analysis of the fault-broadened peaks [2.11] and by an anomalous scattering method [2.38, 75] to extract $\langle \varrho'_{q_{HK}} \rangle$. The two data sets agreed quite well and an iterative procedure was used to extract a set of $\{\beta V_{HK}\}$ from which a potential $V_m(\boldsymbol{r})$ was constructed via (2.2) [2.74].

More recently, a $C_{24}K$ (stage 2 K-GIC) crystal was studied by *Kan* et al. [2.76]. This crystal, as noted earlier, showed a completely random faulting sequence on intercalation and therefore only the $00.L$ and $HK.L$ ($H - K = 3n$) peaks could be integrated for a determination of structure parameters; $HK.L$ ($H - K \neq 3n$) were smeared along L. Figure 2.12 shows the $20.L$ profile fit using a random stacking sequence. By choosing a value of $\langle \varrho'_{q_{20}} \rangle$ over the range where $I_{20}(L)$ in (2.5) goes to zero, a very good fit could be obtained as long as the actual concentration was known from $00.L$ intensities and other data. In this case, $I_{HK}(L)$ is given by the continuous scattering distribution per graphite–alkali–graphite sandwich layer:

$$I_{HK}(L) = |F(HK.L)|^2 \frac{3}{5 + 4\cos 2\pi L/3} \,, \quad (2.5)$$

and it is $|F(HK.L)|^2$ that determines the modulation function and hence $\langle \varrho'_{q_{HK}} \rangle$. Here, as in [2.9, 11], the stacking faults were useful in spreading the intensity smoothly over a range where zeros in $F(HK.L)$ exist. Thus, by a combination of crystallographic methods, values for $\langle \varrho'_{q_{HK}} \rangle$ were determined for the stage 2 K-GIC crystal and a set of $\{\beta V_{HK}\}$ were obtained. Then $V_m(\boldsymbol{r})$ was determined for K and was compared to $V_m(\boldsymbol{r})$ in Rb in a recent review [2.77]. Figure 2.13 presents the plots of $V_m(\boldsymbol{r})$ within a graphite hexagon that were obtained using V_{HK} and (2.3) [2.77]. The barrier height for diffusion in the stage 2 alkali GICs

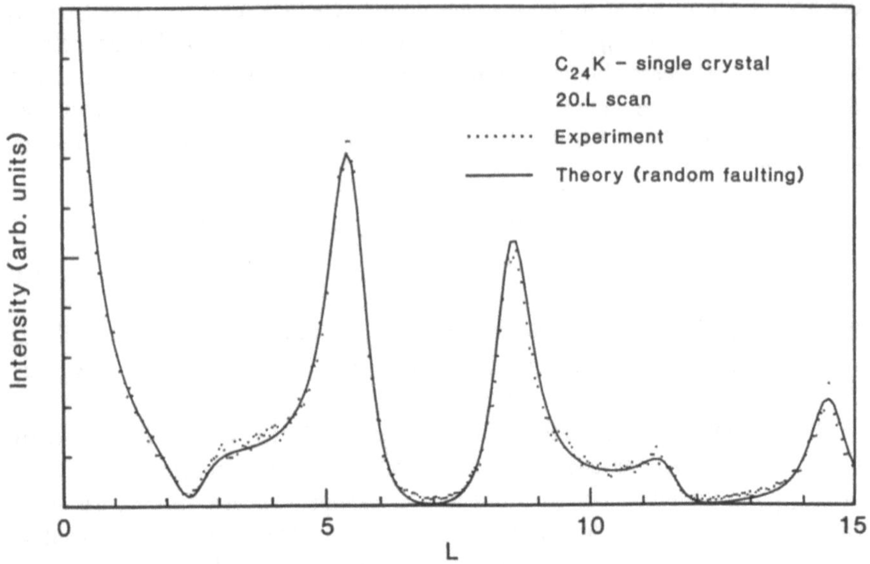

Fig. 2.12. The 20.L X-ray diffraction profile from a stage 2 K-GIC crystal studied at the NSLS synchrotron. The experimental data are corrected for scattering, polarization and geometrical factors and the Compton background has been removed. The theoretical curve is calculated using (2.4) for a completely random faulting in the graphite–K–graphite sandwich sequence (From [2.76])

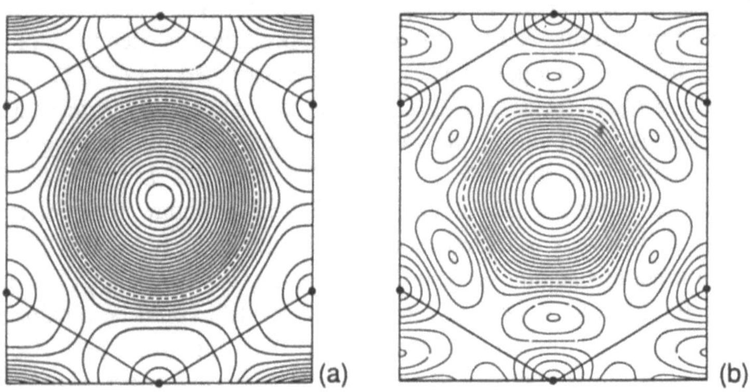

Fig. 2.13. The modulation potential $V_m(\boldsymbol{r})$ for Rb (**a**) and K (**b**) in the stage 2 alkali GICs. The potential at the hexagon center is negative and passes through zero at the dotted contour. For Rb, the weak saddle point minimum occurs at the bond midpoint; for K there is a local minimum on the carbon atom and a respulsive maximum $\sim 2/3$ to the carbon atom from $r = 0$. (From [2.77])

was recently evaluated to be between 0.05 and 0.10 eV [2.78]. The values of 0.092 eV (Rb) and 0.085 eV (K) fall within that range, suggesting that the X-ray method is well-suited to the determination of the modulation potential in these alkali GICs.

2.3.3 Molecular Dynamics Simulations

The use of molecular dynamics (MD) methods to simulate the structure and dynamics of 2d systems has been extensively developed by *Abraham* and co-workers [2.79, 80] in their work on the rare gases. For the alkalis in graphite, Monte Carlo methods have been used by *Plischke* and *Leckie* [2.70] to generate $S(q)$, the circularly averaged liquid structure factor ($S(k)$ in [2.43]) in which they obtained reasonable fits to the available data and observed a Bragg-like contribution from the induced density waves. The modulation potential they employed was of a simple sinusoidal form with 6-fold symmetry. *Di Vincenzo* [2.81] has also used Monte Carlo methods to simulate the temperature-dependent structure in $C_{24}K$. In part he was interested in the phase transition from solid to liquid state in order to evaluate the temperature dependence of the shear constant C_{44} [2.73] and to explore the temperature-dependent domain structure. He also used a $V_m(\boldsymbol{r})$ composed of a 6-fold set of simple sinusoids, as in (2.2) with $\{H, K\} = \{1, 0\}$, together with an unscreened in-plane Coulomb repulsion between alkali ions. The value of $V_m(0)$ in the hexagon center was determined theoretically to be $\sim 0.18\,\mathrm{eV}$ for Rb, which is somewhat larger than the value obtained from experiment of $\sim 0.074\,\mathrm{eV}$. The picture of the disordered state that is displayed by *Di Vincenzo* [2.81] is clearly liquid-like and it evolves on cooling into an arrangement of $(\sqrt{7} \times \sqrt{7})\mathrm{R}19.1^\circ$ ordered domains.

The advantage of MD over the Monte Carlo method is, of course, in its dynamical aspect. In principle, with the proper input potential energy, one may generate the entire $g(\boldsymbol{r}, t)$, the Van Hove space-time correlation function, and its transform $S(\boldsymbol{q}, \omega)$ to compare with both X-ray $[S(\boldsymbol{q}) = \int S(\boldsymbol{q}, \omega) d\omega \rightarrow g(\boldsymbol{r}, 0)]$ and neutron data. *Fan* et al. [2.82] have employed the MD method in a simulation of the structure of stage 2 Rb in graphite and its time evolution. Various 2d sample sizes were used with periodic boundary conditions in which the repulsive Rb-Rb interactions and the in-plane density were the only input parameters. The modulation potential $V_m(\boldsymbol{r})$ was taken directly from experiment [2.74] and the direct Rb-Rb interaction was of a screened coulombic form adapted by *Plischke* [2.69] from *Visscher* and *Falicov* [2.83]. In the final version of this simulation, the MD was run for a 216 atom array with a time step of 0.04 ps [2.84]. Statistics were improved by averaging 280 individual "flash-shots" separated by $100\,\Delta t = 4\,\mathrm{ps}$. Figure 2.14 presents the circular average of $S(\boldsymbol{q}) \Rightarrow S(q)$ to compare with the circularly averaged data of *Thompson* [2.38] on HOPG for the in-plane scattering in the stage 2 Rb-GIC. The Bragg peaks of the experiment were easily removed and they were removed in the MD by an analytical method [2.82]. The agreement is quite good using an adjusted dielectric constant $\varepsilon = 2.35$, compared with $\varepsilon_0 = 1$ for free space, and an in-plane density of $C_{12}Rb$. Because the Rb liquid in graphite is uncorrelated between layers and the graphite hexagons are directly over each other in the bounding planes, the simulation, using the corrugation or modulation of a single substrate, should present an accurate picture of the real intercalant in graphite. Figure 2.15a shows the full $S(\boldsymbol{q})$ to compare with the Rb data of *Rousseaux* et al. [2.64] to be discussed. In the expanded plot of Fig. 2.15b, the

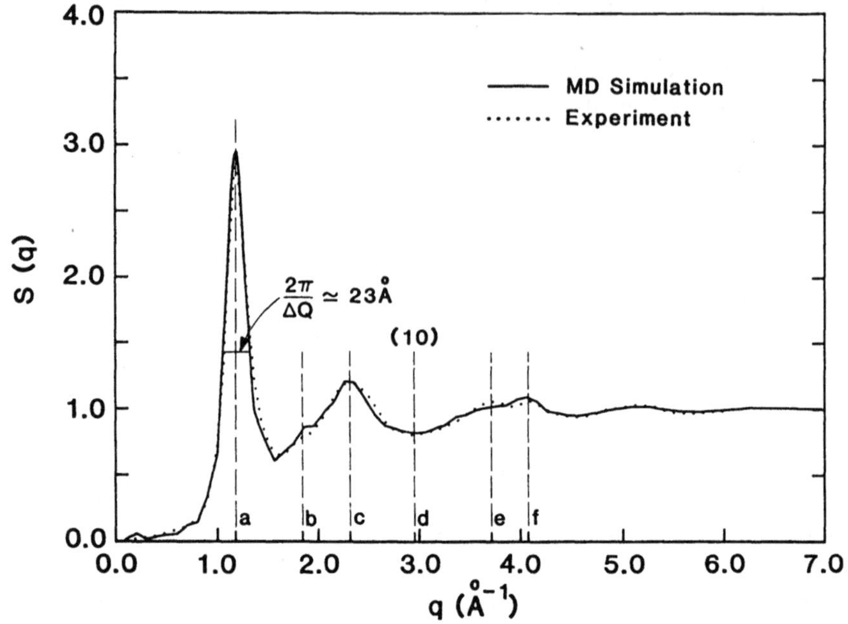

Fig. 2.14. Comparison of a MD simulation with the experimental 2d liquid structure factor for stage 2 Rb in HOPG. The 10.0 and 11.0 graphite reflections were removed (snipped) from the data and were removed analytically from the simulation. Prominent oscillations in the scattering are labeled for Fig. 2.15. (Original data from [2.38]; data reduction and MD simulation from [2.77] and [2.82])

origins of the intensity features at positions a–f in Fig. 2.14 may be seen. In the calculation for Fig. 2.15, a "supercell" method was employed [2.82] which retains the Bragg peak contributions of Rb. (There is, of course, no graphite scattering in any of these plots.)

Fan et al. [2.82] also explored the time evolution of the liquid structure on a scale long compared to an average vibrational period (~ 0.5 ps) but short compared to diffusion hopping times. It was expected that these two time scales would be separable and that the liquid-like component reported by *Zabel* et al. [2.73] would be associated with diffusing atoms that had left their original hexagon centers while the solid-like component would be associated with atoms which remained, for a measurable time, vibrating in their initial hexagons. This indeed seems to be the case and the solid-like fraction estimated by *Zabel* et al. [2.73] to be ~ 0.35 at 300 K was found to be ~ 0.36 by *Fan* et al. [2.82]. Note that while this solid-like fraction gives rise to a measurable phonon density of states at modest momentum transfers, it should not be compared directly with the value extracted by *Ohshima* et al. [2.10] of ~ 0.71 using partial registry and a DWF. The DWF that *Kamitakahara* and *Zabel* [2.71] obtained from their incoherent inelastic scattering was $B \approx 3.8\,Å^2$, whereas *Ohshima* et al. [2.10] found $B \approx 8.1\,Å^2$.

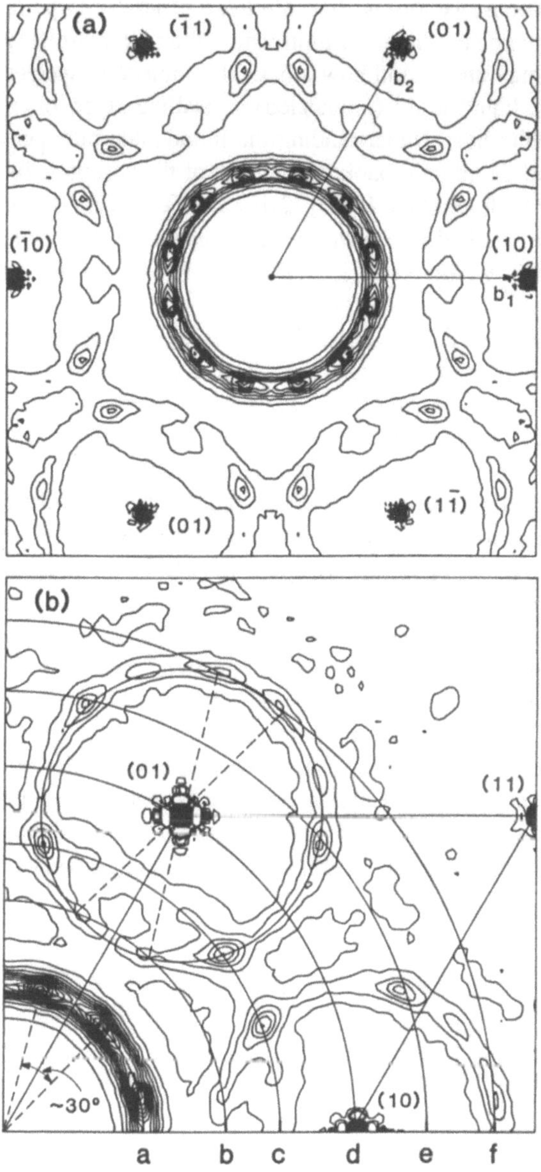

Fig. 2.15. MD simulation of the 2d Rb scattering in a graphite single crystal. (**a**) Overview showing a 6-fold modulation (with split diffuse peaks) and halos about (10) and (01) where b_1 and b_2 are graphite reciprocal lattice vectors. The "Bragg" scattering at (10) and (01) is due to the Rb. (**b**) The origins of the liquid-like oscillations in (**a**). Note the absence of a halo at (11) and the appearance of weak split features in the halos at (10) and (01). (From [2.82])

2.3.4 Acceptor Liquids

Studies of the in-plane structures of acceptor liquids have not advanced as far as the donor work reported above, largely because of the more complex intra- and intermolecular response functions of the molecular species. As we shall discuss in more detail, the melting of the classic acceptor liquid Br_2 [2.45] may be treated within the formalism of 2d melting. We wish here, however, to present the evidence for host modulation effects in the incommensurate acceptor liquids well above their melting transitions; we also wish to discuss a general viewpoint for the formation of these 2d molecular liquids within which their structure may be realized and by which we may understand how a basically molecular species proceeds from molecule to 2d liquid to 2d (modulated) crystalline layer, often incommensurate with its graphite host. Understanding the liquid state may perhaps also help us to understand the often pronounced effects of the acceptors on the host stacking noted earlier with regard to HNO_3 [2.16] and $FeCl_3$ [2.19]. We will concentrate here on HNO_3, studied by *Samuelsen* and reported by *Aberkane* et al. [2.44], on $GaCl_3$, studied by *Nadi* et al. [2.57] and on $FeCl_3$, studied by *Hashimoto* et al. [2.19].

Figure 2.16 from *Aberkane* et al. [2.44] is an $HK.0$ precession photograph of a natural crystal of graphite intercalated with HNO_3 to stage 2 (α-$C_{10}HNO_3$). It shows the in-plane liquid scattering for the crystal whose L-independent liquid scattering is given in Fig. 2.9. The main feature of interest here is the 6-fold modulation of the FSDP, with intensity maxima in the [11.0] directions. [Note the (incomplete) set of $\{10.0\}$ graphite peaks just outside the familiar ring of diffuse scattering that arises in all precession photographs.] This photograph was taken at 297 K; upon cooling *Samuelsen* [2.44] observed a progressive [11.0] enhancement of these azimuthal peaks followed finally by ordering into an incommensurate solid phase at $T_m \simeq 247$ K. It is interesting that a simple construction may be made in Fig. 2.16 in which the first diffuse peak is centered, as a halo, on each of the $\{10.0\}$ graphite points. This produces a [11.0]-type enhancement that nearly overlaps the main maxima on the primary diffuse ring about $q = 0$. In other words Fig. 2.16 reveals not only the anisotropic liquid scattering but also, perhaps, the halos that are seen in the alkalis but may be here somewhat fortuitously overlapping. It would certainly be of interest to extract a modulation potential from the HNO_3 contribution to the graphite peaks. Caution, however,

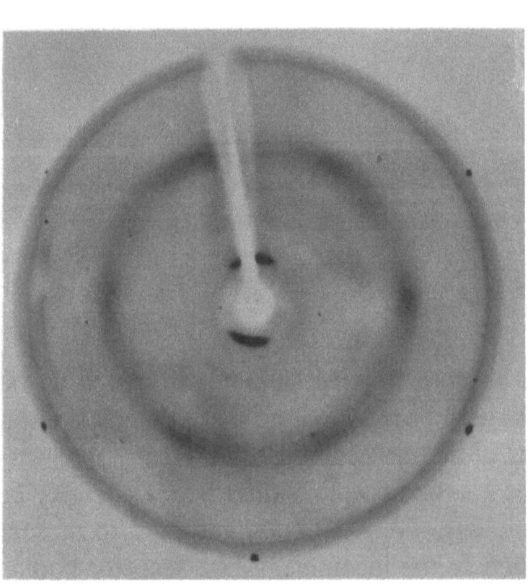

Fig. 2.16. Anisotropic in-plane liquid scattering from the stage 2 HNO_3 single crystal shown in Fig. 2.9. The 6-fold modulation in this $HK.0$ precession photograph is clear; the streaking in Fig. 2.9 is normal to this plane at a diffuse intensity maximum. (From [2.44])

must be used in replacing f_{Rb} in (2.4) with the molecular form factor of the appropriately oriented HNO_3 molecule.

Nadi et al. [2.57] present a single-crystal X-ray study of the order–disorder transition in a stage 1 $GaCl_3$. Their 2d liquid phase at room temperature, in contrast with HNO_3 [2.44], shows continuous rings of diffuse scattering in an $HK.0$ stationary monochromatic Laue photograph taken at the synchrotron at L.U.R.E. (France) with $\lambda = 1.61$ Å. We do not reproduce this photograph here because it is remarkable only in the *absence* of any modulation features. While the phase transition occurs at ~ 245 K into an ordered phase reminiscent of a layer of the crystal, the liquid state is considered to have a mixed constituency of mainly Ga_2Cl_6 molecules, as in the pure gas phase, and a minority phase of $GaCl_4^-$ ions. The evidence for the assignment comes mainly from an analysis of the $00.L$ reflections in which the z-coordinates within the intercalant layer figure in the $00.L$ intensities.

The work of *Hashimoto* et al. [2.19], which provided the L-scans shown in Fig. 2.5, was also directed more recently [2.85] towards an analysis of the in-plane liquid scattering from the stage 4 $FeCl_3$ in HOPG, studied in situ at 350°C. The ordered phase has in the past [2.86] been treated as a layer of the crystal, consisting of an Fe, octahedrally coordinated with six Cl's to form an extended sandwich of Cl-Fe-Cl that fits in the graphite gap. (In pure $FeCl_3$, these layers of Cl-Fe-Cl are stacked along the c-axis of the rhombohedral crystal.) The liquid phase, however, shows a structure factor $S(q)$ in Fig. 2.17 that is more

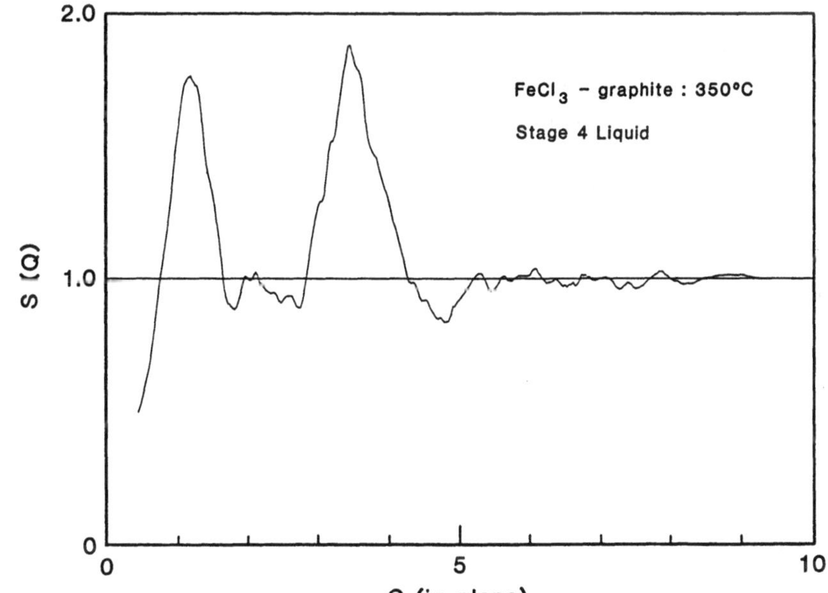

Fig. 2.17. X-ray determined in-plane structure factor $S(Q)$ for stage 4 $FeCl_3$ in HOPG measured in situ at 350°C as in Fig. 2.5. This pattern appears to be more characteristic of an Fe_2Cl_6 molecular liquid than of small islands of crystalline $FeCl_3$. (From [2.19, 86])

reminiscent of a true liquid in the breadth and damping of its diffuse peaks than it is of "micro-islands" of 2d $FeCl_3$. Following this theme, *Hashimoto* et al. [2.19, 85] are modeling the stage 4 $FeCl_3$ liquid in which the intramolecular structure factor is estimated from gas phase data [2.87] and from $00.L$ GIC data; a Percus-Yevick calculation is then used for the intermolecular part [2.59]. This work, currently in progress, takes a similar view of the liquid structure to that of *Nadi* et al. [2.57] and it would be interesting to apply the above modeling scheme to the $GaCl_3$ GIC.

While there are other brief reports of liquid scattering in acceptor compounds, the above studies serve perhaps to highlight the issues in the 2d molecular systems well above their melting transitions. It is clear that modulation effects can vary widely in these liquids.

2.4 Phase Transitions and Ordered States

2.4.1 Overview

There are several aspects of the temperature and pressure dependence of the structure of GICs that warrant our attention. Of principal interest is the development of ordered phases on cooling, i.e., the evolution of the often complex incommensurate ordered state out of the modulated liquid state. There are aspects of dimensionality and crossover effects (in-plane and out-of-plane ordering) involved here, the existence of continuous or second-order transitions versus first-order transitions, and the influence of the stage index or layer separation on the ordering behavior. We also wish to understand the ground state structures adopted by the layered intercalant, especially where they are, by dint of ion size or in-plane density, well removed from the commensurate registered structures that they often assume in stage 1 GICs [2.1–5]. In these strongly incommensurate systems, ordering in 3d may often be frustrated or pinned and unusual temperature dependences of the out-of-plane structure may result.

Modeling of the low temperature structures of both donor and acceptor compounds is becoming an increasingly popular enterprise. The MD work discussed in Sect. 2.3 has naturally been extended to low T to provide insight into the incommensurate ordered state of the alkali GICs. There have also been both Landau and more general phenomenological treatments of the domain-discommensuration states proposed for donors and acceptors and these will be covered as they apply to the available data.

2.4.2 Donors

(a) Temperature-Induced Phase Transitions

We concentrate here on the temperature dependence of the alkali structure in graphite, as the pressure dependence has been well covered in recent reviews

[2.4, 5]. It is well established that the alkali intercalates undergo an order–disorder or melting transformation on heating from low temperatures. The temperature dependence of the structure and the nature of the phase transition(s) have been discussed in a number of papers dealing mostly with X-ray and neutron scattering. These include the early studies by *Nixon, Parry* and co-workers [2.14, 15, 40, 41, 58] followed by investigations reported in [2.7, 8, 21, 27, 32, 33, 42, 43, 63–65, 73, 88–98], many of which have been reviewed in [2.5]. The major issues in these papers, aside from the structure of the disordered or liquid state discussed earlier, have involved the elucidation of the in-plane ordered structures, the transition from 2d to 3d order, and the influence of, and effect on, the graphite stacking.

We first address the cooling of the 2d liquid, whose structure we now believe to be reasonably well understood. Figure 2.11 shows the liquid K-GIC at constant temperature under varying chemical potential. The more common observation is of the liquid cooled at constant in-plane density or composition. *Hastings* et al. [2.7] studied the temperature dependence in $C_{24}K$ and showed that, at the phase transition, in-plane ordering was coincident with a crossover from 2d to 3d correlations among the alkali atoms. The 3d correlations, however, were limited by stacking faults inherited from the heavily faulted graphite. *Zabel* et al. [2.42] studied this 2d–3d crossover in $C_{24}K$ in more detail and demonstrated that the short-range order both above and below the phase transition was 2d, even though the long-range order at the critical temperature was 3d, in analogy with the 2d Ising antiferromagnetic K_2NiF_4. *Zabel* et al. [2.42] also observed that the first sharp peak of the liquid in $C_{24}K$ does not lose its long liquid-like Lorentzian wings on cooling into the ordered phase. *Clarke* et al. [2.43] also measured a 2d–3d crossover in the liquid scattering from $C_{24}Cs$ at $\sim$ 165 K but did not directly correlate this with a sharp in-plane ordering transition. In addition, they studied the intensity in a single crystal around the diffraction ring at $k = 1.16\,\text{Å}^{-1}$ and found an appreciable variation well into the liquid state above 165 K. This modulation effect, as shown in Fig. 2.15 for a stage 2 Rb-GIC, persisted to room temperature in $C_{24}Cs$, where it was first observed by *Parry* [2.58].

Perhaps the most detailed studies of the temperature dependence of the liquid scattering in $C_{24}K$ and $C_{24}Rb$ are by *Rousseaux* et al. [2.63, 64]. Given the success of molecular dynamics [2.77, 82] in accurately simulating the liquid state of a stage 2 Rb-GIC, we show in Fig. 2.18 the MD calculations performed to compare with the companion X-ray photographs of *Rousseaux* et al. [2.64]. The photographs present a distorted picture of the spherical cut (Ewald sphere) of the reciprocal space because the stationary film is in a cylindrical holder surrounding a fixed crystal [2.5, 64]. The monochromatic exposure is then flattened for reproduction. The qualitative agreement is otherwise remarkably good, including a clear indication of ordering in the MD at a temperature between 165 K and 154 K where the reported ordering temperature is between 160 K and 165 K [2.64, 73, 91]. We shall return to the detailed structure of the ordered state and note here only the evolution of the modulation pattern on cooling. The circular average of the simulation of the patterns in Fig. 2.18 [2.82] shows a modest sharp-

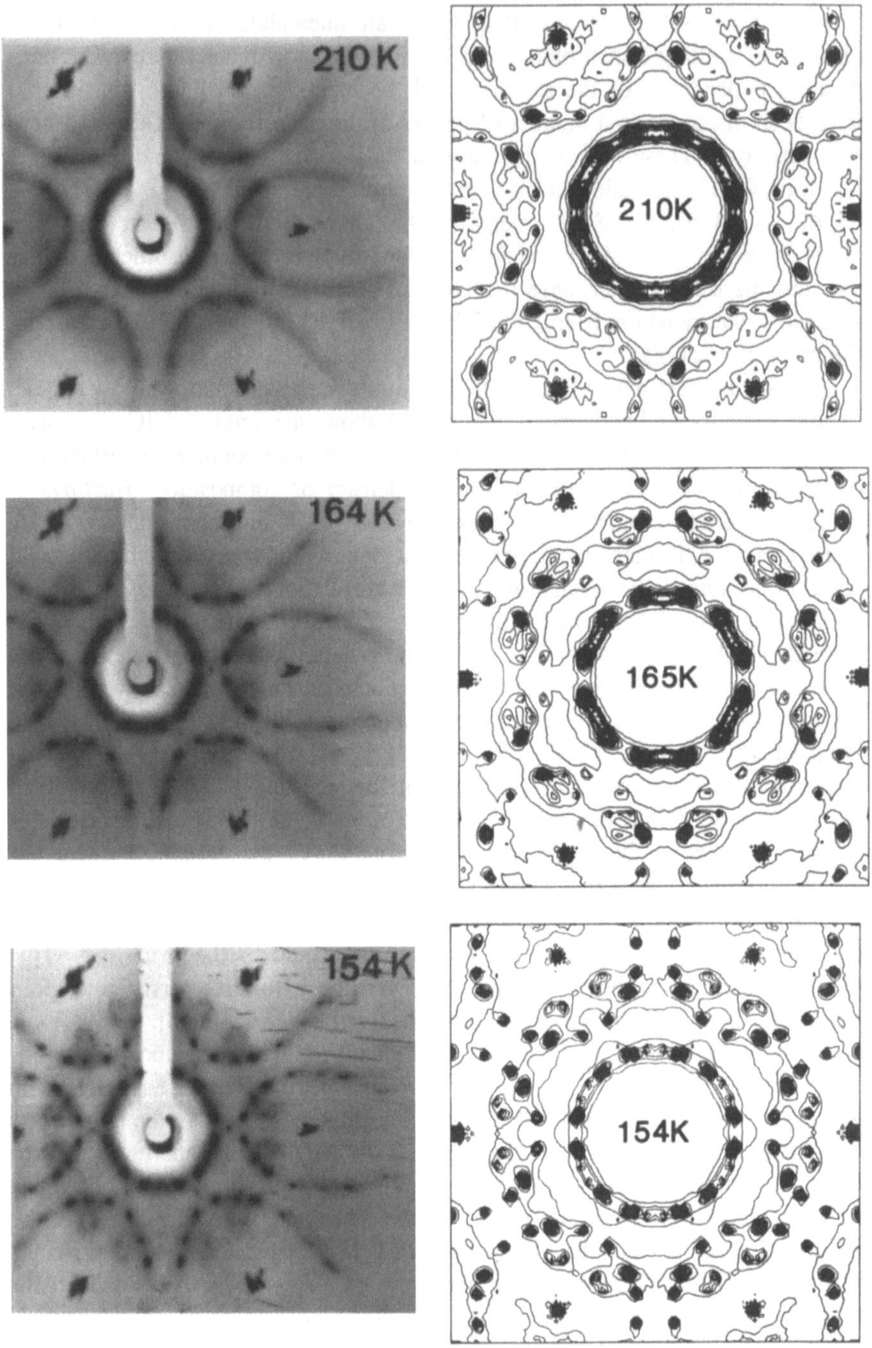

Fig. 2.18. Comparison of the monochromatic Laue data from a stage 2 Rb-graphite crystal [2.64] with the MD simulations [2.82]. The temperatures of the MD calculation are real and a phase transition occurs in the MD at ~ 160 K, in agreement with experiment. The photographic distortion comes from flattening the cylindrical Laue film. Note detailed similarities at 154 K in the weak modulation pattern

ening until quite close to T_c. A qualitative comparison with the photographic data of *Rousseaux* et al. [2.64] is difficult. However, recent counter diffractometry by *Rousseaux* and *Moret* [2.99] shows that the intensity about the FSDP in $C_{24}Rb$ at $q \simeq 1.22\,\text{Å}^{-1}$ has maxima in the graphite [11.0] directions and only a weak shoulder at $\sim \pm 15°$ off these peaks. The MD simulation, however, was done for an experimentally observed FSDP at $q \simeq 1.19\,\text{Å}^{-1}$ [2.38, 74] and shows a maximum in [10.0] with split maxima at $\sim \pm 15°$ off $\langle 10.0 \rangle$ and a second minimum in $\langle 11.0 \rangle$ rather than a maximum. Otherwise, all of the detailed modulation features about, for instance, the 10.0 graphite position are reproduced on cooling.

Rousseaux and *Moret* [2.98] have also studied the ordering in $C_{24}K$ with single-crystal counter diffractometry in which the intensity in the FSDP at $k = 1.26\,\text{Å}^{-1}$ was followed on cooling. The azimuthal variation showed an onset of splitting at ~ 180 K which evolved continually down to the ordering temperature of ~ 123.5 K. Unfortunately, the MD calculations for the stage 2 K-GIC are not yet available for comparison.

In the higher stage GICs, the ordering is somewhat more complicated, particularly with regard to the 2d–3d crossover. In stage 3 and 4 K-GIC *Mori* et al. [2.93] noted an in-plane ordering in HOPG quite similar to that observed in stage 2, albeit at a somewhat higher temperature of ~ 190 K, versus ~ 123 K for stage 2, and at an in-plane density of $\sim 5\,\%$ greater than in the stage 2 compound. The in-plane ordering, however, remained 2d – as determined by X-ray scans along c^* – down to ~ 131 K, where there was an onset of 3d ordering. The change in liquid scattering in this case seemed much more gradual, or continuous, with long Lorentzian wings at the first sharp peak that persisted on cooling well into the ordered state. *Huster* et al. [2.32] observed similar behavior in a stage 4 K-GIC single crystal, which is shown in Fig. 2.19. These profiles directly address the question of the order of the phase transition in the higher stage compounds. The Lorentzian fit is excellent in the liquid state down to ~ 150 K. Below that it is increasingly poor and could perhaps be replaced by a smeared power-law line shape appropriate to a 2d crystal. As the authors note [2.32], it is not clear from these measurements whether a smeared first-order or continuous freezing transition occurs in this compound. For the stage 2 K-GIC, *Hastings* et al. [2.7] and *Zabel* et al. [2.42] found the transition to be continuous and without hysteresis; it was, in fact, possible to extract an exponent of $\beta = 0.18$ for the vanishing of the long-range order. This is to be interpreted with caution because of the prediction by *Bak* and *Domany* [2.100] of a first-order transition in the stage 2 compounds, whose ordered alkali layers are stacked along $\boldsymbol{c}$ in a fcc $\alpha\beta\gamma\alpha\beta\gamma$ array. The *apparent* second-order character has thus been attributed to the high density of stacking faults. It would be very interesting to pursue further the 2d ordering in a stage 4 K- or Rb-GIC single crystal using the angular resolution available at a synchrotron source and good temperature control. The issues of 2d melting noted earlier would be well addressed in these higher stage alkali compounds even with the periodic host.

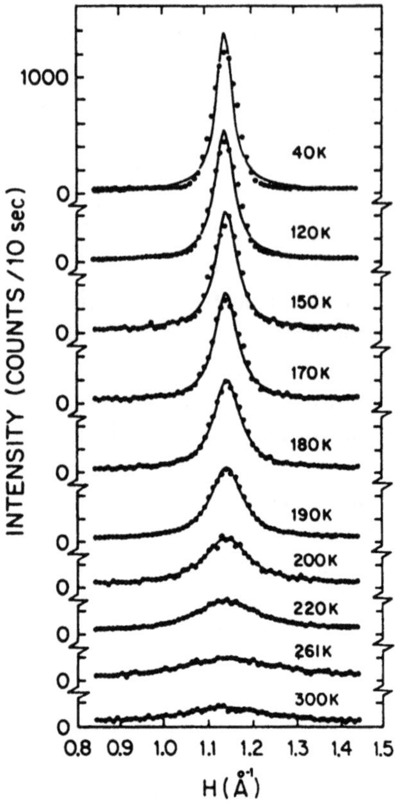

Fig. 2.19. Temperature dependence of the primary potassium (10) peak at $q_M = 1.17\,Å^{-1}$ in a stage 4 K-graphite single crystal. The solid lines are Lorentzian fits to the data indicating a classical liquid shape down to ~ 150 K. (From [2.32])

(b) In-Plane Donor Ordered Structures

The majority of the observations of in-plane ordering in the higher stage ($n \geq 2$) alkali GICs based on both single crystal and HOPG studies support the following:

- The incommensurate modulated liquid state transforms on cooling to an incommensurate 2d solid with 2d–3d crossover at the ordering transition for $n = 2$.
- The diffraction patterns upon which this generalization is based show certain common features including:
  i) A set of six primary alkali reflections rotated off the graphite [10.0] directions by an amount determined by the in-plane density. These reflections and their 6 twins (12 in all) form a primary circle which grows out of the first strong diffraction circle in the liquid as seen in Figs. 2.15 and 2.18.
  ii) The higher-order Fourier components, or Bragg peaks, of the basic alkali lattice fall off rather sharply in the (alkali) reciprocal lattice so that the reflections higher than the {20.0} set are not usually observed. Such a damping implies a strongly modulated system with large displacements off the average alkali lattice as defined by the primary reflections.

iii) The alkali pattern, much as in the liquid case, is repeated in a circle about the graphite reciprocal lattice points as a set of modulation satellites. This modulation pattern also falls off rapidly with increasing graphite reflection order.

The reciprocal lattice geometry as it appears, for example, in Fig. 2.18 in the diffraction pattern of $C_{24}Rb$ at 154 K, may be characterized by the complete set of reflections $(\boldsymbol{q}_G \pm \boldsymbol{q}_M)$ where $\boldsymbol{q}_G = 2\pi(H\boldsymbol{b}_1 + K\boldsymbol{b}_2)$ describes the graphite reciprocal lattice and $\boldsymbol{q}_M$ describes, in a similar fashion, the rotated alkali metal reciprocal lattice. Because the two lattices are incommensurate, this prescription, carried to all orders in both lattices, will fill the entire 2d reciprocal space uniformly with Bragg peaks. In fact, however, it is mainly the lower orders of the alkali set that appear, both directly and as a modulation set about the (lower-order) graphite points. Figure 2.20 from *Mori* et al. [2.21] presents this modulation pattern for $C_{24}K$, where the rotational splitting, 2θ, of a primary pair was found experimentally to be $\sim 15°$. The arrows that are used for both the "potassium" and "graphite modulation" peaks point to the respective graphite reciprocal lattice points for which they are modulation satellites. (The potassium arrows point,

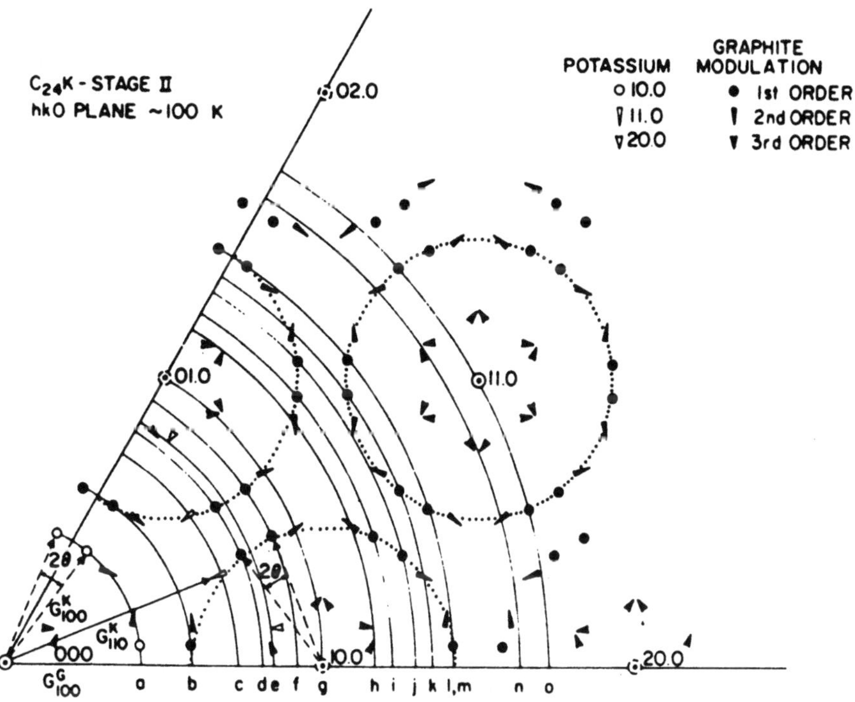

Fig. 2.20. Schematic interpretation of the $HK.0$ single-crystal diffraction patterns from the low-temperature ordered incommensurate alkali GICs drawn here for $C_{24}K$. The primary potassium intensity falls off rapidly with reflection order; the modulation structure also falls off rapidly (by "graphite modulation" we simply mean the pattern that appears about the graphite {10.0} and, weakly, {11.0} positions). The letters **a–o** used in Figs. 2.24 and 2.25 specify the Debye rings arising from the circular averages of these features. (From [2.21])

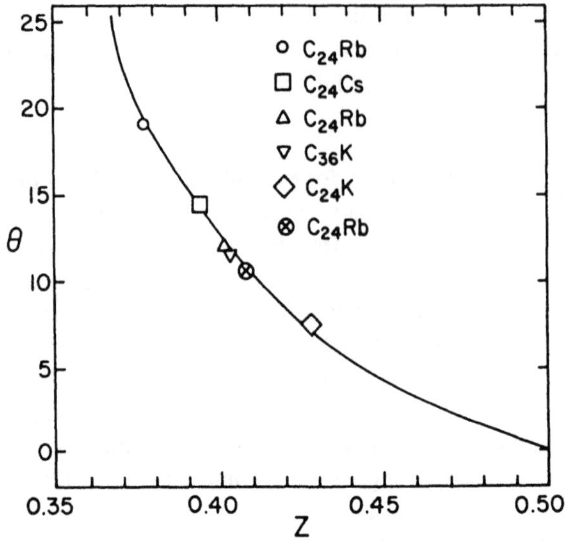

Fig. 2.21. The universal curve of rotation angle, θ, of the alkali lattice (q_M) with respect to graphite [10.0] vs the ratio $z = q_M/q_G (= a_G/a_M)$ for several stage 2 compounds. The curve is given by $\cos(30° - \theta) = (2z^2 + 1)/2\sqrt{3}z$. (From [2.21])

of course, to the origin at 000.) The notations $a, b, \ldots$ etc. refer to the HOPG Debye rings associated with the circular average of these single crystal diffraction features. The question of the L-dependent selection rules for the various peaks is addressed by *Suzuku* et al. [2.8] and *Mori* et al. [2.21].

As noted by *Zabel* [2.101], the positions of the HOPG reflections can be used together with the indexing scheme in Fig. 2.20 to calculate a rotation angle, θ, in the absence of single-crystal data. Using both HOPG and single-crystal results, *Mori* et al. [2.21] assembled all of the known rotation angles and accompanying alkali ion sizes, or primary peak positions, to develop a rather useful "universal curve", shown in Fig. 2.21 for θ vs $z = a_G/a_M$ (a is the lattice parameter). The condition for this universal curve is that the second-order alkali (11.0) peak falls on the first-order modulation circle about, for example, the (10.0) position of graphite. The physical rationale for Fig. 2.21 was subsequently presented by *Yamada* and *Naiki* [2.91] and *Suzuki* and *Suematsu* [2.92], to which we shall return shortly.

While various registered and near-registered structures have been proposed for the in-plane incommensurate ordered state of the alkali GICs, we confine our attention here to those structural proposals for the ground state (and its temperature dependence) of a strongly modulated alkali layer in intimate contact with its host. In principle, the modulation scheme developed for the liquid state [2.60] should extend naturally to the ordered solid. In practice, this is ultimately how one may proceed, i.e., use theory [2.60] and experiment [2.74, 77] to extract a modulation potential; apply that potential to the liquid via a MD simulation, as in Figs. 2.15 and 2.18, to verify its utility; cool the liquid on the computer to simulate the ordered state (and its transform), displaying both the diffraction patterns and the real space positions. Part of this has already been accomplished and will be discussed shortly. We wish first, however, to cover the experiments and proposed models.

In principle, for a weakly modulated system, the relevant diffraction theory may be developed for the case in which the solid 2d alkali lattice is modulated by its graphite host into a rotated pattern as in Fig. 2.20. Such a theory was developed by *Suzuki* et al. [2.8] that naturally reproduced the observed peaks and c-axis selection rules using an $\alpha\beta\gamma\alpha\beta\gamma$ (fcc) stacking of the modulated alkali layers [2.40]. The intensities were not given correctly because, for tractability, a small value of $\boldsymbol{q}\cdot\boldsymbol{\mu}_G$ ($\boldsymbol{q}$ is the diffraction vector, and $\boldsymbol{\mu}_G$ is the Gth component of the induced mass density wave, where $\boldsymbol{G}$ is a graphite reciprocal lattice vector) had to be assumed, which we now know is not correct. The quantity $|\boldsymbol{\mu}_{\mathrm{G}}|$, which is analogous to $\langle \varrho_{q_{HK}} \rangle$ in Sect. 2.3, may be appreciable in these strongly modulated layers; this in turn requires that the diffraction theory be carried out to higher-order (nonlinear) terms in the induced displacements for which the analytic expressions (sums) become rapidly unmanageable.

The apparent failure of this modulated alkali calculation, along with certain distinct features of the observed single crystal intensity distribution, led *Mori* et al. [2.21] to propose a model of alkali-modulated graphite in explanation of Fig. 2.20. While Newton's third law requires both modulations (alkali by graphite and graphite by alkali), and there is, as noted earlier, ample evidence for the rumpling of graphite by the intercalant, we shall not devote more space to this model. The fit achieved by *Mori* et al. [2.21], while quite good, required a degree of graphite deformation that was unreasonably large.

An alternative approach to the ordered alkali structure was suggested by *Dicenzo* [2.102]. In this scheme, the ordered lattice is rotated by the experimentally observed amount and then relaxed into the nearest available hexagon centers, taking care not to bring two alkalis too close. The observed modulation pattern of Fig. 2.20 is thereby reproduced and the "universal curve" in Fig. 2.21 is roughly generated by optimizing θ for each in-plane density. The model represents a way of forcing the observed modulation on the 2d solid alkali and it reveals the natural tendency of the alkali lattice to relax into a $(\sqrt{7}\times\sqrt{7})$R19.11° structure for which the in-plane density is $C_{14}M$ (M = Am). The procedure suggested by *Dicenzo* [2.102] thereby permits a rather practical solution to the structure of the ordered state that does not require any knowledge of the interaction potential or other physical parameters.

In analogy with other incommensurate systems, *Clarke* et al. [2.90] first proposed a domain-discommensuration state for the ordered alkali GICs. This model, which was developed for $C_{24}Cs$, treated the ordered state as a periodic array of $(\sqrt{7}\times\sqrt{7})$R19.11° domains, with domain walls spaced so as to produce the correct average (incommensurate) alkali lattice and rotation angle. The domain walls or discommensurations were proposed to occur along [11.0] directions. This model of course also reproduced the positions of the observed alkali peaks and modulation satellites (all models which permit reflections of the form $\boldsymbol{q}_{\mathrm{G}}\pm\boldsymbol{q}_{\mathrm{M}}$

will do this correctly) and it represented an attractive candidate for the ground state of the modulated alkali layers.

Following *Clarke* et al. [2.90], *Yamada* and *Naiki* [2.91] investigated the thermodynamics of the domain-discommensuration model in close analogy with the incommensurate charge-density-wave ground state in the transition metal dichalcogenides [2.103]. Noting that the mass density wave satisfying a simple commensurability requirement corresponded to the $(\sqrt{7}\times\sqrt{7})$R19.11° structure, *Yamada* and *Naiki* [2.91] considered that structure to be stabilized by the commensurability energy. However, the in-plane density of the $(\sqrt{7}\times\sqrt{7})$ structure is $C_{14}M$ whereas $C_{12}M$ is closer to the actual alkali concentration. The competition engendered by the two types of ordering, $(\sqrt{7}\times\sqrt{7})$ vs close-packed alkali, leads not only to the observed domain-discommensuration ground state but also, via a simple energy argument, to the "universal curve" of *Mori* et al. [2.21].

The thermodynamic treatment by *Yamada* and *Naiki* [2.91] was extended by *Suzuki* and *Suematsu* [2.92] to explain the "universal curve" in more detail and to make a clear prediction both for the relative intensities of the observed X-ray diffraction peaks and for the rotation angle of the large hexagonal domain superlattice off the graphite [10.0] direction (it is not simply in a [11.0] direction). *Suzuki* [2.95] then developed the treatment further to calculate the actual intensities of the observed X-ray reflections to compare with experiment. These papers taken together [2.90–92, 95] have provided essential insight into the probable ground state of the ordered alkali GICs. They have also given us a framework for understanding the disordering process as a domain wall melting, or thickening, and have led *Zabel* and co-workers [2.73] to attribute the measured stiffening of the shear constant, C_{44}, on cooling through the ordering transition in both $C_{24}K$ and $C_{24}Rb$, to domain wall pinning. We shall therefore devote some space to a discussion of the success of this model in its detailed predictions.

In their photographic study of ordering in single crystals of $C_{24}K$ and $C_{24}Rb$, *Rousseaux* et al. [2.96] have reviewed the evidence for a domain-

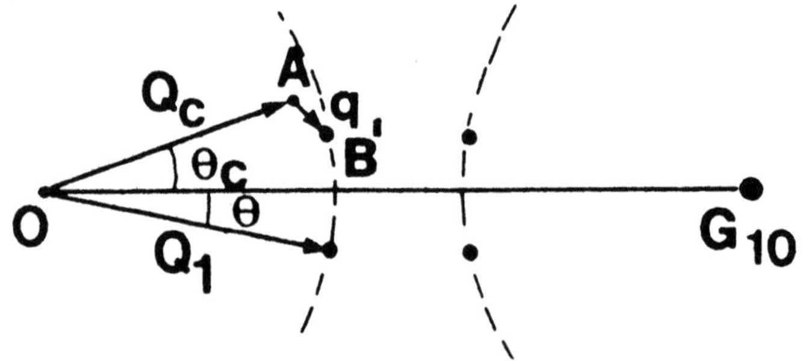

Fig. 2.22. Schematic indication of the prominent reciprocal lattice features in the *Yamada-Naiki* [2.91] and *Suzuki-Suematsu* [2.92] formulations of the $(\sqrt{7}\times\sqrt{7})$R19.11° domain-discommensuration model. Vector OB = $\boldsymbol{Q}_1$ is the observed q_M (a primary alkali reflection) as discussed in the text; θ is the observed rotation angle; OA = $\boldsymbol{Q}_c$ is the $(\sqrt{7}\times\sqrt{7})$R19.11° vector given by $Q_c = (2\pi/\sqrt{7})(1/a_G\cos 30°)$ ($a_G = 2.46$ Å); AB = $\boldsymbol{q}_1$ is a modulation satellite of the registered $(\sqrt{7}\times\sqrt{7})$ lattice which produces the principal (observed) wave vector of the actual incommensurate alkali structure. G_{10} is the graphite 10.0 point. (From [2.96])

discommensuration model for the alkali GICs, using a notation similar to that of *Suzuki* [2.95]. Figure 2.22, from *Rousseaux* et al. [2.96] schematically shows the relevant wave vectors and angles in this model. The basic wave vector, Q_c, at θ_c from the graphite [10.0] direction, is for the $(\sqrt{7}\times\sqrt{7})$R19.11° commensurate lattice. Thus Q_c is given by $Q_c = 2\pi/\sqrt{7}|a^*|$, $a^* = 2/\sqrt{3}a$ ($a = 2.46$ Å), and $\theta_c =$ 19.11°; $\boldsymbol{Q}_1$ is the observed primary alkali reflection at $\boldsymbol{q}_{\mathrm{Am}}$, rotated an amount θ off the graphite [10.0] and G_{10} is the graphite 10.0 point. The vector AB = $\boldsymbol{q}_1$ is one of the three modulation wave vectors of the commensurate $(\sqrt{7}\times\sqrt{7})$ structure that describe the domain-discommensuration state; the normal to the domain wall of this hexagonal domain superlattice is rotated off the graphite [10.0] direction by an angle Ω [2.95], where $\Omega = -60° - \theta_c + \arccos|q_1|/|G_{10}|$; the domain size, L, is given by $L = (1/\sqrt{7})|G_{10}|/|q_1|$ [2.95, 96]. Clearly, as $|q_1| \rightarrow 0$, $L \rightarrow \infty$ and $\Omega = 30° - \theta_c$, which is simply an infinite $(\sqrt{7}\times\sqrt{7})$R19.11° commensurate structure. We may also specify, after *Yamada* and *Naiki* [2.91], a vector $\boldsymbol{l}$ where $\boldsymbol{q}_1 \cdot \boldsymbol{l} = 2\pi$; $\boldsymbol{l}_1$ is thereby a vector connecting domains of the same phase of the mass density wave and $\boldsymbol{l}_1, \boldsymbol{l}_2$ specify a repeating rhombic supercell [2.91].

From Fig. 2.22, we note that $q_1^2 = Q_c^2 + Q_1^2 - 2Q_1Q_c\cos(\theta_c - \theta)$, which permits an estimate of L/a and Ω for the stage 2 alkali GICs $C_{24}K$, $C_{24}Rb$ and $C_{24}Cs$. Tables 2.1 and 2.2 give the values for L/a and Ω calculated from the data. Actually, the data for Rb vary as in Fig. 2.21, and if we choose $|Q_1| = 1.19$ Å^{-1} as observed by *Ohshima* et al. [2.10], which yields $\theta = 11.15°$ from the universal curve, we may calculate $|q_1| = 0.173$ Å^{-1} and $L/a = 6.42$. As *Rousseaux* et al. [2.96] point out, a domain size of $L/a \simeq 11$ contains about 18–19 atoms, whereas a domain with $L/a \simeq 4$ contains only 3–4 atoms and is a rather unlikely ground state for the ordered K-GIC. Even $C_{24}Rb$ at 8–9 atoms per domain is at the border of what could sensibly be called a moderately incommensurate domain-discommensuration state. Figure 2.23 shows the real space density maps, which are the Fourier inversion of the reciprocal space solutions to the free energy

Table 2.1. Values of $Q_1(= q_M)$ in the stage 2 ordered alkali phases; θ, the rotation angle with respect to ⟨10.0⟩; and q_1, the modulation wave vector of the $(\sqrt{7}\times\sqrt{7})$R19.11° structure. (From [2.96])

Metal	Q_1 [Å^{-1}]	θ	q_1 [Å^{-1}]
K	1.26	7.5°	0.280
Rb	1.225	11°	0.197
Cs	1.158	14.5°	0.101

Table 2.2. Domain parameters for the incommensurate donor alkali ordered phases in Table 2.1 calculated from the theory of *Suzuki* [2.95]; L/a is the ratio of domain size to graphite lattice parameter and Ω is the rotation of the domain lattice off the graphite ⟨10.0⟩ direction. (From [2.96])

Metal	L/a	Ω
K	3.98	5.44°
Rb	5.66	7.06°
Cs	11.04	8.93°

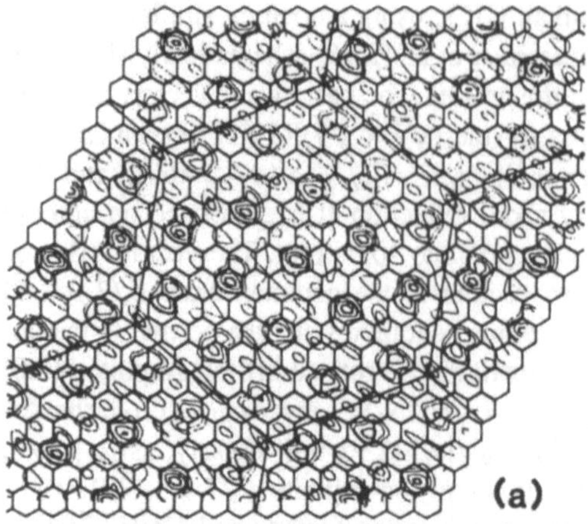

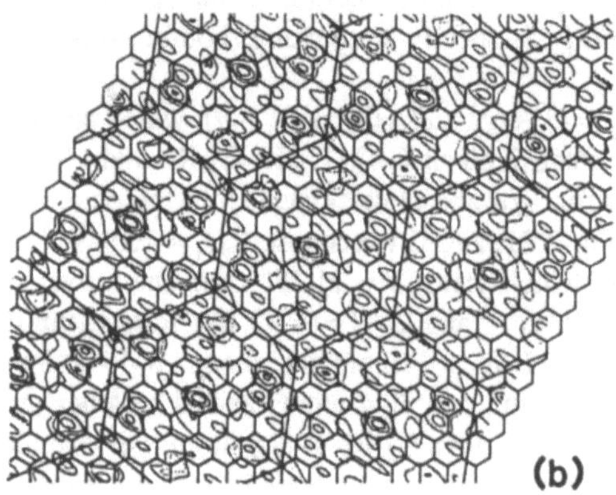

Fig. 2.23. Calculation via the *Suzuki* theory [2.95] of the alkali density distribution in (a) for $C_{24}Cs$ [2.95] and in (b) for $C_{24}Rb$ [2.73]. The rotation angle for $C_{24}Cs$ is larger (closer to 19.11°) and the domain size, as indicated by solid lines, is correspondingly larger than for $C_{24}Rb$ (Tables 2.1 and 2.2). In $C_{24}Rb$ only the central atom appears registered and the domain interpretation becomes somewhat tenuous

equations of *Suzuki* [2.95] as reported in [2.95, 73] in which it is clear that, for Cs, the atoms near a domain center are well registered in a $(\sqrt{7} \times \sqrt{7})$ array while the atoms near the domain wall are rather spread out. For $C_{24}Rb$, only the central atom could be considered registered; the rest seem to belong more to the "wall", being smeared over at least two hexagons.

An important comparison for the domain calculations is with the observed single crystal X-ray intensities. Figure 2.24 is adapted from *Suzuki* [2.95], showing the comparison for $C_{36}Cs$ using data from [2.94], in which the stacking fault smearing has been approximately accounted for. The relative intensities around the modulation circle at the graphite 10.0 position, see Fig. 2.20, are qualitatively accounted for by the domain-discommensuration theory. A better fit, however, was obtained by *Winokur* and *Clarke* [2.94] by first estimating a domain size L from theory [2.91] and placing all the atoms in registered $(\sqrt{7} \times \sqrt{7})$ positions. This domain then served as a box in which the Cs atoms were relaxed under a combined Born-Mayer repulsive interaction and 6-fold sinusoidal substrate attraction. By means of a careful choice of range and strength of these interactions,

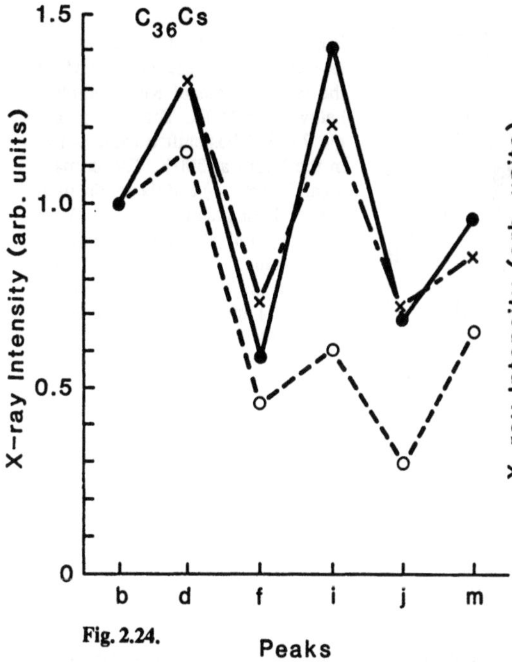

Fig. 2.24.

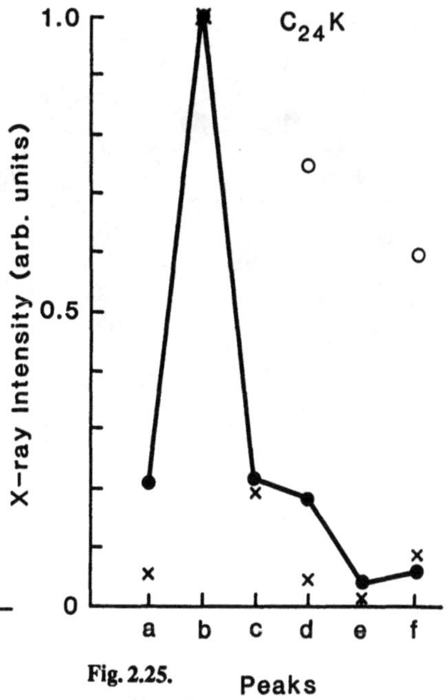

Fig. 2.25.

Fig. 2.24. X-ray intensity for several of the peaks specified in Fig. 2.20 as measured by *Winokur* and *Clarke* [2.94] in single crystal $C_{36}Cs$ and calculated by *Suzuki* [2.95] with peak *b* set in both to 1.0. The domain relaxation calculation by *Winokur* and *Clarke* [2.94] is also included: (•) measured [2.94]; (x) calculated [2.94]; (o) calculated [2.95]

Fig. 2.25. X-ray intensity comparison for recent $C_{24}K$ single-crystal results [2.98] with the calculations of *Suzuki* [2.95] and with the relaxed alkali registry model of *Dicenzo* [2.102]. The notation is again that of Fig. 2.20: (•) measured [2.98]; (x) calculated [2.102, 98]; (o) calculated [2.95]

a good fit to the X-ray intensities was achieved, in which the atoms near the domain wall were found to be rotated with respect to the center. This (somewhat empirical) domain fitting is included in Fig. 2.24.

Figure 2.25 shows recent data of *Rousseaux* and *Moret* [2.98] from a $C_{24}K$ single crystal for which we should not perhaps expect a good fit to the *Suzuki* [2.95] model for the reason given earlier. The data, however, are rather well fit by the scheme suggested by *Dicenzo* [2.102] in which every atom is relaxed to within 0.3 Å of a hexagon center and where the major discrepancies are with the stacking-fault-smeared peaks.

Finally we present in more detail in Fig. 2.26 the diffraction pattern at 154 K for $C_{24}Rb$ (stage Rb-GIC) from the MD simulation [2.82] shown in Fig. 2.18. It is clear that most of the features observed photographically by *Rousseaux* et al. [2.64] appear in this 2d pattern. The presence of the modulation structure along with the damped set of alkali peaks is consistent with all of the data but a quantitative fit remains to be done; it is also clear that the modulation circles do not extend out much beyond the graphite {10.0} positions. Certainly, the intention is to fit the X-ray observations using the MD simulation and then to

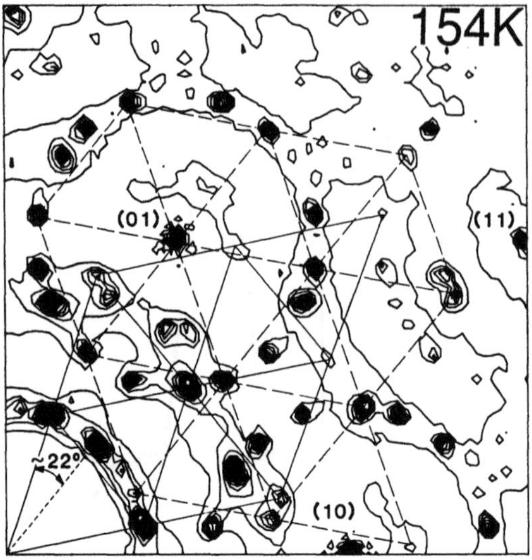

Fig. 2.26. Detailed plot of the single-crystal pattern calculated via MD for the ordered state of Rb in graphite below $T_c = 160\,K$. The rotation angle $\theta = 11°$ has shifted from $\sim 15°$ in the liquid at 300 K. The primary alkali lattice (—) and the modulation lattice about graphite (01) (- - -) are indicated and all intensity maxima are accounted for by these and their equivalents. The modulation lattice about (11) appears to be quite weak. (From [2.82])

display the alkali positions in a way that reveals not only the domain structure but also the domain wall melting through the phase transition. In addition, to the extent that the pure modulation of the alkali lattice by the graphite host potential is insufficient to fit the data, some back modulation may still be required.

The influence of the proposed domain-discommensuration state on the 3d out-of-plane ordering has also been discussed by *Winokur* and *Clarke* [2.94]. They explain the *loss* of *c*-axis correlation on cooling by a pinning effect associated with the incommensurability of the stacked layers, nominally $\alpha\beta\gamma\alpha\beta\gamma$, that have developed a partially registered state with domain walls which become increasingly hard to stack. *Winokur* and *Clarke* [2.104] also studied the *c*-axis correlation in a stage 2 Cs-GIC single crystal in some detail and found that the discrete stacking changes may actually be decomposed into a string of polytypes, whereby the nominally fcc–hcp transition in the alkali stacking may be better characterized via a scheme similar to that given by *Bruinsma* and *Zangwill* [2.105] for the Laves phases. It would perhaps also be interesting to model these L-dependent profiles using a *Hendricks-Teller* scheme [2.28].

In $C_{24}K$, there is a pronounced change of in-plane order at a lower transition temperature of $T_L = 95\,K$ [2.7, 40]. While it has been tempting both theoretically [2.106] and experimentally [2.97] to ascribe this lower transition to a devil's staircase-like fragmentation of commensurate phases, *Rousseaux* et al. [2.5, 64] present rather convincing evidence for a single phase assignment in $C_{24}K$ at low temperatures; they can completely index their high resolution X-ray photographs at 10 K using an oblique unit cell with symmetry related domains and graphite-induced modulations. This is not, however, to suggest that the interesting scheme proposed by *Mitani* and *Niizeki* [2.106] is inappropriate to these intercalation systems; it is only that there does not yet appear to be sound evidence for it, although certainly the observed rotation angles, θ vs z, do not all fall on the smooth "universal curve" in Fig. 2.21.

2.4.3 Acceptors – General Remarks

As noted earlier, our understanding of the phase transitions in graphite–acceptor compounds is more limited because of the complexity associated with their molecular nature; namely, one is faced with intra- and intermolecular electrostatic and elastic interactions which are considerably more difficult to analyze and model than for the single atoms of the alkali metals. For instance, the derivation of the modulation potential of the graphite host for intercalant species such as HNO_3, $FeCl_3$ or $SbCl_5$, and its subsequent application in MD simulations, as developed in Sects. 2.3.2 and 2.3.3 for stage 2 alkali metal GICs, appears to be rather complicated at present. The complexity of the host–intercalant interactions leads to both in-plane and out-of-plane effects. The most typical of the in-plane effects is the frequent low symmetry of the intercalant layer, rectangular as in bromine (Br_2) [2.45, 107] or oblique as in iodine chloride (ICl) [2.108, 109]. Among the out-of-plane effects, there is the possible loss of A/A or A/B stacking for the graphite bounding layers as pointed out in Sect. 2.1 for HNO_3 [2.16] and stage 4 $FeCl_3$ [2.19] GICs.

Another important aspect of acceptor GICs is that the ordered phases of the molecular intercalants retain some of the structural properties of their pristine form, so that the structure of the intercalant layer often resembles a particular layered arrangement that can be isolated in the structure of the pristine solid, as in Br_2 [2.45, 110] or $FeCl_3$ [2.86, 111]. A further consquence is that the intercalant layers are rather "soft", because of the ionic forces, in contrast with the in-plane rigidity of the graphite layers due to the covalent C-C bonds. Temperature and pressure therefore have much stronger effects on the intercalant than on the host, which sponsors a variety of incommensurate–commensurate transitions.

These simple arguments explain the wealth of structural effects and phase transitions which can be expected in acceptor GICs. However, the available data are still restricted to recent studies of a few systems and there remain a number of unexplored ones. Here we shall cover the more advanced studies dealing with the following intercalants: halogens (Br_2, ICl), HNO_3, chlorides ($SbCl_5$, $FeCl_3$, $NiCl_2$, $CoCl_2$) and fluorides.

(a) Temperature-Induced Phase Transitions

Liquid–Solid Transitions. Studies of the liquid-solid transitions induced by cooling have been performed for Br_2, HNO_3 and $SbCl_5$ in graphite. Following the early work of *Eeles* and *Turnbull* [2.107], a lot of attention has been paid to Br_2 over the past few years. In dilute samples, *Bardhan* and *Chung* [2.112, 113] found a continuous melting transition at $T_m \simeq 374\,K$. Detailed in situ diffraction measurements by *Erbil* et al. [2.45] on a stage 4 compound showed that melting occurs at $T_m = 373.41 \pm 0.1$ K as deduced from the peak in the diffuse X-ray scattering (susceptibility) in the immediate vicinity of a Br ordering reflection. The structure of the solid phase is incommensurate (see below). Above T_m, the liquid phase shows an interesting behavior in that it has an anisotropic in-plane character which actually appears to invert as it goes through the transition (the ratio of the correlation lengths in the x and y directions goes from a value of 3 to below 1, within less than 1 K). The study of this crossover, however, was inhibited by intensity and resolution limitations. As noted earlier, a transition from power-law to Lorentzian line shapes was also observed for this stage 4 Br compound, thereby establishing it as a clear example of 2d melting.

A marked anisotropy in the liquid state is observed in C_{5n}-HNO_3, as shown above (Sect. 2.3.4, Fig. 2.16), which is clearly associated with the influence of the graphite potential [2.16, 44]. In contrast with bromine, for this compound *Samuelsen* et al. [2.16] found a liquid–solid transition with a hysteresis of 3–5 K; i.e., for stage 2 the transition is observed at 249 ± 0.5 K on heating and 244 ± 1 K on cooling. The hysteresis may, however, be due to inhomogeneities in the crystal (local variations of in-plane density, for instance) and more work should be done before relating this result to the predictions for 2d melting discussed earlier [2.46, 47]. We note in passing that the motion and diffusion constant of HNO_3 in the liquid phase have been probed by quasi-elastic neutron scattering [2.114]. The authors found a continuous melting transition leading to a lattice liquid with strong orientational correlations, in agreement with the diffraction data. This aspect is treated in more detail by *Magerl* in Chap. 6 of this volume.

At the liquid–solid transition in the HNO_3 GIC, the graphite stacking experiences unusual changes, which were first pointed out by *Nixon* et al. [2.15] and noted above in Sect. 2.1. While in the liquid a classical $A/AB/BA/$stacking is found for stage 2, the symmetric arrangement of the graphite bounding layers is not maintained in the solid phase and two different models have been proposed to account for the intensity of the reflections from the graphite sublattice. *Samuelsen* et al. [2.16] suggest a shift across the intercalant layer by a short vector (on the order of $0.2a_G + 0.05b_G$) leading to a sequence expressed as $A/A'B'/BA$ and related to a further tilt of the NO_3^- ion (the NO_3^- triangle was found to be oriented almost perpendicular to the layers in the liquid [2.115, 116] as in Fig. 2.27). A different model was proposed by *Shaked* et al. [2.117] for a stage 4 sample where they consider shifts of the type $(1/2)(\boldsymbol{a}_G + \boldsymbol{b}_G)$, $(1/2)\boldsymbol{a}_G$ or $(1/2)\boldsymbol{b}_G$ relating the graphite bounding layers, where $\boldsymbol{a}_G$ and $\boldsymbol{b}_G$ are real-space graphite cell vectors $\boldsymbol{a}_1$ and $\boldsymbol{a}_2$. More quantitative studies are obviously needed to understand the mechanism of this stacking effect and its relation to the complex graphite–nitric acid interaction.

In $SbCl_5$-intercalated graphite, where the observation of multiple phases is associated with at least two segregated molecular species, $SbCl_3$ and $SbCl_6^-$, liquid–solid transitions have been studied by *Homma* and *Clarke* [2.118, 119]. The $(\sqrt{39} \times \sqrt{39})$ superstructure of the $SbCl_3$ species is found to melt at about 230 K with a large range of hysteretic behavior (~ 25 K for stage 2) and with stage-dependent effects. Substantial scattering from the ordered phase is, in fact,

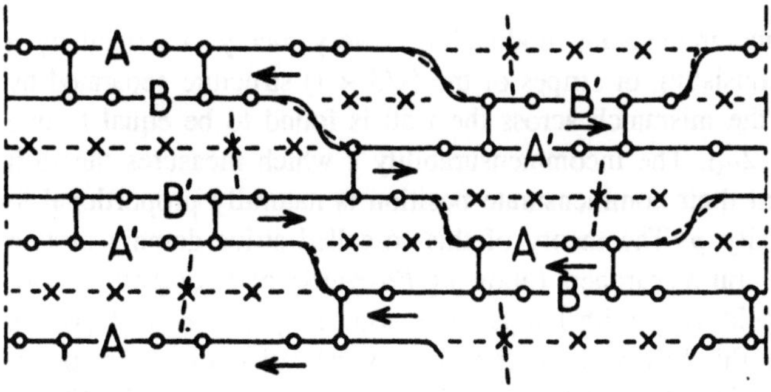

Fig. 2.27. Schematic domain model and graphite stacking scheme for stage 2 $\alpha C_{5n}HNO_3$. The direction of sliding (arrows) and the tilt of the NO_3 triangles (nearly vertical dashed lines) are indicated. (From [2.16])

retained in the liquid phase, especially for low stage compounds. This shows an influence of interlayer forces which could be attributed to the rather large dipole moment of the $SbCl_3$ molecules [2.119]. The second phase arising from the $SbCl_6^-$ regions melts at $T_m \sim 455$ K. The structure of the solid phase below T_m is incommensurate but close to $(\sqrt{7} \times \sqrt{7})$ with respect to the graphite host, while the liquid diffuse scattering is isotropic. A puzzling observation is that the incommensurate $SbCl_6^-$ structure displays reasonably long-range positional correlations (> 100 Å) while a substantial amount of orientational disorder is observed. Moreover the authors report a rotation of the $SbCl_6^-$ layer relative to graphite as the temperature increases towards T_m. It is our feeling that, despite the variety of very interesting phenomena observed in $SbCl_5$ GICs, their multiphase nature puts a rather large obstacle in the way of quantitative studies of quasi-2d liquid–solid transitions.

Commensurate–Incommensurate Transitions. Although a good deal of evidence for incommensurate structures and commensurate-incommensurate (C-I) transitions has been reported for acceptor-GICs, a proper study usually requires single-crystal X-ray diffraction (electron diffraction is not quantitative enough and neutron diffraction is not usable with small natural graphite flakes). This explains the relative scarcity of available quantitative data. The C-I transition which occurs in Br_2-GIC compounds of approximate in-plane concentration C_7Br_2 is the most fully documented transition and has been reviewed recently [2.5, 120]. It was first reported by *Bardhan* et al. [2.121] and was subsequently studied extensively by *Kortan* et al. [2.122], *Erbil* et al. [2.45] and *Mochrie* et al. [2.123] using high-resolution synchrotron diffraction on a stage 4 Kish graphite single crystal sample. At room temperature, the Br_2 layer adopts a commensurate $(\sqrt{3} \times 7)$ centered unit cell whose structure, i.e. the arrangement of the Br_2 molecules, is still controversial (Sect. 2.4.3b). The C-I transition at $T_{C-I} = 69.5°C$ corresponds to a modulation in the sevenfold direction. It is evidenced by a splitting of the reflections of the bromine reciprocal lattice and a temperature-dependent displacement of the resulting satellite reflections in this direction, which is the graphite [110] direction. This 1d incommensurate modulation is very nicely described by a domain-wall model consisting of stripes of the ($\sqrt{3} \times 7$) structure separated by domain walls where the mismatch across the wall is found to be equal to two graphite hexagons or $2a_G$. The incommensurability ε which measures the shift of the reflections from their commensurate position is naturally proportional to the domain wall density ϱ. The theory of this so-called stripe-domain system was developed by several researchers following *Pokrovsky* and *Talapov* [2.124]. It predicts that ε should vary with temperature according to $\varepsilon \propto t^{1/2}$ [where $t = (T - T_c)/T_c$] and this has been verified for several reflections corresponding to different harmonics. A good fit is obtained as shown in Fig. 2.28 [2.45], although a deviation from this power law arises at a higher temperature as the domain-wall density saturates because of repulsive wall–wall interactions. The domain size thereby decreases to a limit of ~ 130 Å (7 or 8 unit cells).

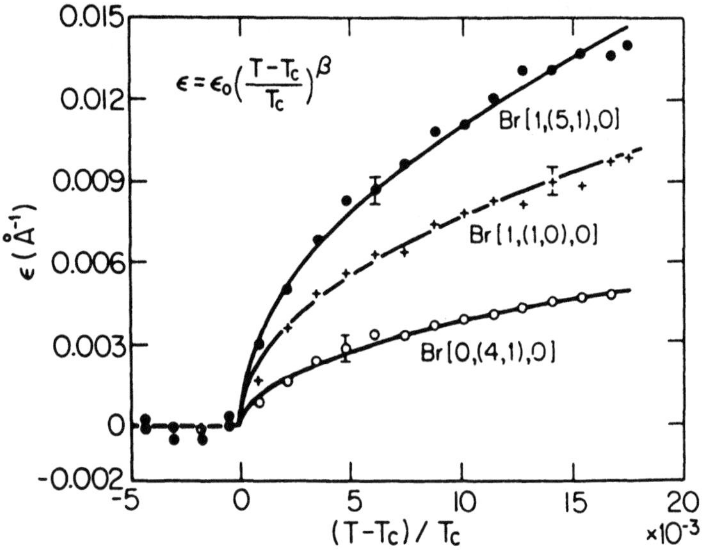

Fig. 2.28. Shift of some principal X-ray peak positions as a function of reduced temperature for the C-I transition in a stage 4 Br_2-GIC crystal. The curves are fits with $\beta = 0.5$ in agreement with predictions of the stripe-domain incommensurate structure. (From [2.45])

A second prediction, useful in the analysis of the stage 4 Br_2-GIC, deals with the line shapes of the reflections in the stripe-domain phase. As noted earlier, in a 2d solid, one expects power law singularities of the form $q^{-2+\eta}$, corresponding to an algebraic form of the order parameter correlation function, rather than the δ functions found in 3d Bragg peaks. For a stripe-domain system, η takes the form $\eta = 2n^2/p^2$, where n is the harmonic index of the reflection and p is the multiplicity of the structure with respect to the underlying lattice, which is 7 in the present case [2.125]. A remarkable aspect of this prediction is therefore that the peak profiles are independent of the temperature and the domain-wall density. A reasonable agreement with the theory was obtained from power-law fits of 5

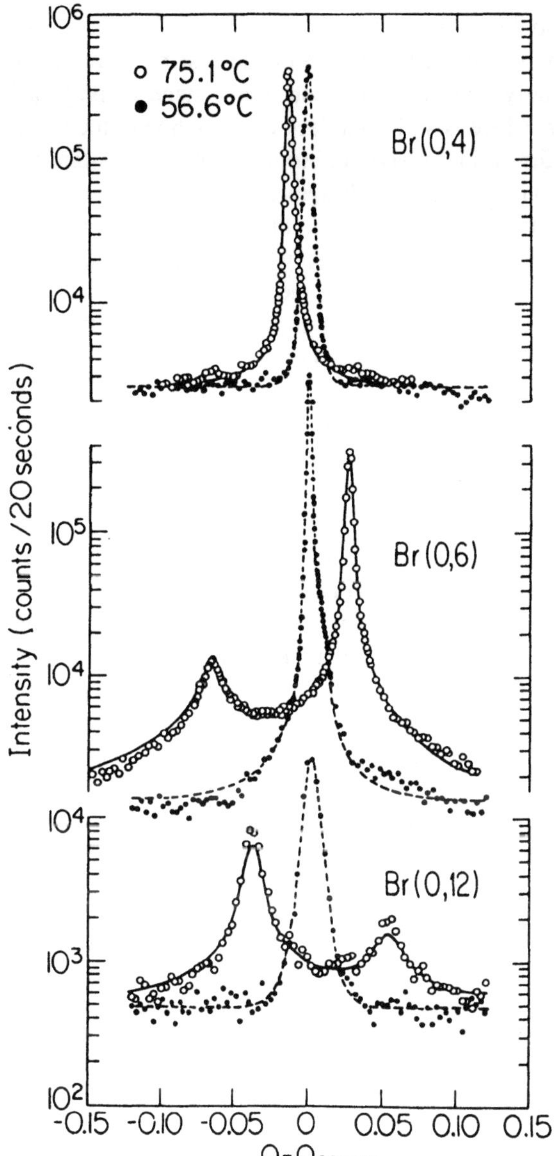

Fig. 2.29. Diffraction profiles of various peaks in the commensurate (•) and incommensurate (o) phases in a stage 4 Br_2-GIC single crystal. The dashed lines are fitted commensurate line shapes while the solid lines are power-law fits calculated from the stripe-domain model. (From [2.123])

different harmonics (within 20%). Typical peak profiles are shown in Fig. 2.29 [2.123]. It was noted, however, that the transition is smeared with a coexistence of the C and I phases over 0.6°C, attributable to heterogeneities in the Kish graphite, so that a Gaussian distribution of critical temperatures had to be added to the mosaic spread of the sample [2.123]. We expect that the use of natural graphite single crystals, which come in similar sizes but have fewer defects, would permit a more conclusive check on the predicted anisotropic scaling of the correlation lengths with the domain-wall density.

Apart from the transition to the stripe-domain phase, several other phases and C-I transitions have been reported in the graphite–bromine system, as reviewed by *Chung* [2.120]. A particularly interesting one concerns a stage 2 compound of in-plane composition C_8Br_2 (there is still some lack of resolution in the stage-composition relationship for the various phases in bromine GICs; for instance, a recent phase diagram by *Simon* et al. [2.126] disagrees with the observation of the above C-I stripe-domain transition for stage 4). In this case, the room temperature structure is incommensurate with the graphite with a 2d oblique unit cell [$a = 4.36$ Å, $b = 8.63$ Å, $\gamma = 99.45°$, with b along the graphite (110) direction] [2.17, 44, 107, 127]. The complex diffraction pattern shows satellite reflections due to a strong modulation effect from the graphite and diffuse streaks in the $(HK0)$ plane; the modulation structure has yet to be solved [2.44]. Significant interlayer correlations are also present at room temperature, where the diffuse rods along c^* are modulated in intensity. At $T_{I-C} \simeq 275$ K, an abrupt, presumably first-order, transition occurs leading to a fully ordered in-plane structure, without diffuse streaks, which has a large commensurate unit cell $(10\sqrt{3} \times 7)$. Moreover the interlayer correlations disappear at the transition and the low-temperature phase is two dimensional [2.129] (Fig. 2.30). This is rather unusual as one would expect a simultaneous improvement of both in-plane and out-of-plane correlations on cooling. It is suggested that the stabilization of a large in-plane commensurate structure (its unit cell area is 140 times that of graphite) is quite unfavorable for a simple stacking order, which would require strong long-range interaction between the bromine layers. Similar losses of interlayer correlations at low temperatures were also noted earlier for the alkali metal GICs [2.94, 99]. These are all interesting examples of frustration and, as such, deserve more theoretical analysis.

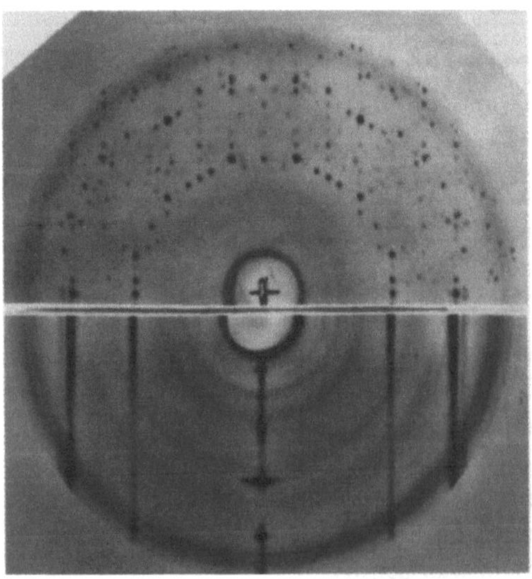

Fig. 2.30. $(HK0)$ (**a**) and $(H0L)$ (**b**) precession photographs of stage 2 C_8Br in the low-temperature commensurate phase at $T \simeq -35°$C. In (**b**) one observes continuous intensity on the reciprocal lattice rods along c^*, indicating no correlation between the bromine layers. (From [2.17, 128])

The solid phase of α-$C_{5n}HNO_3$ (below $T_m \sim 247\,K$) shows an incommensurate structure which was studied in detail by *Samuelsen* et al. [2.44, 129]. The intricate diffraction pattern produced by six equivalent domains of low symmetry (oblique 2d unit cell) can be fully interpreted with a unit cell incommensurate in one direction only, i.e. along a_G,

$$A = (17/12 + \varepsilon)a_G\ , \quad B = 2a_G + 9b_G\ .$$

The incommensurability ε is found to decrease with temperature and in one freshly prepared sample the observation of superlattice reflections indicated that a lock-in to $A = 17a_G$ occurred below about 210 K. However, the lock-in effect degraded with time, although the samples were kept in glass tubes, and it was not observed later on. This is shown in Fig. 2.31, where the angular splitting (ϕ) between 2 reflections from distinct domains is plotted as a function of temperature. At commensurability, $\phi_c = 0.915°$, and one sees that, after a rapid increase down to about 200 K, ϕ saturates below ϕ_c at lower temperatures [2.16]. Other effects of aging were also observed and it is quite understandable that for a high order of commensurability, such as 17, the lock-in gives rise to a very small energy gain and can be easily supressed by the presence of defects.

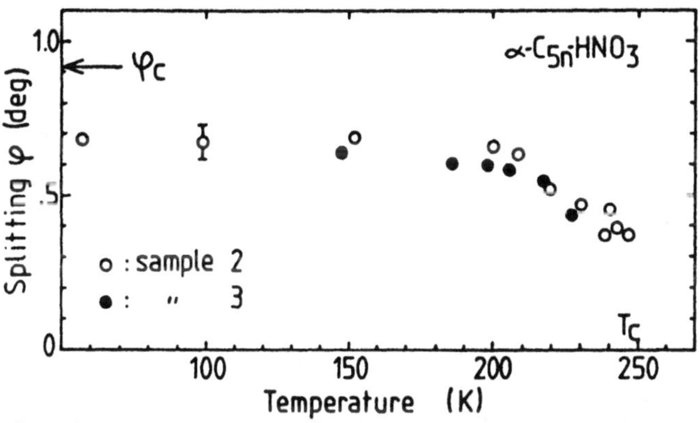

Fig. 2.31. Incommensurability as a function of temperature in stage 2 and 3 HNO_3-GIC. The angular splitting between two selected reflections is very sensitive to the incommensurability. On prolonged aging of these samples, full commensurability (ϕ_c) is still not attained despite an initial large variation below T_c (From [2.16])

(b) Ordered Acceptor Intercalant Structures

We note at the outset that there has to date been no successful determination of the full 3d ordered structure of any acceptor (or donor) GIC by standard methods such as Patterson synthesis or least-squares refinement of atomic positions. This can be partly attributed to the lack (until recently) of good intercalated single crystals and also to problems arising from the anisotropic mosaicity of the samples. Although the in-plane coherence length is generally large (several 1000 Å) the c-axis mosaicity of intercalated crystals often reaches a few degrees, which makes HKL intensity data collection at large angles rather difficult. A more fundamental problem, however, is that these may be strongly modulated (incommensurate) systems, which traditional crystallography is not well equipped to treat.

As a result, our present knowledge of these structures is only approximate and is often restricted to models based on a combination of ($HK0$) and ($00L$) intensity data analyses for the in-plane and out-of-plane coordinates, respectively. This procedure is clearly limited and often inadequate when the unit cell is large or has a low symmetry, or when the intercalant layer is composed of several sublayers. Further progress will be possible if the mosaic spread of the c-axis is improved in the intercalated crystals. For the most actively studied acceptor GICs, we review in the following the current state of structural knowledge of the intercalant species.

Bromine. As pointed out in Sect. 2.4.3, there is some confusion here regarding the number of different room temperature phases and their in-plane concentration, structure and stage dependence [2.120]. Following *Eeles* and *Turnbull* [2.107] in the 1960s, several recent studies have investigated some of these aspects but no clear phase diagram–structure relationship has emerged. The following classification is therefore tentative.

Beginning with the largest in-plane bromine density of C_7Br_2, one obtains an in-plane structure which has been observed for stages 2–4 [2.45, 110]. It can be described using either an oblique 2d unit cell or a centered rectangular cell commensurate with the graphite and defined by $(\sqrt{3} \times 7)a_G$. Two structure

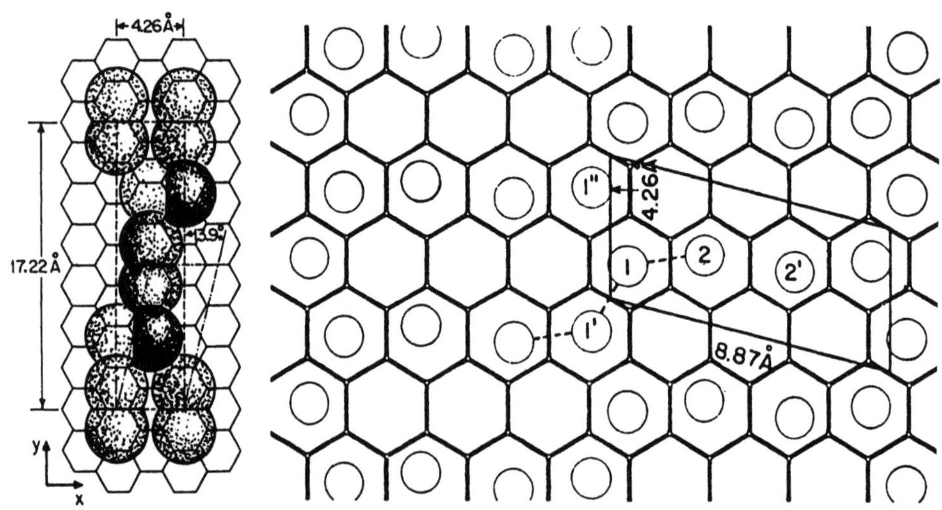

Fig. 2.32. Structure of the bromine layer for the C_7Br_2 in-plane concentration. In (**a**) the centered $(\sqrt{3} \times 7)$ unit cell description is used and the structure can be viewed as a packing of Br_2 molecules. (From [2.45]). In (**b**) an oblique primitive cell is chosen [its area is half that of the $(\sqrt{3} \times 7)$ unit cell] and bromine chains running perpendicular to a_G are suggested by the large Br_2-Br_2' distance of 2.86 Å. (From [2.110])

determinations have been performed. *Erbil* et al. [2.45] used a refinement of the intensity of 9 $(HK0)$ reflections to obtain the structural model of Fig. 2.32a, which can be viewed as a packing of Br_2 molecules and presents similarities with that of pristine solid bromine. However, it is denser [in projection on the (a, b) plane], suggesting that the Br_2 molecules are probably tilted relative to the plane, in agreement with EXAFS results [2.130]. *Ghosh* and *Chung* [2.110] collected $(HK0)$ intensity data using synchrotron radiation and refined 18 reflections. The atomic positions in Fig. 2.32b are significantly different from those of *Erbil* with both short (2.25 Å and 2.38 Å) and long (2.55 Å and 2.86 Å) Br-Br distances. This led the authors to describe the structure as formed by zigzag polymeric chains with a possibility of some out-of-plane corrugation not accessible from the $(HK0)$ reflection data.

The discrepancy between these two results and their incompleteness illustrate the problem encountered in the determination of the structures of acceptor GICs.

When the in-plane concentration is lower, approximately C_8Br_2, one obtains an incommensurate 2d unit cell described above in Sect. 2.4.3a (a = 4.36 Å, b = 8.63 Å, $\gamma = 99.45°$). The diffraction patterns also exhibit satellite reflections and diffuse streaks [2.44, 128]. The satellite reflections can be attributed to a modulation effect due to the graphite sublattice and this provides a good geometrical fit to the experimental diffraction pattern [2.44]. However, the atomic positions and the nature of the modulation are unknown, in spite of a one-dimensional domain-wall analysis by *Ghosh* and *Chung* [2.127] which accounts for a part of the diffraction data. In any case, the arrangement of the bromine atoms is certainly related to that of C_7Br_2 structure, with supplementary features (modulation satellites and diffuse streaks) induced in the lower density layer.

More generally, we suggest that under the influence of the graphite host the structure of the bromine layer is susceptible to modulation effects and, in particular, to domain-wall modulated structures. This has already been shown to be the case for the commensurate–incommensurate transition described in Sect. 2.4.3a with domain walls running perpendicular to an a_G-type direction. Other types of domain walls running parallel to a_G, or even tilted [for instance along the diagonal of the $(\sqrt{3} \times 7)$ unit cell], are also likely to occur as they can accommodate a lowering of the in-plane concentration. Actually, an analogous mechanism was suggested earlier by *Eeles* and *Turnbull* [2.107] to derive the structure of the $C_{14}Br_2$ layer from C_7Br_2 by simply removing alternate chains of Br_2 molecules.

The $C_{14}Br_2$ in-plane composition appears to correspond to a third stable 2d structure (in stage 2 samples for instance). Its diffraction pattern, described by *Eeles* and *Turnbull* [2.107] and *Moret* et al. [2.109], is composed of (i) strong Bragg reflections from a $(\sqrt{3} \times 7)$ unit cell, (ii) weak satellite reflections possibly reflecting a modulation from the graphite host, and (iii) diffuse straight lines at incommensurate positions [2.109]. The two latter effects might well be associated with (stripe) domain wall modulations.

The three structures described above for the in-plane concentrations C_7Br_2, C_8Br_2 and $C_{14}Br_2$ correspond to the ordered structures at room temperature.

There is also the very interesting low-temperature 2d structure obtained from the incommensurate C_8Br_2 type through the I-C transition at 275 K (Sect. 2.4.3a). The large $(10\sqrt{3} \times 7)$ superstructure unit cell, tenfold relative to the $(\sqrt{3} \times 7)$ standard one, is most probably generated by a periodic repetition every 10 unit cells of faults or "domain walls" running along the a_G direction.

Nitric Acid. Two forms of HNO_3 GICs have been characterized. The most studied αC_{5n}-HNO_3 (liquid phase at room temperature, liquid-incommensurate solid at about 250 K, as described in Sect. 2.4.3a) has a relatively large intercalate sandwich distance, $d_s = 7.80$ Å [2.15]. This accounts for a double HNO_3 layer consisting of NO_3 triangles oriented nearly perpendicular to the layer, and pointing either up or down, as found by *Touzain* [2.115] from an analysis of the $(00L)$ reflections. Further studies by *Pinto* et al. [2.116], *Shaked* et al. [2.117] and *Rosenman* et al. [2.131] have concluded that the plane of the NO_3 triangle is actually slightly tilted with respect to the c-axis with a tilt angle that varies from 8° to 25° among the quoted reports. However, the in-plane arrangement of the molecules in the nearly commensurate low-temperature phase is still unknown [2.44].

The second form, so-called β-$C_{8n}HNO_3$, is obtained by a deintercalation of the α variety. It has a smaller intercalate sandwich distance, $d_s = 6.55$ Å, attributed to the existence of a single layer of molecules almost parallel to the graphite layers. In this compound, the HNO_3 molecules are ordered at room temperature but the layers are uncorrelated [2.132]. The complex $(HK0)$ pattern has been unraveled using a rectangular psuedo-unit-cell and a larger hexagonal one [2.133]. From analogies with the structure of nitric acid monohydrate $(HNO_3 \cdot H_2O)$ a convincing structural model was proposed by *Clinard* et al. [2.133]. However, a detailed knowledge of the molecular positions is unfortunately not yet available.

Iodine Chloride. The room temperature ordered structure of ICl intercalated in graphite was first studied in single-crystal X-ray and electron diffraction experiments by *Turnbull* and *Eeles* [2.108]. Based on the most intense Bragg reflections, they proposed an approximate arrangement for the ICl molecules within a pseudo-unit-cell. The full pattern could be interpreted using an oblique incommensurate unit cell together with satellite reflections due to a modulation effect from the graphite host. The agreement between the observed and simulated $(HK0)$ diffraction pattern is excellent as reported in [2.44]. However, no intensity data analysis was performed and the model of *Turnbull* and *Eeles* could not be improved. More recently, *Wortmann* et al. [2.134] performed EXAFS experiments and found, in particular, a significant increase of the ICl bond length in the GIC over that in the solid or gas phases of the pristine compound. The measured value of 2.59 Å enabled them to suggest an improved in-plane arrangement of the ICl molecules compared with that of *Turnbull* and *Eeles* [2.108]. In particular, the positions of the chlorine atoms were modified as in Fig. 2.33. A determination of the in-plane structure of the ICl layer, taking account of the

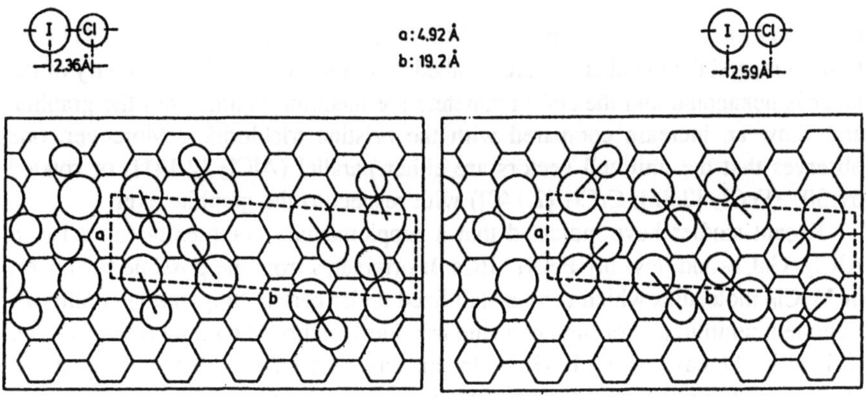

Fig. 2.33. Structure of the ICl layer (a) as proposed by *Turnbull* and *Eeles* [2.3] from X-ray analysis and (b) modified by *Wortmann* et al. [2.37] to account for the large ICl bond length deduced from EXAFS studies. (From [2.134])

modulation effects, has still to be done using quantitative X-ray analyses. This should be feasible given the high quality of the single-crystal diffraction data that can be obtained with this compound.

Dichlorides. Concerning the structure of the dichloride intercalant layer, several studies have developed models based on the analysis of the $(00L)$ reflection intensities and on analogies with the structure of the pristine intercalant. These have resulted in the following general characteristics:

- The metal dichloride layers are composed of a double layer of chlorine atoms with the metal atoms in between adopting either octahedral ($NiCl_2$, $CoCl_2$, $MnCl_2$ [2.135], $CdCl_2$, $HgCl_2$ [2.136], ...) or tetrahedral ($ZnCl_2$ [2.126]) coordination.
- The in-plane unit cells of the intercalant are incommensurate with the graphite host and have either a hexagonal or an oblique symmetry. In the hexagonal case, the orientation of the axes of the intercalant structure relative to those of graphite are found to be either 0 or 30° apart, which is related in a systematic way to the length of the intercalant cell parameters [2.137].
- There is a general tendency or trend for the formation of intercalant domains or islands of typical size 100–200 Å separated by intercalant-free regions [2.135, 138]. This is in agreement with chemical analyses which show an in-plane density of 60%–80% lower than the ideal value [2.135].

Trichlorides. We are particularly concerned here with GICs formed with $FeCl_3$, $AlCl_3$, $InCl_3$, $CrCl_3$ and $GaCl_3$, among which $FeCl_3$ is perhaps the prototype and has already been considered above in Sects. 2.2.2 and 2.3.4.

The structure of the intercalant layer consists generally of a chlorine–metal–chlorine sandwich, where the chlorine atoms are close-packed in a hexagonal

($FeCl_3$) [2.86, 139], cubic ($CrCl_3$) [2.140] or distorted cubic ($AlCl_3$) [2.141] configuration, and the metal atoms are octahedrally coordinated. The symmetry of the layer is hexagonal and the cell parameters are incommensurate with the graphite and show an increase compared with the pristine trichlorides. Moreover, one observes that the unit cell vectors are either parallel ($AlCl_3$ [2.141]) or rotated by 30° ($FeCl_3$ [2.86], $CrCl_3$ [2.140]) with respect to the graphite axes.

Recent studies have indicated that a supplementary complexity can arise in $AlCl_3$, $GaCl_3$ and possibly $FeCl_3$ GICs. It originates from the possible formation of M_2Cl_6 molecules which can be stabilized under certain preparation conditions (high temperature and/or high chlorine pressure), as first shown by *Behrens* et al. [2.141] in the case of $AlCl_3$ GICs. In this case, the $(HK0)$ diffraction patterns could be interpreted with two distinct unit cells. A rectangular commensurate structure was attributed to the Al_2Cl_6 molecules and a hexagonal commensurate structure $(3 \times 3)R0°$ was suggested to be due to an epitaxial arrangement of $AlCl_4^-$ ions [2.141]. Such a model, although quite tentative, may apply to the complex diffraction patterns observed at low temperatures in $GaCl_3$ GICs below the liquid-solid transition at 245 K, if one assumes a coexistence of Ga_2Cl_6 and $GaCl_4^-$ species [2.57]. As noted in Sect. 2.3.4, the stage 4 $FeCl_3$ GIC may also exist as an equilibrium molecular liquid at 350°C, see Fig. 2.17.

$SbCl_5$. The complex structural features of $SbCl_5$ GICs have been briefly discussed in Sect. 2.4.3a. The interpretation of the diffraction data is based on the proposal of a disproportionation of the $SbCl_5$ molecule according to

$$3SbCl_5 + 2e^- \rightarrow 2SbCl_6^- + SbCl_3 .$$

The two resulting species are also assumed to segregate and form $SbCl_6^-$ and $SbCl_3$ domains which behave separately.

Homma and *Clarke* [2.118, 119] have studied the phase transitions and ordered structures in detail. They suggest that the $SbCl_6^-$ ions form a commensurate $(\sqrt{7} \times \sqrt{7})$ structure and are arranged as shown in Fig. 2.34. Figure 2.34a shows a distorted $SbCl_6$ octahedron with a 3-fold axis aligned along the graphite c axis. The proposed close-packing of the $SbCl_6^-$ units in Fig. 2.34b corresponds to an interunit distance of $2\sqrt{3}r_{Cl} = 6.24$ Å (where r_{Cl} is the chlorine ionic radius), which is close to $\sqrt{7}a_G = 6.51$ Å. Superlattices of the $(\sqrt{7} \times \sqrt{7})$ structure, such as (14×14) and (28×28), have been observed in some samples and they are attributed to modulation effects related to the degeneracy of the $(\sqrt{7} \times \sqrt{7})$ structure, which is rotated by $\pm 19.11°$ with respect to the graphite. It is also suggested [2.119] that these very large cells may reflect the tendency, proposed by *Mitani* and *Niizeki* [2.106], toward in-plane commensurate hierarchies.

The minority phase formed by the $SbCl_3$ molecules is thought to give rise to islands 100–200 Å in size. According to *Homma* and *Clarke* [2.119], the $SbCl_3$ molecules would form pairs as shown in Fig. 2.35a. At room temperature, the $SbCl_3$ pairs are in a liquid state and they order at $T = 230$ K into an almost commensurate structure that is slightly compressed (~ 2 %) with respect to the

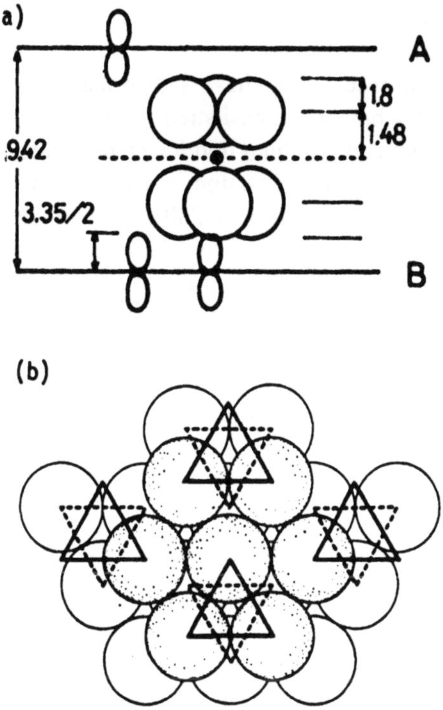

Fig. 2.34. Structural model for the $SbCl_6^-$ phase in a $SbCl_5$ GIC. **(a)** The orientation of the $SbCl_6^-$ ion between the carbon planes A and B. **(b)** The proposed close-packing of these units, which is close to a $(\sqrt{7} \times \sqrt{7})$ superlattice with respect to the graphite. (From [2.119])

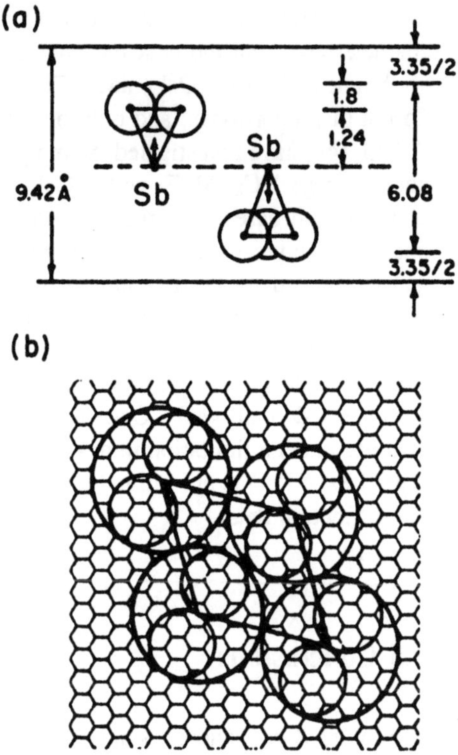

Fig. 2.35. Structural model for the $SbCl_3$ minority phase in a $SbCl_5$ GIC. The $SbCl_3$ molecules form pairs as shown in **(a)**, which can be arranged as in **(b)** to form a $(\sqrt{39} \times \sqrt{39})$ superlattice. (From [2.119])

$(\sqrt{39} \times \sqrt{39})$ superlattice (Fig. 2.35b). Dipole–dipole interactions may be responsible for the stabilization of this relatively large superlattice.

Finally, using the above models and especially the orientation of the $SbCl_6^-$ and $SbCl_3$ units shown in Figs. 2.34a and 2.35a, the authors have fitted the $(00L)$ intensities for stages 2, 4 and 5. They could thus refine for the 3 atomic coordinates, the overall $SbCl_5$ concentration and the relative concentration ratio of $SbCl_6^-$ to $SbCl_3$. One should note that, depending on the sample, the ratio is not always found to be equal to 2 as predicted by the disproportionation reaction but can vary in a range from 1 to 5. The origin of this discrepancy remains to be addressed.

Fluorides. The structure of several metal fluorides MF_5 GICs (such as AsF_5) and MF_6 GICs has been investigated in some detail. Some of these intercalants can undergo a disproportionation reaction similar to that of $SbCl_5$. This is the case with $C_{16}AsF_5$, which was studied recently by *Vaknin* and *Fisher* [2.142] in a stage 2 HOPG sample. The freezing of a liquid-like room temperature phase to about 150 K leads to an incommensurate structure which could not be fully characterized from the neutron scattering data.

Studies of MoF_6, OsF_6 and IrF_6 GICs in recent years have shown that they do not disproportionate and only one species has to be considered. *Kjems* et al. [2.143, 144] have investigated the molybdenum and osmium fluoride compounds, namely stage 1 $C_{8.4}MoF_6$ and C_9OsF_6 using neutron diffraction on HOPG samples. In both cases they found a liquid–solid transition near 280 K. The solid phase is interpreted as a 2d incommensurate structure close to a (2×2) lattice and modulated by the graphite host. A domain-wall model is presented to account for this modulation effect and the size of the domain is found to be on the order of 20–25 graphite units. Both the high intensity of the satellite reflections and the existence of high-order satellites indicate a strongly modulated structure with sharp domain walls. Once again, the intrinsic low resolution of the HOPG data presented a limitation in dealing with complex modulated structures. Single crystal experiments are ultimately needed to confirm the above interpretation of the diffraction data.

2.5 Conclusions

For future X-ray studies, single-crystal graphite of high internal perfection will be required to resolve some of the outstanding issues raised in this review, particularly with regard to complex in-plane ordering.

The structure of the donor alkali liquids seems to be understood in some detail using the concept of a modulated liquid. Bringing this theoretical development into conformity with both static and dynamic scattering data has required the application of X-ray and neutron methods and molecular dynamics. The acceptor liquids are less well understood but are now being actively studied.

The phase transitions have in some cases been very successfully explored, especially as regards higher stage Br-GICs where the predictions of 2d theory and X-ray experiment agree quite well. For the alkalis, although the details of the phase transition(s) need further study, it is clear that there is a regime of in-plane density for which a domain-discommensuration ground state is appropriate. The alkalis are all strongly modulated systems and both in-plane and out-of-plane effects have been studied. More work on the alkali GICs will concentrate on accurate intensity measurements on carefully intercalated crystals of high perfection. In-plane acceptor structures are again more complex and need further research. Molecular dynamics will remain an important adjunct of these studies, especially for the elemental and simpler molecular species.

Finally, the combination of X-ray and neutron scattering will continue to provide insight into these structural issues. GICs remain very interesting examples of the interplay between static and dynamical results in elucidating complex structures. The ordered phases clearly reflect the consequence of competing intercalant–intercalant and host–intercalant interactions and thereby stand as challenging examples of the crystallography of strongly incommensurate systems.

Acknowledgements. We wish to express our deep appreciation to our colleagues George Reiter and Françoise Rousseaux for on-going collaboration and extensive discussions. We wish also to thank the many students and collaborators, referred to herein, with whom we have had the pleasure of working over the past several years. We would also like to thank our numerous colleagues who kindly supplied us with updated reprint sets and new material. Finally, one of us (SCM) gratefully acknowledges support for his program on intercalation compounds at the University of Houston by the National Science Foundation under grant numbers DMR-8214314 and DMR-8603662.

References

2.1 A. Hérold: In *Physics and Chemistry of Materials with Layered Structures*, Vol. 6, ed. by F. Lévy (D. Reidel, Dordrecht 1979)
2.2 M.S. Dresselhaus, G. Dresselhaus: Adv. Phys. **30**, 139 (1981)
2.3 S.A. Solin: Adv. Chem. Phys. **49**, 455 (1982)
2.4 R. Clarke, C. Uher: Adv. Phys. **33**, 469 (1984)
2.5 R. Moret: In *Intercalation in Layered Materials*, ed. by M.S. Dresselhaus, NATO ASI Series, Ser. B: Physics, Vol. 148 (Plenum, New York, 1986) p. 185
2.6 H. Zabel, P.C. Chow: Comments Cond. Mat. Phys. **12**, 225 (1986)
2.7 J.B. Hastings, W.D. Ellenson, J.E. Fischer: Phys. Rev. Lett **42**, 1552 (1979)
2.8 M. Suzuki, H. Ikeda, H. Suematsu, Y. Endoh, H. Shiba, M.T. Hutchings: J. Phys. Soc. Jpn. **49**, 671 (1980)
2.9 F. Rousseaux, D. Tchoubar, C. Tchoubar, D. Guerard, P. Lagrange, A. Hérold, R. Moret: Synth. Met. **7**, 221 (1983)
2.10 K. Ohshima, S.C. Moss, R. Clarke: Synth. Met. **12**, 125 (1985)
2.11 C. Thompson, M.E. Misenheimer, S.C. Moss: Acta Crystallogr. B **44**, 6 (1988)
2.12 X.B. Kan, M.E. Misenheimer, K. Forster, S.C. Moss: Acta Crystallogr. A **43**, 418 (1987)
2.13 J. Maire, J. Méring: In *Chemistry and Physics of Carbon*, Vol. 3 (Marcel Dekker, New York 1970) p. 125
2.14 D.E. Nixon, G.S. Parry: J. Phys. C **2**, 1732 (1969)
2.15 D.E. Nixon, G.S. Parry, A.R. Ubbelohde: Proc. R. Soc. London A **291**, 324 (1966)
2.16 E.J. Samuelsen, R. Moret, H. Fuzellier, M. Klatt, M. Lelaurain, A. Hérold: Phys. Rev. B **32**, 417 (1985)
2.17 F. Aberkane, R. Moret, G. Furdin: In Extended Abstracts, Graphite Intercalation Compounds, MRS Meeting, Boston 1986, ed. by M.S. Dresselhaus, G. Dresselhaus, S.A. Solin (1986) p. 23
2.18 B.E. Warren: *X-Ray Diffraction* (Addison-Wesley, Reading, MA 1969)
2.19 S. Hashimoto, K. Forster, S.C. Moss: Acta Crystallogr. A **44**, 897 (1988)
2.20 N. Daumas, A. Hérold: C.R. Acad. Sci. Ser. C **268**, 373 (1969)
2.21 M. Mori, S.C. Moss, Y.M. Jan, H. Zabel: Phys. Rev. B **25**, 1287 (1982)
2.22 N. Caswell, S.A. Solin, T.M. Hayes, S.J. Hunter: Physica **99B**, 463 (1980)
2.23 J.E. Fischer, H.J. Kim: Phys. Rev. B **35**, 6826 (1987)
2.24 P. Chow, H. Zabel: Synth. Met. **7**, 243 (1983)
2.25 M.E. Misenheimer, H. Zabel: Phys. Rev. Lett **54**, 2521 (1985)
2.26 J.E. Fischer, C.D. Fuerst, K.C. Woo: Synth. Met. **7**, 1 (1983)
2.27 P.A. Heiney, M.E. Huster, V.B. Cajipe, J.E. Fischer: Synth. Met. **12**, 21 (1985)
2.28 S. Hendricks, E. Teller: J. Chem. Phys. **10**, 147 (1942)
2.29 J. Kakinoki, Y. Komura: J. Phys. Soc. Jpn. **7**, 30 (1952)
2.30 R. Nishitani, Y. Uno, H. Suematsu: Phys. Rev. B **27**, 6572 (1983)
2.31 F. Rousseaux, G. Besson, C. Tchoubar, D. Guérard: In Extended Abstracts, Graphite Intercalation Compounds, MRS Meeting, Boston 1986, ed. by M.S. Dresselhaus, G. Dresselhaus, S.A. Solin (1986) p. 13
2.32 M.E. Huster, P.A. Heiney, V.B. Cajipe, J.E. Fischer: Phys. Rev. B **35**, 3311 (1987)

2.33 V.B. Cajipe, J.E. Fischer: Phys. Rev. B **35**, 4854 (1987)
2.34 E. Schrödinger: Phys. Z. **15**, 79, 497 (1914)
2.35 V.J. Emery, J.D. Axe: Phys. Rev. Lett. **40**, 1507 (1978)
2.36 J.M. Cowley: *Diffraction Physics*, 2nd ed. (North-Holland, New York 1981) p. 154 ff.
2.37 S.Y. Leung, M.S. Dresselhaus, C. Underhill, T. Krapchev, G. Dresselhaus, B.J. Wuensch: Phys. Rev. B **24**, 3505 (1981)
2.38 C. Thompson: Ph. D. Thesis, University of Houston (1987)
2.39 J.D. Axe, J.B. Hastings: Acta Crystallogr. A **39**, 593 (1983)
2.40 G.S. Parry, D.E. Nixon: Nature **216**, 909 (1967)
2.41 D.E. Nixon, G.S. Parry: J. Phys. D **1**, 291 (1968)
2.42 H. Zabel, S.C. Moss, N. Caswell, S.A. Solin: Phys. Rev. Lett. **43**, 2022 (1979)
2.43 R. Clarke, N. Caswell, S.A. Solin, P.M. Horn: Phys. Rev. Lett **43**, 2018 (1979)
2.44 F. Aberkane, F. Rousseaux, E.J. Samuelsen, R. Moret: Ann. de Phys., Colloque 2, **11**, 95 (1986)
2.45 A. Erbil, A.R. Kortan, R.J. Birgeneau, M.S. Dresselhaus: Phys. Rev. B **28**, 6329 (1983)
2.46 S.K. Sinha: In *Methods of Experimental Physics*, Vol. 23, Part B, ed. by D.L. Price, K. Skold (Academic, New York 1987) p. 1
2.47 K.J. Strandberg: Rev. Mod. Phys. **60**, 161 (1988)
2.48 L.D. Landau, E.M. Lifshitz: *Statistical Physics* (Pergamon, Oxford 1969) p. 478
2.49 N.D. Mermin, H. Wagner: Phys. Rev. Lett **17**, 1133 (1966)
2.50 B. Jancovici: Phys. Rev. Lett. **19**, 20 (1967)
2.51 Y. Imry: CRC Crit. Rev. Solid State Sci. **8**, 157 (1979)
2.52 P. Dimon, P.M. Horn, M. Sutton, R.J. Birgeneau, D.E. Moncton: Phys. Rev. B **31**, 437 (1984)
2.53 The Kosterlitz-Thouless-Halperin-Nelson-Young (KTHNY) theory is reviewed in [2.47] with a full list of references for the interested reader
2.54 C.A. Murray, D.H. Van Winkle: Phys. Rev. Lett. **58**, 1200 (1987)
2.55 M. Cheng, J.T. Ho, S.W. Hui, R. Pindak: Phys. Rev. Lett. **59**, 1112 (1987)
2.56 J.D. Brock, D.Y. Noh, B.R. McClain, J.D. Litster, R.J. Birgeneau, A. Aharony, P.M. Horn, J.C. Liang: Z. Phys. B **74**, 197 (1989)
2.57 N. Nadi, M. Lelaurain, A. Hérold, F. Rousseaux: Synth. Met. **23**, 75 (1988)
2.58 G.S. Parry: Mat. Sci. Eng. **31**, 99 (1977)
2.59 S.C. Moss, D.L. Price: In *Physics of Disordered Materials*, ed. by D. Adler, H. Fritzsche, S.R. Ovshinsky (Plenum, New York 1985) p. 77
2.60 G. Reiter, S.C. Moss: Phys. Rev. B **33**, 7209 (1986)
2.61 A. Guinier: *X-Ray Diffraction in Crystals, Imperfect Crystals and Amorphous Bodies* (Freeman, San Francisco 1963)
2.62 H. Zabel, Y.M. Jan, S.C. Moss: Physica **99B**, 453 (1980)
2.63 F. Rousseaux, R. Moret, D. Guérard, P. Lagrange, M. Lelaurain: J. de Phys. Lett. **45**, L-1111 (1984)
2.64 F. Rousseaux, R. Moret, D. Guérard, P. Lagrange, M. Lelaurain: Synth. Met. **12**, 45 (1985)
2.65 R. Nishitani, Y. Uno, H. Suematsu, Y. Fujii, T. Matsushita: Phys. Rev. Lett. **52**, 1504 (1984)
2.66 M.J. Huijben, W. van der Lugt: Acta Crystallogr. A **35**, 431 (1979)
2.67 M.J. Winokur, J.H. Rose, R. Clarke: Phys. Rev. B **25**, 3703 (1982)
2.68 A. Bunde, W. Dieterich: In *Physics of Intercalation Compounds*, ed. by L. Pietronero, E. Tosatti, Springer Ser. Solid-State Sci., Vol. 38 (Springer, Berlin, Heidelberg 1981)
2.69 M. Plischke: Can. J. Phys. **59**, 802 (1981)
2.70 M. Plischke, W.D. Leckie: Can. J. Phys. **60**, 1139 (1982)
2.71 W.A. Kamitakahara, H. Zabel: Phys. Rev. B **32**, 7817 (1985)
2.72 H. Zabel, A. Magerl, A.J. Dianoux, J.J. Rush: Phys. Rev. Lett. **50**, 2094 (1983)
2.73 H. Zabel, S.E. Hardcastle, D.A. Neumann, M. Suzuki, A. Magerl: Phys. Rev. Lett. **57**, 2041 (1986)
2.74 S.C. Moss, G. Reiter, J.L. Robertson, C. Thompson, J.D. Fan, K. Ohshima: Phys. Rev. Lett. **57**, 3191 (1986)
2.75 C. Thompson, S.C. Moss, G. Reiter, M.E. Misenheimer: Synth. Met. **12**, 27 (1985)
2.76 X.B. Kan, J.L. Robertson, S.C. Moss, K. Ohshima, C.J. Sparks: Phys. Rev. B **39**, 10627 (1989)
2.77 S.C. Moss, X.B. Kan, J.D. Fan, J.L. Robertson, G. Reiter, O.A. Karim: In *Competing Interactions and Microstructure: Statics and Dynamics*, ed. by R. LeSar, A. Bishop, R. Heffner, Springer Proc. Phys., Vol. 27 (Springer, Berlin, Heidelberg 1988)

2.78 D.P.DiVincenzo, E.J. Mele: Phys. Rev. B **32**, 2538 (1985)
2.79 F.F. Abraham: Phys. Rep. **80**, 339 (1982)
2.80 F.F. Abraham, S.W. Koch, W.E. Rudge: Phys. Rev. Lett. **49**, 1830 (1982)
2.81 D.P. DiVincenzo: Synth. Met. **12**, 111 (1985)
2.82 J.D. Fan, Omar A. Karim, G. Reiter, S.C. Moss. Phys. Rev. B **39**, 6111 (1989)
J.D. Fan: Ph. D. Thesis, University of Houston (1989)
2.83 P.B. Visscher, L.M. Falicov: Phys. Rev. B **3**, 2541 (1971)
2.84 J.P. Hansen, I.R. McDonald: *Theory of Simple Liquids*, 2nd ed. (Academic, New York 1986) p. 45 ff.
2.85 K. Forster: Private communication
2.86 J.M. Cowley, J.A. Ibers: Acta Crystallogr. **9**, 421 (1956)
2.87 M. Hargittai, J. Tremmel, I. Hargittai: J. Chem. Soc., Dalton Trans. **1**, 87 (1980)
2.88 R. Clarke, N. Caswell, S.A. Solin: Phys. Rev. Lett. **42**, 61 (1979)
2.89 H. Suematsu, M. Suzuki, H. Ikeda, Y. Endoh: Synth. Met. **2**, 133 (1980)
2.90 R. Clarke, J.N. Gray, H. Homma, M.J. Winokur: Phys. Rev. Lett. **47**, 1407 (1981)
2.91 Y. Yamada, I. Naiki: J: Phys. Soc. Jpn. **51**, 2174 (1982)
2.92 M. Suzuki, H. Suematsu: J. Phys. Soc. Jpn. **52**, 2761 (1983)
2.93 M. Mori, S.C. Moss, Y.M. Jan. Phys. Rev. B **27**, 6385 (1983)
2.94 M.J. Winokur, R. Clarke: Phys: Rev. Lett. **54**, 811 (1985)
2.95 M. Suzuki: Phys. Rev. B **33**, 1386 (1986)
2.96 F. Rousseaux, R. Moret, D. Guérard, P. Lagrange, M. Lelaurain: Ann. de Phys., Colloque 2, **11**, 85 (1986)
2.97 M.J. Winokur, R. Clarke: Phys. Rev B **34**, 4948 (1986)
2.98 F. Rousseaux, R. Moret: Proc., School on Phase Transitions, Orcieres-Merlette, 1988. Phase Transitions **14**, 275 (1989)
2.99 F. Rousseaux, R. Moret: Phys. Rev. B, still in press
2.100 P. Bak, E. Domany: Phys. Rev. B **23**, 1320 (1981)
2.101 H. Zabel: In *Ordering in Two Dimensions*, ed. by S.K. Sinha (North-Holland, New York 1980) p. 61
2.102 S.B. Dicenzo: Phys. Rev. B **26**, 5878 (1982)
2.103 K. Nakanishi, H. Takatera, Y. Yamada, H. Shiba: J. Phys. Soc. Jpn. **43**, 1509 (1977)
2.104 M.J. Winokur, R. Clarke: Phys. Rev. Lett. **56**, 2072 (1986)
2.105 R. Bruinsma, A. Zangwill: Phys. Rev. Lett. **55**, 214 (1985)
2.106 M. Mitani, K. Niizeki: Solid State Phys. **20**, 1017 (1987); ibid. **21**, 1895 (1988)
2.107 W.T. Eeles, J.A. Turnbull: Proc. R. Soc. London A **283**, 179 (1965)
2.108 J.A. Turnbull, W.T. Eeles: Proc. Int. Conf. Ind. Carbons and Graphite, Soc. Chem. Industry, London (1966) p. 173
2.109 R. Moret, F. Rousseaux, G. Frugin, A. Hérold: Synth. Met. **7**, 289 (1983)
2.110 D. Ghosh, D.D.L. Chung: Mater. Res. Bull. **18**, 1179 (1983)
2.111 F. Rousseaux, R. Vangelisti, A. Plancon, D. Tchoubar: Rev. Chim. Miner. **19**, 572 (1982)
2.112 K.K. Bardhan, D.D.L. Chung: Extended Abstracts Prog. Bien. Conf. Carbon **15**, 113 (1981)
2.113 K.K. Bardhan: Solid State Commun. **44**, 583 (1982)
2.114 F. Batallan, I. Rosenman, A. Magerl, H. Fuzellier: Phys. Rev. B **32**, 4810 (1985);
I. Rosenman, C. Simon, F. Batallan, A. Magerl: Europhys. Lett. **3**, 1013 (1987);
F. Batallan, I. Rosenman, C. Simon, A. Magerl: Synth. Met. **23**, 49 (1988)
2.115 P. Touzain: Synth. Met. **1**, 3 (1979/1980)
2.116 H. Pinto, M. Melamud, O. Smalal, R. Moreh, H. Shaked: Physica **121**B, 121 (1983)
2.117 H. Shaked, H. Pinto, M. Melamud: Phys. Rev. B **35**, 838 (1987)
2.118 H. Homma, R. Clarke: Phys. Rev. Lett. **52**, 629 (1984)
2.119 H. Homma, R. Clarke: Phys. Rev. B **31**, 5865 (1985)
2.120 D.D.L. Chung: Phase Transitions **8**, 35 (1986)
2.121 K.K. Bardhan, J.C. Wu, D.D.L. Chung: Synth. Met. **2**, 109 (1980)
2.122 A.R. Kortan, A. Erbil, R.J. Birgeneau, M.S. Dresselhaus: Phys. Rev. Lett. **49**, 1427 (1982)
2.123 S.G.J. Mochrie, A.R. Kortan, J.R. Birgeneau, P.M. Horn: Phys. Rev. Lett. **53**, 985 (1984)
2.124 V.L. Pokrovsky, A.L. Talapov: Phys. Rev. Lett. **42**, 65 (1979)
2.125 H.J. Schulz: Phys. Rev. B **22**, 5274 (1980)
2.126 C. Simon, I. Rosenman, F. Batallan, G. Pepy, H. Lauter: Synth. Met. **23**, 147 (1988)
2.127 D. Ghosh, D.D.L. Chung: J. de Phys. Lett. **44**, L761 (1983)
2.128 F. Aberkane, R. Moret, G. Furdin: Unpublished

2.129 E.J. Samuelsen, R. Moret, R. Comes, H. Fuzellier, M. Klatt, M. Lelaurain, A. Hérold: Synth. Met. **10**, 13 (1985)
2.130 J.L. Feldman, E.F. Skelton, A.C. Ehrlich, D.D. Dominguez, W.T. Elam, S.B. Qadri, F.W. Lytle: Bull. Am. Phys. Soc. **28**, 346 (1983)
2.131 I. Rosenman, C. Simon, F. Batallan, H. Fuzellier, H.J. Lauter: Synth. Met. **23**, 339 (1988)
2.132 R. Moret, R. Comes, G. Furdin, H. Fuzellier, F. Rousseaux: Proc. MRS Meeting, Vol. 80 (Elsevier, Amsterdam 1983) p. 27
2.133 C. Clinard, D. Tchoubar, C. Tchoubar, F. Rousseaux, M. Fuzellier: Synth. Met. **7**, 333 (1983)
2.134 G. Wortmann, W. Krone, G. Kaindl, R. Schögl: Synth. Met. **23**, 139 (1988)
2.135 F. Baron, S. Flandrois, C. Hauw, J. Gaultier: Solid State Commun. **42**, 759 (1982)
2.136 P. Behrens, H. Beuthien, H.P. Eickhoff, W. Metz, W. Niemann: Synth. Met. **23**, 95 (1988)
2.137 P. Behrens, W. Metz: In *Intercalation in Layered Materials*, ed. by M.S. Dresselhaus, NATO ASI Series, Ser. B: Physics, Vol. 148 (Plenum, New York 1986) p. 229
2.138 S. Flandrois, J.M. Masson, J.C. Rouillon, J. Gaultier, C. Hauw: Synth. Met. **3**, 1 (1981)
2.139 W. Metz, E. Schulze: Z. Kristallogr. **142**, 409 (1975)
2.140 R. Vangelisti, A. Hérold: Carbon **14**, 33 (1976)
2.141 P. Behrens, U. Wiegand, W. Metz: Extended Abstracts, Carbon 86, Baden-Baden (1986) p. 602
2.142 D. Vaknin, J.E. Fischer: Synth. Met. **23**, 101 (1988)
2.143 J.K. Kjems, Y. Yeshurun, D. Vaknin, D. Davidov, H. Selig: Phys. Rev. B **36**, 6981 (1987)
2.144 J.K. Kjems, Y. Yeshurun: Synth. Meth. **23**, 113 (1988)

3. Staging and Kinetics

George Kirczenow

With 17 Figures

When a guest species such as potassium intercalates between the layers of a lamellar host material such as graphite, staged structures are formed. These are periodic sequences of host and guest layers, with n layers of the host material separating neighboring guest layers in a stage n intercalation compound. In some cases, the staged structures are three-dimensionally ordered crystals, but very often this is not the case. Thus staging is commonly regarded as being a quasi-one-dimensional ordering phenomenon. It is characteristic of the very large number of graphite intercalation compounds (GICs) which have been synthesized, and also occurs in many intercalation compounds of the layered transition metal dichalcogenides.

In the past few years, there has been a major experimental and theoretical effort to understand the physics of staging, and many of the developments which have occurred have been the result of an interplay between theoretical ideas and experimental observations. The experimental studies of staging have been directed at analyzing the structure of the staged phases, measuring the staging phase diagrams as a function of the thermodynamic variables, and obtaining an understanding of the stage transitions in which the stage index changes. The role of staging in the intercalation process itself has also been a subject of continuing interest. Recently, attention has focused on the disorder which is frequently observed in the sequence of guest and host layers making up intercalation compounds, and on its role in the equilibrium properties and kinetics.

A few microscopic calculations of the energetics of staging have been carried out, yielding important insights into the form of the interactions responsible for staging, and some semi-quantitative agreement with experimental measurements. However, most theoretical work on staging has been phenomenological. Simple model Hamiltonians describing the thermodynamics of staging have been set up and solved using mean field methods. Significant progress has been achieved in understanding the basic features of the equilibrium staging phase diagrams, and the reasons for the stability of the particular stage-ordered structures which are observed. This phenomenological approach has been extended to treat the problem of stage disorder within the framework of equilibrium thermodynamics, qualitatively explaining many experimental findings. Some kinetic models of stage transformations have been proposed, and recently computer simulations have begun to yield a detailed microscopic picture of the dynamics of intercalate islands in staging phenomena.

This chapter describes the experiments, theoretical studies and computer simulations which have been carried out to understand staged structures, their thermodynamics and kinetics. In addition to the work on GICs, some closely related recent experiments on staging in intercalated transition metal dichalcogenides will also be discussed. Some previous reviews which the reader will find interesting are those of *Hérold* [3.1] and *Dresselhaus* and *Dresselhaus* [3.2] covering the early work on GICs, the review by *Solin* [3.3] of structural properties, the survey of high pressure effects by *Clarke* and *Uher* [3.4], the article by *Solin* [3.5] comparing intercalated graphite, intercalated transition metal dichalcogenides and sheet silicates, the review of electronic properties by *Fischer* [3.6] and by *Pfluger* and *Güntherodt* [3.7], the review of staging domain walls, their structure, and energetics, and their influence on intercalation kinetics by *Ulloa* and *Kirczenow* [3.8], the review of structural and magnetic properties, lattice modes and diffusion by *Zabel* and *Chow* [3.9], the reviews of staging by *Safran* [3.10, 11], and the review of intercalated transition metal dichalcogenides by *Friend* and *Yoffe* [3.12].

3.1 The Energetics of Staging

The principal considerations governing staging energetics are: (a) The charge transfer which occurs between the guest and host species on intercalation. This drives the intercalation process energetically, and is the source of the effective repulsion between different guest layers which is necessary for stage ordering. (b) Elastic effects associated with the distortion of the host by the guest on intercalation, and the related matter of the cohesion between the host layers which must be overcome when the host layers are separated by the intercalating guest. (c) Interactions between guest atoms in the same layer, and between the guest atoms and the adjacent host layers.

3.1.1 Charge Transfer

On intercalation, electrons are transferred from the guest to the host (for donor guest species) or from the host to the guest (for acceptor guests). In graphite, metal guests are donors, while most others are acceptors. The degree of charge transfer is characterized by a parameter f; $f = 1$ corresponds to the transfer of one electron per intercalate atom or formula unit. For typical acceptors in graphite, $f \sim 0.2$–0.5. For alkali metal donors it is larger, typically $f \sim 1$ at least for stages higher than 1. However, the precise values of f are currently the subject of a debate involving complex issues of electronic structure [3.13] and will not be discussed further here.

The simplest description of the charge transfer is in terms of the rigid band approximation for the graphitic layers. In this description, it is assumed that the

π bands associated with the graphitic layers accommodate the electrons transferred from the donor guests and supply the electrons transferred to the acceptor guests, but are otherwise unchanged by the intercalation process. In an intercalation compound with stage $n > 2$, not all of the graphitic layers are equivalent, some being further from the intercalate layers than others. Thus the Coulomb interaction between the charged intercalate layers and the electrons in the graphite π bands causes the transferred charge to be inhomogeneously distributed. The graphitic layers closest to the intercalate layers are the most heavily charged.

A theory of this inhomogeneous charge distribution was given by *Pietronero* et al. [3.14] using a Thomas-Fermi approximation to treat the electronic screening effects. Because of the approximately linear dependence of the density of graphitic π-band states on the energy in the vicinity of the Fermi level [3.15–18], the result obtained was quite different from the exponential screening behavior characteristic of metals. For a single intercalate layer in graphite, the charge density ϱ at a graphitic layer was found to obey an approximate *power law* dependence on the distance z from the intercalate layer: $\varrho(z) \sim (1 + z/z_0)^{-4}$.

Salzano and *Aronson* had suggested earlier that the Coulomb repulsion between the charged intercalate layers is responsible for staging [3.19]. The long-range screening of the intercalate layers by the graphitic electrons found by *Pietronero* et al. [3.14] provides the physical basis for this mechanism. *Safran* and *Hamann* [3.20] pursued this idea by using a Thomas-Fermi approach to calculate the electronic and Coulomb energy of an arbitrary sequence of graphitic and intercalate layers. They found the energy of a stage n-like sandwich of n graphite layers between two intercalate layers to be of the form

$$E(n) \simeq E_\infty[1 + a(1 + n/n_0)^{-5}] , \tag{3.1}$$

where $E(n)$ is the energy per intercalate atom. $E(n)$ decreases with increasing n, which implies an effective "repulsion" between guest layers. For a typical alkali metal GIC with $f = 1$ and stoichiometry MC_{12n} where M stands for the alkali metal, $E_\infty \simeq 1.1\,\mathrm{eV}$, $a \simeq 0.96$, $n_0 \simeq 4.1$. Thus the asymptotic n^{-5} behavior is not reached for actual GICs where $n \lesssim 10$. For low stages, the variation of the "staging energy" $E(n)$ with stage n is close to n^{-1}. Another result of the *Safran Hamann* Thomas-Fermi theory [3.20] is that the effective repulsion between two intercalate layers is very nearly "screened out" by any intervening intercalate layer. This "strong screening" of the interlayer interactions by any intervening intercalate layer is an important ingredient of staging phenomena, as will be explained in Sect. 3.2.

More detailed self-consistent LCAO calculations of the staging energy $E(n)$ have been made by *Safran* and *Hamann* [3.21] and *Ohno* and *Kamimura* [3.22, 23], the latter work making less use of empirical inputs. *Safran* and *Hamann*'s calculation [3.21] agreed well with their earlier Thomas-Fermi theory [3.20]. However, *Ohno* and *Kamimura* [3.22] found the transferred charge to be much more localized to the graphite layers adjacent to the guest layers than did *Safran* and *Hamann* [3.20, 21]. Thus their calculation predicts a much weaker effective repulsion [3.23] between guest layers.

Experimental measurements of the ranges of intercalate chemical potential over which the various stages of GICs are thermodynamically stable can be compared directly with the calculated values of the staging energy $E(n)$, provided that such questions as the influence of stage-dependent stoichiometry and charge transfer can be set aside. Then, at low temperatures where entropy effects can be neglected, the interval of chemical potential over which stage n is stable is given [3.20] by

$$\Delta\mu(n) = n[E(n+1) + E(n-1) - 2E(n)] \,. \tag{3.2}$$

The early experimental work of *Salzano* et al. [3.19, 24] on alkali-metal GICs yielded values of $\Delta\mu(n)$ in fairly good agreement with the results of *Safran* and *Hamann* [3.21]. The more recent experiments of *Nishitani* et al. on K-graphite [3.25] yielded values of $\Delta\mu(n)$ in good agreement with the calculations of *Safran* and *Hamann* for $f = 1/2$. (Note that other measurements of the charge transfer in alkali-metal GICs of stages greater than 1 tend to favor $f \sim 1$ [3.2, 6, 7].) Recent work by *Yosida* et al. [3.26] on $SbCl_5$ graphite yielded results in agreement with the Safran-Hamann theory for $1/4 < f < 1/2$. However, the values $\Delta\mu(n)$ found experimentally are larger than those obtained by *Ohno* and *Kamimura* [3.23] by an order of magnitude.

The discrepancy between the above theories is yet to be resolved. However, there is not much doubt that the energetics of charge transfer provide the repulsive effective interaction between intercalate layers which is responsible for staging, if only because there is no other viable candidate. As is explained in Sects. 3.1.2 and 3.1.3, elastic interactions cannot serve this purpose.

3.1.2 Host Layer Cohesion and Homogeneous Elastic Strains

During intercalation, the cohesion between the host layers must be overcome so that the layers can be separated sufficiently to admit the intercalate atoms between them. In GICs, this increase in the graphite layer spacing on intercalation ranges from 0.36 Å for Li-graphite to > 6 Å for $FeCl_3$, $TiCl_3$ and $AlBr_3$ graphite. A simple estimate of the "separation energy" required to open the host galleries may be obtained using a phenomenological Morse potential to describe the coupling between the graphite layers [3.27]. It has also been estimated from thermodynamic data by *Salzano* and *Aronson* [3.19] for some alkali metal GICs.

The effects of this separation energy on staging were considered by *Millman* and *Kirczenow* [3.28, 29] and by *Dahn* et al. [3.30]. The simplest approximation [3.28, 29] was to treat the host layers as perfectly rigid and to assume the separation energy of two host layers to be independent of the concentration of intercalate between them, provided that the concentration is nonzero. In graphite, this approximation should be adequate for large and moderate-sized intercalate species (which have the effect of nearly cleaving the host crystal) provided that the in-plane density of intercalate is not too low. *Dahn* et al. [3.30], used a simple "plate and spring" model for the guest–host system, to obtain a better expression for the separation energy of the form

$$E_i = \gamma N_i/(N_i/N_0 + \beta) \tag{3.3}$$

where N_i is the number of intercalate atoms in layer i, and N_0 is the total number of sites available to the intercalate atoms in that layer. This form was subsequently used also by *Millman* and *Kirczenow* [3.28, 29].

The derivation of (3.3) given by *Dahn* et al. formally assumes the host layers to be rigid. However, the final expression is proportional to the number N_i of intercalate atoms in the limit of a small intercalate concentration N_i/N_0. Because of this, one can interpret (3.3) much more generally. Let us choose γ to be the energy per intercalate site required to separate *rigid* host layers sufficiently to admit the intercalate, and choose β so that γ/β is the elastic energy of the deformation of the *real* (bendable) host around an isolated intercalate atom in the host matrix. Then (3.3) becomes a valid expression for the separation energy both at high intercalate concentrations, where the host layers can be treated as rigid[1], and at low concentrations of *dispersed* intercalate atoms where the host layers must be treated as *deformable*. The former regime seems to be characteristic of most GICs under conditions of experimental interest. The latter is found in Li-graphite at high temperatures [3.31, 32], where entropy effects completely disorder the intercalate in each host gallery. It also occurs in some intercalated transition metal dichalcogenides [3.33]. In between these two limiting regimes, (3.3) can be viewed as an interpolation formula.

Recently, *Fischer* and *Kim* [3.33] have pointed out that there are noticeable deviations between the predictions of the "plate and spring model" [3.30] and the behavior of the c-axis lattice parameter in stage 1 Li-graphite and Li-TiS_2, and suggested that the bending of the host layers is responsible. However, data obtained by *Dahn* and *Haering* [3.34, 35] suggest that the formation of a disordered stage 2 structure may be an important factor in analyzing the Li-TiS_2 data.

The above discussion concerns the expansion of the host along the c-axis (perpendicular to the layers) on intercalation. The in-plane lattice parameter of the graphite host also changes on intercalation but by a small amount $\sim 1\%$ [3.36–39]. It has been argued by *Pietronero* and *Strässler* [3.39] that this change in the lattice constant is due to the charge transfer between the guest and host. If this is the case, then in higher stage compounds, shear stresses must exist between layers of the host closest to the intercalate (where the transferred charge is concentrated) and those further away. All of the host layers must be elastically strained in order to maintain registry between them. The associated elastic energy should decrease with decreasing stage, resulting in an effective *attraction* between the guest layers. This elastic interaction energy has been estimated by *Safran* and *Hamann* [3.20] and found to be much smaller than the direct effects of charge transfer discussed in Sect. 3.1.1. Because of its weakness and attractive nature, this elastic interaction does not appear to be important in staging energetics. It

[1] This argument assumes β to be small, and ignores the effects of domain boundaries, see Sect. 3.1.3.

is certainly not responsible for the existence of staging, contrary to some early ideas [3.2].

3.1.3 Elasticity, Intercalate Islands and Domains

So far we have considered elastic effects in the "classical model" of staging in which every gallery between the host layers is either uniformly populated with intercalate or empty. Such a picture cannot explain many kinetic staging phenomena, as was pointed out by *Daumas* and *Hérold* [3.40]. They proposed that staged intercalation compounds consist of microscopic domains, as shown schematically in Fig. 3.1. The intercalation compound is stage ordered in every domain, but in different domains the intercalate occupies galleries between different pairs of host layers. The host layers are elastically deformed at the domain boundaries. Within a gallery, the intercalate consists of two-dimensional "islands", each island belonging to a particular domain. Stage transitions in which the stage index n changes to $n \pm 1$ can then occur by movement of intercalate from domain to domain. In this way, *Daumas* and *Hérold* account for the fact that stage transitions happen without the intercalate passing through host layers and without intercalate atoms having to leave the sample in order to vacate some entire galleries and fill others [3.41].

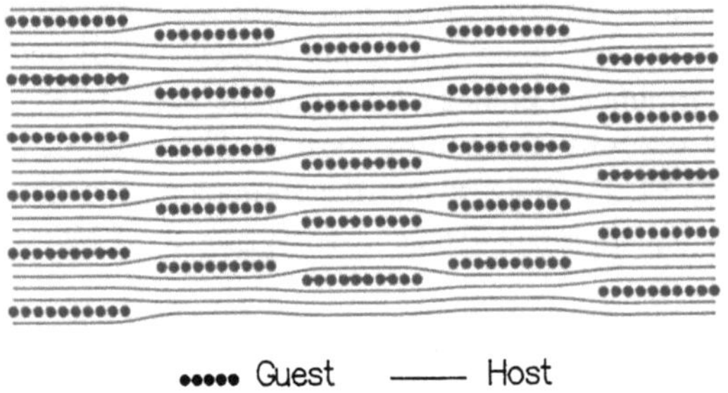

Fig. 3.1. Daumas-Hérold domain model [3.40] for a stage 4 intercalation compound. A slice through five domains is shown. Each domain can be thought of as a stack of two-dimensional islands of intercalate

Numerous experimental studies of stage transformations and of intercalation kinetics have lent indirect support to the Daumas-Hérold model [3.25, 26, 31, 32, 42–51]. Estimates of the in-plane dimensions of the domains range from about 100 Å to more than 10000 Å based on analyses of X-ray line widths [3.47–50, 52–54] and surface profiles of partly intercalated samples [3.51]. Recent theoretical work [3.55] suggests that in-plane residual resistance measurements may provide a relatively simple way of measuring domain sizes in low stage GICs. Estimates of domain sizes obtained in this way [3.55, 56] also fall in the above range.

Recently, electron microscopy has been demonstrated by *Thomas* et al. [3.57] to be capable of resolving the individual guest and host layers viewed edge on at the crystal surface. Such studies have yielded pictures of well-ordered and disordered stacks of guest and host layers, intercalate islands, interpenetrating stages and dislocations in the staged structures [3.57, 58], and also very large, perfectly staged regions [3.59]. An important advance has been the direct observation by *Matsuura* et al. [3.60] and by *Hwang* [3.61] of pleated Daumas-Hérold domain boundaries such as those shown in Fig. 3.1. The structure seen by *Levi-Setti* et al. [3.62] in images of the basal surface of freshly cleaved $SbCl_5$ graphite, using a high-resolution scanning ion microprobe technique, has also been interpreted as being due to intercalate islands arranged in domains.

The intercalate islands form because the elastic energy of the deformed host is relatively low in structures such as that shown in Fig. 3.1. This is because the elastic strains are confined to a small number of domain-wall-type dislocations. The elastic energy would be much higher if the intercalate atoms were randomly scattered throughout the galleries, each atom distorting the host layers in its vicinity. This idea was developed theoretically by *Safran* and *Hamann* [3.63] and *Ohnishi* and *Sugano* [3.64], who treated the host as an anisotropic elastic continuum deformed by embedded intercalate atoms. Their main result was that the effective elastic interaction between guest atoms displaced from one another in a direction parallel to the host layers is attractive, leading to island formation. The elastic interaction between atoms separated by a c-axis displacement was found to be repulsive. This interaction vanishes, however, for uniformly intercalated host galleries and thus cannot be the interlayer repulsion mechanism responsible for staging.

The elastic continuum theory [3.63, 64] does not allow for the fact that the host material is actually made up of discrete layers. *Kirczenow* [3.65] examined the effects of the discreteness of the host layers on the energetics of stage 2 Daumas-Hérold domain walls, by treating the host layers as harmonically coupled plates whose elastic properties are described by a "bending modulus". The dependence of the elastic energy on the domain wall thickness R is shown in Fig. 3.2. For domain walls where the intercalate layers are staggered across the wall (see inset to Fig. 3.2), the energy has a minimum as a function of domain wall thickness (curve a). Thus staggered domains tend to bind to each other. For domains where the intercalate layers match (i.e. occupy the same galleries) across the wall, there is an energy barrier which must be overcome before such domains can merge. The binding of staggered domains should hinder the movement of the intercalate in the host and slow intercalation kinetics, while the energy barrier for matching domains suggests that the Daumas-Hérold domains may be rather stable entities and difficult to "anneal out" of crystals. These calculations were generalized to higher stages by *Hawrylak* et al. [3.66].

A more sophisticated calculation of the domain wall structure and energetics along these lines was made by *Ulloa* and *Kirczenow* [3.27, 67]. A dramatic dependence on the guest species of the strength of the binding energies and energy barriers constraining relative motion of the domains was found. The binding energies are anomalously large for Na, Sm, Ca and Tm, guests which are notoriously difficult to intercalate into graphite. For species such as K and Br_2, which intercalate easily, the binding energies are small or nonexistent. For still others, such as $FeCl_3$, which are intermediate as to ease of intercalation, the binding energies are also intermediate. These results suggest a reasonable explanation of

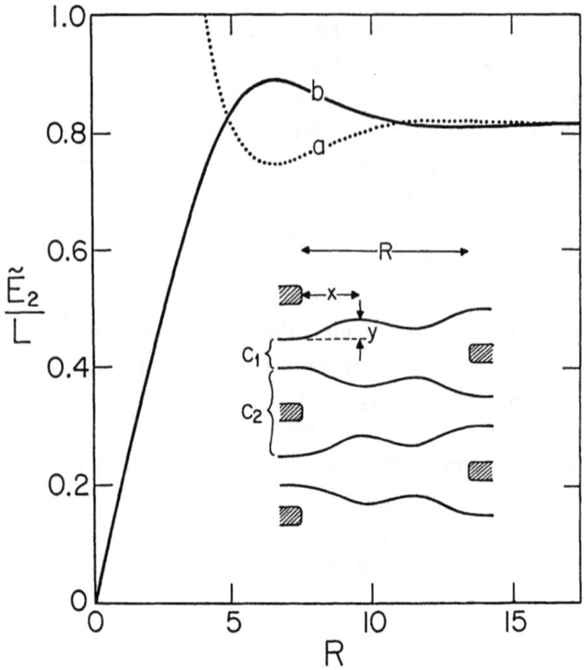

Fig. 3.2. Domain configuration energy $\tilde{E}_2/L$ vs domain wall thickness R for staggered stage 2 domains shown in inset (curve a) and matching domains (curve b). (From [3.65])

the long-standing mystery as to why Na alone among the alkali metals does not intercalate into graphite. Namely, the elastic interactions between Na intercalate islands are such as to greatly hinder the intercalation kinetics. This is consistent also with the findings that Na intercalates much more easily into graphite whose structure has been disrupted by ion implantation [3.68] and into "soft carbons" which are structurally disordered [3.69]. In such systems, the energetics of domain walls, if they exist, would be different.

3.1.4 Intralayer and Guest–Host Interactions

Intercalation compounds exhibit a profusion of structures within the intercalate layers (see [3.1–4, 9] and Chap. 2 of this volume). These can be commensurate with the host, as in many alkali metal GICs, some molecular acceptor compounds such as $C_{7n}Br_2$ for $n \geq 2$, some layered compound acceptors such as $C_{14n}SbCl_5$ and some ternary systems such as C_4KHg. Incommensurate structures are also common. Examples are the transition metal chlorides and bromides in graphite. Commensurate–incommensurate transitions and discommensuration-domain structures have been studied in bromine and alkali-metal GICs [3.9]. These are the result of a competition between the in-plane interactions between the intercalate atoms and the "corrugation potential" imposed on the intercalate by the in-plane periodicity of the graphite host. In addition, various kinds of in-plane *disorder* occur, including lattice-gas-like disorder, liquid-like disorder at high temperatures in some compounds such as KC_{24}, CsC_{24} and RbC_{24} [3.9], and presumably also disorder in the shape of the discommensuration domains.

Progress has been made towards a quantitative treatment of the energetics involved in some of these phenomena in the case of alkali-metal GICs. The strengths of the corrugation potentials in the stage 1 alkali-metal GICs have been calculated by *DiVincenzo* and *Mele* [3.70–72] using a Thomas-Fermi theory, in agreement with experimental diffusion constants [3.73]. Experimental estimates of the Fourier components of the corrugation potential have also been made for stage 2 RbC_{24} based on X-ray scattering data by *Thompson* et al. [3.74] and *Moss* and *Reiter* [3.75]. Microscopic in-plane structure calculations have been made for some cases by *Plischke* [3.76], *DiVincenzo* [3.71] and by *Miyazaki* et al. [3.77], but not in the context of models of staging.

Staging models have treated the in-plane intercalate structure and energetics more simply, using a lattice gas approach [3.28–30, 78], with phenomenological modifications in special situations [3.79, 80]. The details of in-plane structure have usually been considered to be of marginal importance for the *interlayer* interactions responsible for staging. This is because the staging phenomenon is qualitatively very similar in systems having quite different in-plane structures. Physically, the interlayer interactions responsible for staging are mediated by the charge which is transferred to the graphitic π-band states. These have a delocalized (metallic) character in the direction parallel to the host layers, which tends to "wash out" the effects of the details of the in-plane structure on the interlayer forces. The primary effect of the in-plane properties on staging is thus through the allowed values of the local in-plane intercalate density, which are strongly influenced by the in-plane and guest–host interactions, and which in turn control the areal density of the transferred charge.

3.2 Classical Staging Models and Their Implications

3.2.1 Safran's Model of Staging

The first efforts to construct a staging phase diagram based on a model free energy were made by *Safran* [3.78]. *Safran*'s model is a classical staging model in the sense that the distribution of intercalate in every gallery is assumed to be uniform; intercalate islands are not considered. Each gallery is modeled as a lattice gas with N_0 available intercalate sites, N_i of which are occupied in gallery i. The model free energy used by *Safran* can be written as

$$\begin{aligned}\phi =& E(\{N_i\}) - \mu \sum_i N_i - \\ & kT \sum_i [N_0 \ln N_0 - N_i \ln N_i - (N_0 - N_i) \ln(N_0 - N_i)] \,, \end{aligned} \tag{3.4}$$

where $E(\{N_i\})$ is the configurational energy of the set of intercalate layers with occupations $\{N_i\}$, μ is the chemical potential, and the third term represents the entropy of the lattice gas in the simplest mean field approximation. The configurational energy used was

$$E(\{N_i\}) = \sum_i U(N_i) + \sum_{i<j} V_{ij} N_i N_j / N_0 \,. \tag{3.5}$$

The first term represents the intralayer interactions between intercalate atoms, the second term, the repulsion between different intercalate layers. For $U(N_i)$, the Bragg-Williams quadratic form was used:

$$U(N_i) = -U_0 N_i^2 = -\varepsilon\xi N_i^2 / 2N_0 \,, \tag{3.6}$$

where $-\varepsilon$ is the nearest neighbor in-plane interaction energy and ξ is the in-plane coordination number. The interlayer repulsion V_{ij} was taken to be of the power-law form

$$V_{ij} = v_0 |i - j|^{-\alpha} \,. \tag{3.7}$$

The concentration–temperature plane of the phase diagram obtained by minimizing the free energy ϕ with respect to the layer fillings N_i is shown in Fig. 3.3. The concentration x is the ratio of the amount of intercalate present in the crystal to the amount which would be present if all of the galleries were completely filled with intercalate:

$$x = \sum_i N_i / \sum_i N_0 \,. \tag{3.8}$$

The high temperature part of the phase diagram labeled (1) is occupied by a stage 1 phase where all galleries have the same in-plane intercalate density. This

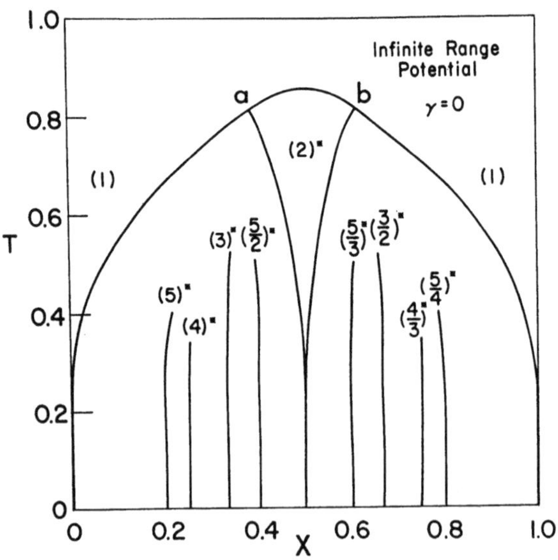

Fig. 3.3. Temperature–density plane of the staging phase diagram obtained from *Safran*'s model free energy. The areas labeled (1) and (2)* and the lines labeled $(n/m)^*$ (which represent narrow areas) are single-phase regions. Only phases with period ≤ 5 are shown. (From [3.28].) Only the phases labelled $(n)^*$ are shown explicitly in the corresponding figure in *Safran*'s original paper [3.78]

ranges from very dilute (at small x) to complete filling. In this regime, entropy effects completely overcome the tendency of the crystal to stage order.

The other pure phases shown in Fig. 3.3 are somewhat different from the ordinary (or *simple*) stage n structures where the period consists of one guest layer followed by n layers of host. In discussing these "complex stages", the term "fractional stage (n/m)" is used to mean a periodic structure where the period consists of n host layers interleaved with m occupied intercalate galleries, the remaining galleries in the period being vacant. [Stage $(n/1)$ is then the simple stage n structure.] In a complex stage $(n)^*$ structure, the period consists of n host layers and n galleries, *all* of the galleries containing some intercalate, but one gallery is special, containing more intercalate than the others. In *Safran*'s model, only that one gallery in the period remains populated in the complex stage $(n)^*$ structure as $T \to 0$. At $T = 0$ the complex stage $(n)^*$ becomes a simple stage n. Similarly, in a stage $(n/m)^*$ structure, all n galleries in the period contain some intercalate, but m of them are more densely populated than the others and only these m remain populated at $T = 0$.

Only structures with periods of up to 5 galleries are shown in Fig. 3.3. However, *all* fractional stages $(n/m)^*$ with $m \leq n$ do exist as stable phases in *Safran*'s model and should be present in the phase diagram, forming a "complete devil's staircase" of phases. This was demonstrated by *Bak* and *Bruinsma* [3.81]. In the regions of the phase diagram between the single-phase regions described above, the different periodic structures coexist in thermodynamic equilibrium.

Of particular interest is the line joining the points a and b which separates the complex stage $(2)^*$ and stage (1) regions in Fig. 3.3. As this line is approached from below, the concentrations of intercalate in the two galleries making up the period of the complex stage $(2)^*$ structure approach each other. They become equal when the line is reached. The result is a stage 1 structure. Thus at the line ab, the transition between stage $(2)^*$ and stage 1 is continuous. All other stage transitions are predicted to be first order.

Safran's model has been studied further by *Schön* et al. [3.82], who obtained results broadly in agreement with those outlined above, although the language used was different. Also, *Alstrøm* [3.83] recently examined some aspects of the "devil's staircase" in more detail.

3.2.2 The Need for Modifications

Safran's model stimulated much of the subsequent interest in staging. However, some of its features are unrealistic and applying it without modification to the interpretation of experimental data, as has at times been done, can be misleading.

One difficulty of the model is its prediction of the complete devil's staircase of fractional stages. Although hundreds of staged intercalation compounds have been studied experimentally, only two examples of a fractional stage have been observed: The stage (3/2) structure found in K-graphite at high pressures [3.48, 84] and the stage (4/3) reported recently in potassium-tetrahydrofuran-graphite (KTHF-graphite) [3.85]. Nothing resembling a complete devil's staircase of fractional stages has been seen[2].

A good reason for the absence of the "devil's staircase" of fractional stages was pointed out by *Safran* and *Hamann* on the basis of their Thomas-Fermi calculation of staging energetics [3.20]. They showed that the interlayer interaction V_{ij} appearing in (3.5) has the property

$$V_{ij} \simeq 0 \quad \text{if any guest layer occurs between guest layers } i \text{ and } j \,, \tag{3.9}$$

i.e., the repulsion between two guest layers i and j is effectively "screened out" by any intervening guest layer, which makes it impossible for structures with complicated periods to be stable. The "strong screening" condition (3.9) was *not* included in *Safran*'s model [3.78].

Another problem of the model is that no evidence of the complex $(n)^*$ and $(n/m)^*$ stages with noticeable amounts of intercalate in the nominally empty galleries has as yet been reported.

3.2.3 Models That Describe Simple Stages

Millman and *Kirczenow* [3.28, 29] modified *Safran*'s model by incorporating the strong screening condition (3.9) into the interlayer repulsion energy (3.7), and added to the configuration energy $E(\{N_i\})$ in (3.5) a term

$$\sum_i \gamma N_0 \theta_i \,, \tag{3.10}$$

where $\theta_i = 0$ if gallery i is empty and 1 if it is not. The contribution (3.10) is the energy that needs to be supplied to overcome the cohesion between the host layers in order to separate them and admit the intercalate in the rigid layer approximation, as discussed in Sect. 3.1.2. This term favors the simple stage n structures over the complex stage $(n)^*$ which are not seen experimentally. For most GICs, the separation energy in γ in (3.10) is of the same order as the other model parameters.

The phase diagram for values of γ typical of GICs is shown in Fig. 3.4a. There are no fractional stages or complex [stage $(n)^*$] structures. The area labeled (1), the narrow loop (2), and the lines (n) are the single-phase stability regions of the respective *simple* stages. The differences between this phase diagram and Fig. 3.3 make it clear that neither the separation energy nor the strong screening condition can be disregarded.

Because the Li^+ ion is smaller than any other guest in graphite, γ, the energy required to separate the graphite layers sufficiently to admit Li between them,

---

[2] *Alstrøm* [3.83] has recently claimed that data on the dilation of graphite during intercalation with AsF_5 supports the existence of the complete devil's staircase, but similar behavior is found in many GICs and X-ray data have consistently ruled out the devil's staircase possibility [3.25, 26, 49, 50].

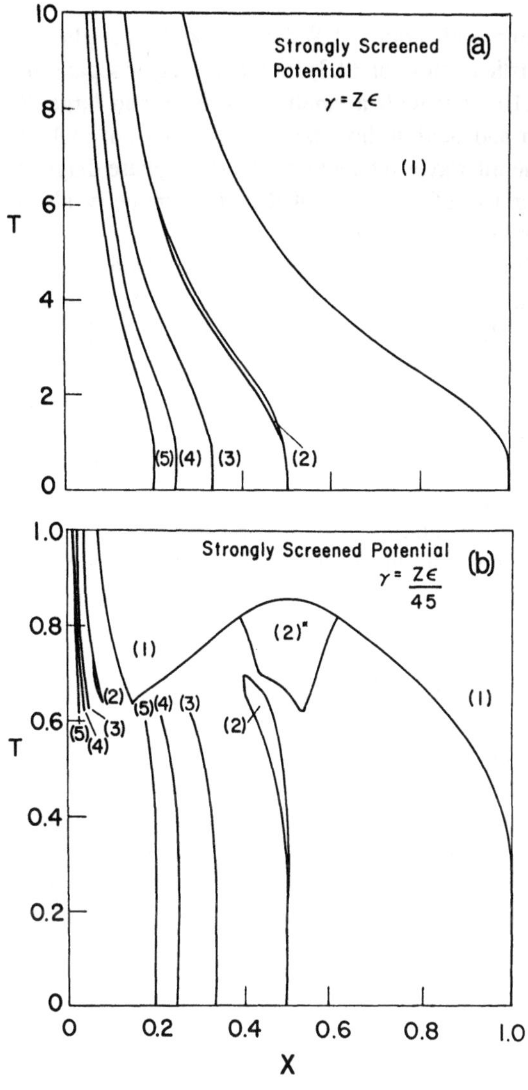

Fig. 3.4a,b. Typical staging phase diagrams obtained by modifying *Safran*'s model to include the strong screening of the repulsion between intercalate layers by any intervening intercalate layer, and the effects of the cohesion between the host layers. **(a)** Moderate values of γ corresponding to most GICs. **(b)** Small values of γ corresponding to small intercalate species (possibly Li) in graphite. The dilute high-stage phases in the upper left corner of the phase diagrams are cut off at high temperatures if the bending of the host layers is allowed for in the model. (From [3.28])

should be relatively small. Some intercalation compounds of transition metal dichalcogenides [3.30, 33, 86] seem also to have small γ values. The staging phase diagrams corresponding to small and intermediate values of γ were studied extensively by *Millman* and *Kirczenow* [3.28, 29]. A typical result for small γ is shown in Fig. 3.4b. Some features of *Safran*'s phase diagram – the re-entrant high temperature stage 1 phase at low x and the complex stage (2)* phase at high temperatures – are recovered for small γ. However, there are also *simple* stage 2 phases present and no fractional staging[3].

[3] Phase diagrams such as that in Fig. 3.4b could occur also for large γ for compounds with strong intralayer bonding. However, the re-entrant stage 1 and complex stage (2)* phases would be confined to unrealistically high temperatures [3.29].

Dahn et al. [3.30] also studied modifications of *Safran*'s model, as applied to Li-TiS_2. They used the more realistic expression (3.3) for the host layer separation energy rather than the form (3.10), but with a small value of γ and large β, corresponding to weak cohesion and flexible host layers. They did not include the "strong screening" (3.9) of the interlayer interactions. Thus the phase diagram which they obtained was similar to *Safran*'s, except that the symmetry about $x = 0.5$ was broken by the separation energy term.

The effects of relaxing the rigidity of the host layers by using *Dahn* et al.'s [3.30] form (3.3) of the separation energy were also studied by *Millman* and *Kirczenow* [3.29]. For the large values of γ characteristic of typical GICs, they found that this had the effect of cutting off the stability regions of all stages higher than 1, in Fig. 3.4a, above a temperature which is most likely too high to be accessible experimentally. For smaller values of γ, relaxing the rigidity of the host truncated the stability regions of stages ≥ 2 seen in the high-temperature low-density corner of the phase diagram of Fig. 3.4b.

These studies resolved the main *qualitative* differences between the predictions of *Safran*'s model and the types of staged phases which are actually observed experimentally.

3.2.4 Towards an Explanation of the Anomalous Fractional Stages

An interesting outstanding problem is to identify the microscopic mechanism(s) responsible for the existence of the exotic stage (3/2) K-graphite and stage (4/3) KTHF-graphite structures reported by *Fuerst* et al. [3.48] and by *Marcus* and *Touzain* [3.85]. If these structures correspond to thermodynamic phases, then they represent a violation of the strong-screening condition (3.9). This implies either a breakdown of the *Safran-Hamann* theory [3.20] or the appearance of an additional microscopic mechanism modifying the form of the interlayer interactions.

Safran [3.11] has suggested that elastic interactions may become sufficiently strong at high pressures in K-graphite to play the role of this additional mechanism. However, this idea has not as yet been developed into a concrete microscopic model.

Another approach to the problem of the high-pressure stage (3/2) structure in K-graphite was pursued by *Hawrylak* and *Subbaswamy* [3.80]. They considered a generalization of *Safran*'s staging model which retained the *unscreened* form (3.7) of the interlayer interactions, but included a separation energy term derived by *Dahn* et al. [3.30]. They also replaced the simple form (3.6) of the intralayer interaction energy $U(N_i)$ by a form having a minimum at a preferred intercalate density, the location of the minimum depending on the thickness of the host–guest–host sandwich. The unscreened interlayer interactions resulted in a stable stage (3/2) phase at high pressures but, remarkably, no stage (3/2) phase was found at zero pressure, reproducing the main feature of the experimental observations [3.48]. The suppression of the fractional stage (3/2) phase at zero pressure appears to be due to the special form of the intralayer interaction energy which was used. The stability of the other fractional stages was not investigated, so that no conclusions can be drawn about the generality of this effect. However, it is clear that, in some cases, effects other than the *Safran-Hamann* strong screening [3.29] are capable of suppressing a fractional stage in a part of the phase diagram.

No microscopic justification for the *unscreened* interlayer interaction used in their model was suggested by *Hawrylak* and *Subbaswamy*. The *completely* unscreened electronic interaction, acting through *many* occupied intercalate layers, which was assumed, seems quite unlikely. However, it is conceivable that the delicately balanced electronic structure [3.13] of the stage-1-like K-graphite guest–host–guest sequences which enter into the stage (3/2) might result in a failure of the Thomas-Fermi arguments of *Safran* and *Hamann* [3.20] and give rise to an *incompletely screened* interlayer interaction of limited range. A precise ab initio total energy calculation would decide this point.

Whether any of these considerations might help to explain the observation of the stage (4/3) KTHF-graphite structure [3.85] is unclear. What is clear is that fractional staged structures are most unusual among intercalation compounds and that highly material-specific ideas will be required to explain each such case.

3.3 Experimental Staging Phase Diagrams

The most popular technique for preparing staged GICs is the two-zone vapor transport method [3.87] where the intercalate and graphite are placed at the opposite ends of an ampoule and sealed under vacuum. The intercalate and graphite are then heated, the temperature T_g of the graphite being kept above the temperature T_i of the intercalate. The saturated vapor of the intercalate reacts with the graphite to form the GIC. The chemical potential of the intercalate is that of the intercalate vapor whose pressure is controlled by T_i but whose temperature is T_g. A great variety of other methods are also used in special situations [3.3]. The stage of the material has been characterized by many methods, including mass uptake of the guest by the host [3.88, 89], chemical analyses, color changes [3.1, 2], c-axis expansion of the samples [3.90], and X-ray and neutron diffraction measurements. At present, in situ diffraction studies are preferred for measuring staging phase diagrams, since they yield precise structural and thermodynamic data.

3.3.1 Typical Graphite Intercalation Compounds

As was noted in Sect. 3.2.3, the energy required to overcome the cohesion between the host layers in order to insert the intercalate is so large in most GICs as to make the low-x re-entrant stage 1 phase and complex stage $(n)^*$ structures nonexistent or unobservable. In situ X-ray diffraction studies of the staging phase diagram have been done on a number of systems [3.25, 26, 49, 50, 91] in this category.

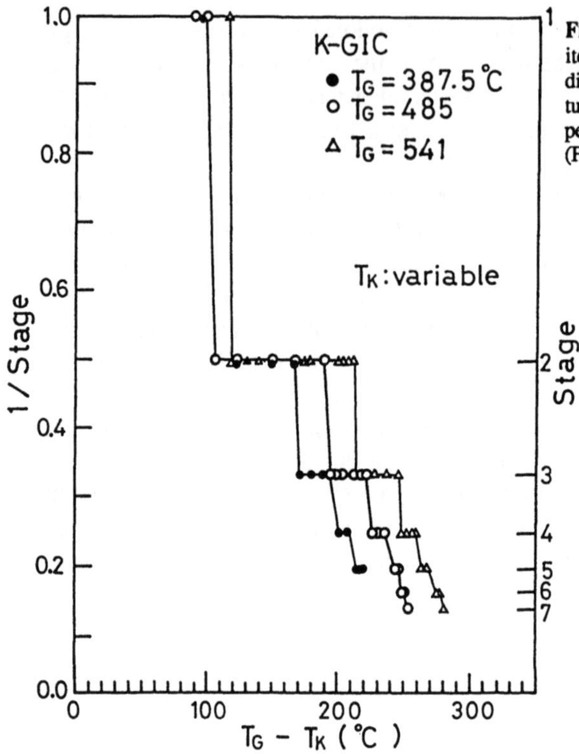

Fig. 3.5. Phase diagram of K-graphite measured by in situ X-ray diffraction. T_G is the temperature of the GIC, T_K is the temperature of the potassium metal. (From [3.25])

The results obtained for K-graphite by *Nishitani* et al. [3.25] using the two-zone vapor transport method are summarized in Fig. 3.5. The stable stage is plotted as a function of the temperature T_K of the potassium intercalate (which determines the K vapor pressure) for three different graphite temperatures. Only simple stages were observed in the X-ray spectra and there was no evidence of a dilute stage 1 phase forming at high sample temperatures and low x (which corresponds to low μ and low T_K). These results agree with another in situ X-ray study of K-graphite by *Misenheimer* and *Zabel* [3.49], who also reported only simple stages up to about stage 7 with no evidence of fractional or complex staging. Similar results were also reported for Rb-graphite by *Nishitani* et al. [3.91], and for the acceptor system $SbCl_5$-graphite by *Yosida* et al. [3.26]. Higher resolution work by *Misenheimer* and *Zabel* on K-graphite confirmed the results of the previous studies but also showed some disorder in the sequence of guest and host layers (i.e. stage disorder), especially close to stage transitions [3.50]. This interesting phenomenon will be discussed further in Sect. 3.4. Earlier ex situ studies of staging in many GICs had yielded results consistent with these findings [3.1, 2]. As was noted in Sect. 3.1.1, the stage stability ranges $\Delta\mu(n)$ obtained from the above data agree fairly well with the *Safran-Hamann* theory [3.20, 21] of the interlayer repulsion responsible for staging.

While the theoretical treatment of the in-plane properties in the models of staging discussed in Sect. 3.2 is very simplified, there is some qualitative agreement with experiment. For example, *Clarke* et al. [3.46] who studied a mixed stage 1–stage 2 Cs-graphite system, observed a decrease of the in-plane intercalate density in the stage 1 Cs-graphite when the temperature was increased, an effect consistent with the shape of the stage 1 stability region at the high-x side of Figs. 3.3 and 3.4. More recently, *Nishitani* et al. [3.92] observed a continuous decrease of the in-plane density of both stage 1 and stage 2 K-graphite with decreasing potassium chemical potential at a constant temperature of 390°C, again in qualitative agreement with the theories. However, the simple lattice gas treatments of staging [3.28–30, 78] predict nearly stage-independent in-plane intercalate densities at low temperatures. Thus they do not describe the change in in-plane density from $MC_{\sim 8}$ to $MC_{\sim 12n}$ which happens between stage 1 and stage $n \geq 2$ for the heavy alkali metals, or the stage dependence of the more delicate superlattice effects, see Chap. 2 and [3.1–4, 9]. It is possible to make some progress by building greater variability of the in-plane densities into staging models phenomenologically [3.79, 80], but at the expense of using increasingly arbitrary parametrizations. It is clear that a more microscopic theoretical treatment of the relationship between staging and the in-plane structure and thermodynamics of the alkali metal GICs would be of interest.

For some acceptor GICs such as $FeCl_3$ graphite, there is evidence that the *local* in-plane intercalate density is indeed only weakly dependent on the stage [3.93]. Also, the intercalate layers have a well-defined lattice structure of their own, not so strongly modified by the graphitic corrugation potential. In such cases, lattice gas models should be appropriate, although improvements to the simple mean field treatment of the in-plane free energy may be needed.

3.3.2 Systems Exhibiting a Reentrant Dilute Stage 1 Phase at High Temperatures

In order to investigate experimentally the exotic high-temperature staging behavior predicted in Figs. 3.3 and 3.4b, it is necessary to study systems with small values of the energy γ needed to separate the host layers sufficiently to admit the intercalate [3.28, 29]. Among GICs, Li-graphite is the best candidate because of the small size of the Li^+ ion.

In situ high-temperature diffraction studies of the Li-graphite staging phase diagram were made by *Woo* et al. [3.31, 32]. Their phase diagram is shown in Fig. 3.6. There is a reentrant dilute stage 1 phase at high T and low x. Two stage 2 phases are present which differ in their in-plane structure. The low T stage 2(S) phase is ordered in-plane and has a nominal stoichiometry of $LiC_{\sim 12}$. The higher temperature stage 2(L) phase is disordered in-plane and has a lower in-plane density. Neither stage 2 phase shows evidence of becoming a Safran-type complex stage (2)* structure in the temperature range which was studied. The very low x, high T part of the phase diagram was not studied so that it is not clear whether the high temperature dilute stages ≥ 2 shown in the upper left corner of Fig. 3.4b are present or completely suppressed in this system by the

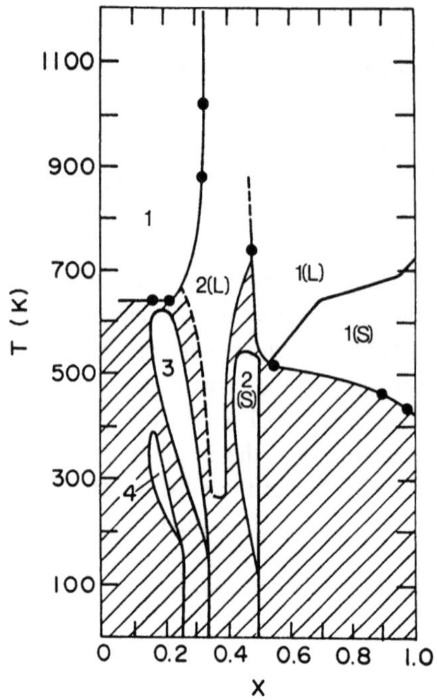

Fig. 3.6. Experimental staging phase diagram of Li-graphite. (From [3.32, 33])

bending of the host layers [3.29]. Two stage 1 phases, 1(S) and 1(L), which differ in their in-plane structure, occur in the high x, high T part of the phase diagram [3.33]. Unfortunately, at temperatures above 750 K, the Li reacts irreversibly with the graphite to form lithium carbide [3.94] so that it was not possible to study the high temperature limit of the stage 2 stability region. Thus the interesting questions as to whether a complex stage (2)* structure forms at high temperatures and eventually makes a continuous transition to stage 1 were left open.

DiVicenzo and *Koch* [3.79] proposed a modification of *Safran*'s staging model [3.78] which reflects the special properties of the in-plane structure of the Li-graphite system. They chose the in-plane spacing of the sites in their lattice gas model to be that of the lithium atoms in the low-tempetature stage 2(S) experimental phase. The in-plane energy $U(N_i)$ in (3.6) was modified to include a slight depression at the in-plane density corresponding to a more dilute commensurate structure having a stage 2 stoichiometry close to that of the experimental stage 2(L) phase. This model reproduced the main qualitative features of the phase diagram of Fig. 3.6: the separate dense and dilute stage 2(S) and 2(L) phases and the high degree of stability of the stage 2 at high temperatures.

Because Li-graphite is unstable at high temperatures, some intercalated transition metal dichalcogenides may be better systems for high T studies of staging phenomena. A study of TiS_2 intercalated with Ag by *Bardhan* et al. [3.86] yielded the high-T staging phase diagram shown in Fig. 3.7. The X-ray diffraction measurements were not carried out in situ but on samples rapidly quenched to room temperature, so that these results must be regarded as tentative. However, the

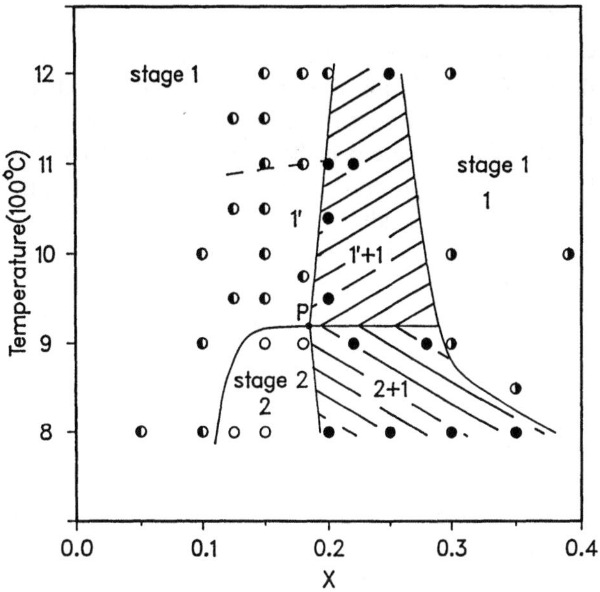

Fig. 3.7. Experimental high temperature staging phase diagram of TiS_2 intercalated with silver. (From [3.86])

kinetics of stage transformations in this system appeared to be very slow, even at the highest temperatures, suggesting that the data obtained from the quenched samples may be representative of the high T phases. The data show a more dilute and a less dilute stage 1 phase which can coexist. Below about 900°C, a pure stage 2 occurs at intermediate values of x. This appears to be a simple stage 2 structure; no evidence of a complex stage (2)* structure was found. At high T and low x the stage 2 terminates in a phase labeled 1′, which is stage-disordered. Its structure is a *disordered* sequence made up of sandwiches of one or two host layers between successive guest layers. The fraction of stage-1-like sandwiches in the sequence increases continuously with increasing temperature. This is quite a different high T termination of stage 2 from that predicted by *Safran*'s model [3.78], but agrees with a picture of stage transitions proceeding via a series of stage-disordered states proposed by *Kirczenow* [3.95]; see Sect. 3.4.1. In situ high temperature studies need to be performed on the Ag-TiS_2 system to obtain a better understanding of these interesting phenomena.

Li_xTiS_2 is another system which needs further study. It exhibits stage 1 structures ranging from very dilute to dense at room temperature [3.30]. There is evidence of a stage-disordered stage 2 structure at intermediate concentrations [3.34, 35] which may be related to the 1′ phase of Fig. 3.7. As has been suggested by *Fischer* et al. [3.32], an ordered stage 2 phase may be stable at lower temperatures, however, experiments are needed to test this hypothesis. Li_xNbSe_2 is yet another promising system for studies of the high temperature staging phase diagram [3.96].

3.3.3 Staging Under Pressure

The alkali-metal GICs have more than one preferred value for the in-plane intercalate density because of the coupling of the alkali metal atoms to the periodicity of the graphite layers. For the heavy alkali metals M, the preferred stoichiometries are roughly $MC_{12\times n}$ and $MC_{8\times n}$ where n is the stage. This property was exploited by *Clarke* et al. [3.47] in a high pressure experiment on K-graphite which introduced an entirely new way of studying staging phenomena.

They found that a sample which was stage 2 (nominally KC_{24}) at zero pressure began to show additional reflections in the X-ray scans at 2.5 kbar and became a pure stage 3 at 6.5 kbar. The transition, although hysteretic, was completely reversible, indicating that the stage 3 was *not* being produced simply by squeezing some potassium out of the sample. The transition takes the form $KC_{12\times 2} \to KC_{8\times 3}$, conserving matter in the sample. The K is more densely packed in-plane into a 2×2 structure commensurate with the graphite lattice and correspondingly occupies fewer galleries in the stage 3 than in the stage 2. The transition is driven energetically by the decrease in volume of the sample under pressure when the stage 2 converts to stage 3. The rearrangement of intercalate which has to take place in this transition provides strong evidence for the Daumas-Hérold domain structure of intercalation compounds described in Sect. 3.1.3.

This work was extended by *Wada* et al. [3.97–100] to other stages and also to Cs-graphite and Rb-graphite, where similar behavior was found. The transitions were of the form $MC_{12\times n} \to MC_{8\times n'}$, the final phase having the dense 2×2 in-plane structure, with intermediate coexisting stages being formed along the way. Obviously, such a reaction can only go to completion and conserve intercalate for special choices of n and n', and this is what was found.

These experiments are consistent with the theoretical results of *Miyazaki* et al. [3.77], who studied the energetics of graphite–alkali metal–graphite sandwiches at 0K at various pressures. They found that for the heavy alkali metals for stages $n \geq 2$, $MC_{14\times n}$ is the most stable stoichiometry at zero pressure while $MC_{8\times n}$ is the most stable at high pressures. The likely reason for the discrepancy between the predicted $MC_{14\times n}$ and observed $MC_{12\times n}$ stoichiometries is that *Miyazaki* et al. [3.77] only studied structures commensurate with the graphite. The observed zero pressure structure is weakly incommensurate, but close to the commensurate $\sqrt{7}\times\sqrt{7}$ structure which would give the $MC_{14\times n}$ stoichiometry, as was pointed out by *Clarke* et al. [3.101].

The pressure experiments of *Fuerst* et al. [3.48] which yielded the stage (3/2) phase of K-graphite discussed in Sect. 3.2.4 are also interesting for other reasons. The starting point at zero pressure was the stage 1 KC_8, which is already a "dense" in-plane structure in the above sense, so that a "superdense" phase must exist for a pressure-driven staging transition to occur. In the experiments, the stage (3/2) is observed together with stage 1 between 15 and 19 kbar and is replaced by stage 2 above 19 kbar. A scheme which is consistent with this behavior [3.48] is the series of transitions

$$6KC_{8\times 1} \to 4KC_{6\times 1} + K_2C_{8\times 3} \to 4KC_{6\times 1} + 4KC_{6\times 2}\,, \tag{3.11}$$

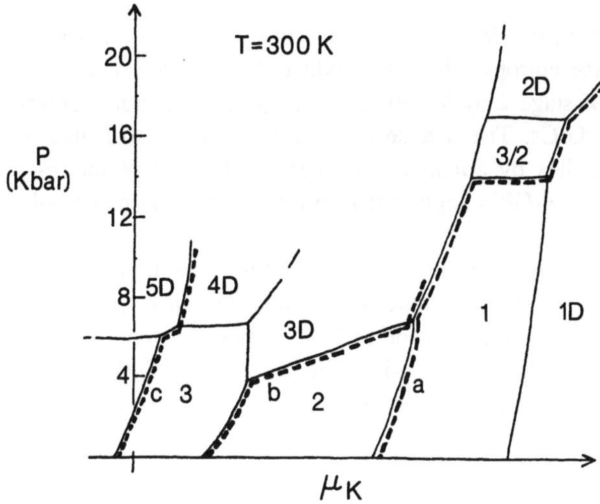

Fig. 3.8. Schematic chemical potential–pressure phase diagram of K-graphite. The pressure scale is determined from experiment, the chemical potential scale is arbitrary and nonlinear. Dashed lines a, b, c represent some experimental trajectories. (From [3.72])

where $KC_{6\times1}$ is a "superdense" stage 1 with a $\sqrt{3} \times \sqrt{3}$ in-plane structure, $K_2C_{8\times3}$ is the stage (3/2) and $KC_{6\times2}$ is a superdense stage 2. X-ray data have confirmed the presence of the in-plane structures appearing in this sequence in the correct pressures ranges [3.102].

In analyzing these and other experiments on the staging transitions that occur in K-graphite under pressure [3.47, 48, 84, 102–105], it is helpful to think in terms of a chemical potential–pressure phase diagram such as shown in Fig. 3.8. This was constructed by *DiVincenzo* [3.72] on the basis of (a) the experimental data, *which is assumed to represent equilibrium states*, (b) the assumption that each stage n comes in a normal density form denoted n and a dense form denoted nD, and (c) a Clausius-Clapeyron relation $dP/d\mu_K = \Delta N_K/\Delta V$ taken along each phase boundary. The change in the number of intercalate atoms per carbon atom across the phase boundary is represented by ΔN_K and ΔV is the change in volume per carbon atom. Dashed line a corresponds to the experiments on the stage (3/2) phase [3.48, 84, 102], b and c to [3.84]. Since the amount of intercalate in the samples is fixed during the experiments, μ_K is not constant but varies freely as the pressure changes. The experimental trajectories are assumed to follow the phase boundaries, reflecting the observed coexistence of the staged phases. Not all of the steps shown in the trajectories have actually been seen in the experiments, but their presence is needed for consistency.

Li-graphite under pressure is again an interesting special case, as was shown by *DiVincenzo* et al. [3.106]. It has a dilute stage $2(L)$ $LiC_{\sim16}$ phase at temperatures $\gtrsim 250$ K (as shown in Fig. 3.6), which can be densified to produce a mixture of the close-packed stage $2(S)$ $LiC_{6\times2}$ and stage 3 $LiC_{6\times3}$ phases either by lowering the temperature at $P = 0$ or by raising the pressure at constant temperature. In the first case, the densification is entropy driven. In the second, the pressure must overcome entropy effects to produce the ordered phases.

Such a densification of stage 2 by lowering the temperature is not observed in the other alkali metal GICs. The precise reasons for this are still unclear, although the extensive studies by *Sundqvist* and *Fischer* [3.105] of the *P–T* phase diagrams of low stage K-GICs suggest that subtle details of in-plane order may be involved.

High pressure studies of acceptor GICs have tended to show transitions associated with molecular rearrangements rather than staging. However, *Sundqvist* and *Lundberg* [3.107] have reported in-plane resistance measurements on SbF_5 graphite and $SbCl_5$ graphite under pressure which are consistent with stage transformations. Diffraction measurements are needed to test this interpretation.

3.4 Stage Disorder

So far we have considered structures with a perfectly periodic arrangement of the guest and host layers along the *c*-axis. However, disorder in the sequence of guest and host layers is also frequently observed. This disorder is referred to as "stage disorder" or "random staging". Most of the structural data on stage disorder has come from diffraction experiments analyzed using variants of the *Hendricks-Teller* [3.108] theory of diffraction from a one-dimensional disordered lattice. Stage disorder appears to be as universal as staging itself, and has been studied in acceptor [3.3, 42, 57, 58, 89] and donor [3.32, 47, 48, 50, 84, 109–113] GICs and in intercalated transition metal dichalcogenides [3.30, 34, 35, 86, 96].

In the early studies, there seemed to be little reason to suspect that stage disorder represents anything other than guest layers in the process of arranging themselves into a stable stage-ordered sequence, i.e., an effect of the kinetics. This is clearly correct in some cases, such as that described recently by *Suematsu* et al. [3.113]. However, in their X-ray study of high-stage K-graphite, *Fuerst* et al. [3.109] found that the degree of stage disorder decreased with decreasing temperature and that this process was reversible. At first sight, this would seem to be strong evidence that the observed stage disorder was *thermal*, however, subsequent work by *Heiney* et al. [3.110] has suggested that it may be premature to draw this conclusion from the data. Another interesting experiment which may be indicative of thermal stage disorder was performed by *Dahn* and *Haering* [3.35]. They made *thermodynamic* measurements of the entropy of TiS_2 intercalated with Li. They found a contribution to the entropy of the Li_xTiS_2 which could not be explained on the basis of a lattice-gas-like disorder, but appeared to be due to stage disorder identified by them also in their X-ray diffraction measurements.

3.4.1 Thermal Theory of Stage Disorder in Daumas-Hérold Domains

These experiments raised the interesting question of whether there can be any *theoretical* basis for *thermal* stage disorder, with an associated well-defined entropy. The models of staging discussed above do not include such effects because they assume that the in-plane density of intercalate in every gallery is spatially uniform. If that is the case, then the entropy of stage disorder is independent of the area of a host layer, whereas the energies involved in staging are proportional to that area. The entropy of stage disorder is then negligible compared to the staging energy for macroscopic crystals (the thermodynamic limit) and no stage disorder is predicted. It is believed, however, that the intercalate layers are not continuous but consist of separate islands stacked up into Daumas-Hérold domains, as is explained in Sect. 3.1.3. The finite island size invalidates the above argument because different domains can have different guest–host sequences. Thus the entropy of stage disorder becomes an extensive quantity (like the energy of staging) and remains important for macroscopic crystals.

A simple equilibrium theory of staging in Daumas-Hérold domains based on the above ideas was proposed by *Kirczenow* [3.95]. Finite domains were assumed and interactions between them were neglected. It was also assumed that the domain size is a parameter imposed by the past history of the sample, i.e., by the kinetics. The theory shows that if the Daumas-Hérold domain model applies, then stage disorder is a direct thermodynamic consequence.

The free energy of a *single* domain was obtained by adding to the staging free energy of *Millman* and *Kirczenow* (Sect. 3.2.3) a term

$$\phi_{\mathrm{DH}} = -k_{\mathrm{B}}T\left\{\left(\sum_i \nu_i\right)\left(\ln\sum_j \nu_j\right) - \sum_i(\nu_i \ln\nu_i)\right\} \tag{3.12}$$

to account for the entropy of stage disorder, where ν_i is the number of stage i units (sandwiches consisting of a pair of guest layers separated by i host layers) in the domain. Since a single domain is being considered, the parameter N_0 in Sect. 3.2 now becomes the number of intercalate sites in a gallery *within a domain*, and is a measure of the in-plane domain size.

The free energy was minimized to obtain the equilibrium distribution of stage units in the domain. A typical result is shown in Fig. 3.9, where the fractions $f_i = \nu_i/(\sum_j \nu_j)$ of the various stage units in the disordered domain are plotted against chemical potential. For large μ, $f_1 = 1$ and the system is pure stage 1. With increasing μ, f_1 decreases continuously to zero in a very narrow range of μ while f_2 increases from zero to 1, whereupon the system becomes pure stage 2. Similar continuous transitions via stage-disordered states are also found between successive higher stages. For high stages ($i > 7$ in Fig. 3.9), pure stages can no longer exist in thermodynamic equilibrium and the domain is always a disordered mixture of two or more different staging units.

For high stages, one can obtain an analytic solution for the equilibrium distribution of stage i units. One finds

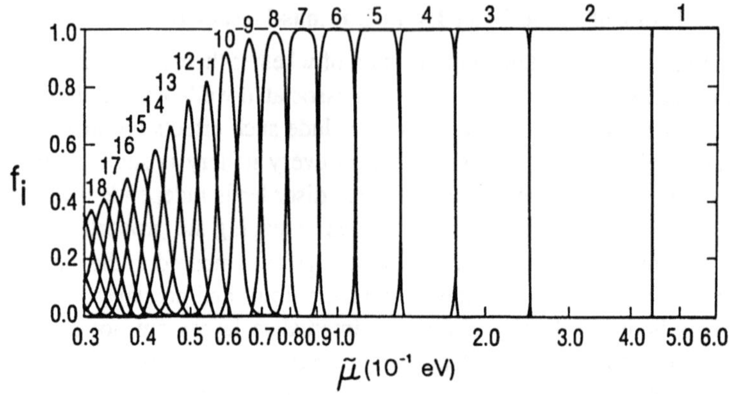

Fig. 3.9. A typical plot of the fraction f_i of stage i units ($1 \leq i \leq 18$) in a stage-disordered Daumas-Hérold domain at equilibrium versus chemical potential. (From [3.95])

$$f_i \simeq \tilde{f} \exp\{-[(i - \tilde{\imath})/\sigma]^2/2\} \tag{3.13}$$

where

$$\begin{aligned}\sigma^2 &= k_B T \left[\left(N_0 \tilde{x}^2 \frac{d^2 u_i}{di^2} \right)_{i=\tilde{\imath}} \right]^{-1} \\ &\simeq \frac{k_B T \tilde{\imath}^{\alpha+2}}{N_0 \tilde{x}^2 v_0 \alpha(\alpha+1)} \, .\end{aligned} \tag{3.14}$$

The distribution is Gaussian about a dominant stage $\tilde{\imath}$. The repulsion between guest layers is given by $u_i = v_0 i^{-\alpha}$; $\tilde{x}$ is the local in-plane density; σ is a measure of the stage disorder, which increases with increasing stage i and temperature T and decreases with increasing domain size N_0, in-plane density $\tilde{x}$ and charge transfer (which is related to v_0). These trends are also obeyed by the stage disorder calculated numerically for low stages. The trends are due to the competition between the interlayer repulsive energy, which favors stage order, and the entropy, which favors stage disorder.

The theory sets a thermodynamic upper limit on the value of the stage index for which a stage-ordered structure can exist. By examining numerical solutions of the minimization of the free energy ϕ, it was found that for combinations of the parameters in (3.14) which yield $\sigma \geq 0.29$, stage ordered structures are always thermodynamically unstable. This corresponds to $i \geq 8$ in Fig. 3.9. In particular, for a given stage i and temperature T, the domain size N_0 must exceed a critical value $\hat{N}_0$ corresponding to $\sigma \simeq 0.29$ for a well-ordered stage i structure to be stable thermodynamically. This result has turned out to be very useful in interpreting the results of computer simulations of staging (Sects. 3.5 and 3.6), which have recently yielded new insights into the nature of stage disorder.

It is plausible that under most conditions the Daumas-Hérold domain size is determined mainly by the kinetics and past history of the samples as was assumed above. However, it is quite possible that in some cases intercalate islands, and even imperfect domain structures, might exist as a *thermal equilibrium* property at high temperatures. Experimental manifestations of this may be the disordered incipient stage 2 structure reported by *Dahn* et al. in Li_xTiS_2 [3.30, 34, 35], the disordered stage 3 in Li_xNbSe_2 [3.30, 96], and the high T disordered stage 2 in Ag_xTiS_2 [3.86]. A high temperature continuous transition from stage 2 to stage 1 via such stage-disordered states is an interesting possible alternative [3.86] to the continuous transition from stage 2 to stage 1 via the complex stage (2)* structure predicted by *Safran*'s model of staging. These ideas require further investigation.

3.4.2 Experiments: General Trends

So far, only qualitative comparisons between the above thermal theory of stage disorder and experiments have been made. One reason is that the domain size N_0 is an important parameter in the theory and should be measured in quantitative experimental studies of stage disorder, but this is difficult and has rarely been done.

The trend towards greater stage disorder with increasing stage which is displayed by the theory was already identifiable in the early experimental work of *Metz* and *Hohlwein* [3.42] on graphite-$FeCl_3$, and was clarified by *Solin* [3.3]. The extensive studies of *Heiney* et al. [3.110] also found this trend in K-graphite for stages ≥ 4. Their finding that the structure progresses from disordered mixtures of units of just two adjacent stages to mixtures of units of many stages with increasing stage also agrees with the theory.

Another interesting trend is that stage disorder is more often seen experimentally, and seen for lower stages, in acceptor GICs than in donor GICs. This may be related to the lower charge transfer in the acceptor GICs, which enters equation (3.14) through the factor v_0. Possible differences in the power α between donors and some acceptors [3.26] might also be involved. However, a systematic study comparing stage disorder in different compounds *and taking proper account of the variation in domain size* from compound to compound and sample to sample is needed to clarify this behavior.

As was mentioned above, *Fuerst* et al. [3.32, 109] and *Heiney* et al. [3.110] observed a marked decrease in stage disorder on cooling high-stage K-graphite samples. This is suggestive of thermal stage disorder, but is not definitive evidence: In *all* of the high stage samples studied, the stoichiometry of the stage-disordered mixture happened to correspond to an "average stage" value slightly *larger* than an integer. *Heiney* et al. [3.110] found that on cooling, there is a slight in-plane *expansion* of the intercalate. Because the studies were not performed in situ but on quenched samples, the amount of intercalate in the system was probably nearly constant. If so, then the in-plane expansion should result in the filling of more galleries with intercalate on cooling, decreasing the "average stage" and hence the degree of stage disorder. Thus the observed stage ordering on cooling might not be driven by entropy effects.

The predictions of the thermal theory of stage disorder for experiments on quenched samples were not discussed in [3.95]. In general, if any parameter such as the temperature or pressure is varied in an experiment where the amount of intercalate is held constant, the chemical potential μ of the intercalate must change in such a way as to remain consistent with the total amount of intercalate present. In the simplest possible case, where *only two different stage units are involved* in a stage-disordered system *and there is no change in the intercalate in-plane density*, this leads to an amusing result: As the temperature is lowered, the stage transitions in Fig. 3.9 narrow, but the chemical potential must move towards the crossover point[4] between the two stages involved in such a way as to maintain the relative amounts of the two stages. If the stage disorder in the system is and stays spatially homogeneous, this means that the degree of stage disorder will remain the same while the system is cooled, even for *perfectly thermal* stage disorder.

This may explain the recent experiments of *Cajipe* and *Fischer* [3.111] on quenched samples of stage-disordered K-graphite which consisted mainly of stage 2 units with small admixtures of stage 3 units. The stage disorder stayed the same on lowering the temperature in a sample in which the stage 2 and 3 units appeared to be homogeneously distributed. In inhomogeneous samples, the predominantly stage 2 regions apparently became less disordered with decreasing temperature by absorbing some intercalate from predominantly stage 3 regions.

The interpretation of the temperature dependence of stage disorder in quenched samples is clearly not simple. However, even in situ measurements may prove difficult to analyze if the crossover value[4] of the chemical potential turns out to be temperature dependent.

No systematic studies of the dependence of stage disorder on domain size have been reported; however, samples used in studies of stage transformations under pressure have generally showed relatively high degrees of stage disorder [3.4, 47, 48, 84]. This may be related to the large numbers of dislocations formed in the high pressure experiments, particularly during stage transitions, and hence to a reduced domain size.

A recent study by *Fischer* and *Kim* [3.112] found that stage disorder increased *smoothly* with the average stage in high stage K-graphite at high pressures, consistent with the previous zero pressure studies [3.110] and with the thermal theory of stage disorder [3.95]. The average stage of the samples was, however, found to increase with increasing pressure, either continuously or abruptly depending on the stage. It appears that in-plane structural changes play an important role in determining this pressure dependence of the average stage.

3.4.3 Experiments: Stage Disorder and Stage Transformations

Some very interesting experiments on stage disorder were performed by *Misenheimer* and *Zabel* [3.50] on low stage K-graphite. Their high-resolution in situ X-ray data showed stage disorder, apparently in thermal equilibrium, for stages

[4] The crossover point is the point where the plots of f_i and f_{i+1} vs μ cross as in Fig. 3.9

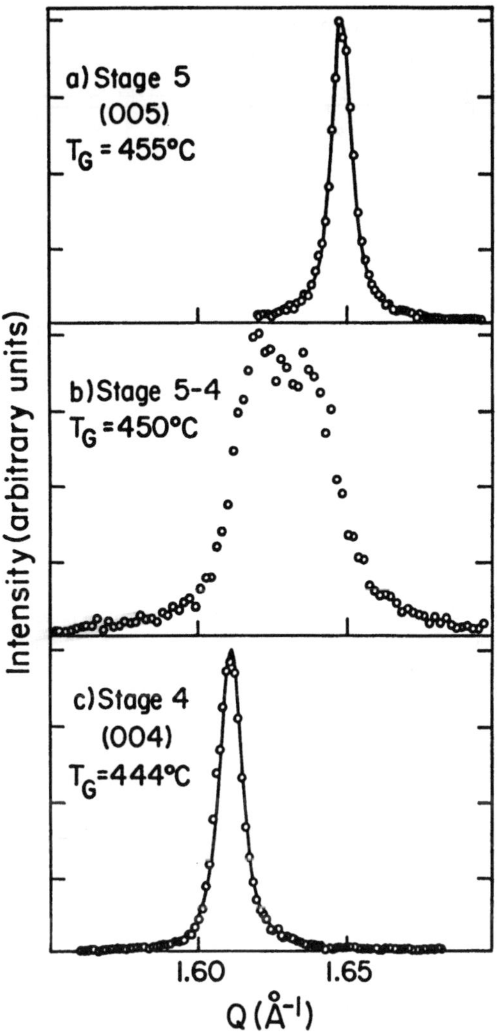

Fig. 3.10a–c. High resolution (00ℓ) X-ray diffraction spectra taken during the stage 5 to stage 4 transition. (a) Stage disordered stage 5 with admixture of stage 4 units. (c) Stage 4 with admixture of stage 5. (b) Coexisting reflections from stage-4-rich and stage-5-rich regions. (From [3.50])

≥ 2. The stage disorder was found to increase close to the transitions between successive stages.

Examples of their diffraction data close to the stage 5–stage 4 transition are shown in Fig. 3.10. All of the lines shown are substantially broader than the instrumental resolution. The solid lines in Fig. 3.10a and c are fits using the *Hendricks-Teller* theory [3.108], assuming disordered mixtures of stage 4 and 5 units. As the stage transition is approached and the degree of stage disorder increases, the disordered stage 5 reflection moves towards the stage 4 position because of the change in the average stage of the sample. Figure 3.10b shows broadened and shifted stage 4 and stage 5 lines together, indicative of the coexistence of disordered stage-4-rich and stage-5-rich regions in the crystal during part of the transition. Figure 3.11 shows the fraction f_n of stage n units as a

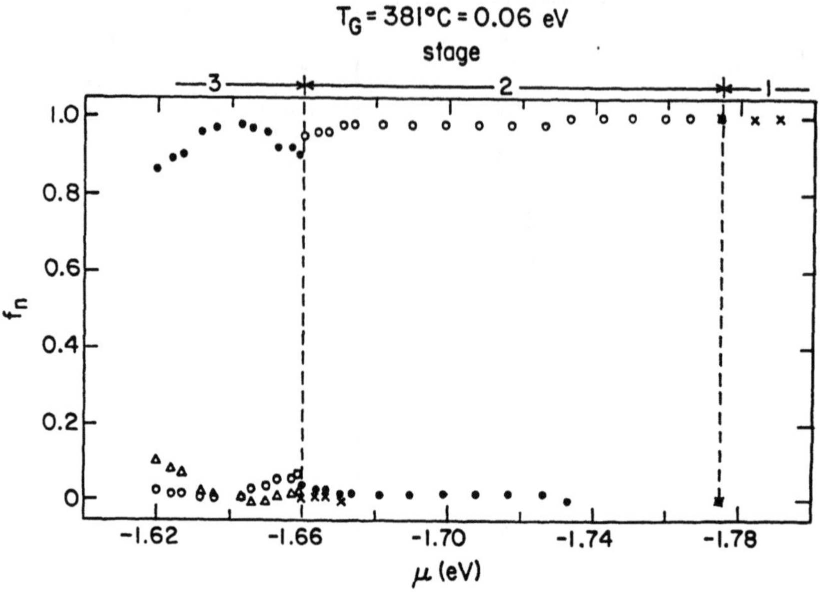

Fig. 3.11. Experimentally measured fraction f_n of stage n units as a function of chemical potential. Arrows at the top indicate the stability region of stage n. The fractions of stage 1, 2, 3 and 4 units are plotted as crosses, circles, dots and triangles respectively. (From [3.50])

function of chemical potential near the stage 3–stage 2 transition. Notice that the stage 3 is never perfect but the fraction of stage 2 units increases as the transition to stage 2 is approached.

The results are in qualitative agreement with the predictions of the thermal theory of stage disorder for stage transitions [3.95]. The predicted buildup of stage disorder near the transitions and the broadening of the transitions with increasing stage are observed. However, the observed transitions appear to be discontinuous, whereas continuous transitions were predicted. This may be due to "miscibility gaps" between different disordered staged phases [3.50], which result from the domain–domain interactions neglected in the theory; these miscibility gaps decrease with increasing stage. It may also be due to some portions of the crystal changing stage earlier than others because of kinetic effects. Another outstanding question is that of domain size. The line widths of the in-plane reflections seen by *Misenheimer* and *Zabel* suggest that the domain sizes in their samples may be larger than those for which the thermal theory [3.95] predicts degrees of stage disorder as large as those which were observed. The nature of stage disorder in large domains has been investigated in the computer simulations discussed in Sect. 3.6.

The experimental data of *Misenheimer* and *Zabel* [3.50] are consistent with the results of *Heiney* et al. [3.110] for stage 4 and stage 5 K-graphite. They are also consistent with the results of *Cajipe* and *Fischer* [3.111] on quenched stage 2–stage 3 K-graphite, which can be interpreted in terms of a miscibility gap that increases with decreasing temperature. However, the studies of K-graphite by

Nishitani et al. [3.25] showed no evidence of stage disorder during stage transitions for stages ≤ 7. One possible explanation is the higher resolution of the spectra taken by *Misenheimer* and *Zabel*. Another is that *Misenheimer* and *Zabel* took their spectra under steady state conditions, making several successive preliminary measurements to check whether the sample had equilibrated. This was not done by *Nishitani* et al. [3.25] who swept their sample through each transition without equilibration since they were studying kinetics. A recent experiment by *Nishitani* et al. [3.114] in which in situ time- and space-resolved Raman studies were performed on stage transitions in H_2SO_4-intercalated graphite, showed that under such nonequilibrium conditions, the stage transition occurs at a narrow front which travels through the sample. Thus at any one time, only a very small part of the sample is at the chemical potential of the stage transition, where stage disorder is favored.

Recently, *Bak* and *Forgacs* [3.115] have suggested that the stage disorder observed by *Misenheimer* and *Zabel* [3.50] is *entirely nonthermal*. They claimed that it is due to slow kinetics of the stage transformations, and that what is being seen is a partly disordered mixture of metastable phases belonging to the complete "devil's staircase" of fractional stages, which is predicted to exist between the integer stages according to Safran's *original* model of staging [3.78]. This suggestion disregards the point that the complete devil's staircase is really a deficiency of Safran's model rather than a prediction applicable to GICs (Sect. 3.2), and that fractional stages only exist in highly exceptional circumstances. The old question remains, however, whether the stage disorder is thermal or kinetic. *Bak* and *Forgacs* [3.115] suggest that studies of temperature dependence and hysteresis should decide this. As explained in Sect. 3.4.2, giving an unambiguous interpretation to the existing temperature-dependent studies is difficult. However, the experimental data [3.110] do not support the view that the stage disorder is *simply* due to "frozen in" defects. Observations of hysteresis in systems with narrow phase transitions and sluggish kinetics are not likely to prove definitive. A better understanding of the nature of stage disorder is more likely to come out of detailed studies of the microscopic aspects of staging kinetics. The recent computer simulations (Sects. 3.5 and 3.6) have made some progress in this direction.

3.5 Staging and the Kinetics of the Intercalation Process

3.5.1 Experiments

Intercalation is a complex phenomenon. In addition to staging, it involves transport of the intercalate species to the host, adsorption and insertion of the intercalate at the host surface and transport within the host. Macroscopic effects such as gross elastic distortions of the host and host polycrystallinity also complicate the behavior. Many of these aspects are not understood in detail.

Early experiments by *Hooley* et al. [3.116, 117] on acceptors intercalating into HOPG (highly oriented pyrolitic graphite) indicated that the diffusion of intercalate in the direction perpendicular to the host layers is not siginificant. They also showed that, at least in thick samples, the intercalation process starts in the vicinity of the basal surfaces where less elastic energy is involved in distorting the sample to accommodate the c-axis expansion associated with intercalation. These results are consistent with such macroscopic intercalation phenomena as the "ashtray effect" [3.118], which is commonly observed during intercalation of HOPG. *Kaluarachchi* and *Frindt* have also convincingly demonstrated the lack of diffusion of the intercalate through the host layers in Ag_xTiS_2 [3.51]. In a recent LEED study of K deposited onto a *highly stepped* graphite basal surface, *Wu* and *Ignatiev* [3.119] observed sequential *layer-by-layer* intercalation of the K atoms, presumably by way of the surface steps. This layer-by-layer intercalation may be related to the tendency of alkali metal guests in graphite to segregate near the basal surface reported by *Laguës* et al. [3.120] and attributed by them to surface electronic screening effects and not macroscopic elasticity. Other recent work by *Polturak* and *Eckstein* [3.121] has suggested that He atoms may be able to penetrate *through* a graphite surface by a tunneling mechanism, apparently making use of structural defects.

A number of experimental studies have been made of the transport of the intercalate within the graphite host. The work of *Metz* and *Siemsglüss* [3.122], on the intercalation of $FeCl_3$ into graphite, found that initially the intercalation rate was limited by entry of the intercalate into the host rather than by the diffusion through the host. *Dowell* and *Badorrek* [3.45, 123] found that intercalation rates of Br_2 and HNO_3 into graphite have the simple $t^{-1/2}$ behavior predicted by Fick's law of diffusion until fairly late in the intercalation process, despite the system passing through gross compositions corresponding to the various stages. The latter point was interpreted qualitatively as evidence for the Daumas-Hérold domain model of staging. More recently, various studies have reported the spatial distribution of stages in partly intercalated samples, diffusion, the presentce of fronts separating different stages in crystals during intercalation, and other related phenomena [3.114, 118, 124–136]. However, none of this work has as yet yielded detailed microscopic structural data on staging comparable in spectroscopic resolution with the quasi-equilibrium studies of *Misenheimer* and *Zabel* [3.50] or *Heiney* et al. [3.110]. Also, most of the systematic space- and time-resolved work has been on macroscopic HOPG samples where it is difficult to separate the influence of polycrystallinity and gross elastic effects from the microscopic staging phenomena which are the present topic. Thus, models which have been developed to describe the observed kinetics have taken a macroscopic point of view even in those cases where staging was specifically included [3.132].

Kaluarachchi and *Frindt* [3.51, 137] have been able to obtain some data on the kinetics of intercalation and staging which is relatively free of the macroscopic complications by studying intercalation of Ag into *individual* microcrystals of TiS_2, a few micrometers thick. Their electron microprobe X-ray fluorescence measurements and radioactive tracer studies showed that when the pristine host is being intercalated to stage 1, a stage 2 front propagates into the crystal followed by a stage 1 front. If intercalation is stopped while there is some pristine host remaining, the stage 1 converts to stage 2 in a few hours, pushing the pre-existing stage 2 material further into the crystal ahead of the originally stage 1 material. Their interpretation of these observations in terms of the Daumas-Hérold domain model is shown in Fig. 3.12. The island size was estimated at 100–200Å. Similarly, it was found that a stage 1 crystal deintercalates by forming a stage 2 region near the a-face of the crystal and the front between the stage 2 and stage 1 then propagates into the crystal. These results show some similarity with the spatially-resolved X-ray data taken by *Metz* et al. [3.124] on single crystals of graphite partly intercalated with $FeCl_3$. They reported overlapping regions of stage 1, stage 3 and pure graphite. More detailed microscopic studies of the kinetics of staging and of the stage disorder in such samples would be of interest.

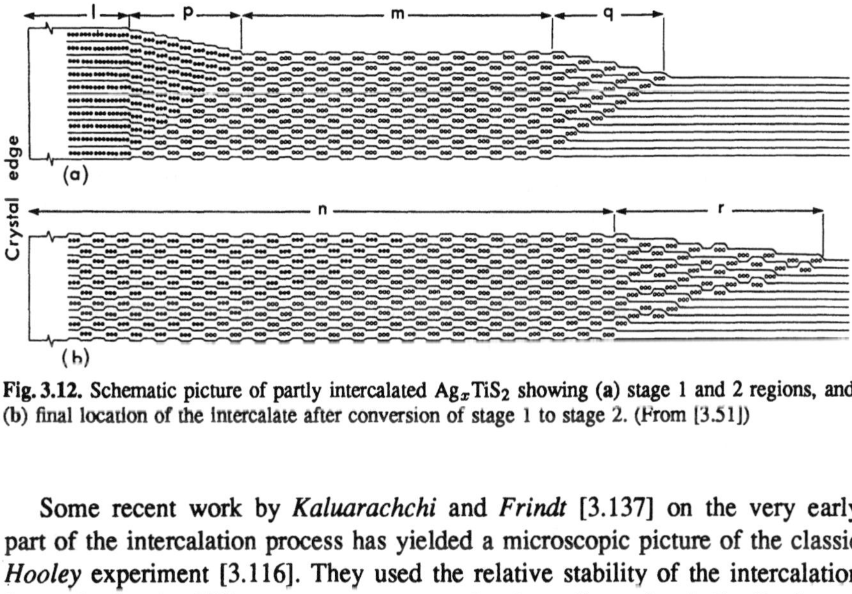

Fig. 3.12. Schematic picture of partly intercalated Ag_xTiS_2 showing (**a**) stage 1 and 2 regions, and (**b**) final location of the intercalate after conversion of stage 1 to stage 2. (From [3.51])

Some recent work by *Kaluarachchi* and *Frindt* [3.137] on the very early part of the intercalation process has yielded a microscopic picture of the classic *Hooley* experiment [3.116]. They used the relative stability of the intercalation fronts in the Ag_xTiS_2 system to measure the three-dimensional distribution of intercalate in the host crystal early in the intercalation process. This was done by measuring the distribution of Ag across the basal planes at different depths in the crystal by removing successive layers by argon ion sputtering followed by Auger analysis of the exposed surfaces. Their results are summarized in Fig. 3.13. The figure shows a section through the crystal, taken perpendicular to the basal planes, which run horizontally. The crystal surface on the left was exposed to the intercalate. In region A, the intercalate density is typical of stage 2. In regions B, C and D it is smaller by about 50 %. It is smaller again in region E and smaller still in region F. The solid lines represent discontinuities in the intercalate density. Region B is a zone of rapid diffusion of the intercalate, presumably because the adjacent basal surface relieves the elastic strains caused by the c-axis expansion.

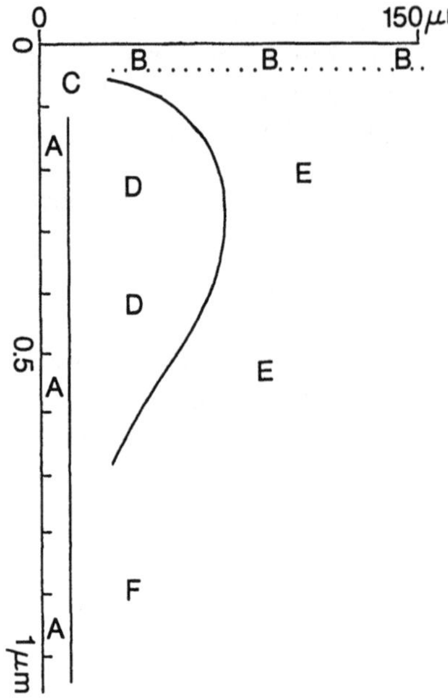

Fig. 3.13. Distribution of Ag in a TiS_2 microcrystal early in the intercalation process. Region A is approximately stage 2, B is a zone of rapid diffusion close to a basal surface extending for a distance of at least 200 μm. (From [3.137])

Another interesting development has been the application of electron spin resonance (ESR) techniques to the study of the kinetics of intercalation. This approach has been used by *Palchan* et al. [3.138] to measure diffusion coefficients characteristic of the very early part of the intercalation process, and it is able to distinguish between intercalation dominated by island growth at the crystal surface and more disordered regimes.

3.5.2 Computer Simulations

As was explained in Sect. 3.1.3, there have been many experiments which indicate that intercalate islands and Daumas-Hérold domains exist and have an important role in the kinetics of intercalation. However, the *specifics* of what happens at this level have so far not been accessible experimentally. A lot is known about the energetics of staging and interactions between intercalate islands and some progress has been made in relating certain aspects of this energetics to the intercalation kinetics of specific GICs by *Ulloa* and *Kirczenow* [3.27] (Sect. 3.1.3). But the complexity of the problem has prevented the development of detailed theoretical models of kinetics of the intercalation process based on microspcopic Hamiltonians.

This complexity can be handled by means of computer simulations, for the case of microscopic single crystals. Three-dimensional computer simulations of intercalation and deintercalation have been carried out by *Kirczenow* [3.139, 140]. In the model used, the intercalate is represented as a collection of small elementary islands (EIs) whose size is determined by the "healing length" of the defor-

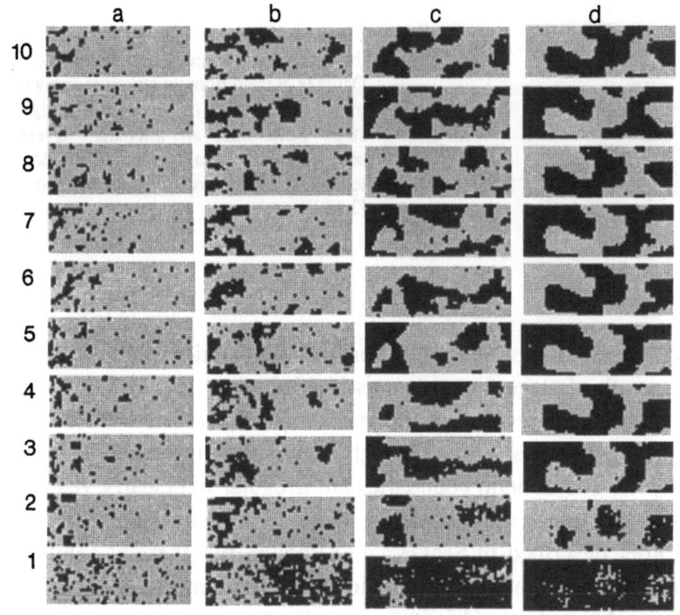

Fig. 3.14. Computer simulation of intercalation of a pristine host to stage 2. The intercalate is shown as black areas in galleries 1–10 of the crystal. Gallery 1 is next to a basal surface. Columns a–d correspond to successive times. The intercalate enters from a reservoir (not shown) at the left of each frame. (From [3.140])

mation of the host by an isolated intercalate atom (~ 10 Å in graphite). The EIs are assumed to interact via repulsive interlayer forces and local elastic strains. They hop in-plane, and between the host crystal and a reservoir, according to a Monte Carlo algorithm. Some surface effects and the physical nature of the reservoir are reflected in the model. Extensive pictorial studies of the behavior of the intercalate islands have been published [3.139–142].

Some "snapshots" of the growth of a stage 2 structure from a pristine host crystal are shown in Fig. 3.14. The intercalate is shown as black areas in galleries 1–10 of the crystal, beginning next to a basal surface. Columns a–d show these galleries at a succession of times. The intercalate diffuses into the crystal from a reservoir (not shown) at the left. The EIs join into islands (column b) which grow and become stage ordered. After some time, a single stage 2 domain (every second gallery filled) forms at the surface next to the reservoir, while an imperfect stage 3 Daumas-Hérold domain structure occupies the rest of the crystal (column c). Finally, the crystal is filled with stage 2 Daumas-Hérold domains (column d). In this model, the intercalate diffuses more rapidly in the surface gallery than in the others and fills it more completely.

It was found to be possible to change the intercalation process in such a way as to fill almost the whole crystal with a *single* stage 2 domain by changing only the physical nature of the intercalate reservoir, while leaving the nature of the guest and of the host and also the chemical potential and temperature the same [3.140]. This clearly showed the sensitivity of the Daumas-Hérold domain size to

Fig. 3.15. Slice taken at right angles to the basal planes through a stage-disordered Daumas-Hérold domain structure formed in a simulation of the growth of a high stage. Solid lines are intercalate layers seen edge on. Vacant galleries are dotted. Host layers are not shown. (From [3.139])

purely kinetic effects. Such single- and multi-domain intercalation regimes were also found for the growth of stage 3 compounds [3.139].

The growth of high stages resulted in *stage-disordered* Daumas-Hérold domains [3.140]. Figure 3.15 shows a slice through such a structure taken at right angles to the host layers. The domain boundaries are irregular but clearly visible. The general structure of these stage-disordered domains is that assumed in the model of stage disorder described in Sect. 3.4.1. Although full thermal equilibrium was not reached in the high stage simulations, the fundamental reason for this stage disorder can be understood in terms of the thermal theory of stage disorder in Daumas-Hérold domains [3.95, 140]: The size of the islands which formed in the high-stage simulations was always smaller than the critical value $\hat{N}_0$ that must be exceeded for stage-ordered structures to be stable thermodynamically. Attempts to grow larger islands in the high stage simulations failed. By contrast, in the low stage simulations, the island size exceeded the critical value $\hat{N}_0$ [from (3.14), $\hat{N}_0 \propto i^{\alpha+2}$ for stage i], and highly stage-ordered structures formed readily.

Deintercalation of a stage 1 crystal [3.140] is shown in Fig. 3.16. First, a deintercalation front traveled through the crystal, leaving behind it a stable pattern of stage 2 Daumas-Hérold domains (Fig. 3.16a, b). Then the crystal passed through a series of increasingly stage-disordered states (Fig. 3.16c) and eventually formed a "residue compound" with some runs of galleries entirely empty of intercalate while others contained moderately large intercalate islands (Fig. 3.16d) whose deintercalation was *kinetically* hindered. A very similar residue structure was observed recently by *Hwang* using electron microscopy [3.61]. Residue compounds, i.e., strongly disordered dilute intercalation compounds which are the result of deintercalation, are often found experimentally [3.1].

A different approach to intercalation kinetics was taken by *Sapoval* et al. [3.142]. They used a lattice gas model to simulate the diffusion of intercalate in a single gallery, the only interaction considered being a hard-core in-plane repulsion which prevents two guest atoms from occupying the same site. They reported the propagation into the host of an intercalation front having a self-similar fractal geometry. The fractal geometry is not seen in the simulations of *Kirczenow* [3.139–141], where it is suppressed by the attractive intralayer interactions and the repulsive interlayer forces, which give rise instead to islands and Daumas-Hérold domains.

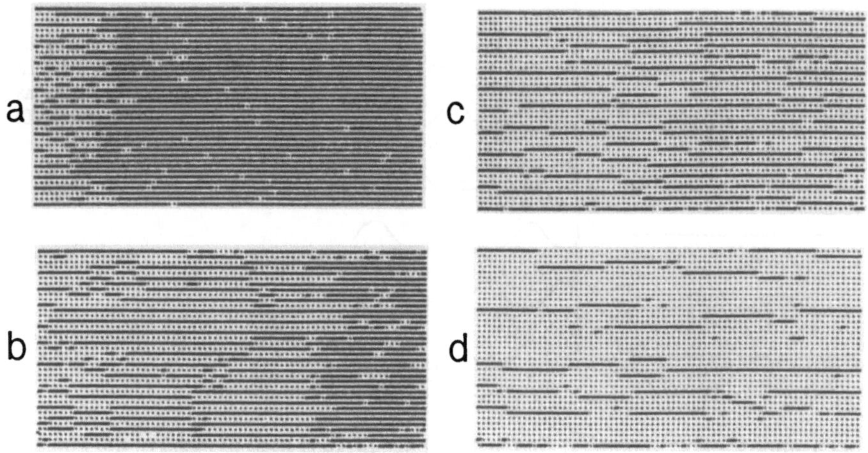

Fig. 3.16. Simulation of the deintercalation of a stage 1 compound. Notation as in Fig. 3.15. (From [3.140])

3.6 The Kinetics of Stage Transformations

3.6.1 Kinetic Models and Experiments

Various models and ideas as to how stage transformations may happen have been proposed. The model of thermal stage disorder in Daumas-Hérold domains [3.95] and the model of partly stage-disordered metastable states [3.115] have already been discussed in Sect. 3.4.

A simple kinetic model describing the transition from a dilute high temperature stage 1 phase to the low temperature stage 2 by spinodal decomposition was proposed by *Safran* [3.10], and a more detailed Landau-Ginzburg theory of this process was developed recently by *Hawrylak* and *Subbaswamy* [3.143]. The free energy density used by *Hawrylak* and *Subbaswamy* was similar to that of the staging model of *Dahn* et al. [3.30] (Sect. 3.2.3) but also included a density gradient term. Nonlinear diffusion equations based on this free energy were solved numerically. The results for the transition from the dilute stage 1 to a stage 2 at lower T are shown in Fig. 3.17. Instabilities develop in the dilute stage 1, and by $t = 200$ a pattern of stage 2 Daumas-Hérold domains has developed in the intercalate density by diffusion of the intercalate into and out of corresponding regions in alternate layers. Quenches from the dilute stage 1 to the low temperature stage 3 stability region resulted in a kinetically stable intermediate stage 2 phase. Studies of the growth of a dense stage 1 from a dilute stage 1 by introducing intercalate from the sample edge resulted in the presence of domains of various stages. This theory does not involve any stochastic component, and so stage disorder does not appear spontaneously during stage transitions.

The time required for the spinodal decomposition during a quench from stage 1 to stage 2 was estimated to be $\sim 10^{-4}$ s [3.143] but its kinetics have not yet

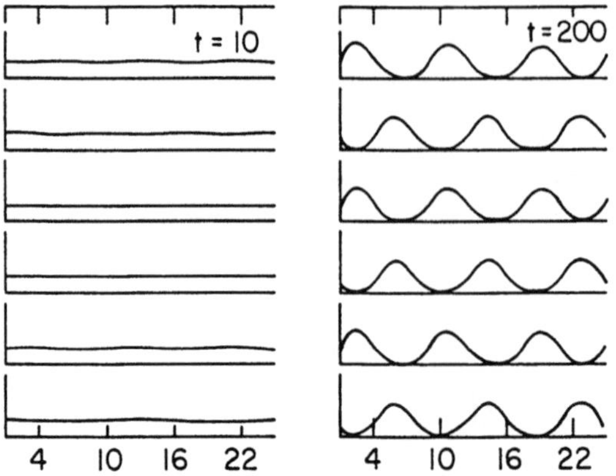

Fig. 3.17. Density profiles of the intercalate in six galleries during the quench from a dilute stage 1 ($t = 10$) to a low temperature stage 2 Daumas-Hérold domain structure ($t = 200$). (From [3.143])

been studied experimentally. *Kim* and *Fischer* [3.104] tried to observe the kinetics of the analogous stage transitions which occur in response to abrupt pressure changes at room temperature, but found that these processes happened too quickly for their time-resolved measurements. They concluded that the pressure-induced stage transitions were nucleation limited.

A different kinetic approach was developed by *Miyazaki* et al. [3.144], who used a one-dimensional stochastic model to study the time evolution of a *single* Daumas-Hérold domain during stage transformations. The stochastic variables were the mean density of intercalate in the different galleries of the domain. The free energy used was that of *Safran*'s staging model [3.78] together with a term which was included to reflect elastic interactions with other domains. This term also partly suppresses the fractional stages introduced by the unrealistic unscreened interlayer interactions of *Safran*'s model. A system of Langevin equations including terms representing movement of intercalate into and out of the domain was solved numerically. A variety of stage transitions was studied under conditions far from equilibrium. They were found to involve considerable stage disorder. The kinetics appeared to be insensitive to the *initial* state of the system. During some stage transitions, narrow peaks, at positions corresponding to the initial and final stages, were simultaneously present in the calculated structure factors. This is in agreement with some kinetic experiments on stage transitions [3.25, 26], but in the experiments the coexisting diffraction lines came from different parts of the sample rather than from a single domain [3.114]. It is not clear what effect the c-axis expansion of the host by the guest, which was not included in the structure factors, would have on this result.

A quite different model of stage transitions was proposed by *Forgacs* and *Uimin* [3.145], who suggested that stage transitions may occur by a *coherent* motion of all of the intercalate islands in the crystal, all of the domain walls tilting in unison away from the direction normal to the host layers, in response to changes in the chemical potential. This model does not include any element of randomness in the arrangement of intercalate islands or their movement. *Misenheimer* and *Zabel* [3.50] found their experimental data on stage transitions to be consistent with stage disorder but not with such domain wall rotations.

Recently, *Hérold* has suggested that the movement of intercalate involved in changes of stage and in the intercalation process, might involve dynamic sliding of intercalate layers in response to thermodynamic forces, rather than the usually assumed diffusion [3.146]. This was motivated by the observation that some intercalation processes occur very quickly. For example, the formation of KC_8 from liquid potassium in contact with graphite occurs *much* more quickly than the purely diffusive replacement of one heavy alkali metal by another in a stage 1 GIC. A calculation of the sliding dynamics, based on a thermodynamic driving force and a viscous force between the guest and host layers, yielded an intercalation rate proportional to $t^{-1/2}$. However, such a time dependence is predicted also by Fick's law of diffusion, so that other considerations such as the temperature dependence of intercalation rates may be more decisive tests of this mechanism. Computer simulations of intercalation kinetics have, in some cases, shown behavior resembling this sliding motion. Examples are the movement of islands during the growth of stage 3 in [3.139], and the "single domain mode" of the growth of stage 2 in [3.140]. However, in these cases the movement of the intercalate actually occurs by *diffusion* of elementary islands or elementary island vacancies. In other cases involving a change of stage, the computer simulations show little or no indication of sliding behavior. Examples are the growth of high stages, and the transition from stage 1 to stage 2 during deintercalation [3.140]. It may also be significant that the conversion from stage 1 to stage 2 observed by *Kaluarachchi* and *Frindt* [3.137] in Ag_xTiS_2 (Fig. 3.12) slowed markedly when samples were cooled to liquid nitrogen temperatures, suggestive of a thermally activated process.

3.6.2 Computer Simulations and More Experiments

Monte Carlo simulations similar to those which have been discussed in Sect. 3.5.2 have also been used to study the stage transformations which are driven by changes in the intercalate chemical potential [3.139, 141, 147]. In the three-dimensional simulations, transitions from stage 2 to stage 1 or from stage 1 to stage 2 are found to proceed by the movement through the crystal of an intercalation or deintercalation front separating the stage 1 and 2 regions. However, the transition from stage 3 to stage 2 is found to be different.

When a stage 3 Daumas-Hérold structure is placed in contact with an intercalate reservoir whose chemical potential is inside the stage 2 stability region, kinetic stage disorder appears at the domain boundaries throughout the crystal [3.139, 141]. The stage disordered regions, which contain mostly stage 3 and stage 2 units, grow and eventually fill the whole crystal, becoming ordered stage 2 domains as they grow [3.139]. None of the simulations have shown any evidence of domain wall rotations [3.145] taking place.

Recently, careful simulations have been made to study the nature of the states which occur in a stage 3 Daumas-Hérold domain structure when it is allowed to reach quasi equilibrium at a chemical potential within the stage 3 stability region, but close to the boundary with stage 2 [3.141]. It was found that under these conditions there is a stage disorder at the domain boundaries and that it is *thermal*. It increases as the stage transition is approached, the domain walls becoming increasingly "rough". It is at present not clear whether the stage-disordered regions completely fill the domains *at equilibrium* right at the stage transition. It is possible that there is a domain wall roughening transition temperature above which this happens. However, strongly stage-disordered domains appear to be at least kinetically stable at the stage 3-2 boundary.

This is a somewhat more complex picture with *spatially inhomogeneous thermal* stage disorder in Daumas-Hérold domains close to stage transitions than is contained in the simple theory discussed in Sect. 3.4, where it was assumed that every gallery in a domain is either completely filled with intercalate or empty. It should be applicable for low stages in the case of large domains, exceeding the critical domain size $\hat{N}_0$ discussed in Sect. 3.4.1. For high stages with domain sizes smaller than $\hat{N}_0$, *only* stage-disordered structures are stable, and as was discussed in Sect. 3.5.2, the simpler picture of *homogeneously* disordered Daumas-Hérold domains appears to be appropriate.

The inhomogeneously stage-disordered Daumas-Hérold domains which are found in the simulations of the stage 3 to stage 2 transition [3.139, 141] have an unusual structure factor, for both thermal and kinetic disorder. Even for quite large amounts of stage disorder ($\sim$ 30 % stage 2 units in the crystal overall), the presence of the stage disorder at the domain walls does not shift the stage 3 (00ℓ) diffraction peaks appreciably towards the stage 2 position, but only reduces their intensity. This means that a Hendricks-Teller analysis, which depends largely on measurements of diffraction line *shifts*, would seriously underestimate the amount of stage disorder present. Changes of absolute diffraction line intensities in layered materials are difficult to measure reliably. Thus experiments aimed at observing the details of such stage transitions should supplement the diffraction data with Raman scattering, which is sensitive to the difference between *individual* stage 2 and stage 3 units [3.2]. Observation of the local vibrational modes predicted by *Hawrylak* and *Williams* [3.148] to be associated with stage disorder may also be a suitable technique.

In the simulations, the kinetics of the stage 3 to 2 transition are different from those of the transition from stage 2 to 1 because the former transition involves a nontrivial rearrangement of the intercalate islands. However, this difference is not apparent in most kinetic experiments on stage transformations [3.25, 26, 49, 91, 114]. For example, in the recent space- and time-resolved in situ study of H_2SO_4-graphite by *Nishitani* et al. [3.114], *both* the stage 2 to 1 *and* the stage 3 to 2 transitions were found to occur in narrow fronts which traveled through the sample. The only difference appeared to be that the stage 3-2 front was somewhat broader.

In comparing such kinetic experiments with the simulations, it should be noted that the experiments were carried out on macroscopic HOPG samples, and that HOPG is a polycrystalline material with many defects. It is well known that HOPG is more difficult to intercalate than single crystal graphite flakes, and that the intercalation process is generally sensitive to crystal imperfections [3.2]. Thus, it is quite possible that the observed front in the stage 3-2 transition in HOPG is a consequence of dislocations and grain boundaries pinning the intercalate islands in the interior of the sample and blocking the stage transition mechanism observed in the simulations. Inhomogeneous elastic strains produced by rapid dilation of thick samples can have a similar effect, as can too rapid a change of stage in any sample. In view of these considerations, it may well be significant that in their spatially resolved ex situ X-ray work on the intercalation of *single crystal* graphite flakes, *Metz* et al. [3.124] found stage 1 and stage 3 in their crystal but not stage 2, which they attributed to the relative ease with which stage 2 is converted to stage 1.

More systematic experimental studies of the kinetics of stage transitions *close to equilibrium in small single crystals*, emphasizing the details of micriscopic structure, stage order and disorder, are clearly needed. However, the very recent work of *Suematsu* et al. [3.113] on K-graphite suggests that studies of stage transitions *far* from equilibrium can, *for higher stages*, also yield interesting information about the *microscopic* aspects of staging kinetics: It was reported that in stage transitions, when changes in the stage index exceeding unity occur, both relatively ordered and *kinetically stage-disordered* intermediate stages can be observed.

Acknowledgements: I would like to thank R.F. Frindt, D. Kaluarachchi and S.E. Ulloa for numerous useful discussions, and also the many people who sent me copies of their articles.

References

3.1 A. Hérold: In *Intercalated Layered Materials*, ed. by F. Lévy, Physics and Chemistry of Materials with Layered Structures, Vol. 6 (Reidel, Dordrecht, Holland 1979)
3.2 M.S. Dresselhaus, G. Dresselhaus: Adv. Phys. **30**, 139 (1981)
3.3 S.A. Solin: In *Advances in Chemical Physics*, ed. by I. Prigogine, S.A. Rice, Vol. XLIX (Wiley, New York 1982)
3.4 R. Clarke, C. Uher: Adv. Phys. **33**, 469 (1984)
3.5 S.A. Solin: J. Mol. Catal. **27**, 293 (1984)
3.6 J.E. Fischer: In *Intercalated Layered Materials*, ed. by F. Lévy, Physics and Chemistry of Materials with Layered Structures, Vol. 6 (Reidel, Dordrecht, Holland 1979)
3.7 P. Pfluger, H.-J. Güntherodt: Festkörperprobleme **21**, 271 (1981)
3.8 S.E. Ulloa, G. Kirczenow: Comments Cond. Mat. Phys. **12**, 181 (1986)
3.9 H. Zabel, P.C. Chow: Comments Cond. Mat. Phys. **12**, 225 (1986)
3.10 S.A. Safran: Synth. Met. **2**, 1 (1980)
3.11 S.A. Safran: In *Solid State Physics*, Vol. 40, ed. by F. Seitz, D. Turnbull, H. Ehrenreich (Academic, New York 1987) p. 183
3.12 R.H. Friend, A.D. Yoffe: Adv. Phys. **36**, 1 (1987)
3.13 For recent discussions see:
H. Kamimura: Ann. de Phys. **11**, 39 (1986);
A. Mansour, S.E. Schnatterly, J.J. Ritsko: Phys. Rev. Lett. **58**, 614 (1987)
3.14 L. Pietronero, S. Strässler, H.R. Zeller, M.J. Rice: Phys. Rev. Lett **41**, 763 (1978); Solid State Commun. **30**, 399 (1979)
3.15 P.R. Wallace: Phys. Rev. **71**, 622 (1947)
3.16 G.S. Painter, D.E. Ellis: Phys. Rev. B **1**, 4747 (1970)
3.17 J.W. McClure: Phys. Rev. **108**, 612 (1957)

3.18 J.C. Slonczewski, P.R. Weiss: Phys. Rev. **109**, 272 (1958)
3.19 F.J. Salzano, S. Aronson: J. Chem. Phys. **45**, 4551 (1966); J. Chem. Phys. **44**, 4320 (1966)
3.20 S.A. Safran, D.R. Hamann: Phys. Rev. B **22**, 606 (1980)
3.12 S.A. Safran, D.R. Hamann: Phys. Rev. B **23**, 565 (1981)
3.22 T. Ohno, H. Kamimura: J. Phys. Soc. Jpn. **52**, 223 (1983)
3.23 T. Ohno, H. Kamimura: Physica B + C **117**, 611 (1983)
3.24 S. Aronson, F.J. Salzano, D. Bellafiore: J. Chem. Phys. **49**, 434 (1968)
3.25 R. Nishitani, Y. Uno, H. Suematsu: Phys. Rev. B **27**, 6572 (1983)
3.26 Y. Yosida, K. Sato, K. Suda, H. Suematsu: Synth. Met. **12**, 167 (1985); J. Phys. Soc. Jpn. **55**, 561 (1986)
3.27 S.E. Ulloa, G. Kirczenow: Phys. Rev. B **33**, 1360 (1986)
3.28 S.E. Millman, G. Kirczenow: Phys. Rev. B **26**, 2310 (1982); S.E. Millman, G. Kirczenow, D. Solenberger: J. Phys. C **15**, L1269 (1982)
3.29 S.E. Millman, G. Kirczenow: Phys. Rev. B **28**, 3482 (1983)
3.30 J.R. Dahn, D.C. Dahn, R.R. Haering: Solid State Commun. **42**, 179 (1982)
3.31 K.C. Woo, H. Mertwoy, J.E. Fischer, W.A. Kamitakahara, D.S. Robinson: Phys. Rev. B **27**, 7831 (1983)
3.32 J.E. Fischer, C.D. Fuerst, K.C. Woo: Synth. Met. **7**, 1 (1983)
3.33 J.E. Fischer, H.J. Kim: Phys. Rev. B **35**, 3295 (1987); Synth. Met. **23**, 121 (1988)
3.34 J.R. Dahn, R.R. Haering: Solid State Commun. **40**, 245 (1981)
3.35 J.R. Dahn, R.R. Haering: Can. J. Phys. **61**, 1093 (1983)
3.36 D.E. Nixon, G.S. Parry: J. Phys. C **2**, 1732 (1969)
3.37 D. Guerard, C. Zeller, A. Hérold: C.R. Acad. Sci. C **283**, 437 (1976)
3.38 S. Flandrois, J. Masson, J. Rouillon, J. Gaultier, C. Hauw: Synth. Met. **3**, 1 (1981)
3.39 L. Pietronero, S. Strässler: Phys. Rev. Lett. **47**, 593 (1981)
3.40 N. Daumas, A. Hérold: C.R. Acad. Sci. C **268**, 373 (1969)
3.41 N. Daumas, A. Hérold: Bull. Soc. Chim. Fr. **5**, 1598 (1971)
3.42 W. Metz, D. Hohlwein: Carbon **13**, 87 (1975)
3.43 G. Schoppen, H. Meyer-Spasche, L. Siemsglüss, W. Metz: Mater. Sci. Eng. **31**, 115 (1977)
3.44 A. Hérold: Mater. Sci. Eng. **31**, 1 (1977)
3.45 M.B. Dowell, D.S. Badorrek: Carbon **16**, 241 (1978)
3.46 R. Clarke, N. Caswell, S.A. Solin: Phys. Rev. Lett. **42**, 61 (1979)
3.47 R. Clarke, N. Wada, S.A. Solin: Phys. Rev. Lett. **44**, 1616 (1980)
3.48 C.D. Fuerst, J.E. Fischer, J.D. Axe, J.B. Hastings, D.B. McWhan: Phys. Rev. Lett. **50**, 357 (1983)
3.49 M.E. Misenheimer, H. Zabel: Phys. Rev. B **27**, 1443 (1983)
3.50 M.E. Misenheimer, H. Zabel: Phys. Rev. Lett. **54**, 2521 (1985)
3.51 D. Kaluarachchi, R.F. Frindt: Phys. Rev. B **28**, 3663 (1983); Phys. Rev. B **31**, 3648 (1985)
3.52 A. Erbil, A.R. Kortan, R.J. Birgeneau, M.S. Dresselhaus: Phys. Rev. B **28**, 6329 (1983)
3.53 S. Flandrois, J.M. Masson, J.C. Rouillon, J. Gaultier, C. Hauw: Synth. Met. **3**, 1 (1981); S. Flandrois, A.W. Hewat, C. Hauw, R.H. Bragg: Synth. Met **7**, 305 (1983)
3.54 F. Baron, S. Flandrois, C. Hauw, J: Gaultier: Solid State Commun. **42**, 759 (1982)
3.55 S.E. Ulloa, G. Kirczenow: Phys. Rev. Lett. **56**, 2537 (1986); Phys. Rev. B **35**, 795 (1987)
3.56 D. Brusilovsky, H. Selig, D. Vaknin, I. Ohana, D. Davidov: Synth. Met. **23**, 377 (1988)
3.57 J.M. Thomas, G.R. Millward, R.F. Schlögl, H.P. Boehm: Mater. Res. Bull. **15**, 671 (1980)
3.58 C. Hauw, J. Gaultier, S. Flandrois, O. Gonzales, D. Dorignac, R. Jagut: Synth. Met. **7**, 313 (1983)
3.59 G. Timp, M.S. Dresselhaus, L. Salamanca-Riba, A. Erbil, L.W. Hobbs, G. Dresselhaus, P.C. Eklund, Y. Iye: Phys. Rev. B **26**, 2323 (1982); G. Timp, M.S. Dresselhaus: J. Phys. C **17**, 2641 (1984)
3.60 M. Matsuura, Y. Murakami, K. Takeda, H. Ikeda, M. Suzuki: Synth. Met. **12**, 427 (1985)
3.61 D.M. Hwang: Unpublished results presented at the March meeting of the A.P.S., Las Vegas (1986)
3.62 R. Levi-Setti, G. Crow, Y.L. Wang, N.W. Parker, R. Mittleman, D.M. Hwang: Phys. Rev. Lett. **54**, 2615 (1985); M. Bretz, R. Clarke: Phys. Rev. Lett. **55**, 2506 (1985); D.M. Hwang, R. Levi-Setti: Phys. Rev. Lett. **55**, 2507 (1985)
3.63 S.A. Safran, D.R. Hamann: Phys. Rev. Lett. **42**, 1410 (1979); Physica (Utrecht) **99B**, 469 (1980)
3.64 S. Ohnishi, S. Sugano: Solid State Commun. **36**, 823 (1980)

3.65 G. Kirczenow: Phys. Rev. Lett. **49**, 1853 (1982)
3.66 P. Hawrylak, K.R. Subbaswamy, G.W. Lehman: Solid State Commun. **51**, 787 (1984)
3.67 S.E. Ulloa, G. Kirczenow: Phys. Rev. Lett. **55**, 218 (1985)
3.68 M.S. Dresselhaus, A. Lusnikov: Synth. Met. **23**, 401 (1988)
3.69 A. Hérold: Private communcation
3.70 D.P. DiVincenzo, E.J. Mele: Phys. Rev. B **32**, 2538 (1985)
3.71 D.P. DiVincenzo: Synth. Met. **12**, 111 (1985)
3.72 D.P. DiVincenzo: In Proceedings of Symposium K, 1986 Fall Meeting of the Materials Research Society, ed. by M.S. Dresselhaus, G. Dresselhaus, S.A. Solin (Materials Research Society, Pittsburgh, PA 1986)
3.73 H. Zabel: In Proceedings of Symposium I, 1984 Fall Meeting of the Materials Research Society, ed. by P.C. Eklund, M.S. Dresselhaus, G. Dresselhaus (Materials Research Society, Pittsburgh, PA 1984)
3.74 C. Thompson, S.C. Moss, G. Reiter, M.E. Misenheimer: Synth. Met. **12**, 57 (1985)
3.75 G. Reiter, S.C. Moss: Phys. Rev. B **33**, 7209 (1986)
3.76 M. Plischke: Can. J. Phys. **59**, 802 (1981)
3.77 H. Miyazaki, Y. Kuramoto, C. Horie: J. Phys. Soc. Jpn. **53**, 1380 (1984)
3.78 S.A. Safran: Phys. Rev. Lett. **44**, 937 (1980)
3.79 D.P. DiVincenzo, T.C. Koch: Phys. Rev. B **30**, 7092 (1984)
3.80 P. Hawrylak, K.R. Subbaswamy: Phys. Rev. B **28**, 4851 (1983)
3.81 P. Bak, R. Bruinsma: Phys. Rev. Lett. **49**, 249 (1982); R. Bruinsma, P. Bak: Phys. Rev. B **27**, 5824 (1983)
3.82 J.C. Schön, D. Adler, G.F. Dresselhaus: In Proceedings of Symposium I, 1984 Fall Meeting of the Materials Research Society, ed. by P.C. Eklund, M.S. Dresselhaus, G. Dresselhaus (Materials Research Society, Pittsburgh, PA 1984);
J.C. Schön, D. Adler, G.F. Dresselhaus: Preprint
3.83 P. Alstrøm: Phys. Scr. **35**, 45 (1987)
3.84 H.J. Kim, J.E. Fischer, D.B. McWhan, J.D. Axe: Phys. Rev. B **33**, 1329 (1986)
3.85 B. Marcus, Ph. Touzain: Synth. Met. **23**, 13 (1988)
3.86 K.K. Bardhan, G. Kirczenow, J.C. Irwin: J. Phys. C **18**, L131 (1985);
K.K. Bardhan, G. Kirczenow, G. Jackle, J.C. Irwin: Phys. Rev. B **33**, 4149 (1986)
3.87 A. Hérold: Bull. Soc. Chim. Fr. **187**, 999 (1955)
3.88 D.E. Nixon, G.S. Parry: J. Phys. C **1**, 291 (1968)
3.89 T. Sasa, Y. Takahashi, T. Mukaibo: Carbon **9**, 407 (1971)
3.90 E.R. Falardeau, L.R. Hanlon, T.E. Thompson: Inorg. Chem. **17**, 301 (1978)
3.91 R. Nishitani, Y. Uno, H. Suematsu: Synth. Met. **7**, 13 (1983)
3.92 R. Nishitani, Y. Uno, H. Suematsu, Y. Fujii, T. Matsushita: Phys. Rev. Lett. **52**, 1504 (1984)
3.93 S.E. Millman: Phys. Lett. **92A**, 441 (1982)
3.94 D. Guerand, A. Hérold: Carbon **13**, 337 (1975)
3.95 G. Kirczenow: Phys. Rev. Lett. **52**, 437 (1984); Phys. Rev. B **31**, 5376 (1985)
3.96 D.C. Dahn, R.R. Haering: Solid State Commun. **44**, 29 (1982)
3.97 N. Wada, R. Clarke, S.A. Solin: Synth. Met. **2**, 27 (1980)
3.98 N. Wada, S.A. Solin: Physica **105B**, 268 (1981)
3.99 N. Wada: Phys. Rev. B **24**, 1065 (1981)
3.100 N. Wada, S. Minomura, J. Pluth: Synth. Met. **12**, 51 (1985)
3.101 R. Clarke, J.N. Gray, H. Homma, M.J. Winokur: Phys. Rev. Lett. **47**, 1407 (1981)
3.102 J.M. Bloch, H. Katz, D. Moses, V.B. Cajipe, J.E. Fischer: Phys. Rev. B **31**, 6785 (1985)
3.103 J.E. Fischer, H.J. Kim: Synth. Met. **12**, 137 (1985)
3.104 H.J. Kim, J.E. Fischer: Phys. Rev. B **33**, 4349 (1986)
3.105 B. Sundqvist, J.E. Fischer: Phys. Rev. B **34**, 3532 (1986); Phys. Rev. B **35**, 8231 (1987)
3.106 D.P. DiVincenzo, C.D. Fuerst, J.E. Fischer: Phys. Rev. B **29**, 1115 (1984)
3.107 B. Sundqvist, B. Lundberg: J. Phys. C **17**, L133 (1984); ibid. **19**, 6915 (1986)
3.108 S. Hendricks, E. Teller: J. Chem. Phys. **10**, 147 (1942)
3.109 C.D. Fuerst, T.C. Koch, J.E. Fischer, J.D. Axe, J.B. Hastings, D.B. McWhan: In *Intercalated Graphite*, Materials Research Society Symposia Proceedings, Vol. 20, ed. by M.S. Dresselhaus, G. Dresselhaus, J.E. Fischer, M.J. Moran (North-Holland, Amsterdan 1983) p. 39
3.110 P.A. Heiney, M.E. Huster, V.B. Cajipe, J.E. Fischer: Synth. Met. **12**, 21 (1985);
M.E. Huster, P.A. Heiney, V.B. Cajipe, J.E. Fischer: Phys. Rev. B **35**, 3311 (1987)
3.111 V.B. Cajipe, J.E. Fischer: Phys. Rev. B **35**, 4854 (1987)
3.112 J.E. Fischer, H.J. Kim: Phys. Rev. B **35**, 6826 (1987)

3.113 H. Suematsu, K. Suda, N. Metoki: Synth. Met. **23**, 7 (1988)
3.114 R. Nishitani, Y. Sasaki, Y. Nishina: J. Phys. Soc. Jpn. **56**, 1051 (1987)
3.115 P. Bak, G. Forgacs: Phys. Rev. B **32**, 7535 (1985)
3.116 J.G. Hooley, W.P. Garby, J. Valentin: Carbon **3**, 7 (1965)
3.117 J.G. Hooley: Mater. Sci. Eng. **31**, 17 (1977)
3.118 See, for example, K.K. Bardhan, D.D.L. Chung: Carbon **18**, 313 (1980)
3.119 N.J. Wu, A. Ignatiev: Solid State Commun. **46**, 59 (1983)
3.120 M. Laguës, D. Marchand, C. Fretigny: Solid State Commun. **59**, 583 (1986); Ann. de Phys. **11**, 49 (1986);
C. Fretigny, D. Marchand, M. Laguës: Phys. Rev. B **32**, 8462 (1985);
M. Laguës, D. Marchand, C. Fretigny: In Proceedings of Symposium K, 1986 Fall Meeting of the Materials Research Society, ed. by M.S. Dresselhaus, G. Dresselhaus, S.A. Solin (Materials Research Society, Pittsburgh, PA 1986)
3.121 E. Polturak, Y. Eckstein: Phys. Rev. Lett. **58**, 2680 (1987)
3.122 W. Metz, L. Siemsglüss: Mater. Sci. Eng. **31**, 119 (1977)
3.123 M.B. Dowell: Mater. Sci. Eng. **31**, 129 (1977)
3.124 W. Metz, P. Josuks, U. Kleinman: Synth. Met. **7**, 319 (1983)
3.125 A. Hamwi, P. Touzain, C. Riekel: Synth. Met. **7**, 23 (1983)
3.126 A. Hamwi, P. Touzain, L. Bonnetain, A. Boeuf, A. Freund, C. Riekel: Mater. Sci. Eng. **57**, 161 (1983)
3.127 K.K. Bardhan, J.C. Wu, J.S. Culik, S.H. Anderson, D.D.L. Chung: Synth. Met. **2**, 57 (1980)
3.128 T. Dziemanowicz, W. Forsman, R. Vangelisti, A. Hérold: Carbon **22**, 53 (1984)
3.129 S.H. Anderson, D.D.L. Chung: Synth. Met. **7**, 107 (1983)
3.130 Y.Y. Huang, D. Stump, S.A. Solin, J.P. Heremans: Solid State Commun. **61**, 469 (1987)
3.131 C. Riekel, A. Hamwi, Ph. Touzain: Synth. Met. **15**, 345 (1986)
3.132 S.H. Anderson Axdal, D.D.L. Chung: Carbon **25**, 191, 211, 377 (1987)
3.133 Ph. Touzain: Ann. de Phys. **11**, 23 (1986)
3.134 K. Okabe, S. Tanuma: Synth. Met. **23**, 61 (1988)
3.135 A. Metrot, M. Tihli: Synth. Met. **23**, 19 (1988)
3.136 Ph. Touzain, A. Hamwi: Synth. Met. **23**, 127 (1988)
3.137 D. Kaluarachchi: Ph. D. Thesis, Simon Fraser University (1987)
3.138 I. Palchan, D. Davidov, V. Zevin, G. Polatsek: Phys. Rev. B **32**, 5554 (1985); Synth. Met. **12**, 413 (1985); and unpublished work
3.139 G. Kirczenow: Synth. Met. **12**, 143 (1985); Phys. Rev. Lett. **55**, 2180 (1985)
3.140 G. Kirczenow: Can. J. Phys. **66**, 39 (1988)
3.141 G. Kirczenow: Synth. Met. **23**, 1 (1988)
3.142 B. Sapoval, M. Rosso, J.-F. Gouyet, J.-F. Colonna: Solid State Ionics **18 & 19**, 21 (1986);
B. Sapoval, M. Rosso, J.-F. Gouyet: Solid State Ionics **18 & 19**, 232 (1986)
3.143 P. Hawrylak, K.R. Subbaswamy: Phys. Rev. Lett. **53**, 2098 (1984)
3.144 H. Miyazaki, T. Watanabe, C. Horie: Phys. Rev. B **34**, 5736 (1986);
H. Miyazaki, C. Horie: Synth. Met. **12**, 149 (1985)
3.145 G. Forgacs, G. Uimin: Phys. Rev. Lett. **52**, 633 (1984)
3.146 A. Hérold: Synth. Met. **23**, 27 (1988)
3.147 S. Miyazima: Synth. Met. **12**, 155 (1985)
3.148 P. Hawrylak, M.L. Williams: Phys. Rev. B **33**, 1503 (1986)

4. Lattice Dynamics I: Neutron Studies

Hartmut Zabel

With 29 Figures

This chapter reviews inelastic neutron scattering investigations of the lattice dynamics of graphite intercalation compounds. Pristine graphite and its intercalation compounds exhibit phonon frequencies as high as 200 meV. Most existing neutron sources allow the investigation of phonon frequencies only up to about 100 meV. It is therefore the lower part of the phonon dispersion of graphite intercalation compounds which is discussed here. Complementary studies by infrared and Raman techniques are reviewed by *Solin* in Chap. 5 of this volume.

Graphite intercalation compounds (GICs) with a regular sequence of alternating intercalate layers and graphite basal planes (also termed "graphene layers") represent a unique facility for the exploration of the lattice dynamics of layered materials. By changing the intercalate species, the stage of the compound, or the intercalate in-plane stoichiometry, the symmetry and interaction parameters can be altered in a systematic fashion, allowing detailed investigations of the relation between structure and phonons. In addition, the intercalated layers in graphite display a rich variety of ordered and disordered structures, commensurate and incommensurate phases, melting transitions on periodic substrates, and metal–insulator transitions. In all these cases, examination of the lattice dynamics is of prime interest for obtaining interatomic potentials, which in turn leads to accurate modeling of the structures, dynamics, and phase transitions in these compounds.

The lattice dynamics of graphite and of GICs can be classified as high-frequency *intralayer* modes and low-frequency *interlayer* modes. The different frequency regimes are schematically sketched in Fig. 4.1. The low-lying modes, which are obtainable by inelastic neutron scattering techniques, comprise the acoustic and optic branches of the $[00q]$ longitudinal (L) modes (so-called layer breathing modes) and the $[00q]$ transverse (T) or "layer shear modes". Both modes propagate perpendicular to the graphite and intercalate layers. The layered structure of the compounds leads, in addition, to very soft $[q00]\mathrm{T}_{\perp}$ modes which propagate parallel to the layers with polarization perpendicular to the graphene and intercalate planes. These modes are commonly referred to as "layer bending" or "rippling" modes. Finally, we find modes specific to the intercalate atoms or molecules. These comprise either simple in-plane longitudinal and transverse modes, in the case of monatomic intercalate layers, or more complex intramolecular modes, in the case of molecular intercalates.

A full exploration of the lattice dynamics of GICs via inelastic neutron scattering is, however, hampered by the lack of sufficiently large single crystals. Therefore, all phonon data from neutron scattering have been collected using

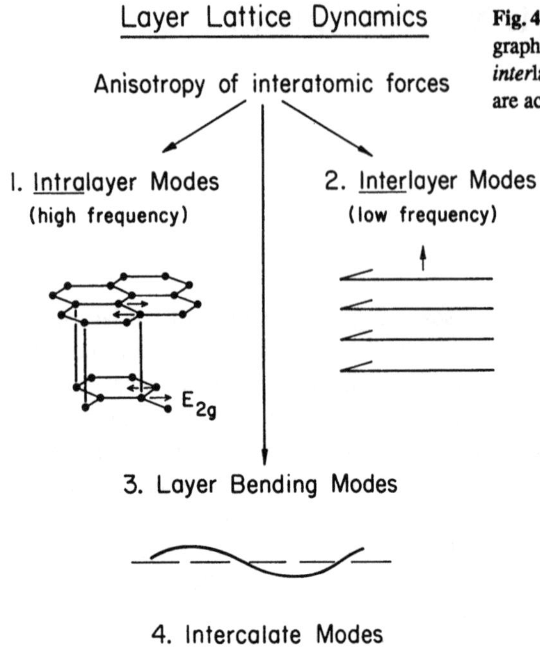

Fig. 4.1. Schematics of layer dynamics in graphite intercalation compounds. Only the *inter*layer, bending, and intercalate modes are accessible to inelastic neutron scattering

intercalated highly oriented pyrolytic graphite (HOPG) or even lower grade materials. The two-dimensional powder nature of this material has no effect on the $[00q]$L phonons which propagate along the oriented c-axis. In fact, the $[00q]$L modes are the most frequently measured phonon modes of GICs. The $[q00]T_{\perp}$ modes are also rather well defined, in spite of the powder averaged $(HK0)$ plane, owing to the close elastic isotropy of the graphite and intercalate planes. All other modes are more difficult to determine, and have been measured in only a few cases.

Pristine graphite was one of the first materials to have its phonon dispersion measured after the installation of the first neutron triple-axis spectrometer at Chalk River, Canada. In their paper, *Dolling* and *Brockhouse* [4.1] describe the extremely simple dispersion relation of the graphite $[00q]$L breathing modes as the dispersion "in which undistorted hexagonal planes of carbon atoms move relative to each other". Thus, the $[00q]$L phonon dispersion of the rigid graphite planes is equivalent to the dispersion of a simple linear chain of single atoms, in which the "atoms" represent the graphite basal planes. Later, *Nicklow* et al. [4.2] reported in a seminal paper a more complete neutron scattering study of the phonon dispersion of pristine graphite, enabling the first critical test of lattice dynamical models to be conducted. This study and similar lattice dynamical investigations on other layered compounds are described in an excellent review by *Wakabayashi* and *Nicklow* [4.3].

Inelastic neutron scattering experiment on GICs started in 1977 with the work by *Ellenson* et al. on RbC_8 [4.4], followed by a systematic investigation of $[00q]$L modes of a number of alkali-metal GICs by *Zabel* and *Magerl* [4.5]. Meanwhile, the phonon dispersions of a large number of GICs have been determined, including the bending modes and the phonon density of states of intercalate modes. Short reviews on the lattice dynamics of GICs have appeared previously [4.6–8], but this is the first comprehensive review of this subject.

The remainder of this chapter is organized as follows: In Sect. 4.1, the theory of the lattice dynamics of layered compounds is outlined, and experimental techniques for obtaining the various phonon branches are discussed. Because most of the experimental work is focused on the $[00q]\mathrm{L}$ phonon modes, these will all be discussed and compared in Sect. 4.2. Although this procedure is a bit unusual, it appears here to be more practical, considering that for most of the compounds only the $[00q]\mathrm{L}$ branches are known at present. In Sect. 4.3, the $[q00]\mathrm{T}_\perp$ bending modes are reviewed, followed by a discussion in Sect. 4.4 of the intercalate phonon density of states. Finally, in Sect. 4.5, thermal properties related to the phonon measurements are discussed, including specific heat, Debye-Waller factors, and thermal expansion measurements.

4.1 Lattice Dynamics and Inelastic Neutron Scattering

4.1.1 Lattice Dynamical Models

For the basic theory of the vibrational dynamics of crystal lattices we refer to [4.9]. Here we will only outline some vibrational properties that are required for later discussion of the lattice dynamics of graphite intercalation compounds.

The equation of motion in the harmonic approximation, assuming plane waves for the lattice vibrations, is [4.9]

$$\omega^2 e_\alpha(d;\boldsymbol{q}) - \sum_{\beta d'} D_{\alpha\beta}(d,d';\boldsymbol{q}) e_\beta(d';\boldsymbol{q}) = 0 \,, \tag{4.1}$$

and couples the displacement amplitude $e_\alpha(d;\boldsymbol{q})$ $\{\alpha = x,y,z\}$ of each atom d in a unit cell, to the displacement of another atom d', independent of the cell parameter, where $\boldsymbol{q}$ is the wave vector of a particular phonon mode, and the coupling term or dynamical matrix $D_{\alpha\beta}(d,d';\boldsymbol{q})$ is

$$D_{\alpha\beta}(d,d';\boldsymbol{q}) = \frac{1}{\sqrt{M_d M_{d'}}} \sum_{\ell} \phi_{\alpha\beta}\begin{pmatrix} 0\ell \\ dd' \end{pmatrix} \mathrm{e}^{-\mathrm{i}\ell\boldsymbol{q}\cdot\boldsymbol{R}} \,. \tag{4.2}$$

Here, M_d and $M_{d'}$ are the respective masses of the atoms d and d' and $\phi_{\alpha\beta}$ is the force constant connecting the atom d in cell 0 at the origin with the atom d' in cell ℓ at distance $\ell|\boldsymbol{R}|$, where $\boldsymbol{R}$ is the position vector of the unit cell. By the ansatz of planar waves, the original equation of motion, comprising $3Nr$ coupled differential equations, is reduced to a set of equations of order $3r$, where r is the number of atoms per unit cell and N is the total number of unit cells in a crystal. Nontrivial solutions to the $3r$ eqautions of motion are found if the determinant of the amplitudes vanishes:

$$|D_{\alpha\beta}(d,d';\boldsymbol{q}) - \omega^2\delta_{\alpha\beta}\delta_{dd'}| = 0 \, . \tag{4.3}$$

The $3r$ eigenvalues $\omega_j(q)$ $\{j = 1,2,\ldots,3r\}$ for each wave vector q represent the dispersion relation, or the phonon dispersion. In a three-dimensional crystal with r atoms per unit cell, there are 3 acoustic branches with $\lim[\omega_j(q \to 0)] = 0$. The remaining $3r-3$ optical branches have finite values at $q = 0$. Longitudinal phonon modes are characterized by parallel polarization and propagation vectors ($\boldsymbol{e}\|\boldsymbol{q}$), while for transverse modes polarization and propagation vectors are perpendicular to each other ($\boldsymbol{e} \perp \boldsymbol{q}$).

The lattice dynamics of GICs and pristine graphite are closely related to each other. Therefore, we will start with a brief discussion of the lattice dynamical models which have been proposed in recent years for graphite, followed by a discussion of the phonon models for graphite intercalation compounds. The graphite structure is shown in Fig. 4.2 together with a labeling of the neighbors in the same and the adjacent plane with respect to the carbon atom 0 at the origin. The graphite unit cell with two planes stacked in an $ABA\ldots$ fashion contains 4 atoms.

Most of the lattice dynamical models proposed for graphite and graphite intercalation compounds are of the Born–von Kármán type and differ mainly in

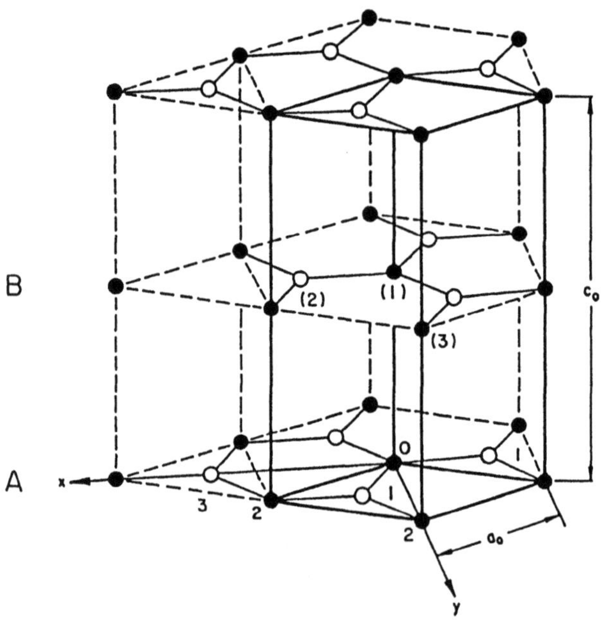

Fig. 4.2. Crystal structure of graphite comprising two graphene planes in $ABA\ldots$ stacking sequence. The thick solid lines indicate the unit cell. Numbers in plane A indicate the order of neighbor to the atom at the origin 0, and numbers in parentheses refer to the order of out-of-plane neighbors in the B plane. The lattice parameters are $a_0 = 2.46$ Å and $c_0 = 6.7$ Å

the number and symmetry of the force constants taken into account. The models of *Nicklow* et al. [4.2], of *Maeda* et al. [4.10] and of *Al-Jishi* and *Dresselhaus* [4.11] for pristine graphite are of this type, as are also the models of *Horie* et al. [4.12], and of *Al-Jishi* and *Dresselhaus* [4.13] for graphite intercalations. Simplified force constant models were developed by *Batallan* et al. [4.14, 15] for the description of stage 2 Br compounds. A different approach is taken by *Gupta* and co-workers [4.16–19], who derive the lattice dynamics of graphite and of the intercalation compounds from a Born-Mayer type potential, and *Pietronero* and *Strässler* [4.20], who employ a jellium-like model for describing the intercalate dynamics parallel to the layers. In the following, we will discuss these lattice dynamical models in more detail.

Nicklow et al. [4.2] described the lattice dynamics of graphite by an axially symmetric Born–von Kármán (AS-BvK) model, consisting of radial and tangential force constants $\phi_{\mathrm{ir}}^{\mathrm{i}n}$ and $\phi_{\mathrm{t}}^{\mathrm{i}n}$, respectively, up to the third nearest neighbor in the plane, and one respective radial and tangential force constant $\phi_{\mathrm{or}}^{\mathrm{o}1}$ and $\phi_{\mathrm{t}}^{\mathrm{o}1}$, coupling adjacent graphene layers. Here, we are using a force constant notation in which the superscript n indicates the order of the neighbor, and the subscripts r and t distinguish radial and transverse force constants. The additional superscripts i and o specify in- and out-of-plane neighbors, whereas the subscripts i and o indicate the in- and out-of-plane direction in which the force constants are acting. If i or o is not specified, the force constants are taken to be equal in both directions. In the AS-BvK model, for instance, $\phi_{\mathrm{it}}^{\mathrm{i}n} = \phi_{\mathrm{ot}}^{\mathrm{i}n} = \phi_{\mathrm{t}}^{\mathrm{i}n}$.

In the AS-BvK model, the high anisotropy of graphite is taken into account by the anisotropy of the in- and out-of-plane interactions. The anisotropy of the C-C bond bending in and out of the plane is, however, neglected by virtue of the assumption of axial symmetry. The force constants were fit to available neutron, Raman, and elastic data, and are listed in Table 4.1. The calculated phonon dispersion, reproduced in Fig. 4.3, displays all the essential features of the lattice dynamics of a layered crystal. These are the extremely steep slopes of the in-plane [q00]L acoustic (A) and [q00]L optic (O) modes which are due to the very strong covalent bonds between the carbon atoms in the plane. The low-frequency layer shear modes [00q]T and the quadratic dispersion of the layer bending modes [q00]TA$_{\perp}$ are characteristic of the weak interlayer shear interaction in layered crystals. Finally, the layer breathing modes [00q]L probe the interplanar radial or stretching force constants $\phi_{\mathrm{or}}^{\mathrm{o}n}$. The high-lying intralayer modes due to the strong covalent bonds in graphite are further characteristic signs. Some of these high-frequency modes appear almost degenerate in pairs, because they involve the weak interlayer shear constant $\phi_{\mathrm{t}}^{\mathrm{o}1}$. The AS-BvK model also predicts a Raman active A_{2u} out-of-plane mode at 174 meV. Later work by *Nemanich* et al. [4.21], however, showed that this mode actually occurs at 109 meV instead, revealing the inadequacy of the As-BvK model for describing the intralayer modes.

Maeda et al. [4.10] subsequently modified the AS-BvK model to account for the proper location of the A_{2u} mode. They improved the fit by distinguishing between bond bending force constants in and out of the plane, $\phi_{\mathrm{it}}^{\mathrm{i}n}$ and $\phi_{\mathrm{ot}}^{\mathrm{i}n}$,

Table 4.1. Force constants of graphite [10^5 dyn/cm] (= [10^2 N/m])

	Nicklow et al. [4.2]	*Maeda* et al. [4.10]	*Al-Jishi* and *Dresselhaus* [4.11]	*Gupta et al.* [4.16, 19]
ϕ^{i1}_{ir}	3.62	3.066	3.125	4.972
ϕ^{i2}_{ir}	1.33	1.363	1.2092	0.761
ϕ^{i3}_{ir}	−0.037	-	0.27978	-
ϕ^{i4}_{ir}	-	-	−0.25508	-
$\phi^{o1\,a}_{or}$	0.029	0.02895	0.02747	0.029
ϕ^{o2}_{or}	-	-	0.00595	-
ϕ^{o3}	-	-	−0.00473	-
ϕ^{o4}	-	-	0.002	-
ϕ^{i1}_{it}	1.99	2.810	2.675	0.765
ϕ^{i1}_{ot}	1.99	0.856	0.86545	0.765
ϕ^{i2}_{it}	−0.52	−0.527	−0.63731	−0.123
ϕ^{i2}_{ot}	−0.52	−0.142	−0.09312	−0.123
ϕ^{i3}_{it}	0.288	-	0.19	-
ϕ^{i3}_{ot}	0.288	-	0.12695	-
ϕ^{i4}_{it}	-	-	0.09488	-
ϕ^{i4}_{ot}	-	-	−0.05498	-
$\phi^{o1\,a}_{ot}$	0.00385	0.00359	−0.05934	0.0031
ϕ^{o2}_{ot}	-	-	0.01271	-
ϕ^{o3}_{ot}	-	-	0.00473	-
ϕ^{o4}_{ot}	-	-	−0.00927	-

[a] force constants per carbon atom.

respectively, while leaving all other force constants basically the same as in the AS-BvK model. *Al-Jishi* and *Dresselhaus* [4.11] extended the model of *Maeda* and co-workers to provide a better description of neutron inelastic data between the points Γ and M and also to obtain a better fit to experimental elastic constants. Their model includes up to the fourth neighbor intra- and interplane interactions, and it assumes axially asymmetric in-plane but symmetric out-of-plane tangential force constants, similar to the model of *Maeda* et al. [4.10]. With a total of 20 force constants, *Al-Jishi* and *Dresselhaus* obtained an excellent agreement with experimental data.

A different route was taken by *Gupta* et al. [4.16], who proposed a Born-Mayer-type two-body potential approach for the lattice dynamics of graphite,

$$\Phi(\mathcal{R}) = A(\mathcal{R} - \mathcal{R}_a)^2(\mathcal{R} - \mathcal{R}_c)^2 \mathrm{e}^{-\mathcal{R}/\varrho} , \tag{4.4}$$

from which the radial and tangential force constants are derived via the derivatives

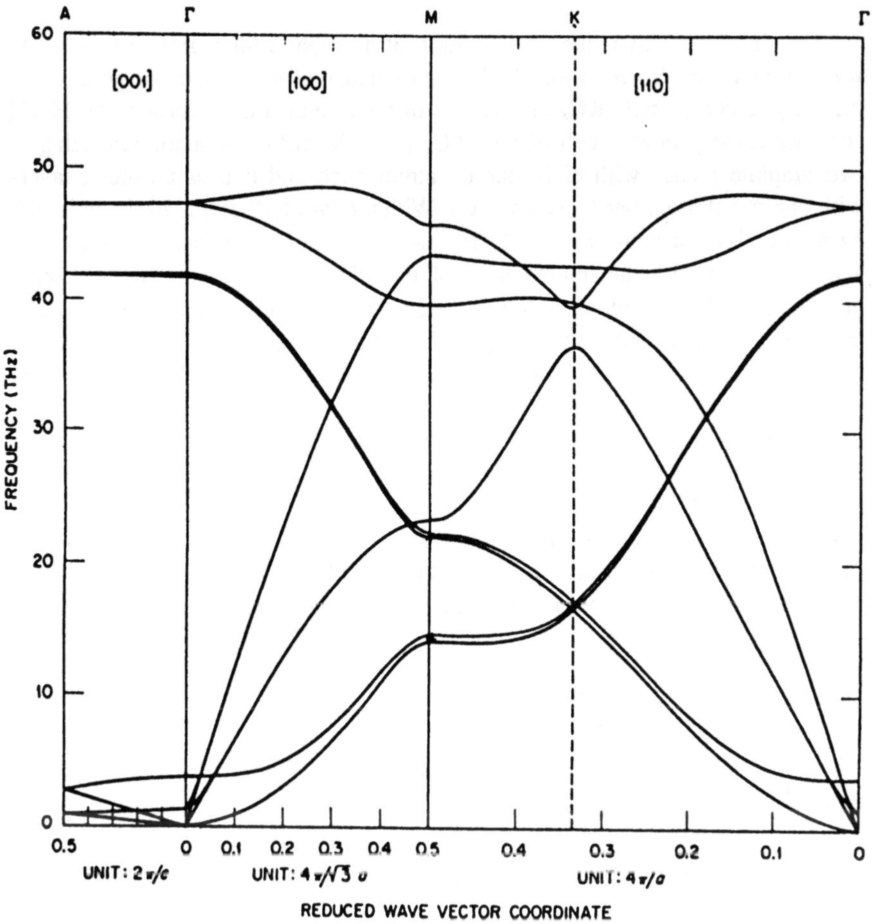

Fig. 4.3. Phonon dispersion relations for graphite from calculations of *Nicklow* et al. [4.2]

$$\phi_{\mathrm{r}}^{n} = \left(\frac{d^2\Phi}{d\mathcal{R}^2}\right)_n , \quad \phi_{\mathrm{t}}^{n} = \frac{1}{\mathcal{R}}\left(\frac{d\Phi}{d\mathcal{R}}\right)_n . \tag{4.5}$$

Here, $\mathcal{R}_a$ and $\mathcal{R}_c$ are the equilibrium in- and out-of-plane bond lengths, and A and ϱ are Born-Mayer-type constants. In this model, *Gupta* et al. [4.16] have taken into account first and second neighbor interactions in the plane, but only one first neighbor interaction along the c-axis. The axial symmetry of the in- and out-of-plane interactions resembles the AS-BvK model. The main advantage of the model of *Gupta* and co-workers is, however, that the force constants are derived from a potential, requiring a fit of only 4 potential parameters instead of 8 force constants in the AS-BvK model or 20 force constants in the BvK model calculated by *Al-Jishi* and *Dresselhaus*. With these four parameters ($A, \varrho, \mathcal{R}_a, \mathcal{R}_c$), *Gupta* et al. [4.16] are able to describe the graphite phonon dispersion with reasonable accuracy.

The first lattice dynamical calculation of a graphite intercalation compound was presented by *Horie* et al. [4.12], who calculated the phonon dispersion for the stage 1 compounds KC_8 and RbC_8. In their calculation, *Horie* et al. [4.12] considered the primitive cell of the MC_8 ($M = K, Rb$) compound, consisting of two graphite planes with eight carbon atoms each and two alkali-metal layers with one metal atom per layer unit cell. With a total of 18 atoms in the unit cell, 54 phonon branches are produced, from which only six branches are really new and associated with the intercalation of the alkali-metal layers, the rest being graphite-like. In this model, only first neighbor radial and transverse metal-metal in-plane force constants $\tilde{\phi}^{i1}_{ir}$ and $\tilde{\phi}^{i1}_{it}$, respectively, were considered and those were taken to be the same as in the free alkali metals. (The tilde indicates that these force constants refer to the intercalate layer.) The intercalate and graphene layers were coupled by first neighbor radial and tangential force constants, ϕ^{IC}_{r} and ϕ^{IC}_{t}, respectively. Numerical values of force constants used in the model are listed in Table 4.2. Although the calculated phonon energies do not agree well with experimental data, the model has almost all the ingredients for a proper description of the lattice dynamics of stage 1 GICs and a new fit of the force constants with current experimental data would certainly improve the agreement.

Table 4.2. Force constants for intercalates in stage 1 compounds used in model calculations by *Horie* et al. [4.12], *Al-Jishi* and *Dresselhaus* [4.13], and *Gupta* et al. [4.17, 19]. All force constants are in units of dyn/cm

	LiC_6	KC_8	RbC_8	CsC_8	Ref.
$\tilde{\phi}^{i1}_{ir}$	–	1400	1300	1300	[4.13]
		2500			[4.12]
$\tilde{\phi}^{i1}_{it}$	–	50	−50	−50	[4.13]
		100			[4.12]
ϕ^{IC}_{r}	29330				[4.17]
		1660	1600	1600	[4.13]
		10000			[4.19]
		10000			[4.12]
ϕ^{IC}_{t}	1340				[4.17]
		−120	−100	−100	[4.13]
		1000			[4.19]
ϕ'_{r}	–	100			[4.13]

The model of *Horie* et al. [4.12] is limited to the symmetry of stage 1 compounds, and a simple extension to higher stage compounds is not possible. A more general model was proposed by *Leung* et al. [4.22] and later in a refined version by *Al-Jishi* and *Dresselhaus* [4.13]. These authors derive the dispersion of a stage n alkali-metal GIC from appropriate c^*-axis folding and replacement of graphene layers by intercalate layers. The calculation proceeds in several steps. In the first step, either the lattice dynamical model of *Maeda* et al. [4.10] or the more precise model of *Al-Jishi* and *Dresselhaus* [4.11] is used to describe the dispersion of the graphite layers with a stacking sequence $ABAB\ldots$. The second step requires a layer-by-layer expansion of the c-axis repeat distance, which causes a simultaneous folding of the c^*-axis. For instance, in a graphite lattice with periodicity mI_c, where $I_c = 3.35$ Å is the distance between adjacent graphene layers, the smallest reciprocal lattice vector is $G = \pi z / mI_c$, and the new phonon frequencies are given by

$$\omega(q) = \omega(nG + q) \; ; \quad n = 1, 2, \dots, m-1 \, . \tag{4.6}$$

Here, z is the unit vector pointing along the c^* direction. Next, a proper stacking of the graphite layers is assumed and some of them are replaced by intercalate layers at appropriate positions for a stage n compound. For a stage 1 compound, the resulting stacking sequence is $AIAIA\dots$; for stage 2, $ABIABIABI\dots$; for stage 3, $ABAIABAIA\dots$, etc. Although these stacking arrangements do not represent the crystallographically correct stacking sequences of the intercalation compounds, they are adopted to facilitate the computation. *Al-Jishi* and *Dresselhaus* [4.13] note that this simplification, however, does not affect the quality of the calculation. A unit cell of a stage n compound then contains n graphene layers with two carbon atoms per layer and one intercalate layer containing a single intercalate mass unit. With this stacking arrangement, the Brillouin zone along the c^* direction of a stage n compound is reduced by a factor $2/(n+1)$ as compared to pristine graphite.

Finally, *Al-Jishi* and *Dresselhaus* [4.13] use appropriate effective masses for the carbon and intercalate atoms, which allow for the proper layer stoichiometry, and introduce new force constants for the in- and out-of-plane intercalate interaction, including a new interplanar force constant, ϕ_r', connecting the bounding graphene layers across an intercalate layer.

For the radial force constants in the c direction, *Al-Jishi* and *Dresselhaus* have made various assumptions, to take charge transfer effects into account. They scaled the force constant ϕ_r^{IC}, connecting the graphite layers adjacent to an alkali-metal layer, according to

$$\phi_r^{IC} = \phi_{or}^{ol}(1.04 - 0.04/n) \, , \tag{4.7}$$

while for interior layers not adjacent to intercalate layers they assumed

$$\phi_r^{cc} = \phi_{or}^{ol}(1 - 0.025/n) \, , \quad n \geq 2 \, . \tag{4.8}$$

The tangential interplanar force constants were adjusted to match the experimental observation that the shear elastic constant C_{44} is softer in higher stage compounds than in graphite (Sect. 4.3).

The lattice dynamical model of *Al-Jishi* and *Dresselhaus* [4.13] is very powerful and can be applied to various stages and compounds merely by adjusting the effective intercalate masses. Calculated phonon dispersions for the stages 1 through 4 of Rb intercalated in graphite are shown in Fig. 4.4. The only serious deficiency of their model is the assumption of equal spacings between all graphite–graphite and graphite-intercalate layers, which leads to fits with inproper force constants.

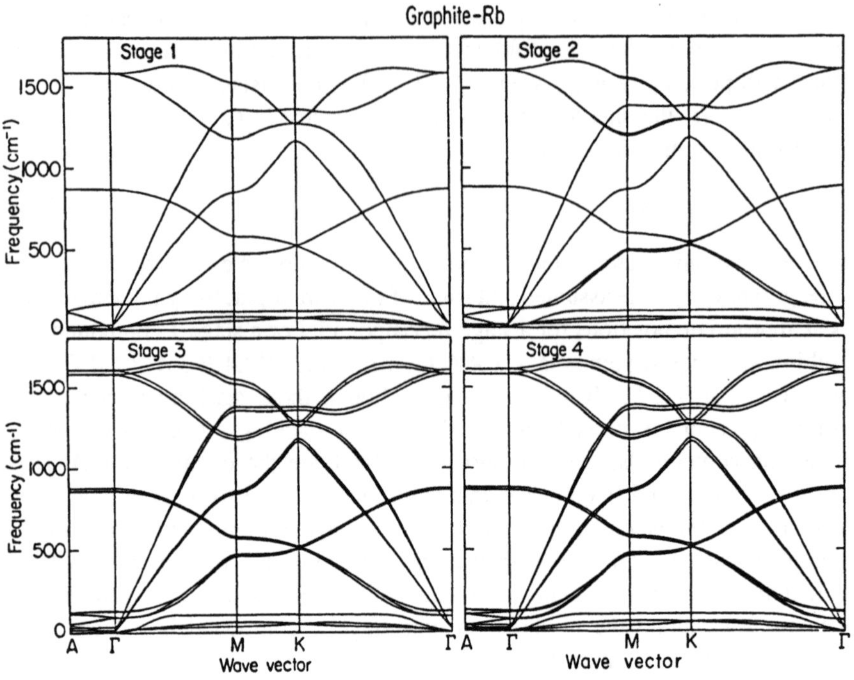

Fig. 4.4. Calculated phonon dispersion curves for stages 1 to 4 of Rb intercalated in graphite. (From [4.13])

Based on the force constant models described above, *Al-Jishi* and *Dresselhaus* [4.13] and *Kimura* et al. [4.23] calculated the phonon density of states (PDS) of a number of alkali-metal GICs with very similar results. Both groups find that the PDS of MC_8 (M = K, Rb, Cs) is similar to pristine graphite, except for two peaks in the lower part of the frequency spectra, both of which are split. The lower split peak is due to the metal in-plane transverse and longitudinal vibrations, and appears superimposed on a featureless background of the graphite phonon spectrum. These peaks scale with the mass of the intercalate atom, as one would expect, and are only weakly stage dependent, according to *Al-Jishi* and *Dresselhaus* [4.13]. These features are in excellent agreement with measured PDSs of the alkali-metal in-plane vibration, which are discussed in more detail in Sect. 4.5. The upper peak comes from vibrations of the graphite and intercalate planes along the c-axis.

Gupta et al. [4.17–19] have calculated the phonon dispersion of a number of stage 1 compounds, applying the two-body pseudopotential approach described before. The intercalate-graphene interlayer interaction is again modeled by one radial and one transverse force constant, but the intercalate intralayer interaction is neglected. The pseudopotential model of *Gupta* et al. [4.17–19] is rather sucessful for a description of the phonon dispersion of stage 1 compounds. However, it cannot easily be extended to higher stage compounds.

The force constants for stage 1 compounds used in the model calculations described above are listed in Table 4.2.

In an early lattice dynamical calculation, *Pietronero* and *Strässler* [4.20] proposed a jellium-like model for the description of the dynamics of the intercalate species. In this two-dimensional model it is assumed that the dominant interaction between intercalate atoms or molecules is the Coulomb repulsion arising from their partial ionization due to the charge transfer. The graphite layers merely play the role of providing a charged background. Within the frame of this model, *Pietronero* and *Strässler* [4.20] obtain for the plasma frequency of the intercalate layer [4.22]

$$\omega^2 = \frac{\varrho f^2}{M_{\mathrm{I}}} \langle\alpha\rangle q_{\|}^2 , \tag{4.9}$$

where ϱ is the intercalate in-plane volume density, f is the charge per intercalate atom or molecule, M_{I} is the intercalate mass, and $q_{\|}$ is the phonon wave number with polarization parallel to the layers. The z-dependent polarizability which depends on the screening effects of the graphite layers is described by $\langle\alpha\rangle$. While a simple 2d BvK-type model predicts two in-plane modes, the jellium model yields only one, longitudinal, mode. The other, transverse, in-plane mode has to come from a different assumption, as shown below.

Kamitakahara et al. [4.24,25] described the in-plane vibration of the alkali-metal atoms in stage 1 RbC_8 in a similar fashion to the model of *Pietronero* and *Strässler* [4.20]. Applying this model to the measured PDS of the intercalate in-plane modes, *Kamitakahara* and co-workers noticed that the upper, longitudinal, peak occurs close to the ion plasma frequency for the bare Coulomb interaction, indicating that screening effects can be neglected. A second peak at lower frequencies could be described as arising from the interaction of the intercalate atoms with the graphite substrate potential.

4.1.2 Inelastic Neutron Scattering

The coherent differential cross section for one-phonon neutron scattering is [4.26,27]

$$\begin{aligned} \frac{d\sigma}{d\Omega dE'} = & \frac{k'}{k}\frac{(2\pi)^3}{v_0} \sum_{j,G,Q} F^2(Q,q_j)\frac{1}{2\omega_j(q_j)} \left[n(\omega_j) + \tfrac{1}{2} \mp \left(-\tfrac{1}{2}\right)\right] \\ & \times \delta(\omega \mp \omega_j(q))\delta(Q \mp q - G) \end{aligned} \tag{4.10}$$

with the dynamical structure factor

$$F(Q,q_j) = \left| \sum_d \frac{b_d}{\sqrt{M_d}} \mathrm{e}^{\mathrm{i}Q\cdot r_d}[Q \cdot e_d(q_j)]\mathrm{e}^{-W_d(Q)} \right| . \tag{4.11}$$

Here, $n(\omega_j) = [\exp(\hbar\omega_j/k_{\mathrm{B}}T) - 1]^{-1}$ is the Bose occupation number, b_d is the coherent scattering length of atom d, W_d is the Debye-Waller factor, k and

k' are the neutron wave numbers of the incident and scattered neutron beam, respectively, and v_0 is the unit cell volume. One-phonon scattering occurs for

$$\frac{1}{\hbar}(E' - E) = \pm\omega_j \tag{4.12}$$

and

$$(\boldsymbol{k}' - \boldsymbol{k}) = \boldsymbol{Q} = \boldsymbol{G} \pm \boldsymbol{q} \,, \tag{4.13}$$

which guarantees energy and momentum conservation. Here, E is the initial and E' the final neutron energy, $\boldsymbol{Q}$ is the scattering vector, $\boldsymbol{G}$ is a reciprocal lattice vector, and $\boldsymbol{q}$ is a phonon wave vector. The scattering triangle for neutron inelastic scattering is shown in Fig. 4.5. In (4.10), the + sign stands for phonon annihilation, and the minus sign for phonon creation.

Phonon dispersions are usually measured with a triple-axis neutron spectrometer. A schematic diagram showing the essential parts of such a spectrometer is shown in Fig. 4.6. The monochromator angle $2\theta_M$ determines the magnitude of the incident wave vector $|\boldsymbol{k}|$, ϕ is the scattering angle, Ψ measures the angle between the scattering vector $\boldsymbol{Q}$ and a crystallographic axis, and $2\theta_A$ is the analyzer angle which defines the final wave number $|\boldsymbol{k}'|$. Phonon dispersions are usually measured in the constant $\boldsymbol{Q}$ (or $\boldsymbol{q}$) mode, in which $\boldsymbol{Q} = \boldsymbol{G} \pm \boldsymbol{q}$ is kept fixed while scanning the energy. This can be done either with fixed initial energy and variable final energy or vice versa.

One important consideration in the study of phonon dispersions is the selection of a unique polarization vector for the phonon mode in question. Since the dynamical structure factor F is proportional to the scalar product $(\boldsymbol{Q} \cdot \boldsymbol{e})$ between the scattering vector and the phonon polarization, longitudinal and transverse

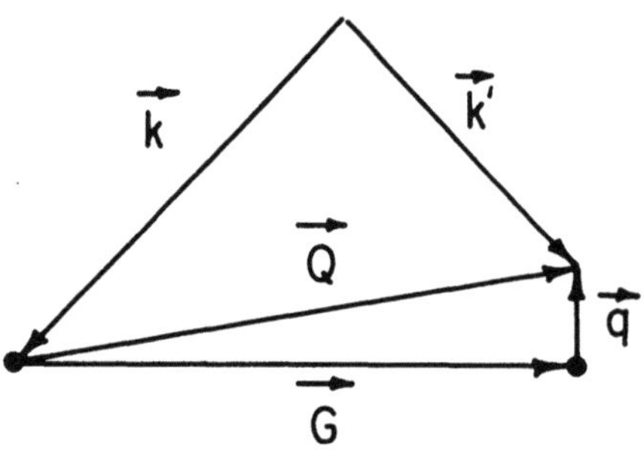

Fig. 4.5. Schematic of the scattering triangle for inelastic neutron scattering. $\boldsymbol{k}$ and $\boldsymbol{k}'$ designate the incident and final neutron wave vectors, $\boldsymbol{Q}$ is the scattering vector, $\boldsymbol{G}$ is a reciprocal lattice vector, and $\boldsymbol{q}$ refers to a phonon wave vector. The scattering geometry shown is used for measuring transverse phonons

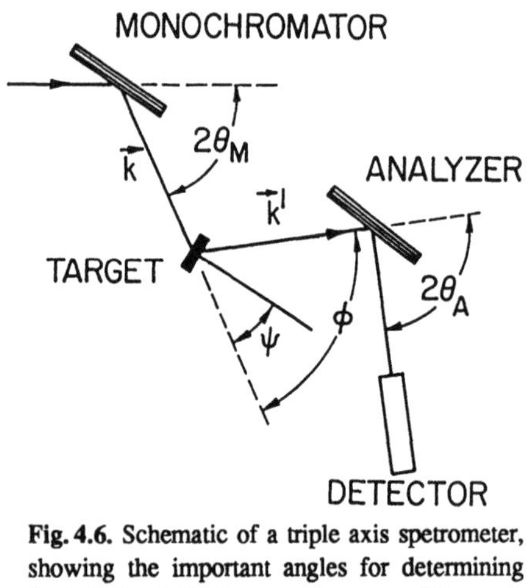

Fig. 4.6. Schematic of a triple axis spetrometer, showing the important angles for determining the initial and final neutron energy and the position of the scattering vector $\boldsymbol{Q}$ in relation to the crystallographic orientation of the sample

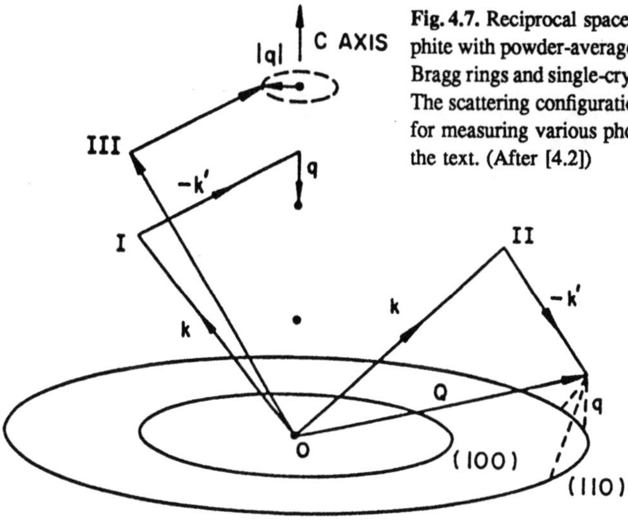

Fig. 4.7. Reciprocal space diagram for pyrolytic graphite with powder-averaged in-plane (100) and (110) Bragg rings and single-crystal-like ($00L$) Bragg spots. The scattering configurations indicate scan directions for measuring various phonon branches described in the text. (After [4.2])

modes can be easily separated out. For longitudinal modes, $e \parallel q$ and therefore the scalar product has a maximum when $Q \parallel q$. On the other hand, transverse modes are defined by $e \perp q$, and therefore require a configuration with $Q \parallel e$, while $Q \perp q$.

The scattering configurations for measuring the most important modes of graphite and its intercalation compounds are depicted in Fig. 4.7, which is adapted from [4.2]. The $[00q]$L modes are obtained along the c^*-axis (configuration III), and the $[q00]\mathrm{T}_\perp$ modes are measured with respect to any of the (00ℓ) reflections, to meet the polarization requirement (configuration I). It is easy to see that in this configuration the mixing of modes steadily increases when the angle between the scattering vector Q and the c^* direction becomes larger, which is an unavoidable feature in the measurement of these modes. The circles represent the powder averaged ($hk0$) in-plane reflections in pyrolytic graphite. The layer shear modes $[00q]$T can be measured away from these circles in the configuration II, however, an average over several phonon wave vectors q is taken for each energy scan, as indicated by the three arrows, and the measured dispersion has to be corrected for these effects [4.2]. Any measurement of the in-plane longitudinal mode $[q00]\mathrm{L}_\parallel$ suffers from the same difficulties, and the in-plane transverse mode $[q00]\mathrm{T}_\parallel$ cannot be measured at all using pyrolytic graphite.

In order to test the validity of lattice dynamical models, it is necessary to establish the dynamical structure factor of individual phonon modes. Also, in order to set up a neutron scattering experiment, it is necessary to have a rough idea of the dynamical structure factor, which determines the intensity of phonon groups in different Brillouin zones. This is particularly important for measuring the $[00q]$L modes of graphite intercalation compounds. For these modes, the dynamical structure factor is

$$|F(Q,q_j)/Q|^2 = \left|\sum_d \frac{b_d}{\sqrt{M_d}} e_d(q_j)\exp(\mathrm{i}Qr_d)\right|^2 , \qquad (4.14)$$

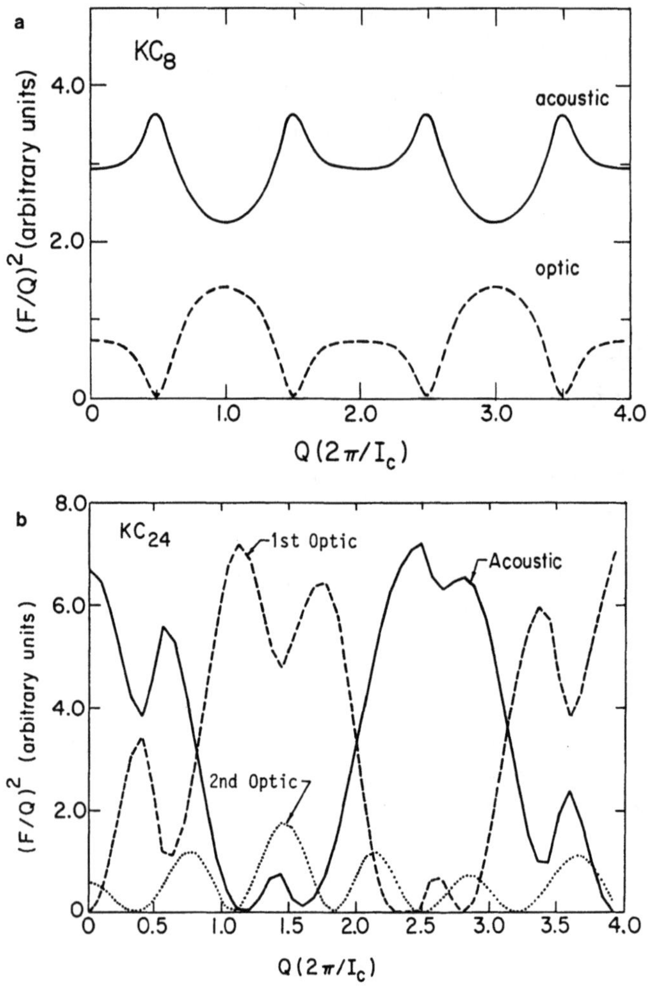

Fig. 4.8a,b. Dynamical structure factor $|F/Q|^2$ from (4.14) calculated for acoustic and optic [00q]L branches of potassium intercalated in graphite: (**a**) for KC_8; (**b**) for KC_{24}

where the displacements $e_\alpha(q_j)$ follow from a lattice dynamical model. For stage 1 compounds, it is easy to guess the functional dependence of $|F/Q|^2$ which is periodic in Q, as shown in Fig. 4.8a for KC_8. However, in compounds with $n \geq 2$, the layer spacings r_d are not rational fractions of the c-axis repeat distance, and therefore the dynamic structure factor becomes nonperiodic with intensities which vary strongly from Brillouin zone to Brillouin zone. Therefore, the dynamic structure factor has to be calculated beforehand in order to conduct a proper experiment. An example of the dynamic structure factor is given in Fig. 4.8b for stage 2 KC_{24}, based on a simple one-dimensional nearest neighbor BvK model discussed in the next section.

4.2 [00q]L Phonon Dispersion

4.2.1 Lattice Dynamical Models for [00q]L Modes

The [00q]L phonon dispersion of GICs can be described as the dispersion of linear chains of mass points, representing the graphene and intercalate layers, as discussed in the previous section. Therefore, the dispersion depends on the areal mass densities of the intercalate layers, the force constants between the layers, and the number of layers within a repeat unit. In general, a stage n compound contains $n+m$ layers: n graphene layers between consecutive intercalate layers and an intercalate package with m sublayers. Then the dispersion consists of $n+m$ branches, 1 acoustic branch and $n+m-1$ optic branches. In the simple binary alkali-metal GICs, $m=1$, while for ternary trilayer and acceptor compounds with transition metal chlorides, $m=3$. All these compounds exhibit characteristic stacking sequences of the graphene and intercalate layers which may extend over several repeat distances. The [00q]L modes propagating normal to the layers are, however, insensitive to these stacking arrangements, and therefore a simplified structure model is sufficient for the description of the [00q]L phonon dispersion. In this description, only the simple repeat distance I_c between consecutive intercalate layers is taken into account:

$$I_c = d_1 + (n-1)I_G \tag{4.15}$$

where $I_G = 3.35\,\text{Å}$ is the layer spacing between graphene planes in GICs, which is usually identical to the layer separation in pristine graphite. An approximate sandwich thickness d_1 containing the intercalate layer in higher stage compounds can be obtained from stage 1 compounds.

In general, the [00q]L dispersion of a stage n compound follows from a $(n+m)\times(n+m)$ dynamical matrix, with matrix elements determined from (4.2). For example, the dynamical matrix of a stage 2 compound with $m=2$ using the force constant notation, masses, and layer separations, shown in Fig. 4.9a, is

$$\begin{bmatrix} -\dfrac{2\phi_r^{IC}}{M_I} & \dfrac{\phi_r^{IC}}{\sqrt{M_I M_C}} & \dfrac{\phi_r^{IC}}{\sqrt{M_I M_C}}e^{iqI_c} \\ & -\dfrac{\phi_r^{IC}+\phi_r^{CC}}{M_C} & \dfrac{1}{M_C}(\phi_r^{CC}+\phi_r' e^{iqI_c}) \\ \text{c.c.} & & -\dfrac{\phi_r^{IC}+\phi_r^{CC}}{M_C} \end{bmatrix} . \tag{4.16}$$

Here, M_C is the graphite mass in atomic mass units (M_C = 12 a.u.) and M_I is the intercalate effective mass per layer and per carbon atom: $M_I = M_A/\xi$, where M_A is the atomic or molecular mass in atomic mass units and ξ is the layer stoichiometry. Also, ϕ_r^{IC} describes the radial force constant connecting the intercalate and the bonding graphene layers, ϕ_r^{CC} is the force constant between inner graphene layers, and ϕ_r' is a second-neighbor force constant connecting

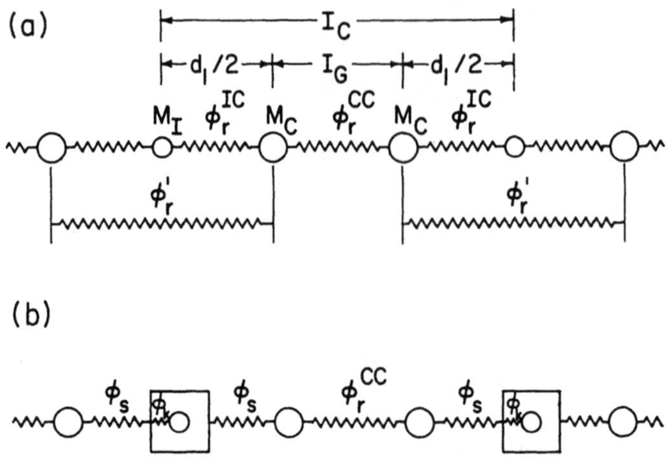

Fig. 4.9a,b. Linear chain models for the calculation of $[00q]$L modes in stage 2 graphite intercalation compounds. (a) Born–von Kármán model with first and second nearest neighbor force constants. (b) Mixed Born–von Kármán and shell model where the alkali ion core is coupled to a rigid electronic shell via the force constant ϕ_k and the shell is coupled to the graphite planes via ϕ_s. The interaction of the inner graphite layers is simply described by the radial force constant ϕ_r^{CC}, as in the Born–von Kármán model

graphene layers across an intercalate layer. All force constants are per carbon atom.

This model takes into account only one second neighbor force constant, but it can easily be extended to include further second and third neighbor interactions. However, the present form has been particularly successful for describing the $[00q]$L modes, as we will show below. Clearly, higher stages follow simply by adding more force constants ϕ_r^{CC} for the interaction of the inner graphene layers. For stages $n \geq 4$, one may have to distinguish between interactions of graphite layers that are next to intercalate layers and those that are not. However, at present there are not enough data on higher stage compounds to justify such a distinction.

An alternative mixed ion-shell and BvK model was proposed by *Magerl* and *Zabel* [4.5, 28] to account for the charge transfer and the ionic character of the intercalate layers. In their model the interaction between the intercalate and bounding graphene layer is mediated by an electronic shell whose mass is assumed to be zero [4.29]. In akali-metal GICs, the shell is coupled to the core of the alkali ion by a force constant ϕ_k and to the bounding graphite layers by the shell force constant ϕ_s. Then we have $n+3$ equations of motion for a stage n compound instead of $n+1$ in the BvK model. The two extra equations describe the shell motion on either side of the intercalate layer. But since their mass is zero, they serve to eliminate the two unknown displacement parameters for the shell and the matrix is again reduced to the dimension $(n+1) \times (n+1)$. In the notation of Fig. 4.9b, the dynamical matrix for the mixed shell–BvK model is

$$
\begin{bmatrix}
-\dfrac{2\phi_s\phi_k}{M_I(2\phi_s+\phi_k)} & \dfrac{\phi_s\phi_k}{\sqrt{M_I M_C}(2\phi_s+\phi_k)} & \dfrac{\phi_s\phi_k}{\sqrt{M_I M_C}(2\phi_s+\phi_k)}e^{-iqI_c} \\
 & -\dfrac{1}{M_C}\left(\dfrac{\phi_s\phi_k}{\phi_k+2\phi_s}+\phi_r^{CC}\right) & \dfrac{1}{M_C}\left(\phi_r^{CC}+\dfrac{\phi_s^2}{2\phi_s+\phi_k}e^{-i2qI_c}\right) \\
\text{c.c.} & & -\dfrac{1}{M_C}\left(\dfrac{\phi_s\phi_k+\phi_s^2}{\phi_k+2\phi_s}+\phi_r^{CC}\right)
\end{bmatrix} . \tag{4.17}
$$

This mixed ion-shell and BvK model can also easily be modified for higher stages. Inspection of the two matrices (4.16) and (4.17) shows that they are equivalent if we set

$$
\phi_r^{IC} = \frac{\phi_s\phi_k}{\phi_k+2\phi_s} \quad \text{and} \tag{4.18}
$$

$$
\phi_r' = \frac{\phi_s^2}{\phi_k+2\phi_s} . \tag{4.19}
$$

The equivalence of these two particular models was pointed out by *Funahashi* et al. [4.30] and holds for all stages. The electronic shell can always be replaced by the first neighbor force constant connecting the intercalate layer with the graphite bonding layers and by a second neighbor force constant connecting the bonding graphite layers across the intercalate layers. Increasing the stage merely adds more force constants for the internal graphene–graphene interaction represented by ϕ_r^{CC}. Although the BvK model is conceptionally simpler than the ion-shell model, the latter more clearly reveals the nature of the intercalate-graphene bonding. For instance, in the alkali-metal GICs the shell force constant ϕ_k plays an important role, reflecting the compressibility of the metal electronic shell, whereas in acceptor compounds ϕ_k appears to be infinitely stiff such that $\phi_s = \phi_r^{IC}$ in most cases.

The force constants can also be related to elastic constants. In the elastic limit ($q \to 0$) one obtains from the lattice dynamical models described above the compressional elastic constant C_{33}^n for a stage n compound

$$
C_{33}^n = \frac{4I_c}{a^2\sqrt{3}}\left(\frac{2}{\phi_r^{IC}+2\phi_r'}+\frac{n-1}{\phi_r^{CC}}\right)^{-1} . \tag{4.20}
$$

Here, $a = 2.46$ Å is the graphite in-plane lattice parameter. The elastic constant C_{33} can also be determined directly from the slope of the acoustic branch via

$$
C_{33} = v^2\varrho , \tag{4.21}
$$

where $v = \Delta E/\hbar\Delta q$ is the sound velocity, and

$$
\varrho = \frac{4p(nM_c+M_I)}{\sqrt{3}I_c a^2} \tag{4.22}
$$

is the volume density, p being the proton mass. Conversely, if C_{33} is known, one can estimate the force constants. For a rough calculation it is safe to assume that $\phi_r^{CC} = \phi_{or}^{ol}$ and $\phi_r'/\phi_r^{IC} \approx 0.1$–$0.2$ in donor compounds with alkali metals, whereas $\phi_r' \approx 0$ in most acceptor compounds.

4.2.2 Experimental Results

(a) Donor Compounds

The donor compounds are ideally suited for testing lattice dynamical models and for the investigation of zone-folding and mode-splitting effects. This is because of the simple periodic arrangement of the layers, the large number of stages which are available with high stage purity, and last but not least, the large sample sizes which can be prepared with reasonable crystal quality, essential for neutron inelastic studies.

The effect of the intercalate mass on the phonon energies is best demonstrated by the series of stage 1 compounds KC_8, RbC_8, and CsC_8, the phonon dispersions of which are reproduced in Fig. 4.10. (The phonon energies are given in units of meV. Other common units are frequencies [Hz], and wave numbers

Table 4.3. Elastic data and best fit force constants from $[00q]$L modes of graphite intercalation compounds from neutron scattering experiments. All force constants are per carbon atom and have typical accuracies of 5%–10%

	Stage	ϕ_r^{IC} [dyn/cm]	ϕ_r^{CC} [dyn/cm]	ϕ_r' [dyn/cm]	C_{33} [10^{11}dyn/cm^2]	v [10^5cm/s]	Ref.
Graphite			2880		3.65 ± 0.1	4.0	[4.2]
Donor Compounds:							
LiC_6	1	7445	–	527	8.86 ± 0.3	6.28 ± 0.1	[4.34]
LiC_{12}	2	10040	4070		3.71	5.56	[4.36]
LiC_{18}	2	5740	3000		7.04	4.07	[4.36]
KC_8	1	4140	–	354	4.85 ± 0.14	4.91 ± 0.07	[4.5]
KC_{24}	2	2734	3000	477	3.75 ± 0.1	4.36 ± 0.05	[4.5]
KC_{36}	3	2674	2885	418	3.53 ± 0.1	4.19 ± 0.05	[4.5]
$K(ND_3)_{4.3}C_{24}$	1	2090	–		1.71 ± 0.1	3.22 ± 0.1	[4.51]
$K(ND_3)_{3.1}C_{22}$	1	2120	–		1.62 ± 0.1	3.18 ± 0.1	[4.51]
$K(ND_3)_{2.3}\ C_{28}$	2				2.94 ± 0.09	3.90 ± 0.06	[4.51]
$KHgC_4$ [a]	1	2404	1682	323	2.27 ± 0.05		[4.38]
RbC_8	1	3850		354	4.84 ± 0.1	4.35 ± 0.1	[4.54]
RbC_{24}	2	2505	2950	422	3.55 ± 0.1	4.09 ± 0.05	[4.5]
		2540	2940	360			[4.64]
CsC_8	1	3917		651	5.83 ± 0.12	4.36 ± 0.1	[4.5]
CsC_{24}	2	2380	3065	520	3.10 ± 0.13	3.61 ± 0.07	[4.5]
CsC_{36}	3	2715	2809	180			[4.30]
Acceptor Compounds:							
$C_{13.2}FeCl_3$	2	2030	2800		2.81		[4.43]
$C_{8.36}CoCl_2$	2	2350	2800		4.019	3.82	[4.45]
$C_{16}AsF_5$	2				3.64		[4.46]
$C_{28.2}SbCl_5$	2	1400	2800		2.84		[4.44]
$C_{56.4}SbCl_5$	4	1400	2800		3.082		[4.44]
$C_{14}Br$	2	945	2560		2.81		[4.15]
$C_{20}D_2SO_4$	1	1970		-10			[4.47]

[a] Force constants listed are in the sequence ϕ_1, ϕ_2, and ϕ_3 and are defined in the inset of Fig. 4.14.

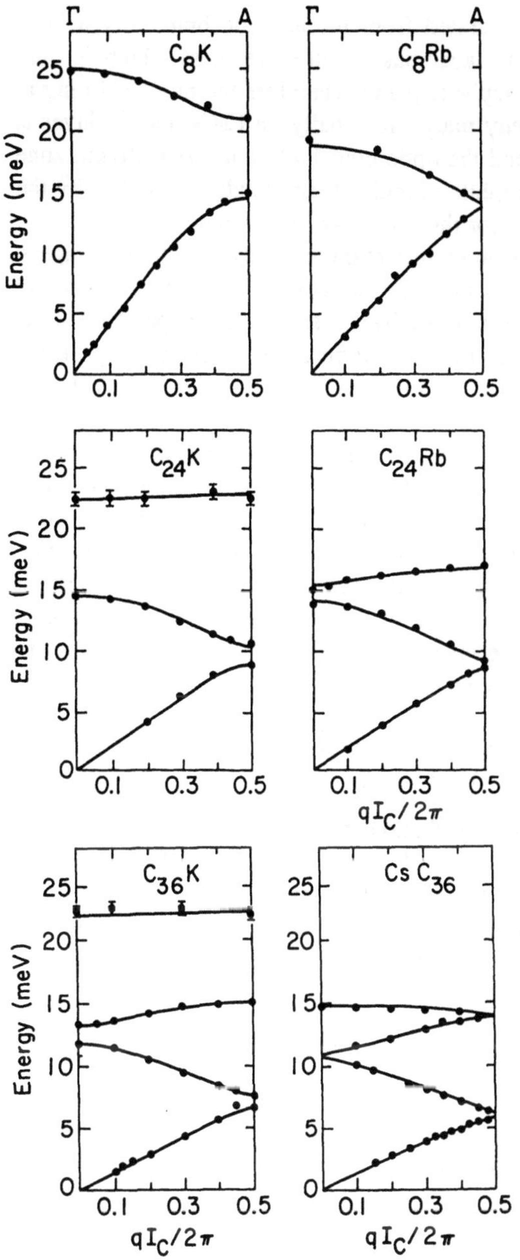

Fig. 4.10. Measured phonon energies of [00q]L modes in alkali-metal graphite intercalation compounds at room temperature. The solid lines represent best fits to the phonon energies with one-dimensional lattice dynamical models described in the text. KC_8, CsC_8, KC_{24}, RbC_{24}, CsC_{24} and KC_{36} from [4.5]; RbC_8 from [4.4, 24]; CsC_{36} from [4.30]

[cm^{-1}], which are related via 1 meV = 0.242×10^{12} Hz = 8.065 cm^{-1}.) For all three compounds, the symmetry along the c^*-axis and the number ratio of intercalate to carbon atoms is the same, and there are only small differences in the force constants, see Table 4.3. The dominant effect is attributable to the in-plane effective mass of the intercalate layer. In KC_8, with the lightest intercalate mass, the optic branch is well separated from the acoustic branch by a large frequency gap. The Rb intercalate layer has an effective mass which is only slightly smaller than that of the graphene plane. Therefore the mode splitting at the Brillouin zone boundary is only marginal. Finally, in CsC_8, the Cs layer is heavier than the graphene layer and the optic branch close to the Brillouin zone boundary is determined by the graphite vibrational amplitude rather than by the intercalate vibration as is the case for the other two compounds.

The result of staging on the phonon dispersion is clearly seen by the stage 1, 2 and 3 K-GICs, whose phonon energies are reproduced in the first column of Fig. 4.10. An increase of the stage number by 1 results in the appearance of an additional optic branch. For all stages, zone-folding and mode-splitting effects are clearly visible. The highest optic branch is predominantly determined by the vibrational amplitude of the planes with the smallest effective mass. These are the potassium layers in all three K-GICs. It should be noted that with increasing stage number the dispersion of the top optic branch becomes flatter, indicating that the coupling between the alkali-metal planes weakens from stage 1 to 3.

The Einstein-like behavior of the top optic mode in high stage compounds can be modeled assuming a single alkali-metal layer intercalated in an otherwise pure graphite crystal. Coupling this layer to the adjacent graphene layer by a force constant ϕ_r^{IC} and the graphite layers via the usual force constant ϕ_{or}^{ol}, this model predicts a localized mode frequency of

$$\omega_{loc}^2 = \omega_{max}^2 \left[\frac{M_C \phi_r^{IC}}{2 M_I \phi_{or}^{ol}} \left(\frac{\phi_r^{IC}(2M_C - M_I) + M_I \phi_{or}^{ol}}{2\phi_r^{IC} M_C + M_I \phi_{or}^{ol}} \right) \right] , \qquad (4.23)$$

where $\omega_{max} = 15.7$ meV is the highest phonon frequency of the graphite $[00q]$L modes. Inserting force constant values ϕ_r^{IC} from stage 3 K-GIC and ϕ_{or}^{ol} for pristine graphite, (4.23) predicts for potassium layers, a localized mode at 22.225 meV, compared to the experimental value of (22.2 ± 0.5) meV for the top branch in KC_{36}. The vibrational amplitude of the defect mode is exponentially damped with a damping constant

$$\alpha = \ell n \left(\frac{2\phi_r^{IC} M_C - \phi_r^C M_I}{M_I \phi_r^{IC}} \right) , \qquad (4.24)$$

which is essentially proportional to the mass difference $M_C - M_I$. Figure 4.11 shows the vibrational amplitude for isolated K, Rb, and Cs layers in graphite as a function of the number of graphite planes s away from the intercalate layer. Due to the large mass mismatch, the K vibrational amplitude is most strongly attenuated and the wave function of the localized mode does not extend much further than one graphite layer from the origin. This explains the weak dispersion of the K optic branch in the stage 3 K-GIC. On the other hand, the Cs layers in compounds with $n \geq 2$ have an effective mass of 11.08 a.u., which is closely matched to M_C. Therefore, the local vibration can propagate over several graphite planes. Consequently, the top optic branch of CsC_{36} should exhibit more dispersion, which is indeed the case.

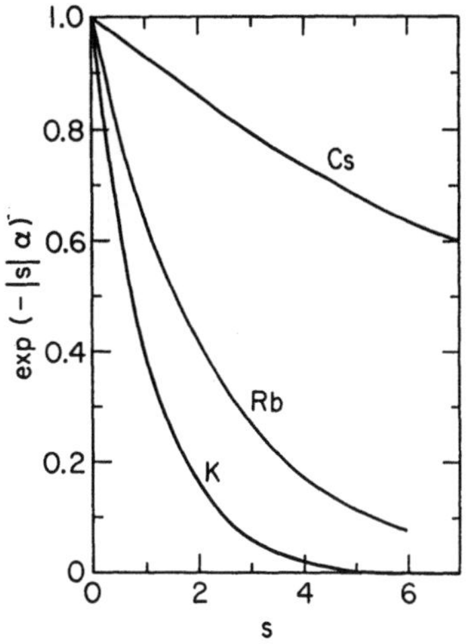

Fig. 4.11. Amplitude of longitudinal vibrations of isolated intercalate layers in graphite vs the number of graphite layers away from the intercalates at the origin

The effect of phonon confinement, which is nicely demonstrated here for the alkali-metal GICs, has also been observed and studied in great detail in semiconductor superlattices [4.31].

All solid lines in Fig. 4.10 represent fits of the data to the one-dimensional lattice dynamical models discussed previously. The best fit force constants are listed in Table 4.3. The fits are usually rather good, the only significant deviation occurring in stage 1 CsC_8 at the zone center and boundary. This deviation may be caused by the large compressibility of the Cs electronic shell [4.32], which is not properly taken into account by the present simple lattice dynamical models.

The Li-GICs occupy a special place among the alkali-metal compounds because of the very small effective mass of the intercalate atoms, the strong interaction of the Li atoms with the carbon atoms in the direction perpendicular to the planes, and the variation of the layer stoichiometry which is possible to a greater extent in these compounds than it is in the heavy alkali-metal GICs. Phonon dispersion measurements of the stage 1 LiC_6 compound were reported by *Rossat-Mignod* et al. [4.33], and a more complete study was provided by *Zabel* et al. [4.34]. The $[00q]$L phonon dispersion as measured by *Zabel* and co-workers is shown in Fig. 4.12, and reflects strong metal–graphite interlayer interaction. The optic branch was located at the very high energy of about 56 meV and was verified by several experimental methods, including careful structure factor measurements.

The stage 2 compounds with Li exhibit the unique property of a tunable in-plane intercalate concentration, which enables one to study the phonon energies as a function of the intercalate concentration. The $[00q]$L dispersion of LiC_{12} and LiC_{18} are shown in Fig. 4.13 from the measurements of *Woo* et al. [4.36]. The compound with the smaller intercalate number density, LiC_{18}, exhibits the softer phonon energies, in line with the expectation that the strength of the interplanar interaction scales with the number of metal atoms per unit area. For both compounds the highest optic branch has not been determined, making

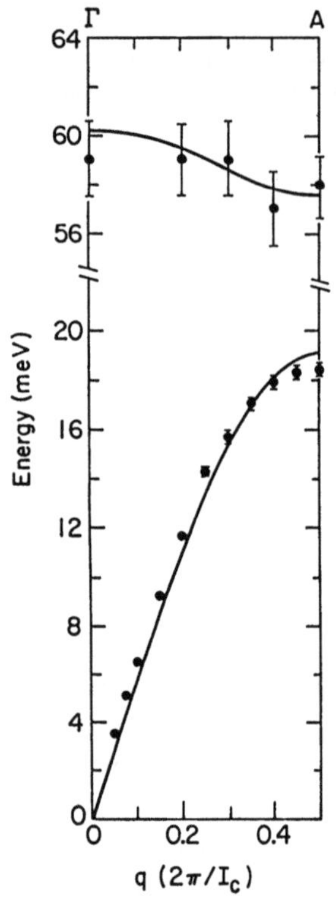

Fig. 4.12. [00q]L phonon dispersion of LiC_6. The solid line is a best fit from a model calculation. (From [4.34])

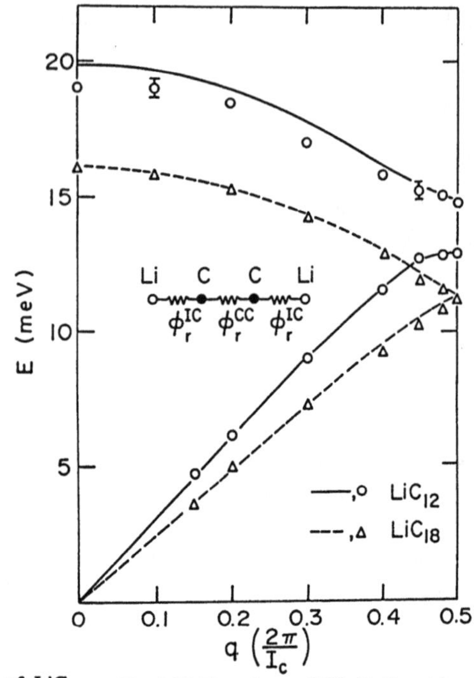

Fig. 4.13. Experimental [00q]L dispersion of stage 2 Li intercalated in graphite with different in-plane concentrations. (o) LiC_{12}; (Δ) LiC_{18}. (From [4.36])

a reliable fit to the data essentially impossible. *Woo* et al. [4.36] used only two force constants for fitting the acoustic and first optic branches, which is clearly not sufficient in the case of donor compounds, and discrepancies between measured and calculated phonon energies are noticeable in particular for LiC_{12}.

All force constants obtained from fits to the measured phonon energies are listed in Table 4.3. This table also contains experimental values for the elastic constant C_{33} and sound velocities $\mathcal{V}$ as determined from the acoustic branches.

(b) Trilayer Compounds

A particularly interesting class of layered structures is formed when the amalgams Hg, Tl, Bi are intercalated together with alkali-metal atoms. This co-intercalation process results in the formation of trilayer ternary intercalation compounds, consisting of heavy metal layers, sandwiched between layers of alkali-metal atoms in a layer sequence $CAHAC\ldots$, where A refers to the alkali-metal and H to the heavy metal layers. Trilayer compounds have also been reported in which monatomic hydrogen layers sit in the midplane between alkali-metal layers [4.36]. The synthesis, and the structural and physical properties of these trilayer GICs

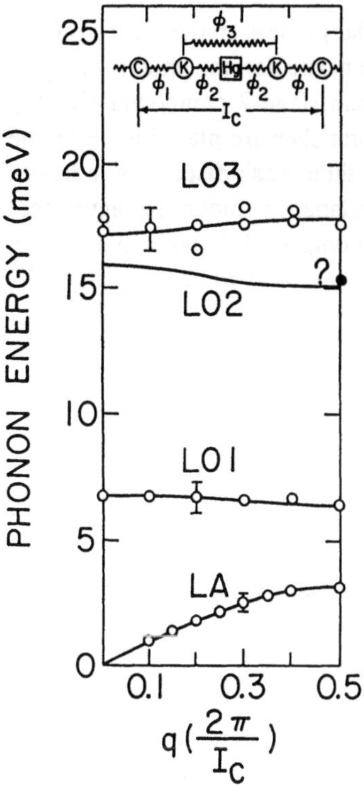

Fig. 4.14. Phonon dispersion for [00q]L modes in the trilayer stage 1 $KHgC_4$ compound. The inset shows the one-dimensional lattice dynamical model used for calculating the phonon energies. (From [4.38])

have recently been reviewed by *Solin* and *Zabel* [4.37] and are also described in Chap. 8 of this volume by *Lagrange* and *Setton*.

Kim et al. [4.38] conducted a detailed investigation of the [00q]L dispersion of the stage 1 trilayer compound $KHgC_4$. The c-axis repeat distance comprises four layers, giving rise to one acoustic and three optic branches. As shown by the data in Fig. 4.14 from *Kim* et al. [4.38], all these branches were determined, apart from the second optic branch for which only a weak indication from a zone center scan was found. The acoustic branch lies much lower than is usually the case in the donor compounds because of the large Hg contribution to this mode. This mode is well separated from the high-frequency optic modes at 16 meV and 17 meV, in which the K displacements dominate, and a graphite-like branch at 7 meV. The solid lines represent best fits of a BvK model to the data using force constants ϕ_1 and ϕ_2 between nearest neighbor C-K and K-Hg layers, and a second neighbor force constant ϕ_3 joining the potassium planes on either side of a mercury layer. These force constants are schematically shown in the inset of Fig. 4.14, and their values are listed in Table 4.3. The values of ϕ_1 and ϕ_2 are roughly consistent with the idea of an equal share of charge transferred from the potassium layers to the neighboring C and Hg layers. On an absolute scale, ϕ_1 is about half the value determined for the C-K interaction in KC_8, indicating a softening of the interlayer interaction in the ternary compound. *Kim* et al. [4.38] describe this interaction as being more metallic and less ionic than the C-K interaction in the binary alkali-metal compounds.

Kamitakahara et al. [4.39, 40] carried out similar phonon studies of the [00q]L modes in K-Hg trilayer compounds, comparing, in particular, the phonon energies in the stage 1 compound $KHgC_4$ with those in the stage 2 compound $KHgC_8$. These authors find that the potassium-like optic branches are placed at about the same energy of 17.7 meV in both compounds and their weak dispersion indicates a small force constant ϕ_3 describing the K-K interplanar interaction, in agreement with the results of *Kim* et al. [4.38] on the stage 1 compound. In both compounds, the zone boundary acoustic phonon has the same low energy of about 2.5–3 meV, which is due to the heavy mass of the Hg layer.

It is interesting to compare the vibrational energies of the Hg layer for displacements in and out of the plane. The out-of-plane vibrational energy can be estimated from the zone boundary acoustic frequency as about 3 meV, whereas the in-plane vibrational energy is about 9 meV, as determined from phonon density-of-states measurements [4.39]. The ratio of the in-plane to the out-of-plane vibrational energy is opposite to the general observation in binary alkali-metal GICs, where this ratio is about 0.6–0.7. We will come back to this point later in Sect. 4.4. Neutron scattering experiments on the Tl and Bi trilayer compounds have not been performed to date, but they would be of considerable interest for comparative studies.

Some phonon properties of the hydrogen-derived trilayer compound $KC_4H_{0.8}$ were recently investigated by *Kamitakahara* et al. [4.41]. From measurements of the acoustic [00q]L branch, they inferred a weakening of the interlayer interaction as compared to KC_8 (Table 4.3). *Kamitakahara* et al. also measured the density of states for vibrations of the intercalated hydrogen atoms parallel and perpendicular to the layers and compared those with the density of states of free KH. Due to the light mass of the hydrogen atoms, the intercalate modes are located at the very high energies of 93 and 71 meV, respectively [4.41], comparable to vibrational energies of hydrogen atoms in transition metals [4.42]. In free KH the corresponding hydrogen modes are at 110 and 66 meV [4.41]. The observed energies and anisotropies of the hydrogen vibration in both materials are quite close, indicating that the K-H potential and the local symmetries are similar in the intercalated compound and in the ionic solid K^+-H^-, with the former providing a slightly less tetragonal potential.

(c) Acceptor Compounds

In acceptor compounds, the intercalate layers often have an internal structure similar to the trilayer compounds discussed above. For instance, in transition metal chloride compounds, the intercalate consists of three layers, i.e., a transition metal layer bonded on either side to a layer of chlorine ions. It is customary to consider the trilayer structure chlorine–metal–chlorine as one rigid unit, which reduces the dynamical matrix to a $(n+1)\times(n+1)$ matrix as for the alkali-metal GICs. Only the compounds with the halogen atoms I, Cl, and Br form a simple monolayer structure similar to the alkali-metal compounds.

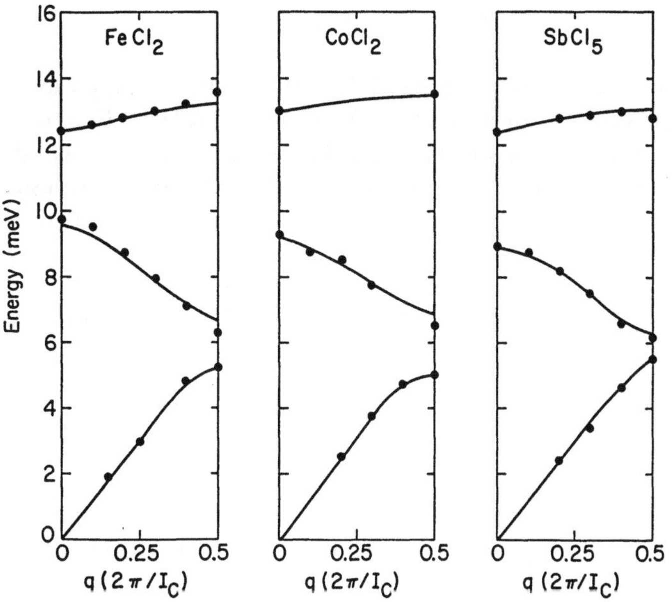

Fig. 4.15. Phonon dispersions for [00q]L modes of stage 2 acceptor graphite intercalation compounds. $FeCl_3$ from [4.43], $CoCl_2$ from [4.45] and $SbCl_5$ from [4.44]

Dresselhaus et al. [4.43] investigated the [00q]L modes of a stage 2 $FeCl_3$ GIC, *Eklund* et al. [4.44] studied the same modes in the stages 2 and 4 of an $SbCl_5$ GIC, and recently *Zabel* et al. [4.45] determined the [00q]L dispersion of a stage 2 $CoCl_2$ GIC. The dispersion of all three stage 2 compounds, which are reproduced in Fig. 4.15, exhibit the $n+1$ phonon branches as expected from a rigid intercalate unit. In none of the measurements have the internal dynamics of the metal-chlorine intercalate layers been revealed. This is a bit surprising since, for instance, in $FeCl_3$ GIC the modes with dominantly Fe-Cl displacements should be located at 17.5 meV and 22.5 meV, as estimated from the $FeCl_3$ crystal. Theses energies are easily attainable by inelastic neutron scattering.

Lattice dynamical calculations of the [00q]L dispersion with one-dimensional BvK models show that the force constant ϕ_r^{IC} is consistently softer in these metal-chloride intercalation compounds than in the alkali-metal GICs, while the force constants for the inner graphene layers are roughly the same and close to the value accepted for pristine graphite. The reduced values of ϕ_r^{IC} reflect the fact that the charge transfer in these acceptor compounds is considerably smaller than in the donor compounds. In all these calculations it was assumed that the metal-chloride trilayer structure moves as a rigid unit, which eliminates the upper two optic branches. This assumption seems not to harm the quality of the fit. A striking difference as compared to the donor compounds is the fact that the fits do not require a second neighbor force constant ϕ_r' connecting the graphite bounding layers across the intercalate layers. This indicates that the electronic shells in the transition metal chlorides are more rigidly bound to the ion core than in the alkali-metals. The best fit force constants and elastic constants are listed in Table 4.3.

The only phonon investigation of a metal-fluoride GIC was recently carried out by *Vaknin* and *Fischer* [4.46] on the stage 2 compound $C_{16}AsF_5$. Since only two out of three branches were observed, a complete lattice dynamical analysis was not possible. However, the elastic constant C_{33} derived from the acoustic branch confirms the overall trend that the c-axis interaction is in general softer in acceptor compounds than in donor compounds.

Eklund et al. [4.47] conducted a detailed investigation into the structural and vibrational properties of D_2SO_4 intercalated in graphite. These compounds are unique among the acceptor compounds since they allow a continuous variation of the charge within a given stage. They therefore provide an excellent opportunity to study the role of charge transfer on the physical properties and in particular on the lattice dynamics. The intercalation was carried out in situ in an electrochemical cell placed on a neutron spectrometer. Phonon spectra were recorded at either extreme of charge transfer allowed in these stage 1 and 2 compounds, and they are reproduced in Fig. 4.16. The phonon energies renormalize in both stages in the expected way: with increasing charge transfer the c-axis shrinks and the phonons stiffen. A 1d BvK model was used to describe the phonon dispersion, again assuming rigid intercalate sheets. *Eklund* et al. [4.47] showed that relaxation of this assumption does not significantly degrade the quality of the fit to the two lower phonon branches; ϕ_r^{CC} was found to be nearly identical to ϕ_{or}^{o1} in pristine graphite, while the values for ϕ_r^{IC} in both stage compounds turned out to be almost the same for the same amount of charge transfer.

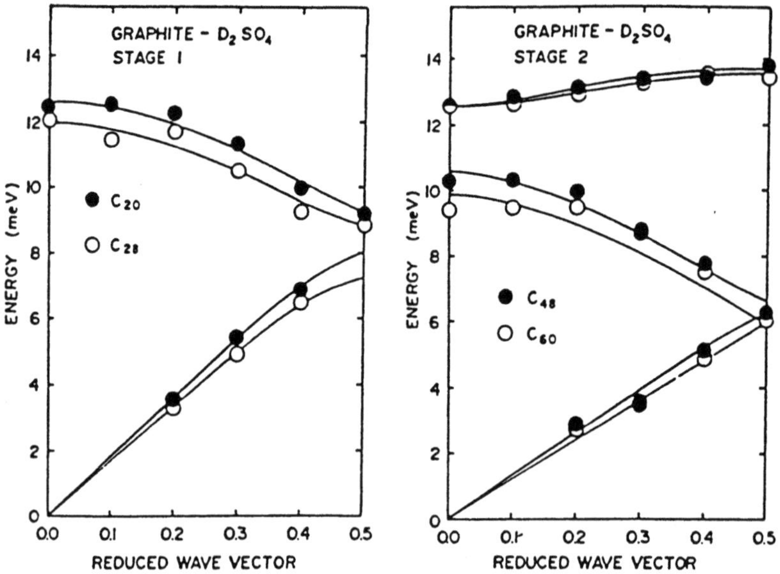

Fig. 4.16. [00q]L phonon dispersions of stage 1 and 2 D_2SO_4 showing the effect of charge transfer on the phonon energies. (From [4.47])

The phonon dispersion of the stage 2 halogen intercalation compound $C_{14}Br$ was investigated by *Simon* et al. [4.15] and analyzed in the usual way. The interlayer force constants not only confirm the general trend of a reduced charge transfer in acceptor compounds, but also uncover an extremely weak C-Br bonding of 945 dyne/cm, the weakest bonding measured yet for any intercalation compound.

(d) Substitutional Compounds

Ammonia molecules are soluble in K-GICs, forming stage 1, 2 and even higher stage compounds depending on the ammonia concentration [4.48, 37]. Recent X-ray [4.49] and neutron [4.50] structural investigations have revealed that the $K\text{-}NH_3$ intercalate layer is a 2d liquid at room temperature and that the liquid structure factor can be reasonably well described by a model in which four ammonia molecules are loosely associated with a single K^+ ion while the rest of the ammonia molecules can be considered free.

The [00q]L phonon modes of stage 1 $K(ND_3)_{4.3}C_{24}$ were measured by *Neumann* et al. [4.51] and are shown in Fig. 4.17. These measurements were carried out with deuterated ammonia to reduce the incoherent scattering and enhance the coherent cross section, thereby allowing the measurement of the phonon dispersion. Three branches are clearly observed, the acoustic branch obviously being split by the hybridization with an Einstein-like mode at 7 meV. *Neumann* et al. [4.51] associated this mode with the libration of the ammonia molecules within the host gallery. A zone center scan reveals a broad feature centered at 7 meV due to the additional excitation. Thus the librational mode is essentially

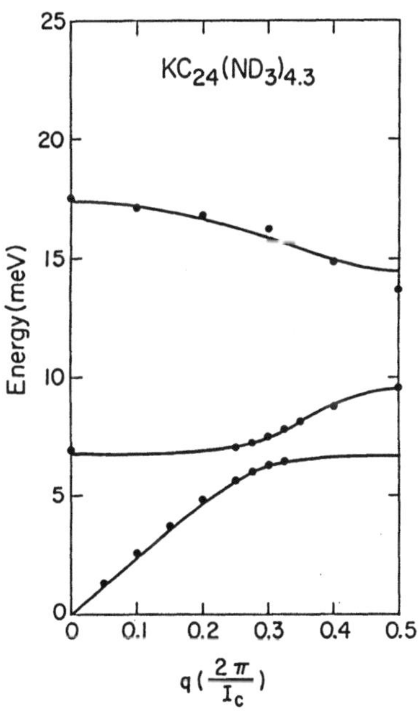

Fig. 4.17. The [00q]L phonon dispersion of the stage 1 compound $KC_{24}(ND_3)_{4.3}$. The solid lines represent a fit of the data to the one-dimensional Born–von Kármán model including a phonon–libron coupling term describing the hybridization of the acoustic branch with a libronic excitation at 7 meV. (From [4.52, 53])

dispersionless along the c-axis, indicating a very weak interaction between ND_3 molecules in adjacent layers. This finding is in accordance with the complete absence of intercalate interlayer correlations as manifested by structural investigations [4.49,50].

Neumann et al. [4.51] modeled the observed dispersion relation by a 3×3 dynamical matrix that includes a virtual crystal, one-force-constant BvK model, an Einstein oscillator of energy α representing the libration, and a term containing the coupling constant γ:

$$\left(\begin{array}{cc:c} \dfrac{2\phi_r^{IC}}{M_C} & \dfrac{-\phi_r^{IC}}{\sqrt{M_I M_C}}(1+e^{iqI_c}) & -\gamma\alpha\left(1-e^{i2qI_c}\right) \\ \dfrac{-\phi_r^{IC}}{\sqrt{M_I M_C}}\left(1+e^{-iqI_c}\right) & \dfrac{2\phi_r^{IC}}{M_I} & 0 \\ \hdashline -\gamma\alpha\left(1-e^{-i2qI_c}\right) & 0 & \alpha^2 \end{array} \right) . \qquad (4.25)$$

All masses and force constants have their usual meaning. The upper left corner, divided off by the dotted line, represents the uncoupled phonon dispersion. The best fit parameters corresponding to the dispersion relations shown as the solid lines in Fig. 4.17 are $\phi_r^{IC} = 2090$ dyn/cm, $\alpha = 6.7$ meV, and $\gamma = 0.70$ meV. The interlayer force constant ϕ_r^{IC} is smaller by a factor of two than the value obtained for the binary stage 1 compound KC_8, see Table 4.3. This decrease is mostly due to the c-axis layer expansion in the ternary compound as compared to the binary compound and partially also to the back transfer of electrons from the graphite π^* band to the intercalate bands [4.48].

The actual physical motion corresponding to the observed librational excitation has not been unambiguously determined yet. However, it is most likely that oscillations of the C_3 molecular symmetry axis in the direction parallel to the c-axis are responsible for the mode splitting. *Neumann* et al. [4.51] demonstrated, by considering a simple model of an oscillating dipole, that in order to account for a librational energy of 7 meV the center of rotation must be located 1.4 Å away from the center of the ND_3 molecular mass, placing the axis of rotation between the potassium and ammonia. The observed excitation was therefore tentatively named a K-ND_3 "bond-bending" mode [4.51]. *Resing* et al. [4.52] pointed out that including a charged graphite layer above and below the ammonia molecule would increase the vibrational frequency by about 10 %. The complete dynamics of the ammonia molecules including rotational and diffusional excitations have recently been reviewed by *Zabel* and *Neumann* [4.53], and by *Magerl* in Chap. 6 of this volume.

(e) Disorder

Graphite intercalation compounds provide a unique testing ground for the investigation of the effect that disorder has on the lattice dynamics of layered structures. We have to distinguish between two kinds of disorder; a lateral and a c-axis-type disorder. Lateral disorder is introduced by mixing two different atoms or one atom and a vacancy within an intercalate layer while maintaining a proper stage. The second kind of disorder occurs by mixing the stage of a particular intercalation compound.

Neumann et al. [4.54] investigated the lattice dynamics in the presence of lateral disorder by studying the [00q]L dispersion in the mixed stage 1 compound $K_{1-x}Rb_xC_8$. *Chow* and *Zabel* [4.55, 56] recently argued that these ternary compounds can be regarded as two-dimensional binary alkali alloys with complete substitutional solubility and site disorder on a (2 × 2)R0° superlattice, interleaved by graphene layers. These compounds have generated much interest in recent years because of the anomalies which occur as a function of the binary alkali-metal composition x. *Billaud* and *Hérold* [4.57] and later *Chow* and *Zabel* [4.55, 56] investigated the composition dependence of the interlayer spacing and found a pronounced deviation from Vegard's law with an anomalous steep slope occurring at $x = 2/3$. Strong paramagnetic signals were observed at the same composition [4.58]. Striking correlations between the repeat distance and the susceptibility have also been seen for the other two ternary stage 1 compounds $Rb_{1-x}Cs_xC_8$ and $K_{1-x}Cs_xC_8$.

In $K_{1-x}Rb_xC_8$, the interlayer force constants of the end members ($x = 0$ and 1) are almost identical (Table 4.3), their main difference being the different effective intercalate masses. *Neumann* et al. [4.54] measured the phonon dispersion for the compositions $x = 0, 0.3, 0.45, 0.65, 0.92$ and 1.0 and found that for all compositions investigated the optic and acoustic phonon branches showed a normal dispersion. The [00q]L dispersion for the composition $x = 0.45$ is reproduced in Fig. 4.18. No signs of local or gap modes have been observed for

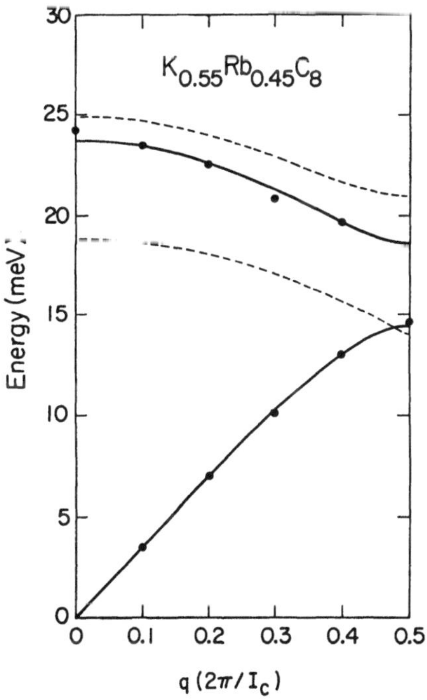

Fig. 4.18. Phonon dispersion curves of the [00q]L modes of the ternary compound $K_{0.55}Rb_{0.45}C_8$. The solid lines represent fits to a one-dimensional Born-von Kármán model with composition-dependent force constants $\phi_r^{IC}(x)$ and $\phi_r^{CC}(x)$ as shown in the inset of Fig. 4.19. The dashed lines show the position of the optic branches in KC_8 (upper curve) and RbC_8 (lower curve). (From [4.54])

any of the compositions. Clearly, the difference of the two alkali-metal masses is rather small, and the development of a localized state would only be marginal.

Based on their experiments, *Neumann* et al. [4.54] described the phonons in the mixed compound by a virtual crystal model, in which the vibrational Hamiltonian is replaced by its average using an average effective mass for the alkali-metal layers:

$$M(x) = [(1 - x)M_K + xM_{Rb}] . \tag{4.26}$$

Using a Born–von Kármán model with first and second nearest neighbor interactions as discussed previously, *Neumann* and co-workers were able to fit the measured dispersion and to determine the composition dependence of the force constants $\phi_r^{IC}(x)$ and $\phi_r'(x)$. The salient feature of their results shown in Fig. 4.19 is the strong composition dependence of both force constants. From the best fit force constants as well as from the initial slope of the acoustic branch, *Neumann* et al. [4.54] derived the layer compressional elastic constant C_{33} as a function of the composition and found an anomalous softening at $x \cong 2/3$, which coincides with the structural and magnetic anomalies mentioned before. They attributed

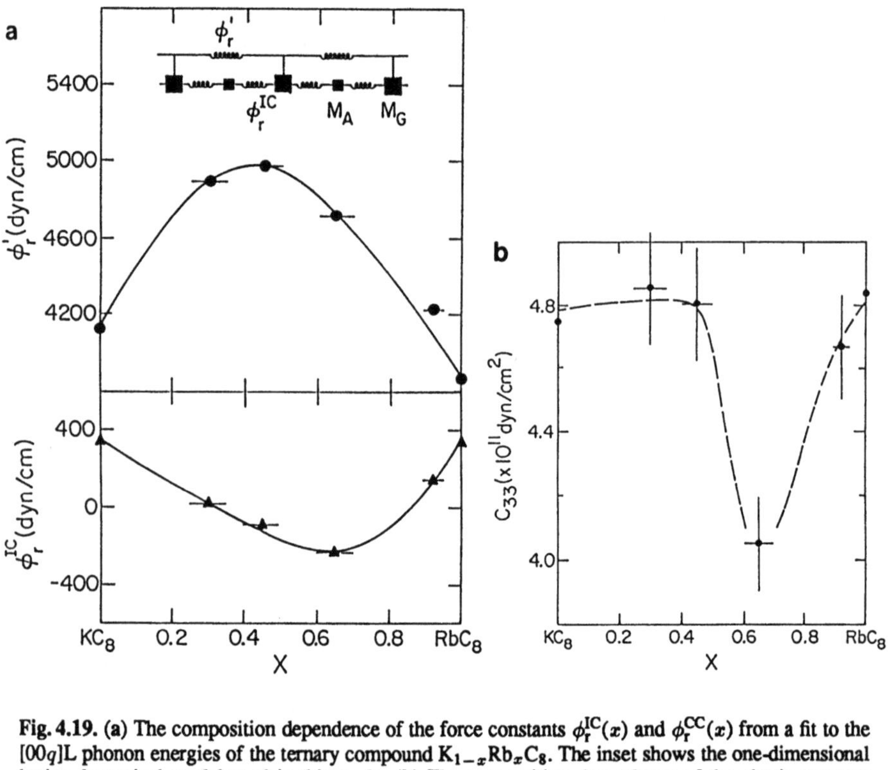

Fig. 4.19. (a) The composition dependence of the force constants $\phi_r^{IC}(x)$ and $\phi_r^{CC}(x)$ from a fit to the $[00q]$L phonon energies of the ternary compound $K_{1-x}Rb_xC_8$. The inset shows the one-dimensional lattice dynamical model used in this study. (b) The composition dependence of the elastic constant C_{33}, calculated from the initial slope of the acoustic branch and from the force constants, exhibits a strong anomaly at the composition $x = 0.67$. The dashed line is a guide to the eye. (From [4.54])

the force constant variation and the elastic anomaly to a composition-dependent charge transfer between the alkali-metal intercalate layer and the graphene planes. More recently, *Akera* and *Kamimura* [4.59] have explained these anomalies in terms of a cusp-shaped Van Hove singularity that occurs in the electron density of states at the Fermi level for the composition in question.

The *c*-axis-type disorder requires a random mixing of various stages and this can be generated in a controlled fashion close to staging transitions from stage n to $n \pm 1$ [4.60, 61] (see also Chap. 3). Stage disordered compounds provide model systems for the investigation of phonons in random linear chains [4.62]. The dispersion relation and the localization of phonons in stage disordered compounds has recently been investigated theoretically by *Hawrylak* and *Williams* [4.63], but at present experimental data are absent. There have been several reports of unintentional neutron measurements on mixed stage compounds, but none of them focused on the specific nature of the lattice dynamics in disordered linear chains.

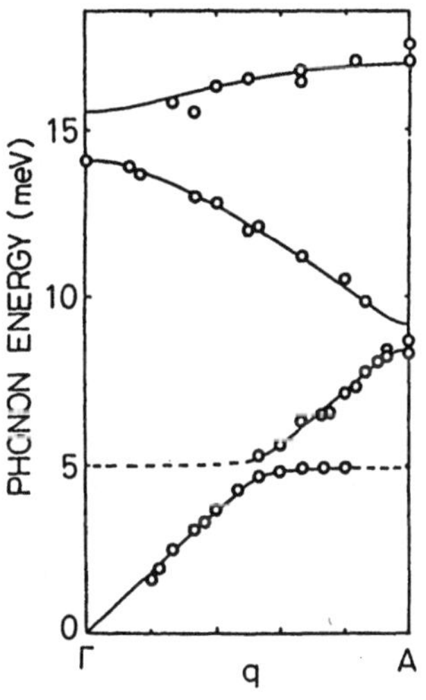

Fig. 4.20. Phonon dispersion of the [00*q*]L modes in RbC_{24} at 80 K. The solid lines represent a fit to a one-dimensional Born–von Kármán model including a transverse Einstein oscillation of the Rb atoms, causing the splitting of the acoustic branch. (From [4.64])

(f) Temperature Dependence

The temperature dependence of the [00*q*]L modes has been studied in only a few cases. *Funahashi* et al. [4.64, 65] measured the [00*q*]L phonon dispersion of RbC_{24} between 40 K and 400 K, and noticed a splitting of the acoustic branch at 5 meV in the temperature range from 40 K to 130 K, which closes at about 250 K. The phonon dispersion at 80 K is reproduced in Fig. 4.20. *Magerl* et al. [4.66] reported similar temperature-dependent studies of the layer breathing modes in KC_{24}, where they observed a mode splitting of the acoustic branch at 8 meV for measurements taken at 78 K and 208 K. No splitting was seen at 300 K, in agreement with earlier room temperature measurements of the same phonon dispersion. There is a striking resemblance of these low temperature mode splittings to those observed earlier of the acoustic bending mode at roughly the same energy in KC_{24} [4.67]. This suggests that the splitting is due to a localized mode hybridizing all phonon dispersions having a displacement vector perpendicular to the planes.

A further hint to the origin of the splitting comes from considering the ratio of the splitting energies in KC_{24} and RbC_{24}, which is close to the square root of the mass ratio (M_{Rb}/M_K), indicating that it may be induced by the intercalate atoms. It is therefore very likely that in-plane vibrations of the alkali-atoms couple to the out-of-plane components of the phonon dispersion. A physical picture for this coupling comes from the low temperature structure of the intercalate atoms. At low temperatures, the alkali-metal intercalates order in a discommensuration domain structure (Chap. 2), in which the atoms close to the domain boundaries are shifted off the centers of carbon hexagons provided by the graphene substrate potential. Alkali-atoms which are located over carbon–carbon bonds may oscillate between potential wells on either side of the bond, thereby inducing a displacement perpendicular to the planes. Above the order–disorder transition temperatures, which are 123 K and 165 K in KC_{24} and RbC_{24}, respectively, the intercalate layers transform steadily into liquids with diffusional motion of the alkali-metal intercalate atoms dominating over vibrational motions. At those temperatures, we expect the in- and out-of-plane coupling to cease, in agreement with the observations. Einstein-like modes in stage 2 Rb and Cs but not in K compounds have been predicted by *Suganuma* et al. [4.68] in order to explain specific heat anomalies. The specific heat data were described by the usual electronic and vibrational contributions. However, a harmonic-oscillator-type excitation with an Einstein temperature of 51 K or 4.4 meV for RbC_{24} had to be taken into account in order to fit the data. This frequency is very close to the mode splitting energy of 5 meV.

Funahashi et al. [4.64, 65] modeled the phonon dispersion of RbC_{24} in the usual way using three force constants ϕ_r^{CC}, ϕ_r^{IC} and ϕ_r', and the splitting was introduced by coupling the out-of-plane displacement to an Einstein-like in-plane mode of frequency $\omega = 5\,\text{meV}/\hbar$ via a coupling constant γ. The phonon dispersion was then obtained by solving a 4×4 dynamical matrix which includes the 3×3 matrix reproduced in (4.16) and a term containing the Einstein oscillator frequency. The fit to the data points is shown in Fig. 4.20 by the solid lines and reproduces very well the hybridization of the acoustic branch.

Both groups also investigated the low temperature behavior of the $[00q]$L phonon dispersions in the stage 3 compounds KC_{36} and CsC_{36} [4.66, 30], but could not identify any mode splittings. This may be taken as a sign of an intercalate in-plane structure which is more in registry with the substrate in the stage 3 than in the stage 2 compounds, see Chap. 2.

4.2.3 Interlayer Force Constants

From the available data on the radial interlayer coupling constants in graphite intercalation compounds, the following general conclusions can be drawn:

1. The constant ϕ_r^{CC} is almost always equal to ϕ_{or}^{ol} in pristine graphite, independent of the stage of the compound and of whether the compound is a donor or an acceptor compound. The fluctuations observed are still within the error bars of the force constant determination and seem not to indicate systematic trends. Therefore ϕ_r^{CC} is rather charge insensitive unlike the *intra*planar force constants of graphite, whose charge sensitivity has been manifested by infrared and Raman scattering experiments described in Chap. 5. An exception seems to be the graphite coupling constant in stage 2 LiC_{12}, which appears unusually high. But this may be the result of an inaccurate fit.

2. Donor compounds: In stage 1 compounds, $\phi_r^{IC} > \phi_{or}^{ol}$ and drops with increasing stage. In some cases, ϕ_r^{IC} becomes even smaller than ϕ_{or}^{ol}. Many of the changes observed are merely due to changes of the intercalate layer stoichiometry. Evaluation of the force constants per intercalate atom or molecule rather than per carbon atom does indeed reveal a smooth variation of the interlayer interaction with compound and stage, as can be seen from Table 4.4. The Cs compounds, however, behave anomalously, since ϕ_r^{IC} is larger in the third stage than in the first stage. The force constants for the stage 2 Li compounds are not reliable for reasons mentioned before.

Table 4.4. Force constant for alkali-metal–graphene interlayer interaction from neutron scattering experiments and calculated per alkali-atom

Compound	$\phi_r^{IC}\xi$[dyn/cm]	Compound	$\phi_r^{IC}\xi$[dyn/cm]
LiC_6	44 670	RbC_8	30 800
LiC_{12}	60 240	RbC_{24}	30 060
LiC_{18}	51 660		
KC_8	33 120	CsC_8	31 336
KC_{24}	32 808	CsC_{24}	28 560
KC_{36}	32 088	CsC_{36}	32 580

The ratio ϕ_r'/ϕ_r^{IC} in stage 1 compounds increases steadily from LiC_6 to CsC_8, indicating that the electronic shell becomes softer with increasing atomic radius. The same trend can be observed in the stage 2 compounds. The exception is CsC_{36}, for which the ratio $\phi_r'/\phi_r^{IC} = 0.066$ appears too small.

3. Acceptor compounds: ϕ_r^{IC} is typically much softer in acceptor compounds than in donor compounds, reflecting the reduced charge transfer. Particularly soft interlayer force constants are found in $SbCl_5$ and Br-GICs, where ϕ_r^{IC} is less than half the interlayer force constant of graphite. In contrast to donor compounds, ϕ_r' is essentially zero in acceptor compounds, reflecting a more rigid electronic shell in the latter compounds.

A word of caution may be in order here. The force constant determination is only reliable when *all* $n+1$ branches have been determined. This is particularly important for donor compounds, for which the force constants ϕ_r' and ϕ_r^{IC} become highly correlated in the fits. Because of this correlation, the accuracy of the force constants is on the order of 5%–10%. There is therefore a danger of over-interpretation of small changes in the force constants.

Woo et al. [4.36] explain the different c-axis compressibilities of stage 2 donor compounds by a combined elastic and electrostatic model. They assume that the contribution of the inner graphene layers to the compressibility is the same as in pristine graphite and is given by K_G, while the contribution from a C-I-C sandwich is represented by an electrostatic term K_e. The total compressibility then follows from the combination

$$K_{tot} = (I_G K_G + d_1 K_e)/(I_G + d_1) , \tag{4.27}$$

where I_G and d_1 have their usual meanings and $K_{tot} = 1/C_{33}$. The electrostatic contribution was estimated from the Coulomb energy per sandwich and was found to be proportional to the inverse square of the areal charge density σ : $K_e \propto \sigma^{-2}$. Combined with the composition law stated above (4.26), this model predicts a linear relationship between K_{tot} and σ^{-2} with an intercept given by the compressibility of the graphite material. An analysis of the compressibility data of several intercalation compounds and stages indicates that the electrostatic contribution represents an important part of the interlayer interaction between the graphene and alkali-metal intercalate layers.

In a more fundamental way, *DiVincenzo* and *Mele* [4.69] have calculated the structural and elastic properties of stage 1 alkali-metal compounds from a density functional calculation within a Thomas-Fermi approximation of the electronic structure of the intercalation compound. They have modeled the carbon–alkali-metal interaction by a competition between the attractive Coulomb and exchange-correlation interaction and hard core repulsion between the ion cores. The *intra*planar alkali–alkali interaction was simply described by a classical Coulombic interaction. Assuming full charge transfer and no hybridization between metal and carbon states, this ansatz is simple enough to make the underlying physics transparent and still complex enough to be sensitive to the structural and energetic details of the compounds in question. The structural and elastic parameters that *DiVincenzo* and *Mele* [4.69] calculated from this theory are in general agreement with experimental data and reproduce the correct trend. The calculated compressional elastic constants C_{33} for the four stage 1 alkali-metal compounds are plotted in Fig. 4.21 and compared to experimental data. The general trend

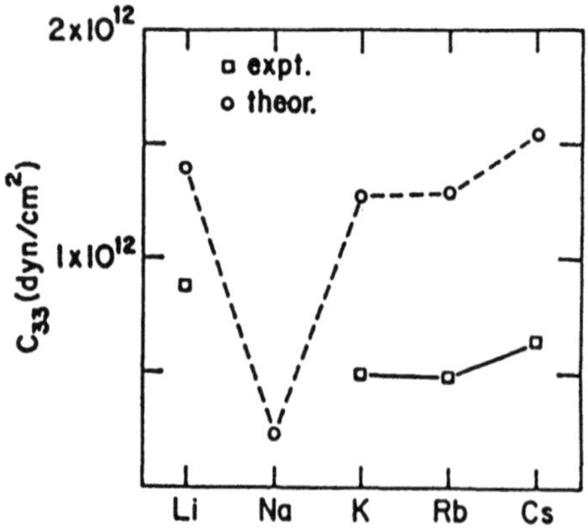

Fig. 4.21. Compressional elastic constant C_{33} calculated for stage 1 alkali-metal compound (○) and comparison with experimental values (□). (From [4.69])

is very well reproduced, in particular, note that C_{33} is largest for LiC_6, drops for KC_8 and RbC_8, and increases again slightly in CsC_{24}. The ionic radius of Li is much smaller than that of the other three metals, such that direct C-plane–C-plane interaction becomes important. The compactness of the Li core is also expressed by the very large core elastic constant ϕ_k, and the enhanced C-plane interaction by a large ϕ_r'. The increase of C_{33} in CsC_8 is also accompanied by an increase in ϕ_r'; however, *DiVincenzo* and *Mele* [4.69] cannot offer a physical explanation for this behavior. They have also calculated C_{33} for stage 1 NaC_8 and found it anomalously small, indicating a very weak interlayer interaction. *DiVincenzo* and *Mele* speculate that this may be one of the reasons that NaC_8 does not form.

4.3 $[q00]T_\perp$ Phonon Dispersion

4.3.1 Introductory Remarks

One of the most intriguing features of the $[q00]T_\perp$ modes is their parabolic dispersion relation at small phonon wavenumbers q, characteristic of layered materials with weak interlayer shear interaction. As already shown by *Lord Raleigh* [4.70], single isolated sheets of homogeneous material exhibit a dispersion for bending modes $\omega \propto q^2$. In reference to these transverse modes, the $[q00]T_\perp$ modes are also referred to as "bending modes" or "rippling modes". Increasing the shear interaction between neighboring layers causes the dispersion to have a nonzero slope at $q \to 0$, so that the small q dispersion of the $[00q]T_\perp$ mode can be written

$$\omega^2 = Aq^2 + Bq^4 \ . \tag{4.28}$$

The slope $(A)^{1/2}$ at $q \to 0$ is given by $(C_{44}/\varrho)^{1/2}$, where C_{44} is the layer shear elastic constant and ϱ is the volume density. This is the same elastic constant which also determines the slope of the acoustic $[00q]$TA branch. While it is very difficult to measure the extremely soft $[00q]$T modes by inelastic neutron scattering, the analysis of the bending modes also yields information on the layer shear interaction. In fact, neutron scattering provides the only reliable source for the shear elastic constant C_{44} in GICs, since ultrasonic methods often are not applicable and Brillouin scattering suffers from the small penetration depth of light in surface regions of the compound which may not represent bulk structural and compositional properties. The term B is the parameter determining the resistance of the layer to bending, and is usually referred to as the "bending stiffness constant", akin to the spin stiffness constant in ferromagnets, determined by the exchange operator J.

Both parameters, B and C_{44}, can be obtained from the acoustic dispersion relation by plotting ω^2/q^2 as a function of q^2. The bending stiffness constant B, deduced from the slope of such a plot, represents the combined bending stiffness of the graphite and intercalate layers, and for a stage n compound is given by

$$B^n = \frac{1}{n+1}\sum_{i=1}^{n+1} B_i \,, \tag{4.29}$$

where B_i is the bending stiffness constant of the individual layer i. The shear elastic constant C_{44} follows from the intercept and is determined by the shear (or transverse) force constants between the graphite and intercalate layer (ϕ_t^{IC}) and between the inner graphene layers (ϕ_t^{CC}):

$$C_{44}^n = \frac{4I_c}{a^2\sqrt{3}}\left(\frac{2}{\phi_t^{IC}} + \frac{n-1}{\phi_t^{CC}}\right)^{-1} . \tag{4.30}$$

Here only nearest neighbor shear force constants have been taken into account. All other parameters have their usual meaning.

Since all experimental data on $[q00]T_\perp$ modes in GICs have been obtained using intercalated HOPG material, constant Q scans are not capable of distinguishing between phonons propagating along the [100] or [110] directions. However, owing to the approximate elastic isotropy of the graphite and intercalate planes, we assume that no essential difference exists for the two directions. Therefore, while we continue with the notation $[q00]T_\perp$, in fact an in-plane orientational average is implied.

The orientational average in HOPG material has the effect that the Brillouin zone boundary is not well defined. This complicates the determination of the dispersion relation at higher q-values, in addition to the fact that for increasing phonon wave numbers the selection rule for determining the transverse mode becomes less well defined, as already alluded to in the introductory chapter.

In order to apply the continuum approximation for $q \to 0$, acoustic phonons at small wave numbers have to be collected. This causes a serious experimental problem in the presence of a large mosaic, since then the proximity of the Bragg peak dominates the intensity at small q values. For high quality $[q00]T_\perp$ data it is therefore of paramount importance to use samples with very good c-axis mosaicity, equal to or better than 1°. Nowadays, HOPG material with a c-axis mosaic distribution 0.5° or less is available. Unfortunately the mosaicity increases somewhat during the intercalation process and great care has to be taken to keep it to a tolerable level, either by intercalating the sample very slowly or by pressure treatment after completing the intercalation to the desired stage.

Early measurements of layer bending modes have been carried out on pristine graphite by *Nicklow* et al. [4.2] and subsequently on some transition metal dichalcogenides, in particular on 2H-MoS_2 by *Wakabayashi* et al. [4.71]. However, only the GICs with a large variety of intercalates and stages permit systematic exploration of these modes and their dependence on phase transitions and charge transfer effects.

4.3.2 Experimental Results

(a) Room-Temperature Measurements

The $[q00]T_\perp$ modes in donor compounds have been investigated by *Kamitakahara* et al. [4.24, 25], *Zabel* et al. [4.67, 34], and *Fischer* [4.72]. Experimental results from room-temperature measurements of stage 1 compounds are shown in Fig. 4.22. In all cases, the quadratic dispersion of the acoustic branch is clearly visible. Even in stage 1 LiC_6, which is the compound with the least layered character among the GICs, the upward curvature of the $[q00]T_\perp$ is still very pronounced. Recently, *Vaknin* and *Fischer* [4.46] and *Zabel* et al. [4.45] determined the bending modes of the stage 2 acceptor compounds with AsF_5 and $CoCl_2$, respectively, which also revealed the characteristic parabolic shape of the

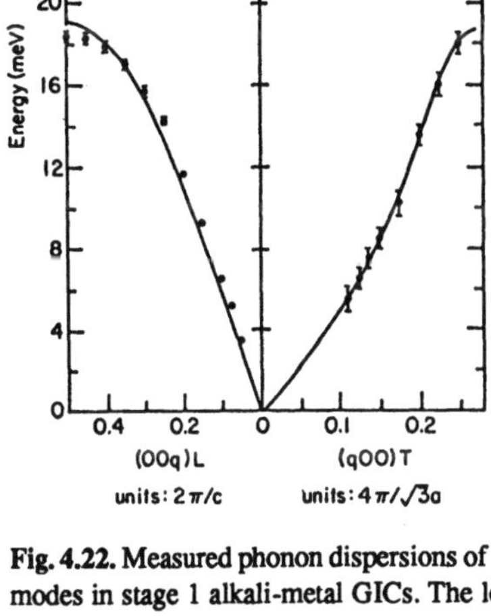

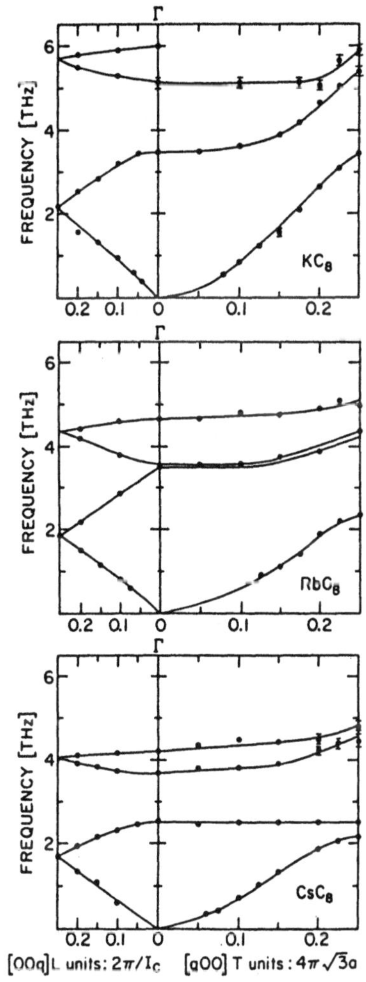

Fig. 4.22. Measured phonon dispersions of $[q00]T_\perp$ modes in stage 1 alkali-metal GICs. The left panels reproduce the $[00q]L$ modes, and the right panels show the bending modes. The figures are compiled from the following references: KC_8 and CsC_8 from [4.67], RbC_8 from [4.23], LiC_6 from [4.34]

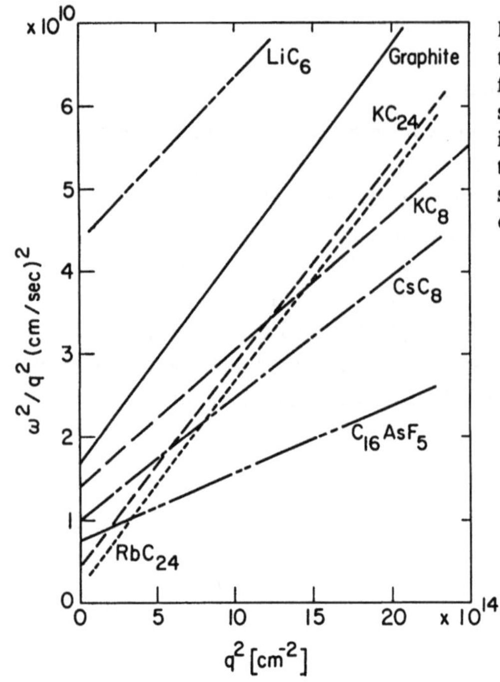

Fig. 4.23. Acoustic phonon frequencies plotted as ω^2/q^2 vs q^2 for pristine graphite and for various donor and acceptor GICs. The slope of the lines is determined by the bending stiffness constant of the layers B, and the intercept by the layer shear elastic constant C_{44}. For the experimental values of C_{44} and B refer to Table 4.5

bending modes. Transverse modes of the stage 2 $C_{14}Br$ compound were reported by *Simon* et al. [4.15].

Plots of ω^2/q^2 as a function of q^2 from the acoustic branches are shown in Fig. 4.23 for most compounds measured so far, and the experimental values for B and C_{44} are listed in Table 4.5. There are a few important trends to notice. In most compounds, C_{44} is reduced relative to its value for pristine graphite. The only exception is LiC_6, where an increase by a factor of 3–4 is seen [4.34, 72]. The most dramatic softening of C_{44} is observed in the stage 2 alkali-metal [4.67] and Br-GICs [4.15]. The softening of the shear elastic constant in the stage 2 donor

Table 4.5. Shear elastic constants and bending modulus of graphite and graphite intercalation compounds

	Stage	$C_{44}[10^{10}\text{dyn/cm}^2]$	$B[10^{-5}\text{cm}^4/\text{s}^2]$	Ref.
Graphite		3.84 ± 0.4	2.55 ± 0.15	[4.2]
LiC_6	1	10.0 ± 1.0	2.06 ± 0.2	[4.34]
		17.3	1.5	[4.72]
KC_8	1	2.82 ± 0.2	1.66 ± 0.1	[4.67]
CsC_8	1	3.07 ± 0.15	1.48 ± 0.05	[4.67]
KC_{24}	2	0.89 ± 0.2	2.45 ± 0.15	[4.67]
RbC_{24}	2	0.44 ± 0.2	2.48 ± 0.1	[4.67]
BrC_{14}	2	0.08		[4.48]
AsF_5C_{16}	2	1.9 ± 0.01	1.21	[4.46]

compounds is in good agreement with Raman results on the [00q]T shear modes by *Wada* et al. [4.73], and has been explained by the incommensurate structure of the intercalates in these compounds [4.74]. This aspect will be discussed in the next section.

DiVincenzo and *Mele* [4.69] have calculated the shear elastic constant from a density functional approach within a Thomas-Fermi approximation. With their simple assumption about the electronic structure, they can successfully reproduce the shear elastic constant C_{44} in stage 1 compounds. For the shear elastic constant they obtain the expression

$$C_{44} = \frac{\pi^2}{\sqrt{3}} \frac{E_1 I_c}{a^4} , \tag{4.31}$$

which relates the shear modulus to the in-plane corrugation energy E_1. All other parameters have their usual meanings. Their theoretical values are compared in Fig. 4.24 with experimental data for the shear modulus. Although there is some disagreement with respect to the numerical values, the overall trend follows the experimental data. In particular, the large shear elastic constant of LiC_6 is correctly predicted by the density functional theory and is a result of the compact compound, yielding a high corrugation energy E_1 of 1.3 eV. *DiVincenzo* and *Mele* also calculate C_{44} for the imaginary stage 1 compound NaC_8, which behaves normally, in contrast to the compressional elastic constant C_{33}, which they found anomalously soft.

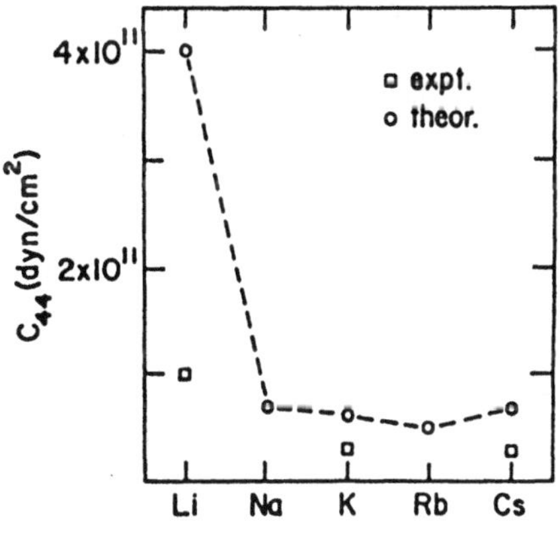

Fig. 4.24. Layer shear elastic constant C_{44} calculated for stage 1 akali-metal compounds (○) compared with experimental values (□). (From [4.69])

The bending stiffness constant appears to be dominated by the resistance of the graphite layer against bending, while the intercalate layers seem more "floppy". Using (4.29), it follows that in the ordered stage 1 compounds B_I is only about 20 % of B_G, whereas in stage 2 compounds the intercalate layers exhibit bending stiffness constants which are more comparable to those of graphite. These estimates are based on the assumption that B_G is the same before and after intercalation, independent of charge transfer effects. Independent estimates come from an analysis of the optic modes, as detailed below. While the dispersion of the acoustic branch is determined by the bending stiffness constant of the graphite *and* intercalate layers, the dispersion of the optic branches is sensitive to the bending modulus of the individual layers [4.67]. For instance, for stage 1 compounds, the dispersion of the optic $[q00]\mathrm{T}_\perp$ modes at small q values can be written

$$\omega^2(q_{x,y}, q_z) = \omega^2(q_z) + Aq_{x,y}^2 + \tfrac{1}{2}(e_\mathrm{G}B_\mathrm{G} + e_\mathrm{I}B_\mathrm{I})q_{x,y}^4 \;, \qquad (4.32)$$

where $q_{x,y}$ and q_z are the wave-vector components for phonons propagating in the (a^*, b^*) plane and along the c^* direction, respectively, e_G and e_I are the displacement amplitudes of the graphite and intercalate planes for the $[00q]\mathrm{L}$ modes, and $\omega(q_z)$ is the frequency of the $[00q]\mathrm{L}$ mode. At the zone center, Γ, the $[00q]\mathrm{L}$ and $[q00]\mathrm{T}_\perp$ modes exhibit the same symmetry. Therefore, the frequency of any transverse mode having a finite q-component is composed of layer breathing and layer bending modes. For instance, in stage 1 compounds at $q_z = 0.5$, which is at the Γ-point in a zone folded scheme, the lower energy is due to vibrations of the layer with the larger mass, and increasing the in-plane component $q_{x,y}$ causes this particular layer to bend in addition, while the other, lighter layer remains unbent. In $\mathrm{CsC_8}$, the intercalate layer is heavier than the graphite layer and therefore the quadratic component of the dispersion of the first optic branch is given by the bending modulus of the Cs layer. The fact that the dispersion of this branch is negligibly small lends itself to the suggestion that B_I of the metal layer is vanishingly small. Conversely, the dispersion of the second optic branch is determined by B_G and exhibits the same dispersion as the acoustic branch. Similar arguments can be made for $\mathrm{KC_8}$, however, in reversed order, since the K layers are lighter than the graphene layers. Therefore, the dispersion of the second optic branch is alkali-metal-like and is flat until the branch is required to bend due to the hybridization with the graphite-like first optic branch. An intermediate case can be found in $\mathrm{RbC_8}$ due to the similar effective masses of the graphene and Rb layers, such that a distinction between intercalate and graphene dispersion cannot clearly be drawn. All these observations taken together show that the quadratic dispersion of the acoustic $[q00]\mathrm{T}_\perp$ mode is predominantly due to the resistance to the bending of the graphite layers rather than of the alkali-intercalate layers, for which the bending stiffness constant is essentially zero.

Similar arguments can be presented for the dispersion relations in stage 2 compounds; however, here the intercalate and graphite modes are more mixed at the zone center.

In acceptor compounds with weak in-plane interaction between the intercalate molecules, we also expect the bending modulus B_I to be small. An indiciation of this is seen in the results of *Vaknin* and *Fischer* on stage 2 $\mathrm{AsF_5}$ [4.46]. In some cases, for instance the transition metal chlorides, the intercalate preserves its crystal integrity, which should result in considerably larger B_I values. This effect has not been investigated yet, partly because of the large mosaicity of most of these compounds.

The partial ionicity of the alkali-metal GICs along the c-axis has been interpreted by *Neumann* et al. [4.75] as a hint of a possible Lyddane-Sachs-Teller (LST)-type splitting between the zone-center optic $[00q]$L and $[q00]\mathrm{T}_{\perp}$ modes, which are polarized parallel to the c-axis. Careful measurements of the optic branches close to the Γ point in CsC_8, however, did not support the idea of such a LST splitting, indicating that it has been quenched by screening from free c-axis carriers.

(b) Temperature Dependence

As already noted in the previous section, there is a striking difference between the shear elastic constant C_{44} of stage 1 and stage 2 alkali-metal GICs. It has been conjectured that this difference may be caused by the structural (epitaxial) relationship between the intercalate layer and the graphite substrate. At room temperature, stage 1 compounds exhibit an ordered commensurate in-plane structure, while the alkali-metal intercalate layers in stage 2 compounds can be described as incommensurate liquids with only a small fraction of alkali ions sitting in the dips of the graphite potential at any time [4.76]. This fraction, however, increases with decreasing temperature, and below the ordering temperature a discommensuration domain structure is formed with intercalate atoms placed in approximate registry with the substrate potential, and phase slips between the domains [4.77–79]. *Hardcastle* et al. [4.80] and *Zabel* et al. [4.74, 81] showed that the intercalate ordering in KC_{24} and RbC_{24} induces a stiffening of the shear elastic constant, and that the temperature dependence of C_{44} closely follows the temperature dependence of the order parameter for this phase transition, see Fig. 4.25. According to *Zabel* et al. [4.74], the softening of the layer shear elastic constant C_{44} is induced by the melting of the discommensurate intercalate layer into an incommensurate liquid via unpinning of domain walls. Monte Carlo simulation of the layer shear interaction by *DiVincenzo* [4.82] support this picture.

Recently, *Vaknin* and *Fischer* [4.46] also observed a softening of C_{44} in stage 2 $C_{16}AsF_5$, again in a temperature range which carries the intercalate layer from a discommensurate solid to an incommensurate liquid at high temperatures. It is most likely that a similar mechanism of shear softening through unpinning of discommensuration domain walls is also responsible for these results.

Another crucial test of the shear softening through melting was recently carried out on LiC_6 by *Fischer* [4.72]. The Li atoms form a commensurate $(\sqrt{3} \times \sqrt{3})\mathrm{R}30°$ structure at room temperature, and from structural data [4.83] and the study of the Li in-plane diffusivity [4.84] it has been inferred that the melting of the intercalate layer at 740 K leads to a commensurate lattice liquid above the melting temperature. In this case, one should not expect a softening of C_{44} through the melting transition, except for a more or less linear decrease due to lattice anharmonicity effects. The striking results of *Fischer* [4.72], however, indicate that in this case too $C_{44}(T)$ roughly follows the order parameter of the order–disorder transition. In order to reconcile these results with previous experiments, we have to assume that either another mechanism is also capable of

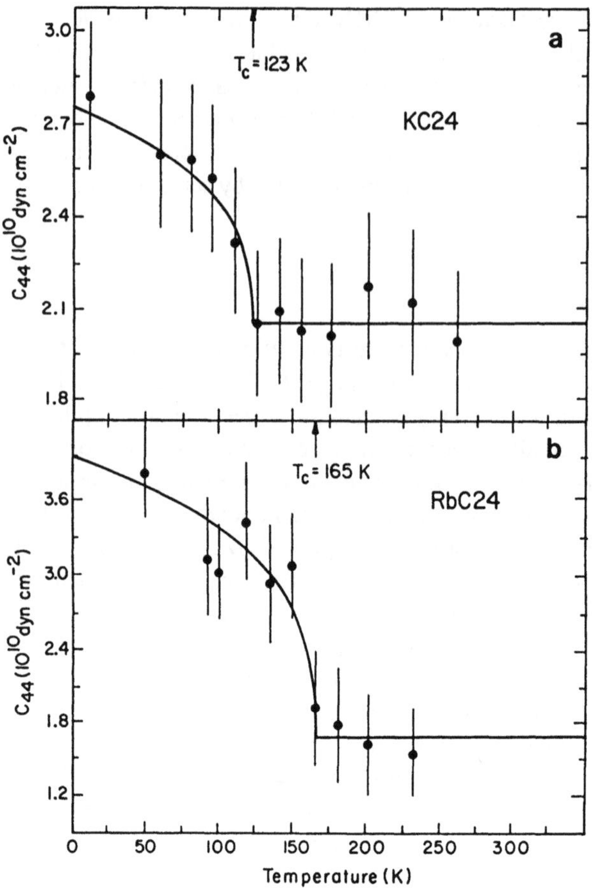

Fig. 4.25. Temperature dependence of the layer shear elastic constant C_{44} in (a) KC_{24} and (b) RbC_{24}, determined from the initial slope of the $[q00]T_{\perp}$ acoustic mode. The full lines are calculated on the assumption of a power law $[(T_c - T)/T_c]^{2\beta}$ with $\beta = 0.165$, suggesting a power law behavior similar to the order parameter for the quasi-two-dimensional discommensuration–domain transition. (From [4.74])

causing a mode softening, or that the melting leads indeed to an incommensurate Li liquid above T_c, despite conclusions from structural data. This difficult but important question requires further careful investigation.

4.4 Phonon Density of States of Intercalate Modes

4.4.1 Introductory Remarks

While for most of the phonon modes discussed so far, the intercalate and host graphite displacements are coupled, this is the case to a much lesser extent for the intercalate in-plane modes with propagation and polarization parallel to the planes. In the plane, the extremely strong C-C interaction leads to $[q00]T_{\parallel}$ and $[q00]L_{\parallel}$ dispersions with very steep slopes. The bonds between the intercalate atoms or molecules are usually much weaker and their vibration is therefore essentially decoupled from the graphite modes. Because of the lack of single crystals, the dispersion of the intercalate in-plane modes is difficult to measure. Only the longitudinal modes could, in principle, be measured with some certainty in a scattering geometry similar to the configuration II sketched in Fig. 4.7, with the modification, however, that the scattering vector $\boldsymbol{Q}$ has to be placed parallel to the $HK0$ plane. The dispersion of the transverse in-plane mode cannot be measured at all within the 2d powder averaged sample geometry. It is therefore a better practice to determine both the longitudinal and transverse modes via PDS measurements. In general, PDSs are obtained via incoherent mixing and averaging of phonon modes, usually achieved by using nuclei with predominant incoherent scattering cross sections for thermal neutron scattering [4.26]. For alkali-intercalate atoms, the scattering lengths are, however, exclusively coherent [4.85]. The same effect of incoherent mixing is then achieved by adding and averaging the coherent phonon signal over many Brillioun zones. This procedure is termed the *incoherent approximation* [4.26, 86].

Here, we are interested in the partial PDSs of the intercalate modes with propagation and polarization parallel to the layers. This implies that no graphite modes and no out-of-plane modes should be contained in the scattered intensity. Clearly, because of the mosaicity of the sample and the vertical instrumental resolution, some out-of-plane components of the intercalate vibration may contribute to the in-plane PDS, however, this contribution should only play a role at small wave vectors, if at all, whereas the averaging of the phonon intensities is usually carried at large scattering vectors. The partial PDS of the intercalate in-plane modes can then be obtained via the dynamical scattering function in the incoherent approximation [4.26]:

$$S(Q,\nu) = \frac{KQ^2\sigma}{M_I\nu}[n(\nu)+1]e^{-2W_I}g(\nu) + L(Q,\nu)\,. \qquad (4.33)$$

Here, $n(\nu)$ is the Bose-Einstein occupation number, M_I and σ are the mass and the cross section of the intercalate species, respectively, K is a constant, W_I is the Debye-Waller factor for the intercalate in-plane vibration, and $\nu = 2\pi/\omega$ is the vibrational frequency. The first term in (4.33) represents the one-phonon dynamical scattering function, from which the phonon density of states $g(\nu)$ can be calculated. The second term $L(Q,\nu)$ collects all contributions from multiphonon processes, which is usually a smooth function of ν and not easily distinguished from a uniform background. Nonetheless, the PDS data shown below have been corrected for two-phonon contributions and for the graphite contributions via procedures described in [4.87].

The PDS of a two-dimensional lattice is expected to increase linearly with ν at small frequencies [$g(\nu) \propto \nu$] and should exhibit two peaks at larger frequencies from the flat parts of the transverse and longitudinal dispersion close to the Brillouin zone boundary. This is shown in Fig. 4.26 from a model cal-

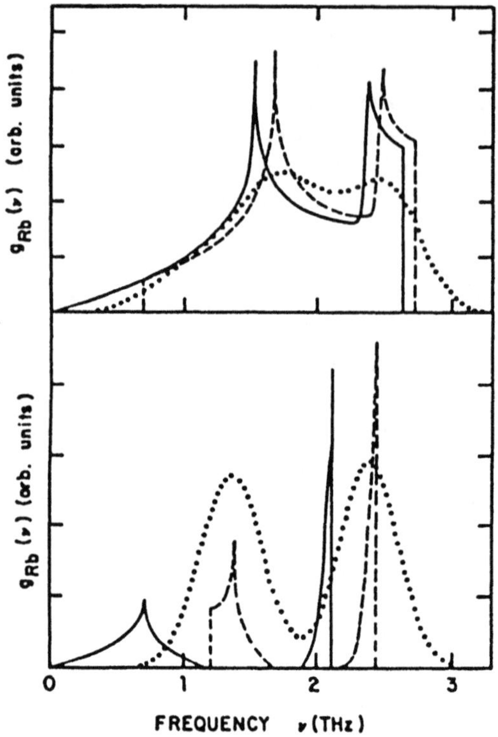

Fig. 4.26. (*Top*): Phonon density of states for a 2d Born–von Kármán model of a triangular lattice with only nearest neighbor interaction (—), compared with the same model but including interaction with the graphite substrate (- - - -), and broadening effects due to instrumental resolution and anharmonicity (· · ·). (*Bottom*): Phonon density of states for a 2d lattice assuming a Coulomb force model. The solid, dashed, and dotted lines have the same meanings as in the top panel. (From [4.24])

culation by *Kamitakahara* et al. [4.24], which takes into account a triangular lattice and nearest neighbor BvK force constants for the radial and angular coupling. Turning on the substrate potential causes a gap to occur in the PDS of the size $\nu_0 = (1/2\pi)\sqrt{\phi_{IC}/M_I}$, where ϕ_{IC} represents the intercalate–graphite in-plane coupling constant, which is proportional to the curvature of the corrugated substrate potential. Therefore, measuring the PDS yields important information on the intercalate-intercalate in-plane interaction as well as on the strength of the substrate potential.

4.4.2 Experimental Results

(a) Stage 1 Alkali-Metal Compounds

Experimental results of PDS measurement for the alkali-metal in-plane modes in the stage 1 compounds KC_8, RbC_8, and CsC_8 are shown in Fig. 4.27. For each sample, constant Q scans over an energy range $0 < E < 20$ meV ($0 < E < 4.8$ THz) were carried out and repeated for several Q values in the range of $3.5 < Q < 4.5$ Å^{-1}. This Q range was chosen to lie between the (100) and (110) graphite peaks at 2.94 Å^{-1} and 5.01 Å^{-1}, respectively, ensuring that the averaging in the $HK0$ plane extends over many Brillouin zones without contamination by signals from the graphite host. In the energy region scanned,

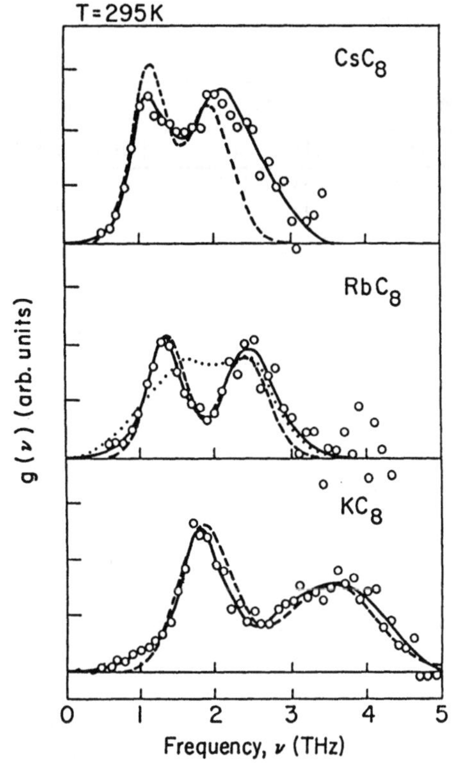

Fig. 4.27. Measured phonon density of states of alkali-metal in-plane modes in stage 1 compounds. Dashed curves are derived from a Coulomb force model; the dotted line in the middle panel is the result of a 2d BvK model applied to RbC_8. The model spectra have been broadened to take instrumental resolution effects into account. The solid line is a guide to the eye. (From [4.24, 87])

the corresponding PDSs of the graphite host do not have any characteristic feature and are expected to contribute only to a gently sloping background intensity.

The partial PDSs of the alkali-metal in-plane modes consistently show a double peak structure of about equal intensity, and the positions of both peaks scale roughly with the square root of the intercalate mass. Although this kind of PDS shape is expected for a two-dimensional solid interactiong with a substrate potential as discussed above, *Kamitakahara* et al. [4.24] could show that such a model is inappropriate for the description of the measured PDSs of alkali-metal GICs. The dotted line in the middle panel of Fig. 4.27 represents the PDSs calculated from a two-dimensional BvK model applied to RbC_8 including a cut-off frequency ν_0. A much better description of the observed PDS was achieved by a model in which the metal–substrate interaction is retained, but the alkali–alkali in-plane interaction is replaced by a long range Coulomb interaction between the bare M^+ ions. In the absence of the I-C interaction, the Coulomb model predicts that the upper peak occurs very close to the plasma frequency $\nu_p = e/(\pi M_I V)^{1/2}$, where V is the volume per M^+ ion. Both effects combined predict a frequency $\nu_c = (\nu_0^2 + \nu_p^2)^{1/2}$ for the upper peak. The calculated spectra for all three stage 1 compounds based on the Coulomb model are shown in Fig. 4.27 by the dashed lines. The only fit parameter, ν_0, is listed in Table 4.6 and is chosen slightly below the frequency of the lower peak. This small shift is due to the weak alkali–alkali

Table 4.6. Values for ν_0 and the ion-plasma frequency ν_p from [4.24, 87]. Calculated and observed values for the upper peak, $\nu_c = (\nu_0^2 + \nu_p^2)^{1/2}$ and ν_{obs} are also shown

	T	ν_0		ν_p		ν_c		ν_{obs}	
	[K]	[THz]	[meV]	[THz]	[meV]	[THz]	[meV]	[THz]	[meV]
KC_8	296	1.60	6.61	3.20	13.22	3.58	14.79	3.65	15.08
RbC_8	296	1.20	4.95	2.10	8.67	2.42	10.0	2.50	10.33
CsC_8	10	1.00	4.13	1.64	6.77	1.92	7.93	2.6	10.74

in-plane shear interaction. Table 4.6 also contains the values for ν_p, ν_c and ν_{obs}. The calculated spectra have been broadened to take instrumental resolution and anharmonicity effects into account, and in particular for KC_8 and RbC_8 they reproduce the measurements rather well. Some discrepancy is noticeable for CsC_8, possibly due to short range interactions not properly taken into account by the Coulomb model. It is interesting to note that CsC_8 is also the compound which presents the most difficulty in fitting the $[00q]$L modes.

Another interesting feature of the Coulomb model is that no conduction electron screening is required for the prediction of the upper phonon frequency. In general, for metals, the electron–phonon interaction leads to strong screening effects at larger phonon wave number q and results in a down-scaling of the maximum frequency. Typical ratios are $\nu_{obs}/\nu_p = 0.2$ to 0.5 [4.88]. The fact that for GICs this ratio is close to 1 implies that the conduction electron screening for small wave numbers must be ineffective, i.e., the screening length of the conduction electrons must be larger than typical $M - M$ distances in the intercalate layer, in agreement with calculations by *DiVincenzo* and *Mele* [4.69].

(b) Stage 2 Alkali-Metal Compounds

Stage 2 compounds differ from stage 1 compounds in many ways, as suggested previously and as is described in more detail in Chap. 2. The I-C interaction is weaker, the alkali-metal areal mass density is reduced, and in most cases the alkali layers form incommensurate modulated structures that melt at low temperatures. The intercalate in-plane modes are expected to be a sensitive probe of the discommensuration domain structure and of the melting transition. For instance, the first peak which is proportional to the strength of the I-C in-plane interaction should depend crucially on the location of the intercalate atom with respect to the substrate.

In the ordered low temperature state, a two-peak structure is again observed in the PDS, similar to the results on the stage 1 compounds. This is shown for KC_{24} and RbC_{24} in Fig. 4.28. In the low temperature phase below 123 K and 162 K for KC_{24} and RbC_{24}, respectively, the lower frequency peak is surprisingly well defined, in spite of the fact that many of the alkali atoms in the discommensuration domain structure do not sit precisely on commensurate sites with respect to the graphite substrate and should therefore experience a distribution of I-C interactions. Upon heating above the melting temperature, some anharmonicity

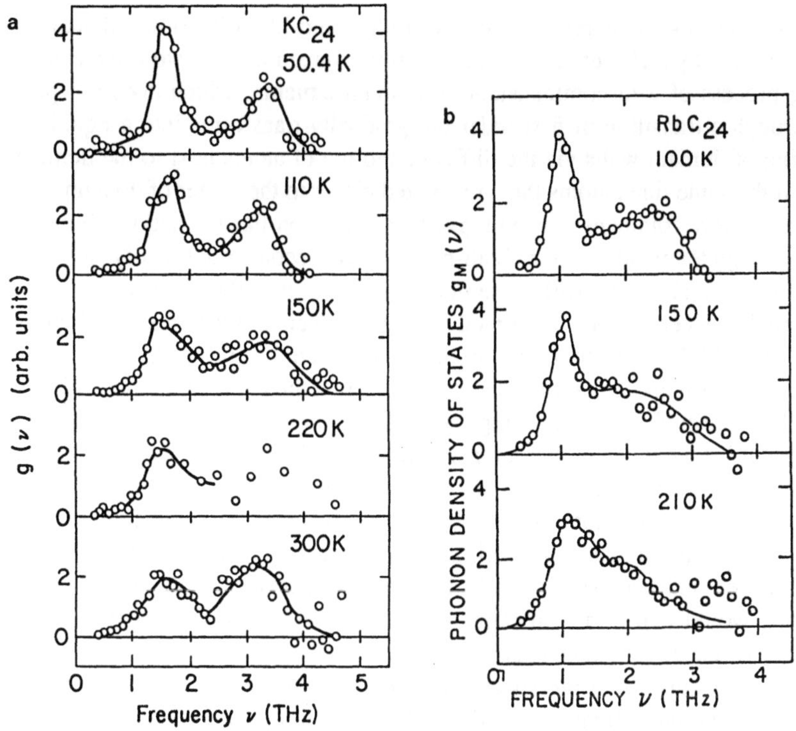

Fig. 4.28. Phonon density of states of intercalate in-plane vibrations in stage 2 KC_{24} (a), and in RbC_{24} (b) below and above their respective temperatures for the discommensuration-domain transition at 123 K and 162 K. (From [4.87])

becomes noticeable, but nothing dramatic happens on increasing the temperature in excess of T_c.

The observed behavior of the PDS above T_c is fundamentally different from what one would have expected if the high temperature state were describable by a simple liquid or glassy state. In the first case, diffusive motion and vibrational excitation become indistinguishable in the region of phonon wave vectors probed here, leading to a broad quasi-elastic line shape of the dynamical scattering function centered at $\nu = 0$ [4.88]. In the second case, rather ill-defined phonon excitations may exist, as shown for glassy metals by *Suck* et al. [4.89], along with a central peak which is resolution limited. In the present case, however, both types of excitations coexist, the solid-like vibrational and the liquid-like diffusional. The diffusional part giving rise to a quasi-elastic peak cannot be seen with the limited instrumental resolution of the triple-axis spectrometer using thermal neutrons but was unraveled using higher resolution time-of-flight spectrometers [4.90], and is described in more detail in Chap. 5 of this volume. The phonon results discussed here indicate that the I-C in-plane interaction responsible for the lower peak in the PDSs remains important above T_c and that collective excitations giving rise to the upper peak persist upon melting. The fact that these solid-like features can coexist with liquid-like diffusional dynamics was interpreted by *Zabel* et al. [4.74] as a strong signature for a two-dimensional melting process of a discommensuration domain structure which does not proceed through a transition of first order but gradually transforms into a liquid by unpinning of domain walls via the diffusive motion of atoms next to the domain walls. At the same time, atoms that are located closer to the center of domains are structurally more commensurate and give rise to phonon-like excitations. Finally, at high enough temperatures, the alkali-metal layers become rotationally isotropic 2d liquids when all atoms within the domains take part in the diffusion process. This view has recently been confirmed through molecular dynamics calculations by *Fan* et al. [4.91], which are described in more detail in Chap. 2.

Kamitakahara and *Zabel* [4.87] also investigated the temperature dependence of the PDSs through the disordering transition of the intercalate layer in RbC_8, which occurs at the much higher temperature of 745 K. Unlike the case of stage 2 compounds, they observed a rather discontinuous change of the phonon spectrum through the phase transition. While the intensity increases at small frequencies, possibly due to quasi-elastic scattering from alkali-metal diffusion, the phonon peaks broaden and become less well defined. For a lattice liquid, on the other hand, one would expect a Lorentzian peak centered at $\nu = 0$ from the planar diffusivity of the alkali atoms coexisting with a peak at ν_0. How well defined the oscillator peak appears in the spectrum depends on the ratio of the vibrational frequency to the jump frequency and will be a function of the temperature. The details of the dynamical behavior of 2d commensurate solids transforming to 2d lattice liquids has not yet been investigated in detail. The stage 1 compounds with heavy alkali-metals represent ideal test systems for these studies and much more work needs to be done.

4.4.3 Anisotropy of Intercalate Interaction

Combining the results of the $[00q]L$ layer breathing modes with those of the intercalate in-plane modes, important information can be obtained about the anisotropy of the intercalate interaction parallel and perpendicular to the intercalate planes. If the effective mass of the intercalate layer is smaller than that of the graphite basal planes, the top $[00q]L$ optic branch is predominantly determined by the vibrational energy of the intercalate perpendicular to the layers. In that case, a rough estimate for the vibrational anisotropy is derived by comparing the phonon energies of the top $[00q]L$ optic branch and the plasma frequency ν_p, reflecting the strength of the in-plane I-I interaction. A few energy ratios E_{out}/E_{in} are listed in Table 4.7, showing that in binary donor compounds this ratio is roughly 3/2, independent of the alkali-metal and stage of the compound, while in the trilayer compounds the anisotropy for the mercury and hydrogen vibrational energies is reversed, being larger in than out of the plane.

The anisotropy of the alkali-metal bonding in binary GICs is opposite to the bonding of the host graphite atoms, which is stronger in the plane than out of the plane.

Table 4.7. Anisotropy of the vibrational energy for intercalate vibrations parallel and perpendicular to the intercalate plane. $E_\perp$ is estimated from the top optic branch of the $[00q]$L dispersion, and $E_{\|} \simeq h\nu_p = h(\nu_{obs}^2 - \nu_0^2)^{1/2}$

	$E_\perp$[meV]	$E_{\|}$[meV]	$E_\perp/E_{\|}$	Ref.
KC_8	21	13.55	1.55	[4.5, 87]
RbC_8	14	9.10	1.54	[4.5, 87]
CsC_8	15	9.90	1.51	[4.5, 87]
KC_{24}	22	12.6	1.75	[4.5, 87]
RbC_{24}	15	9.27	1.61	[4.5, 87]
H in $KH_{0.8}C_4$	71	93.0	0.76	[4.41]
Hg in $KHgC_4$	3	9.2	0.32	[4.39]

A more precise estimate of the vibrational anisotropy is obtained by calculating the thermal energy of the intercalate atom directly via integration of the corresponding phonon spectrum

$$E_{\text{therm}}^{\text{I}} = \frac{\int_0^{\nu_m} S_{\text{I}}(\nu) g_{\text{I}}(\nu) h\nu n(\nu) d\nu}{\int_0^{\nu_m} g_{\text{I}}(\nu) d\nu} , \tag{4.34}$$

where $g_{\text{I}}(\nu)$ is the intercalate phonon density of states, S_{I} is the energy fraction shared by the intercalate atom in each vibrational mode, and ν_m is the maximum phonon frequency occurring in the spectrum. For the in-plane intercalate modes, $S_{\text{I}}(\nu) = 1$ because of the decoupling between the intercalate and graphite vibration, and the integration can be carried out immediately over the measured phonon density of states $g_{\text{I}}(\nu)$. For the out-of-plane modes, the density of states has to be calculated from the $[00q]$L dispersion and the fractional contribution of the intercalate atoms in the mode energy is given by $S = E_{\text{I}}/(E_{\text{I}} + E_{\text{C}})$, where E_{I} and E_{C} are the energies carried by the intercalate and graphene planes. This calculation has been carried out by *Moreh* et al. [4.92] for KC_8, and for $T = 0$K they obtain for the K vibrational energy an anisotropy of 1.79 as compared to the 1.55 derived from the rough estimate. In addition, *Moreh* et al. [4.92] found that the K bonding in graphite, parallel as well as perpendicular to the layers, is significantly stronger than in metallic K. Compared to metallic potassium, the binding energies of K in KC_8 are $E_{\|}/E_{\text{met}} = 1.8$ and $E_\perp/E_{\text{met}} = 3.25$.

4.5 Thermal Properties

In this section we will give a short overview of some thermal properties which relate to the phonon measurements discussed in the previous section.

4.5.1 Specific Heat

Specific heat measurements have been reported for a number of intercalation compounds [4.68, 93, 94]. The data were analyzed using the standard expression for the specific heat:

$$C_p = \gamma T + \alpha T^3 + \Delta C \ . \tag{4.35}$$

The linear term is (4.35) is due to electronic contributions, the second term results from the thermal energy, and the last term collects any Einstein-like excitations. From the coefficient α, an orientationally averaged Debye temperature $\theta_D(0)$ for zero temperatures can be deduced via $\alpha = 12k_B\pi^2/5\theta_D^3(0)$, where k_B is Boltzmann's constant.

Plots of C_p/T versus T^2 have, in many cases, exhibited deviations from a linear slope, from which Einstein-like modes were inferred. Support for this picture was provided from low temperature neutron scattering experiments [4.64], see Fig. 4.20, which showed a mode splitting in the acoustic $[00q]$L branch of RbC_{24}, coinciding with the Einstein oscillator energy as determined from specific heat measurements [4.68]. In contrast, specific heat measurements of KC_{24} did not indicate any extra contribution [4.93]. Nevertheless, a mode splitting of the $[00q]$L mode was also seen here [4.66].

We believe that the coincidence in RbC_{24} is not based on physical grounds, and that the analysis of the specific heat data has not been carried out properly. In general, for a d-dimensional lattice and a phonon dispersion which, at small frequencies, is proportional to $\nu \propto q^n$, the specific heat is

$$C_P = \gamma T + \alpha T^{d/n} + \Delta C \ . \tag{4.36}$$

For three-dimensional crystals ($d = 3$) and a linear dispersion, one obtains the usual T^3 dependence at low temperatures. However, in layered materials with $2 \leq d \leq 3$ and with bending modes having quadratic dispersion relations, the lattice vibrational contribution to the specific heat may well deviate from the usual T^3 behavior. A T^2 dependence has indeed been observed at low temperatures in pristine graphite [4.2]. In graphite intercalation compounds, additional contributions are expected from the intercalate in-plane modes. For instance, the alkali-metal intercalates exhibit low frequency in-plane vibrations ν_0 which are due to the metal–substrate interaction. These frequencies match rather well with Einstein frequencies obtained from the analysis of the specific heat data and with the mode splitting energies observed in KC_{24} and RbC_{24} at low temperatures. A re-analysis of the specific heat data in light of the experimentally determined phonon dispersions may be required to obtain reliable parameters from these measurements, such as Debye temperatures and Einstein frequencies.

4.5.2 Debye-Waller Factor

Debye-Waller factors of graphite intercalation compounds have been measured by a variety of methods, including X-ray scattering, Mössbauer spectroscopy, nuclear resonance photon scattering, and from the temperature dependence of the phonon density of states. All four methods yield slightly different information and direct comparison is sometimes difficult.

With X-ray scattering the mean square amplitude $\langle u^2 \rangle$ of atomic vibrations is measured. In a complex system with several different atoms per unit cell, many Bragg reflections need to be measured to obtain the mean square amplitudes of all individual atoms and for the different directions. In this regard, the most complete work on GICs is that by *Ohshima* et al. [4.95], who analyzed $\langle u^2 \rangle$ of Rb in RbC_{24}. Mean square amplitudes for alkali-metal vibrations perpendicular to the intercalate plane in stage 1 compounds were also determined by means of X-ray scattering by *Leung* et al. [4.96].

The Debye-Waller factor of Mössbauer spectroscopy is sensitive to the mean square displacement of the γ emitting nuclei, and is therefore specific to a subset of atoms in the unit cell. Thus, the relevant information is more direct and does not need to be extracted from a number of Bragg reflections, as in X-ray scattering. The main disadvantage of this method is the limited number of Mössbauer atoms available. The Mössbauer effect has been successfully applied to the study of the Debye-Waller factor of Cs in CsC_8 by *Campbell* et al. [4.97].

Nuclear resonance photon scattering (NRPS) shares with the Mössbauer effect selectivity for specific nuclei. But instead of measuring the recoil-free fraction of γ transitions, the Doppler broadening of the nuclear levels is determined. Therefore NRPS is sensitive to the mean square velocity $\langle V^2 \rangle$ instead of the mean square displacements. Thus the NPRS method weights more effectively the high frequency modes, while the Debye-Waller factors derived from X-ray scattering or Mössbuaer spectroscopy are more tuned into the low frequency modes. In a detailed study, *Moreh* et al. [4.92] examined the Doppler broadening of K nuclei in KC_8 by the NPRS method and determined the anisotropy of the K interaction in this compound as compared to metallic potassium. In their analysis of the experimental data, *Moreh* and co-workers obtain an anisotropy factor which is very close to the calculated value from integrating over the relevant phonon energies. Expressed in terms of Debye temperatures, *Moreh* et al. [4.92] obtained for the K vibration parallel and perpendicular to the planes the values $\theta_{\parallel} = 173\,\mathrm{K}$ and $\theta_{\perp} = 378\,\mathrm{K}$, respectively, whereas $\theta_D = 91\,\mathrm{K}$ for metallic K. This is a manifestation of the predominantly ionic forces governing the K-K and K-C interactions in the intercalation compound, which are anisotropic and much stronger than in the metal. For the anisotropy of the C-C interaction, *Moreh* et al. [4.92] find values which are in good agreement with the graphite phonon dispersion as measured by *Nicklow* et al. [4.2] and with Debye temperatures as determined by *Albinet* et al. [4.98].

The determination of the Debye-Waller factor from phonon measurements is perhaps the most straightforward method, since it is nuclei as well as mode specific. Nonetheless, this method has been applied to date only to the stage 1 alkali-metal compounds [4.87]. From those measurements, reliable mean square amplitudes for the metal vibrations parallel to the intercalate plane are available, including their temperature dependence. This is shown in Fig. 4.29. The agreement between the mean square displacements derived from Mössbauer spectroscopy and PDS measurements is excellent in the case of Cs intercalate atoms in CsC_8, as can be seen from Table 4.8.

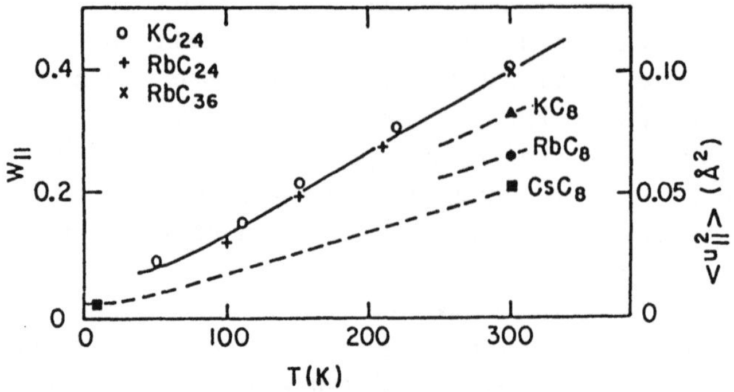

Fig. 4.29. In-plane components of the Debye-Waller exponent $W_{\|}$ and the mean square displacement $\langle u_{\|}^2 \rangle$ for alkali-atoms in stage 1 compounds, determined from the intercalate in-plane vibrational density of states. (From [4.87])

Table 4.8. Mean square displacements for alkali-atoms in graphite parallel and perpendicular to the intercalate plane. All values are for room temperature

Compound	$\langle u_{\perp}^2 \rangle$[Å^2]	$\langle u_{\|}^2 \rangle$[Å^2]	Method	Ref.
K in KC_8		0.082	neutron – PDS	[4.87]
	0.0416		X-ray	[4.96]
Rb in RbC_8		0.065	neutron – PDS	[4.87]
	0.0208		X-ray	[4.96]
Cs in CsC_8		0.055	neutron - PDS	[4.87]
	0.0157[a]	0.0574[a]	Mössbauer	[4.97]
	0.0019		X-ray	[4.96]
Rb in RbC_{24}	0.015	0.12	X-ray	[4.95]
Graphite	0.0067[b]	0.00175[b]		[4.98]

[a] Extrapolated from 4.2 K to 300 K.
[b] Calculated from Debye temperatures.

4.5.3 Thermal Expansion

The thermal expansion is related to the specific heat via the Grüneisen relation:

$$\frac{\Delta V}{V}(T) = \gamma B^{-1} E_{\text{therm}}(T) \,, \tag{4.37}$$

where B is the bulk modulus and γ is the Grüneisen parameter, defined by

$$\gamma = \frac{V\delta\omega}{\omega\delta V} \,. \tag{4.38}$$

If the specific heat or the thermal energy is known, thermal expansion measurements provide information on the lattice anharmonicity. A series of measurements of the thermal expansion of alkali-metal GICs have been carried out by *Hardcastle* and *Zabel* [4.99]. Their measurements show that the c-axis thermal expansion depends systematically on the effective mass of the intercalate and the stage of the compound, with a tendency to increase with decreasing $M_{\rm I}$ and stage number n. The a-axis thermal expansion is negative in pristine graphite for temperatures between 10 and 300 K and does not significantly change upon intercalation with alkali-metals.

Hardcastle and *Zabel* [4.99] have analyzed the c-axis thermal expansion using (4.37) and calculating the c-axis thermal energy from the known [00q]L phonon dispersions. This way, they found that the large increase of the thermal expansion is mainly due to an increase in the anharmonicity of the alkali-metal–graphite interlayer interaction rather than due to an increase in the thermal energy.

Thermal expansion measurements have also been carried out recently on Hg- and Bi-based trilayer compounds by *McRae* et al. [4.100] between 4 and 300 K. They measured large thermal expansion coefficients, but could not identify any systematic trend with respect to the mass of the intercalated guest or the stage of the compound, in contrast to the binary alkali-metal GICs.

4.6 Conclusion

The available experimental data on layer breathing modes provide a rather complete understanding of the longitudinal interlayer interactions in the graphite intercalation compounds. One-dimensional lattice dynamical models have proven to provide adequate descriptions of the phonon energies. The general trend for the transverse modes and the intercalate in-plane modes is also visible from the present data; however, more detailed and systematic work is still required for many compounds. The study of molecular rotations, librations, and translations is still in its infancy, and the *intra*molecular dynamics of molecular intercalates has not yet been the focus of attention. Another point which requires more consideration in the future is the dynamics of quasi-two-dimensional lattice gases and lattice liquids, for which the stage 1 alkali-metal intercalation compounds provide perfect model systems.

Acknowledgements. I wish to express my appreciation to my colleagues as well as to my excellent present and former students for fruitful collaboration and beneficial discussions. Much of the work described here is due to major contributions by P.C. Chow, W.A. Kamitakahara, A. Magerl, D.A. Neumann, R.M. Nicklow, J.J. Rush, S.M. Shapiro, and S.A. Solin. A critical reading of the manuscript by D.A. Neumann and D.G. Wiesler, and the dynamical structure factor calculations by X.-M. Zhu are also highly appreciated. I would also like to thank J.E. Fischer for communicating experimental results prior to publication. The support of the National Science Foundation through grant DMR 86-05565 is gratefully acknowledged.

References

4.1 G. Dolling, B.N. Brockhouse: Phys. Rev. **18**, 1120 (1962)
4.2 R.M. Nicklow, N. Wakayabashi, H.G. Smith: Phys. Rev. B **5**, 4951 (1972)
4.3 N. Wakayabashi, R.M. Nicklow: In *Electrons and Phonons in Layered Crystal Structures*, ed. by T.J. Wieting, M. Schlüter (Reidel, Dordrecht, Holland 1979) p. 409–464
4.4 W.B. Ellenson, D. Semmingsen, D. Guérard, D. Onn, J.E. Fischer: Mater. Sci. Eng. **31**, 137 (1977)
4.5 H. Zabel, A. Magerl: Phys. Rev. B **25**, 2463 (1982)
4.6 H. Zabel: Physica **136B**, 1 (1986); In *Phonon Physics*, Proc. 2nd Int. Conf. on Phonon Physics, Budapest, ed. by J.Kollár, N. Kroó, N. Menyhárd, T. Siklós (World Scientific, Singapore 1985) p. 336
4.7 S.A. Solin: In *Intercalation in Layered Materials*, Series B, Vol. 148, ed. by M.S. Dresselhaus (Plenum, New York 1986)
4.8 C. Stassis: In *Methods of Experimental Physics: Neutron Scattering*, Vol. 23, Part A, ed. by K. Sköld, D.L. Price (Academic, San Diego 1986) p. 369
4.9 M. Born, K. Huang: *Dynamical Theory of Crystal Lattices* (Oxford University Press, Oxford 1954);
A.A. Maradudin, E.W. Montroll, G.H. Weiss, I.P. Iopatova: *Solid State Physics*, Suppl. 3, 2nd ed. (Academic, New York 1971)
4.10 M. Maeda, Y. Kuramoto, C. Horie: J. Phys. Soc. Jpn. **47**, 337 (1979)
4.11 R. Al-Jishi, G. Dresselhaus: Phys. Rev. B **26**, 4515 (1982)
4.12 C. Horie, M. Maeda, Y. Kuramoto: Physica **99B**, 430 (1980)
4.13 R. Al-Jishi, G. Dresselhaus: Phys. Rev. B **26**, 4523 (1982)
4.14 F. Batallan, I. Rosenman, Ch. Simon, H. Lauter: Synth. Met. **7**, 361 (1983)
4.15 Ch. Simon, F. Batallan, I. Rosenman, H. Lauter, F. Furdin: Phys. Rev. B **27**, 5088 (1983)
4.16 H.C. Gupta, J. Malhotra, N. Rani, B.B. Tripathi: Solid State Commun. **57**, 263 (1986)
4.17 H.C. Gupta, J. Malhotra, N. Rani, B.B. Tripathi: Mater. Sci. Eng. **85**, 187 (1987)
4.18 H.C. Gupta, R.S. Narayanan, N. Rani, B.B. Tripathi: Synth. Met. **7**, 347 (1983)
4.19 H.C. Gupta, J. Malhotra, N. Rani, B.B. Tripathi: Phys. Rev. B **33**, 7285 (1986)
4.20 L. Pietronero, S. Strässler: Solid State Commun. **32**, 1337 (1979)
4.21 R.J. Nemanich, S.A. Solin, D. Guérard: Phys. Rev. B **16**, 2965 (1977)
4.22 S.Y. Leung, G. Dresselhaus, M.S. Dresselhaus: Phys. Rev. B **24**, 6083 (1981)
4.23 T. Kimura, R. Yamamoto, M. Doyama: J. Phys. F **12**, 1907 (1982)
4.24 W.A. Kamitakahara, N. Wada, S.A. Solin: Solid State Commun. **44**, 297 (1982)
4.25 W.A. Kamitakahara, N. Wada, S.A. Solin, L.M. Seaverson: Phys. Rev. B **28**, 3457 (1983)
4.26 W. Marshall, S.W. Lovesey: *Theory of Thermal Neutron Scattering* (Oxford University Press, Oxford 1971)
4.27 S.W. Lovesey: *Theory of Neutron Scattering from Condensed Matter* (Oxford University Press, Oxford 1984)
4.28 A. Magerl, H. Zabel: In *Physics of Intercalation Compounds*, ed. by L. Pietronero, E. Tosatti, Springer Ser. Solid State Sci., Vol. 38 (Springer, Berlin, Heidelberg 1981) p. 180
4.29 B.G. Dick, A.W. Overhauser: Phys. Rev. **112**, 90 (1958)
4.30 S. Funahashi, T. Kondow: Synth. Met. **6**, 101 (1983)
4.31 C. Colvard, T.A. Gant, M.V. Klein, R. Merlin, R. Fischer, H. Morkoç, A.C. Gossard: Phys. Rev. B **31**, 2080 (1985)
4.32 K. Gschneider, Jr.: Solid State Phys. **16**, 275 (1964)
4.33 J. Rossat-Mignod, D. Fruchart, M.J. Moran, J.W. Milliken, J.E. Fischer: Synth. Met. **2**, 143 (1980)
4.34 H. Zabel, A. Magerl, J.J. Rush: Phys. Rev. B **27**, 3930 (1983)
4.35 A. Hérold, D. Billaud, D. Guérard, P. Lagrange, M. El Makrini: Physica **104B**, 253 (1981)
4.36 K.C. Woo, A.W. Kamitakahara, D.P. DiVincenzo, D.S. Robinson, H. Mertwoy, J.W. Milliken, J.E. Fischer: Phys. Rev. Lett. **50**, 182 (1983)
4.37 S.A. Solin, H. Zabel: Adv. Phys. **37**, 87 (1988)
4.38 H.J. Kim, H. Mertwoy, T. Koch, J.E. Fischer, D.B. McWahn, J.D. Axe: Phys. Rev. B **29**, 5947 (1984)
4.39 W.A. Kamitakahara, L.E. DeLong, P.C. Eklund, R.M. Nicklow: Phys. Rev. B **29**, 460 (1984)

4.40 W.A. Kamitakahara, P.E. Eklund, L.E. DeLong, M.H. Yang: Proc. of the 17th Int. Conf. on Low Temperature Physics, ed. by U. Eckern, A. Schmid, W. Weber, H. Wühl (North-Holland, Amsterdam 1984) p. 861
4.41 W.A. Kamitakahara, G.L. Doll, P.C. Eklund: Proc. of the 1986 Fall Meeting of the Materials Research Society, ed. by M.S. Dresselhaus, G. Dresselhaus, S.A. Solin (Materials Research Society, Pittsburgh, PA 1986) p. 69
4.42 T. Springer, D. Richter: In *Methods of Experimental Physics: Neutron Scattering*, Vol. 23, Part B, ed. by D.L. Price, K. Sköld (Academic, San Diego, CA 1987) p. 131
4.43 G. Dresselhaus, R. Al-Jishi, J.D. Axe, C.F. Majkrzak, L. Passell, S.K. Satija: Solid State Commun. **40**, 229 (1981)
4.44 P.C. Eklund, V. Yeh, H.G. Smith, R.M. Nicklow, R. Al-Jishi, G. Dresselhaus: Phys. Rev. B **29**, 2138 (1984)
4.45 H. Zabel, S.M. Shapiro, D.G. Wiesler: Unpublished
4.46 D. Vaknin, J.E. Fischer: Synth. Met. **23**, 101 (1988)
4.47 P.C. Eklund, E.T. Arakawa, J.L. Zarestky, W.A. Kamitakahara, G.D. Mahan: Synth. Met. **12**, 97 (1985)
4.48 B.R. York, S.A. Solin: Phys. Rev. B **31**, 8206 (1985)
4.49 X.W. Qian, D.R. Stumpp, S.A. Solin: Phys. Rev. B **33**, 5756 (1986)
4.50 Y.B. Fan, S.A. Solin, D.A. Neumann, H. Zabel, J.J. Rush: Phys. Rev. B **36**, 3386 (1987)
4.51 D.A. Neumann, H. Zabel, Y.B. Fan, S.A. Solin: Phys. Rev. B **37**, 8424 (1988)
4.52 H.A. Resing, R.M. Fronko, T. Tsang, S.A. Solin: Synth. Met. **23**, 297 (1988)
4.53 H. Zabel, D.A. Neumann: Can. J. Chem. **66**, 666 (1988)
4.54 D.A. Neumann, H. Zabel, J.J. Rush, N. Berk: Phys. Rev. Lett. **53**, 56 (1984)
4.55 P.C. Chow, H. Zabel: Synth. Met. **7**, 243 (1983)
4.56 P.C. Chow, H. Zabel: Phys. Rev. B **38**, 12837 (1988)
4.57 D. Billaud, A. Hérold: Bull. Soc. Chim. Fr. **11**, 2407 (1974)
4.58 G. Furdin, B. Carton, A. Hérold, C. Zeller: C.R. Acad. Sci. B **280**, 753 (1975)
4.59 H. Akera, H. Kamimura: J. Phys. Soc. Jpn. **55**, 2326 (1986)
4.60 G. Kirczenow: Phys. Rev. Lett. **52**, 437 (1984); see also Chap. 3 in this volume
4.61 M.E. Misenheimer, H. Zabel: Phys. Rev. Lett. **54**, 2521 (1985)
4.62 P. Dean: Rev. Mod. Phys. **44**, 127 (1972)
4.63 P. Hawrylak, M.L. Williams: Phys. Rev. B **33**, 1503 (1986)
4.64 S. Funahashi, T. Kondow, M. Iizumi: Solid State Commun. **44**, 1515 (1982)
4.65 S. Funahashi, T. Kondow, M. Iizumi: Physica **120B**, 305 (1983)
4.66 A. Magerl, H. Zabel, J.J. Rush: Synth. Met. **7**, 339 (1983)
4.67 H. Zabel, W.A. Kamitakahara, R.M. Nicklow: Phys. Rev. B **26**, 5919 (1982)
4.68 M. Suganuma, T. Kondow, U. Mizutani: Phys. Rev. B **23**, 706 (1981)
4.69 D.P. DiVincenzo, E.J. Mele: Phys. Rev. B **32**, 2538 (1985)
4.70 Lord Raleigh: *Theory of Sound*, Vol. 1 (Macmillan, London 1984) p. 352
4.71 N. Wakayabashi, H.G. Smith, R.M. Nicklow: Phys. Rev. B **12**, 659 (1975)
4.72 J.E. Fischer: Extended Abstracts, Materials Research Society Symposium, Boston 1988, ed. by M. Endo, M.S. Dresselhaus, G. Dresselhaus (Materials Research Society, Pittsburgh, PA 1988) p. 1
4.73 N. Wada, M.V. Klein, H. Zabel: J. de Phys. Colloq. **42**, C6–350 (1981)
4.74 H. Zabel, S.E. Hardcastle, D.A. Neumann, M. Suzuki, A. Magerl: Phys. Rev. Lett. **57**, 2041 (1986)
4.75 D.A. Neumann, H. Zabel, D.P. DiVincenzo: Phys. Rev. B **31**, 1136 (1985)
4.76 S.C. Moss, G. Reiter, J.L. Robertson, C. Thompson, J.D. Fan, K. Ohshima: Phys. Rev. Lett. **57**, 3191 (1986)
4.77 M. Suzuki, H. Suematsu: J. Phys. Soc. Jpn. **52**, 2761 (1983)
4.78 M.J: Winokur, R. Clarke: Phys. Rev. Lett. **54**, 811 (1985)
4.79 M. Suzuki: Phys. Rev. B **33**, 1386 (1986)
4.80 S.E. Hardcastle, D.A. Neumann, H. Zabel, J.J. Rush: Extended Abstracts, Materials Research Society Symposium I, ed. by P.C. Eklund, M.S. Dresselhaus, G. Dresselhaus (Materials Research Society, Pittsburgh, PA 1984) p. 115
4.81 H. Zabel, M. Suzuki, D.A. Neumann, S.E. Hardcastle, A. Magerl, W.A. Kamitakahara: Synth. Met. **12**, 105 (1985)
4.82 D.P. DiVincenzo: Synth. Met. **12**, 111 (1985)
4.83 D.S. Robinson, M.B. Salamon: Phys. Rev. Lett. **48**, 156 (1982)
4.84 A. Magerl, H. Zabel, I.S. Anderson: Phys. Rev. Lett. **55**, 222 (1985)

4.85 V.F. Sears: In *Methods of Experimental Physics: Neutron Scattering*, Vol. 23, Part A, ed. by K. Sköld, D.L. Price (Academic, San Diego, CA 1986) p. 521
4.86 M.M. Bredov, B.A. Kotov, N.M. Okuneva, V.S. Oskotskii, A.L. Shakh-Budagov: Sov. Phys. – Solid State **9**, 214 (1967)
4.87 W.A. Kamitakahara, H. Zabel: Phys. Rev. B **32**, 7817 (1985)
4.88 J.R.D. Copley, J.M. Rowe: Phys. Rev. A **9**, 1656 (1974)
4.89 J.B. Suck, H. Rudin, H.J. Güntherodt, H. Beck: Phys. Rev. Lett. **50**, 49 (1983)
4.90 H. Zabel, A. Magerl, A.J. Dianoux, J.J. Rush: Phys. Rev. Lett. **50**, 2094 (1983)
4.91 J.D. Fan, G. Reiter, S.C. Moss: Extended Abstracts, Materials Research Society Symposium, Boston 1988, ed. by M. Endo, M.S. Dresselhaus, G. Dresselhaus (Materials Research Society, Pittsburgh, PA 1988) p. 13
4.92 R. Moreh, W.C. Sellyey, D.C. Sutton, H. Zabel: Phys. Rev. B **35**, 821 (1987)
4.93 U. Mizutani, T. Kondow, T.B. Massalski: Phys. Rev. B **17**, 3165 (1978)
4.94 M.G. Alexander, D.P. Goshorn, D.G. Onn: Phys. Rev. B **22**, 4535 (1980)
4.95 K. Ohshima, S.C. Moss, R. Clarke: Synth. Met. **12**, 125 (1985)
4.96 S.Y. Leung, M.S. Dresselhaus, C. Underhill, T. Krapchev, G. Dresselhaus, B.J. Wuensch: Phys. Rev. B **24**, 3505 (1981)
4.97 L.E. Campbell, G.L. Montet, G.J. Perlow: Phys. Rev. B **15**, 3318 (1977)
4.98 G. Albinet, J.P. Biberian, M. Bienfait: Phys. Rev. B **3**, 2015 (1971)
4.99 S.E. Hardcastle, H. Zabel: Phys. Rev. B **27**, 6363 (1983)
4.100 E. McRae, J.F. Mareche, A. Bendriss-Rerhrhaye, P. Lagrange, M. Lelaurain: Ann de Phys. Colloq. 2, Suppl. 2, **11**, 13 (1986)

5. Lattice Dynamics II: Optical Studies

Stuart A. Solin

With 46 Figures

In the previous chapter, *Zabel* reviewed the study of the lattice dynamics of graphite intercalation compounds (GICs) using neutron scattering as the probe technique. Neutron scattering is of course the most powerful technique for examining the vibrational excitations of solids since one can examine the phonon dispersion curves throughout the wave-vector range of the Brillouin zone (BZ). In contrast, first-order infrared absorption/reflection spectroscopy and Raman scattering optical studies of the lattice dynamics which use photons as the probing particles access only a very small region near the BZ center [5.1]. Nevertheless, these optical studies provide important information complementary to the neutron measurements.

While photons access a very small wave-vector region of the BZ, they can be used to study high energy optical phonon modes which are well beyond the region where neutron sources provide useful flux [5.2]. This is especially important for GICs since they exhibit the highest energy phonons of any known solid [5.3]. Moreover, whereas bulk specimens of typical volume $1\,cm^3$ are required for neutron scattering measurements, sample regions as small as $4\,\mu m^2 \times 0.1\,\mu m = 1 \times 10^{-13} cm^3$ can be probed with Raman techniques. This allows one to monitor the phonon properties with moderately high spatial resolution. Optical probes typically exhibit considerably higher energy resolution than can be achieved with inelastic neutron scattering. This feature can be very useful in line shape/lifetime analyses. Finally, polarization analysis is a special advantage of the Raman technique owing to the second-rank tensor character of the scattering process. [5.1]. Polarization studies of IR reflection or absorption can also be quite useful when one is dealing with highly oriented or single crystal specimens.

In view of the wealth of complementary lattice dynamical information that can be obtained using optical techniques it is not surprising that the literature on GICs is replete with reports of numerous studies of the optical phonons in these novel materials. To date, there have been several reviews of this subject, the first of which by the present author appeared in 1980 [5.4]. Subsequently, excellent treatises of the optical phonons and lattice dynamics of GICs were prepared by *Dresselhaus* and *Dresselhaus* [5.5, 6], and by *Eklund* [5.7], while *Solin* [5.8–11] has written several short reviews on this topic since the publication of his 1980 article.

All of the above cited review articles dealt exclusively with binary GICs (BGICs) whereas new studies of the optical phonons of ternary GICs (TGICs) have revealed fascinating new physical phenomena [5.12]. In addition, new measurement techniques [5.13–16] have been developed and exploited heavily since the last review on this subject was prepared. These techniques provide heretofore inaccessible information which significantly enhances our understanding of the physics of GICs. Finally, optical phonon studies of GICs have become a routine analytical technique for monitoring a host of phenomena, including staging dynamics [5.15], structural disorder [5.17], diffusion, kinetics [5.15], and intercalation chemistry [5.18], all of which postdate the last comprehensive review on this subject. Clearly, a treatise of the optical studies of the lattice dynamics of GICs which integrates the extensive recent developments with past activities in the field is timely and warranted. The purpose of this chapter is to provide such a review.

In order to avoid redundancy, this review will emphasize the most recent developments. However, to give the nonexpert reader a proper scientific perspective and to provide the background material necessary for the appreciation of the above cited recent developments, some "generic" material must be included. These tasks are accomplished with the following organization of this chapter. Following this introduction, the theory, including group theoretical methods, of first- and second-order Raman scattering in ordered and disordered solids are discussed in Sect. 5.1. In Sect. 5.2, both traditional and new experimental methods for acquiring optical spectra are addressed. Section 5.3 covers pristine graphite. Section 5.4 constitutes the major part of this chapter since it extensively discusses detailed studies of a variety of specific BGICs. The recent work on TGICs is reviewed in Sect. 5.5. In Sect. 5.6, the applications of optical phonon spectroscopy to the study of various systems, including electrochemically intercalated GICs and intercalated fibers, are considered. The final section contains concluding remarks.

5.1 Theoretical Considerations

The theory of IR and Raman scattering from solids of various types has been treated extensively in review articles [5.19, 20], textbooks [5.21, 1], etc., so a lengthy review of these phenomena is inappropriate here. Moreover, IR absorption and reflection spectroscopy are such pervasive techniques in solid-state physics and chemistry that it is reasonable to assume that all of the readers of this book are familiar with the basic physics on which those techniques are based. In contrast, Raman scattering is still a somewhat esoteric phenomenon. Therefore, we provide here a brief theoretical description of the Raman process following an analysis presented by *Hayes* and *Loudon* [5.1]. We will also briefly discuss the selection rules which govern the Raman process.

5.1.1 First-Order Raman Scattering

(a) The Raman Cross-Section

We present here a synopsis of the macroscopic theory of inelastic light scattering following *Hayes* and *Loudon* [5.1]. For additional details and for information on the microscopic theory, the reader is referred to the literature [5.1]. Raman scattering is a two-photon/two-step process. In the first step, an incident photon is absorbed in a real (in the case of resonant Raman scattering) or virtual transition to an intermediate state. The second step involves a transition from the intermediate state to a real final state accompanied by the emission of a scattered photon and the creation or destruction of a phonon. The overall process must conserve energy and crystal momentum, constraints which have important consequences as discussed below. If the initial state in step 1 is the ground state, then a phonon will be created and the scattered photon will occur at lower energy than the incident photon. This is called Stokes scattering. However, if the initial state is a thermally populated phonon level then the scattered photon will have a higher energy than the incident photon. This is called an anti-Stokes process or anti-Stokes scattering. In semiclassical language, the scattering results from polarization fluctuations in the solids which are induced by the applied photon field and modulated by the phonons. These fluctuations generate time-dependent fluctuations in the dielectric function which scatter light inelastically.

Assume that the solid in question is isotropic and sustains a number of non-polar phonon excitations, each of which is characterized by a space- and time-dependent normal coordinate $W_\sigma(\boldsymbol{r},t)$, where

$$W_\sigma(\boldsymbol{r},t) = \sum_{\boldsymbol{q}} \{W_{\sigma\boldsymbol{q}}(t)\exp(\mathrm{i}\boldsymbol{q}\cdot\boldsymbol{r}) + W^*_{\sigma\boldsymbol{q}}(t)\exp(-\mathrm{i}\boldsymbol{q}\cdot\boldsymbol{r})\} . \tag{5.1}$$

Here σ is a phonon branch index and $\boldsymbol{q}$ is the phonon wave vector. We express $W_\sigma(\boldsymbol{r},t)$ in terms of its Fourier components using

$$W_\sigma(\boldsymbol{r},t) = \int W_{\sigma\boldsymbol{q}}(\omega)\exp(-\mathrm{i}\omega t)d\omega . \tag{5.2}$$

The ith component of the fluctuation-induced polarization which produces Stokes Raman scattering is given by

$$P^i_{\mathrm{S}} = \varepsilon_0 \sum_\sigma \chi^{ij}_\sigma(\omega_{\mathrm{I}},0) W_{\sigma\boldsymbol{q}} E^j_{\mathrm{I}} , \tag{5.3}$$

where we define $\chi^{ij}_\sigma = \sum_{\boldsymbol{q}} \chi^{ij}_{\sigma\boldsymbol{q}}$, the indices i and j correspond to the coordinates x, y and z, E^j_{I} is the jth component of the incident photon field whose frequency is ω_{I}, ε_0 is the permittivity of free space and

$$\chi^{ij}_{\sigma\boldsymbol{q}}(\omega_{\mathrm{I}},0) = \frac{\partial \chi^{ij}_{\sigma\boldsymbol{q}}(\omega_{\mathrm{I}})}{\partial W^*_{\sigma\boldsymbol{q}}} \tag{5.4}$$

is the susceptibility derivative.

One can express the differential scattering cross section per unit solid angle, Ω, per unit frequency as

$$\frac{d^2\sigma}{d\Omega d\omega} = \sum_\sigma \sum_q \frac{\omega_I \omega_S^3 bV \eta_S |\varepsilon_0 e_S^i e_I^j \chi_{\sigma q}^{ij}(\omega_I, -\omega_{\sigma q})|^2}{(4\pi\varepsilon_0)^2 c^4 \eta_I} \langle W_{\sigma q} W_{\sigma q}^* \rangle_\omega . \tag{5.5}$$

In (5.5), I and S refer to the incident and scattered photons, η is the refractive index of the medium, e is the polarization vector, b is the volume of the primitive cell, and V is the scattering volume, while the angular bracketed term is the power spectrum of the fluctuations in the normal coordinates and corresponds to the dynamic structure factor used in the terminology of neutron scattering, see Chap. 4. Since, for phonon-induced scattering $\omega_I \sim \omega_S$, the scattering cross section is essentially proportional to ω_I^4, as one expects for classical dipolar scattering [5.22]. The power spectrum in the scattering cross section can be obtained using the fluctuation dissipation theorem [5.23], from which one finds

$$\langle W_{\sigma q} W_{\sigma q}^* \rangle = (h/2N\omega_{\sigma q})[n(\omega_{\sigma q}) + 1] g_\sigma(\omega) , \tag{5.6}$$

where N/V is the density of the scattering medium, n is the Bose-Einstein thermal factor, and $g_\sigma(\omega)$ is a normalized (Lorentzian) line shape function.

The anti-Stokes cross section can be obtained in a similar way and is given by (5.5) and (5.6) above with $S \to AS$, $-\omega_{\sigma q} \to \omega_{\sigma q}$, and $[n(\omega_{\sigma q}) + 1] \to n(\omega_{\sigma q})$. These transformations are valid because the susceptibility derivatives obey the relation

$$\chi_\sigma^{ij}(\omega_I, 0) = \chi_\sigma^{ji}(\omega_S, 0) , \tag{5.7}$$

where we have made the approximation that $\omega_{\sigma q} \ll \omega_I, \omega_S$ so $\omega_{\sigma q} \sim 0$.

(b) The Raman Tensor and Polarization Effects

As can be seen from (5.3), the susceptibility derivatives form a second-rank tensor or Raman tensor $\underset{\sim}{\chi}$ which governs the scattering process. Under nonresonant scattering conditions, i.e., when neither ω_I nor ω_S correspond to an allowed electronic transition in the scattering medium, this tensor is symmetric and contains only six independent components. The σth phonon in the medium is said to be Raman active if one or more of its Raman tensor components is nonzero. The Raman tensor also governs the polarization selection rules, since the scattering cross section in (5.5) is proportional to

$$|e_S^i e_I^j \chi_\sigma^{ij}|^2 \to |e_S \cdot \underset{\sim}{\chi} \cdot e_I|^2 . \tag{5.8}$$

Thus by judiciously choosing the experimental configuration (see below) one can measure the square (but not the sign) of the nonzero tensor components of each of the σ Raman active phonons in the solid. Some of these components may be vanishingly small yet in principle nonzero while others may be identically zero as a result of symmetry selection rules, which can be group theoretically derived as we now briefly describe.

(c) Group Theory and Selection Rules

For a crystalline lattice, the symmetry properties of the Raman tensor and therefore of the scattering cross section are determined by its space group, which is the group of all invariant translation, rotation and reflection operations [5.24]. In first-order Raman scattering from crystals, momentum conservation constrains the observable phonons to the BZ center as we will show shortly. In this case, the relevant symmetry group is the point group, of which there are 32 distinct types for all possible crystal symmetries. Associated with each zone-center phonon excitation of the crystal is an irreducible representation, say Γ_σ, that characterizes its spatial symmetry properties relative to the point group. Now the incident, scattered and polarization fields which characterize the scattering process are all polar vectors whose spatial properties are represented by the polar vector representation Γ_{PV}. But (5.3) must be invariant under all of the appropriate symmetry operations of the crystal, i.e., the point group operations for first-order scattering. This condition imposes symmetry selection rules on the scattering process. In particular, the only symmetry allowed Raman active phonons are those whose irreducible representations Γ_σ are contained in the decomposition of the group theoretical direct product $\Gamma^*_{\mathrm{PV}} \times \Gamma_{\mathrm{PV}}$. As an additional consequence, the relation between the tensor components of the Raman allowed excitations is further constrained beyond the requirement that the tensor be symmetric. Thus for the Raman allowed phonons, some of the tensor components are required to vanish while others are related through equalities. The Raman-allowed zone-center phonon irreducible representations for each of the 32 crystal point groups have been tabulated, together with their Raman tensors, by many authors [5.1]. Thus by assessing the crystal point group symmetry, one can easily determine the form of the Raman tensor for each of the Raman active phonons and identify those phonons experimentally using the polarization selection rules.

(d) Conservation of Momentum and Energy

We now discuss the momentum and energy conservation requirements which limit first-order Raman scattering to the zone-center phonons. Let $\boldsymbol{K}_\alpha$, $\alpha = \mathrm{I, S}$, be the wave vector of the incident or scattered photons in the first-order Raman process. Then

$$|\boldsymbol{K}_\alpha| = 2\pi\eta_\alpha/\lambda_\alpha\,, \quad \alpha = \mathrm{I, S}\,, \tag{5.9}$$

where λ_α is the wavelength of the incident or scattered photon. Conservation of crystal momentum defined in the usual way as $\boldsymbol{P}_\alpha = h\boldsymbol{K}_\alpha$ yields the conditions

$$\boldsymbol{K}_{\mathrm{S}} = \boldsymbol{K}_{\mathrm{I}} - \boldsymbol{q}_\sigma \tag{5.10}$$

and

$$K_{AS} = K_I + q_\sigma \tag{5.11}$$

for Stokes and anti-Stokes scattering, respectively, from the σth phonon. But when visible photons are used to excite the Raman scattering as is usually the case, $|K_\alpha| \sim 2\pi/5000\,\text{Å} \ll |q_{ZB}| \sim 2\pi/3\,\text{Å}$, where $|q_{ZB}|$ is the magnitude of a typical zone-boundary phonon. Therefore, if (5.10) or (5.11) is to be satisfied, $|q_\sigma| \sim 0$ and only zone-center phonons participate in first-order scattering. The conservation-of-energy conditions corresponding to (5.10) and (5.11) are

$$\omega_\sigma = \omega_I - \omega_S \tag{5.12}$$

and

$$\omega_\sigma = \omega_{As} - \omega_I \,. \tag{5.13}$$

So the first-order Raman spectrum of a crystalline solid will contain a series of pairs of sharp or discrete phonon side bands displaced equally in energy from the incident photon at ω_I by an amount $\pm\omega_\sigma$. The amplitude of each of these bands will depend on the polarization configuration in which they are measured and on the Raman cross section given in (5.5)

(e) Disorder and Defect-Induced Scattering

Topological and/or chemical disorder in a nominally crystalline solid can destroy its long range order [5.25]. So too can a finite size which obtains when a crystallite contains too few atoms to be reasonably described by cyclic boundary conditions [5.26]. In such cases, the phonon wave vector is no longer a "good" quantum number with which to describe the vibrational excitations and the momentum conservation conditions given in (5.8) and (5.9) are no longer valid. Scattering is not then limited to zone-center phonons and the first-order Raman spectrum may be a continuum upon which are superposed certain sharp features characteristic of singularities in the phonon density of states. A detailed discussion of disorder-induced scattering is beyond the scope of this chapter, so the reader is referred to the literature for additional details [5.25, 26].

5.1.2 Second-Order Raman Scattering

(a) The Raman Cross Section

Higher-order Raman scattering processes in which multiple phonons are created and/or destroyed in various linear combinations are also possible. Since second-order scattering is usually the highest-order process which can be observed for graphite or its intercalation compounds, we briefly describe the theory for this process here. Because of the thermal population factors in the scattering process, the second-order Raman spectrum is typically dominated by Stokes scattering from overtones or combinations of phonons. Using an analysis similar to that for first-order scattering, it can be shown [5.1] that the second-order differential cross section for Stokes scattering is

$$\frac{d^2\sigma}{d\Omega d\omega} = \frac{\omega_\mathrm{I}\omega_\mathrm{S}^3 bV\eta_\mathrm{S}}{(4\pi\varepsilon_0)^2c^4\eta_\mathrm{I}}\frac{h^2}{4N^2}\sum_{\sigma\sigma'}\sum_{q}|\varepsilon_0 e_\mathrm{S}^i e_\mathrm{I}^j\chi_{\sigma\sigma' q}^{ij}(\omega_\mathrm{I},0)|^2$$
$$\times(\omega_{\sigma q}\omega_{\sigma' q})^{-1}[n(\omega_{\sigma q})+1][n(\omega_{\sigma' q})+1]\delta(\omega-\omega_{\sigma q}-\omega_{\sigma' q}) \,. \quad (5.14)$$

In (5.14) the susceptibility derivative for second-order scattering is defined as

$$\chi_{\sigma\sigma' q}^{ij}(\omega_\mathrm{I},0) = \frac{\partial^2\chi^{ij}(\omega_\mathrm{I})}{\partial W_{\sigma q}^*\partial W_{\sigma' q}^*} \,. \quad (5.15)$$

Now the joint or combined phonon density of states, $\varrho_2(\omega)$, is simply given by

$$\varrho_2(\omega) = \sum_{\sigma\sigma'}\sum_{q}\delta(\omega-\omega_{\sigma q}-\omega_{\sigma' q}) \quad (5.16)$$

so that the second-order spectrum will be a continuum weighted joint density of states whose sharp features represent Van Hove singularities in $\varrho_2(\omega)$.

(b) The Raman Tensor and Polarization Effects

As in the case of first-order scattering, the polarization properties of the second-order scattering are obtained from $|e_\mathrm{S}\cdot\underset{\sim}{\chi}\cdot e_\mathrm{I}|^2$ where the components of the Raman tensor are now defined by (5.15) and the fluctuation-induced second-order Stokes polarization is given by

$$P_\mathrm{S}^i = \varepsilon_0\sum_{\sigma\sigma'}\sum_{q}\chi_{\sigma\sigma' q}^{ij}(\omega_\mathrm{I},0)W_{\sigma q}^*W_{\sigma',-q}^*E_\mathrm{I}^j \,. \quad (5.17)$$

(c) Group Theory and Selection Rules

We will show in the next subsection that the momentum selection rules require the phonon pairs participating in second order scattering to have equal and opposite wave vectors. If $\Gamma_{\sigma\sigma'}$ is the excitation symmetry of the phonon pairs then

$$\Gamma_{\sigma\sigma'} = \Gamma_{\sigma q}\times\Gamma_{\sigma',-q} \,, \quad (5.18)$$

where the right hand side of (5.18) is a direct product [5.24] of the irreducible representations of the space group of the crystal. Raman-allowed two-phonon pair scattering is defined by those pairs for which the irreducible representations of $\Gamma_{\sigma\sigma'}$ are contained in the direct product of the polar vector representation [5.27] $\Gamma_\mathrm{PV}^*\times\Gamma_\mathrm{PV}$ as in the case of first-order scattering. It can be shown that for general phonons with q values not corresponding to high symmetry points in the BZ, $\Gamma_{\sigma\sigma'}'$ contains all zone-center Raman active symmetries [5.27]. Therefore, there are no selection rules on the participation of these phonons in second-order scattering. In contrast, there are selection rules which may apply for scattering involving phonons from high symmetry points in the zone. These symmetry points are the most likely source of Van Hove singularities, so some of the features associated with these singularities may be selection-rule suppressed in the second-order spectrum.

(d) Conservation of Momentum and Energy

The generalization of (5.10) for second-order Stokes scattering is

$$\boldsymbol{q} + \boldsymbol{q}' = \boldsymbol{K}_{\mathrm{I}} - \boldsymbol{K}_{\mathrm{S}} \,. \tag{5.19}$$

But the phonon wave vectors are typically much larger than the photon wave vectors and the right hand side of (5.19) is ~ 0. Thus the momentum selection rules becomes $\boldsymbol{q}' = -\boldsymbol{q}$ and all phonon pairs for which that condition is satisfied can in principle contribute to second-order scattering. This is the reason for the second-order spectrum being a continuum. The energy conservation condition corresponding to (5.12) can be written as

$$\omega_{\sigma \boldsymbol{q}} + \omega_{\sigma' \boldsymbol{q}} = \omega_{\mathrm{I}} - \omega_{\mathrm{S}} \,, \tag{5.20}$$

where we have used the fact that $\omega_{\sigma' \boldsymbol{q}} = \omega'_{\sigma - \boldsymbol{q}}$.

5.2 Experimental Techniques

In this section we review experimental spectroscopic techniques which have been successfully applied to the study of graphite and GICs, with emphasis on new techniques that have not been addressed in previous review articles. For a general discussion of the methodology and apparatus employed in Raman and IR spectroscopy, the reader is referred to the excellent texts by *Hayes* and *Loudon* [5.1] and by *Muller* and *Rothschild* [5.21], respectively.

5.2.1 Traditional Methods

(a) The Reflection Geometry

c-Face Measurements. Unless one prepares extremely thin specimens, graphite and GICs are opaque to the incident photons normally used in IR and Raman spectroscopy. Moreover, since graphitic materials easily cleave parallel to the layer planes, it is the basal surface or *c*-face that is normally exposed to the incident radiation in an optical experiment. When this is the case the Raman spectra are measured using the "backscattering" reflection geometry depicted in Fig. 5.1. In such a configuration, the incident and scattered photons sample a

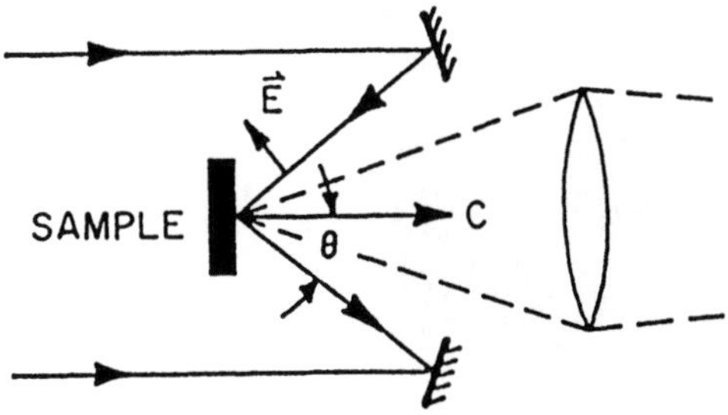

Fig. 5.1. Backscattering geometry for Raman scattering experiments on the c-face. The incident photon is polarized in the scattering plane. This is the Brewster angle configuration. (From [5.7])

shallow region within $\sim$ 1000 Å of the surface. Therefore, to increase the coupling of the incident photon to the sample, the incident beam is usually polarized parallel (||) to the scattering plane while the scattered beam is analyzed parallel (||) or perpendicular ($\perp$) to the scattering plane. This is called the Brewster angle backscattering configuration. Then, using the ||, || and ||, $\perp$ polarization configurations one can probe the diagonal and off-diagonal components of the Raman tensor, e.g., χ^{xx} and χ^{xy}. However, because the index of refraction at the incident wavelength is very large, photons incident on the basal surface at essentially any angle propagate along the c-axis in the specimen. Therefore, with the above specified configuration, tensor components such as $\chi^{z\alpha}$, $\alpha = x, y, z \parallel c$ cannot be studied. [It is conventional in polarized Raman scattering studies to use either the above designations or to designate the scattering configuration with the label $i(jk)\ell$, where $i(\ell)$ is the direction of the incident (scattered) photon and $j(k)$ are the corresponding polarizations. Thus a $z(xy)z$ experiment is a "backscattering" measurement of the off-diagonal tensor component χ^{xy} with the incident and scattered photons propagating along the c-axis.] For IR studies, the near-normal incidence reflection geometry is normally used. Since IR experiments constitute a one-photon vector process, in contrast to the two-photon tensor process or Raman scattering, polarization analysis of the c-face studies of graphite or GICs provides little if any useful information. It is not surprising then that no such studies have been reported to date.

The detection of the high- and low-frequency external modes of the host lattice in c-face IR and Raman measurements of graphite and GICs is quite straightforward, possibly because the scattering from these modes is resonantly enhanced when visible incident photons are used [5.28]. However, in the case of molecular intercalates which form acceptor GICs, only rarely have the internal molecular modes been deteceted in c-face measurements. In one such example, bromine-graphite [5.29], detection of the internal modes was clearly facilitated by resonance enhancement of the Raman signal. Other reported observations of internal modes in $FeCl_3$ graphite [5.30] using nonresonant c-face scattering are still controversial.

***a*-Face Measurements.** While the preparation of c-faces of graphite and GICs for optical studies is straightforward, the preparation of damage-free optical quality a-faces using, e.g., HOPG is much more difficult. But for IR measurements of modes with displacements along the c-axis, the so-called out-of-plane modes,

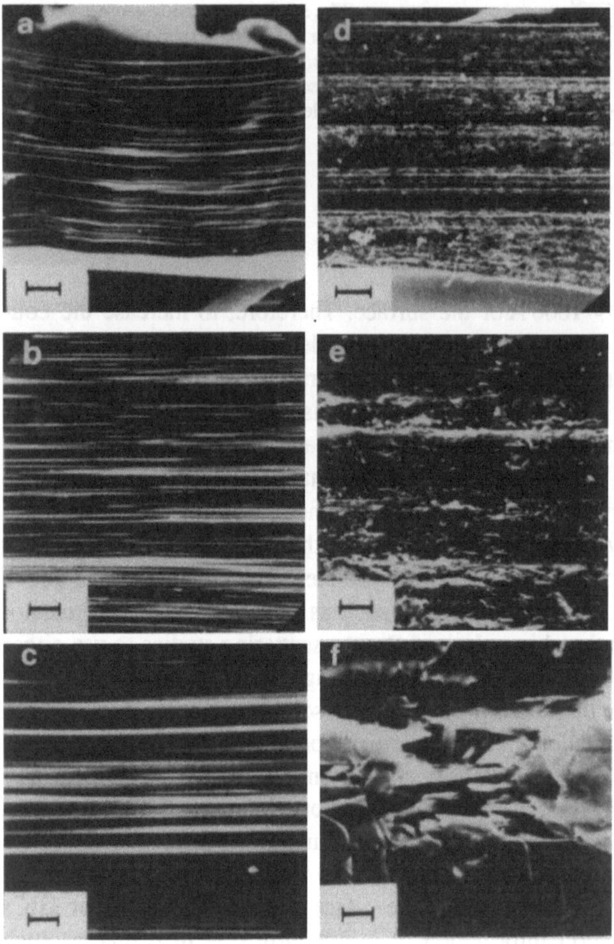

Fig. 5.2. Scanning-electron-microscope pictures of the a-faces of two different samples. (a)–(c) The surface of a torn sample. (d)–(f) Pictures with different scales of a knife-cut sample. The lines in the pictures indicate the scale; 100, 10 and 1 μm in (a) and (d), (b) and (e), and (c) and (f), respectively. (From [5.16])

the use of a reflection a-face geometry is desirable. *Nemanich* et al. [5.31] prepared a-faces of HOPG for their IR studies of pristine graphite by mechanical polishing. These faces were heavily damaged on a microscopic scale, primarily due to bending and interleaving of the crumpled layer edges. Others have used sputtering techniques to prepare even better a-faces on both pristine graphite and GICs [5.32, 33]. Surprisingly, the best a-faces prepared to date [5.16] were obtained by tearing a clamped piece of HOPG along a direction parallel to the layers. Scanning electron micrographs of a torn edge a-face of such samples are shown in Fig. 5.2 and compared with the micrographs of the a-face of a knife-cut specimen. As can be seen, the torn edge a-face is clearly superior and possesses layers that are undamaged all the way to the surface, which is smooth in the direction parallel to the layers but rough on a scale of about 1000 Å in the c-axis direction.

As noted above, the observation of nonresonantly enhanced internal molecular modes is extremely difficult, if not impossible, using c-face excitation [5.16]. This limitation results in part from the very small penetration depth sampled by photons incident on the c-face. Attempts to observe molecular modes in acceptor GICs by selecting incident photons with energies in the reflectivity minimum were unsuccessful [5.16, 34]. However, *Ohana* and *Yacoby* [5.16] were able to obtain Raman spectra of the internal molecular modes of acceptor GICs by using a backscattering reflection geometry on torn edge a-face specimens with the incident field polarized parallel to the c-axis. For reasons which will be amplified upon below, the penetration depth under those conditions is not limited by electronic absorption or reflection. As a result, the scattering volume is sufficiently enhanced to render the nonresonant internal molecular modes observable.

(b) The Transmission Geometry

Although transmission IR studies of thin pristine graphite and BGIC specimens have been reported [5.35], the phonon modes were not probed in these investigations. Transmission Raman measurements have, to our knowledge, not yet been reported for graphite or any of its intercalation compounds.

5.2.2 Novel Techniques

All of the techniques to be discussed below were carried out using the reflection backscattering geometry. Moreover, with the exception of certain micro-Raman measurements on fibers, photons were incident and scattered from the c-face.

(a) Micro-Raman Spectroscopy

In recent years, micro sampling techniques have been developed for Raman spectroscopy and these techniques have proved to be very useful when applied to graphite and GICs [5.14], especially when the host material is in the form of a thin fiber [5.36]. The optical arrangement typically used to acquire laser Raman spectra with a spatial resolution of $< 5\,\mu$m is shown in Fig. 5.3. As can be seen from the figure, a backscattering geometry is typically employed with a very large collection angle for the scattered light. This large collection angle seriously degrades polarization analysis so one normally studies only polarization unanalyzed spectra. By accurately locating the specimen at the small focal spot of the micro-Raman apparatus using a micropositioner, a one- or two-dimensional spatial profile of the Raman spectrum can be obtained. Measurements of this type have been made for both pitch-based and polyacrylonitrile (PAN)-based fibers [5.37] in both the pristine and intercalated form. Note however that pitch fibers exhibit a morphology in which the graphitic planes are directed radially outward from the fiber axis while in PAN fibers these planes adopt a cylindrical morphol-

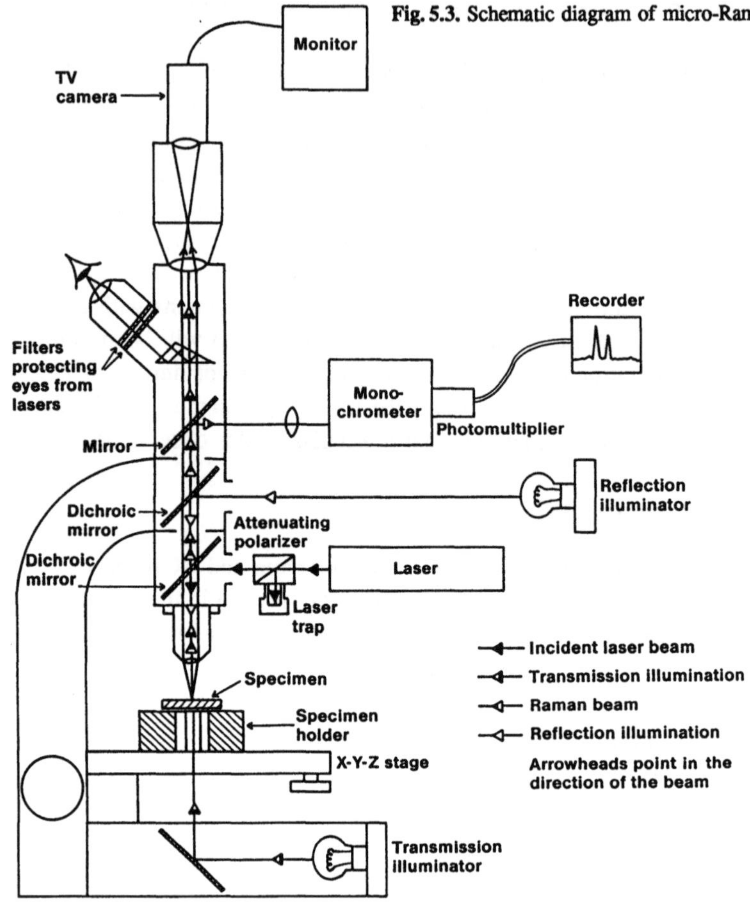

Fig. 5.3. Schematic diagram of micro-Raman apparatus

ogy about the fiber axis [5.37]. Thus for a reflection backscattering geometry with light incident normal to the fiber axis, PAN fibers yield primarily a c-face response while pitch fibers contain a major portion of the a-face response.

(b) High-Pressure Measurements

Diamond anvil optical cells capable of generating true hydrostatic pressures up to several Mbar have been used for some time in the study of the optical properties of solids [5.38] but have been employed only recently by *Wada* [5.13] to study the pressure and stage dependence of the first-order Raman spectra of GICs. Because these materials are highly reactive to air and to fluids commonly used as pressure transmitting media [5.38], special sample handling techniques were developed to facilitate the high pressure optical studies of GICs [5.13]. For a detailed discussion of the use of the diamond anvil technique, the reader is referred to the excellent treatise by *Hazen* and *Finger* [5.39].

(c) In Situ Raman Studies of Intercalation

While in situ X-ray studies of the intercalation process are quite straightforward, corresponding Raman studies, especially with the specimen at elevated temperature, can be plagued by the blackbody background emission. Nevertheless, *Caswell* and *Solin* did successfully employ a lock-in electronic detection technique to study, in situ, the in-plane intercalant order–disorder melting transition in Cs-graphite [5.39]. More recently, *Eklund* and co-workers [5.40] pioneered room-temperature in situ Raman studies of intercalation using an electrochemical process. They were able to monitor with relatively coarse spatial resolution the Raman spectrum of H_2SO_4 graphite as a function of charge exchange to the host lattice. Following *Eklund* et al., *Nishitani* et al. [5.15] monitored the spectral properties of the same intercalation system as a function of position in real time. They were thus able to probe directly the kinetics of intercalation as well as the kinetic effect on staging.

5.3 Pristine Graphite

5.3.1 Structure and Symmetry

Graphite, the prototypical layered solid, is composed of stacked layers of hexagonally arranged carbon atoms. The most common form of graphite is the hexagonal form, which possesses an *ABAB* stacking arrangement as shown in Fig. 5.4 [5.41]. But there also exists in nature a rhombohedral form with a stacking sequence *ABCABC* [5.41]. Since only the hexagonal form of graphite has been the subject of intercalation studies, we will concentrate here exclusively on that

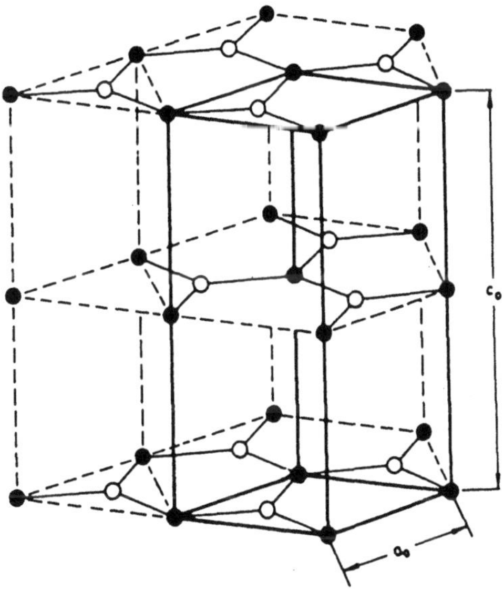

Fig. 5.4. Crystal structure of graphite, a_0 = 2.45 Å, c_0 = 6.70 Å. (From [5.41])

form. The highly anisotropic physical properties of hexagonal graphite and its intercalation compounds result in part from the large difference in the in-plane C-C distance (1.42 Å) relative to the interplanar distance (3.35 Å) as can be seen in Fig. 5.4. Also, note from the figure that hexagonal pristine graphite crystallizes in the D_{6h}^4 space group and contains four atoms in the biplanar primitive cell.

5.3.2 First-Order Optical Spectra

In order to understand the optical spectra of graphite and its intercalation compounds, we must of course refer to the relevant phonon dispersion curves. *Zabel*, in Chap. 4, has reviewed the various models that have been advanced to explain the phonon dispersion of pristine graphite and has pointed out that the 20-parameter Born–von Kármán model employed by *Al-Jishi* and *Dresselhaus* [5.42] provides the best agreement with all of the available experimental data. It is interesting to note that the multitude of previous models [5.43], all of which provided reasonable agreement with the available neutron scattering results that were acquired in the low-frequency region, yielded conflicting results for the high-frequency intralayer modes. This conflict was not resolved until optical data [5.31] on some of the key high-frequency phonons became available.

The phonon dispersion curves for pristine graphite, based on the *Al-Jishi–Dresselhaus* calculations [5.42] are shown in Fig. 5.5 and the hexagonal BZ in Fig. 5.6. Both of these figures are labeled with conventional notations for the key points and lines in a hexagonal BZ. Consider first the zone-center Γ-point phonons. Using well-known group theoretical techniques [5.24, 44], one can show that there are 12 such modes of vibration associated with the three degrees of freedom of the four atoms in the primitive cell. These modes can be labeled according to the following irreducible representations of the D_{6h} point group:

$$\Gamma_{\text{vib}} = 2A_{2u} + 2B_{1g} + 2E_{1u} + 2E_{2g} \tag{5.21}$$

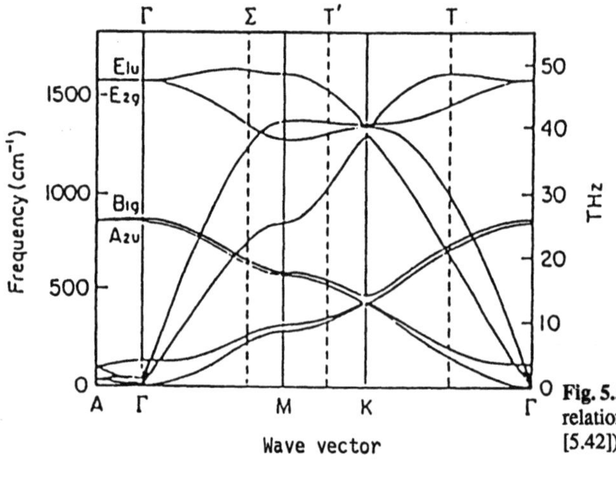

Fig. 5.5. The phonon dispersion relations of pristine graphite. (From [5.42])

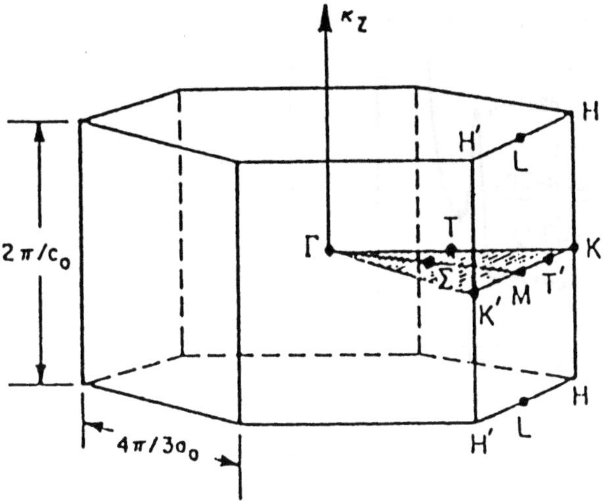

Fig. 5.6. The hexagonal Brillouin zone of pristine graphite

and they are indicated in Fig. 5.5. Three of these modes, the A_{2u} singlet and the E_{1u} doublet are acoustic vibrations which correspond to pure crystal translation. They are not observable in the optical spectra. The two B_{1g} singlet excitations are silent and so are also unobservable optically. Of the remaining zone-center excitations, the $2E_{2g}$ doubly degenerate excitations yield four Raman active modes while the A_{2u} and E_{1u} excitations yield three IR active modes. Thus one expects to observe a total of four spectroscopically detectable optic phonons for pristine graphite. All four of these phonon modes have indeed been observed [5.4] as shown in Fig. 5.7. Figure 5.7 also shows the specific eigenvectors corresponding to each optically detectable vibration. The $E_{2g}^{(2)}$ intralayer mode is the highest-energy zone center excitation exhibited by any solid [5.45] and is evidence that, for intraplanar displacements, graphite is in fact "harder" than diamond [5.46]. The companion IR active E_{1u} mode shown in Fig. 5.7 is very close in energy and exhibits identical intraplanar motions but interplanar motions of opposite phase relative to those of the $E_{2g}^{(2)}$ mode. Notice that the $E_{2g}^{(1)}$ mode involves motions devoid of intralayer bond stretching and that adjacent layers move as whole units in opposite directions. The energy of such rigid-layer shear modes is determined by the interlayer shearing interaction and is thus very low, since this interaction is much weaker than the intralayer interaction. Of the observable optical modes, only the A_{2u} vibration involves out-of-plane motions and therefore intermediate energies. Indeed, prior to the work of *Nemanich* et al. [5.31], the values for the energy of this mode based on various models of the phonon dispersion curves of graphite ranged from $600\,\mathrm{cm}^{-1}$ to $1400\,\mathrm{cm}^{-1}$ [5.4]. Thus the determination of the A_{2u} mode energy provided a significant additional constraint on the lattice dynamical models.

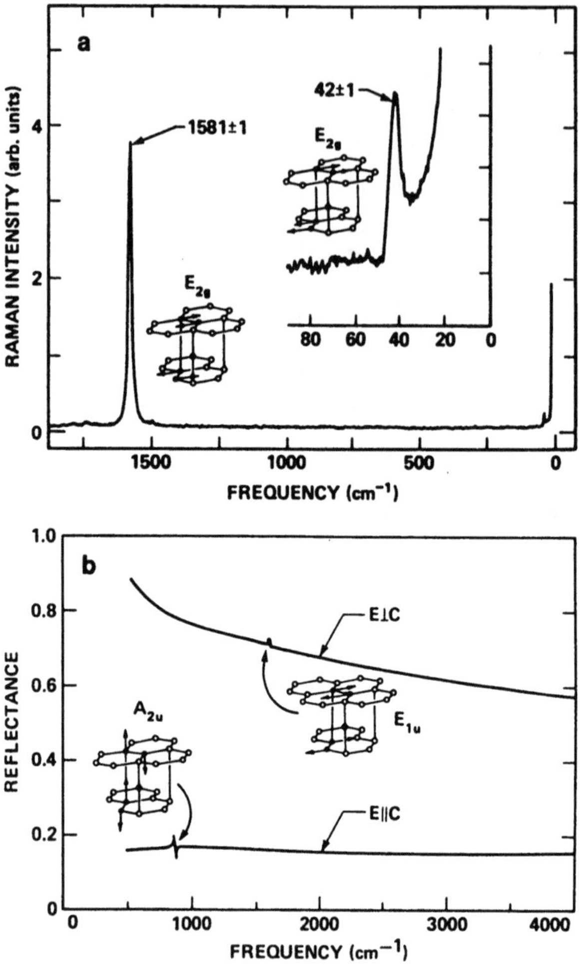

Fig. 5.7. (a) The Raman spectra obtained from a cleaved surface of graphite, and the atomic displacements of the Raman active E_{2g} modes. (b) The $\boldsymbol{E} \perp \boldsymbol{c}$ and $\boldsymbol{E} \parallel \boldsymbol{c}$ reflectance of graphite, and the atomic displacements of the IR active modes. The $\boldsymbol{E} \perp \boldsymbol{c}$, spectrum has been normalized for surface damage. (From [5.4])

5.3.3 Second-Order Raman Scattering

The second-order Raman scattering from HOPG has been extensively studied by *Nemanich* and *Solin* [5.3, 47] and more recently by *Al-Jishi* and *Dresselhaus* [5.42]. The polarized spectra reported by the former [5.47] are shown in Fig. 5.8. The key features of those spectra were attributed to overtone scattering associated with peaks in the two-phonon density of vibrational states calculated on the basis of the phonon dispersion curves deduced by *Maeda* et al. [5.48]. Such a calculation is of course only approximate. The second-order spectrum is in fact a weighted density of states where the weighting function is proportional to the

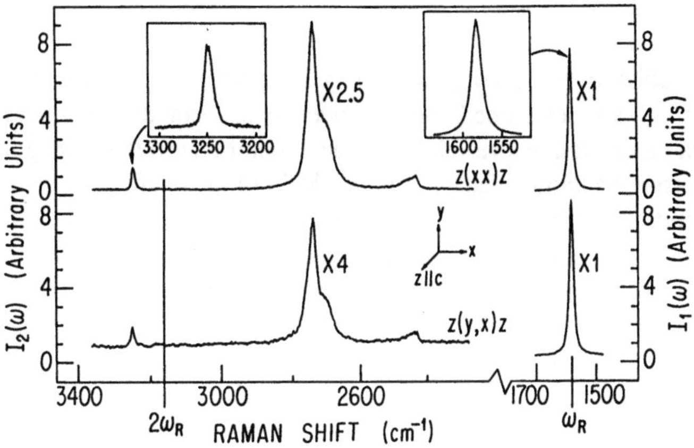

Fig. 5.8. Polarized first (I_1) and second (I_2) order room temperature Raman spectra of highly oriented pyrolytic graphite. The spectra were excited with 550 mW of 4880 Å argon ion laser radiation and were recorded with spectral slit widths of 6.2 cm^{-1} and 5.5 cm^{-1} respectively. Laboratory axes are labeled x, y, and z with z parallel to the crystal "c" axis. The abscissa is marked in wave numbers but is linear in wavelength. (From [5.47])

square of the relevant second-order Raman tensor for the scattering polarization in question. *Al-Jishi* and *Dresselhaus* [5.42] have calculated the second-order spectrum including the weight function using their own phonon dispersion curves. Their calculation confirms the band assignments reported by *Nemanich* and *Solin* [5.3, 47] and in addition accounts reasonably well for the relative intensities of all of the observed spectral features. Moreover, the energies of the doublet at 2435 cm^{-1} and 2468 cm^{-1} are correctly reproduced. This indicates that the phonon dispersion curves reported by *Al-Jishi* and *Dresselhaus* [5.42] are superior to those of *Maeda* et al. [5.48], since calculations based on the latter do not yield the correct doublet energies.

A key feature of more current lattice dynamical models of graphite that distinguishes them from most previous models is the unusual rise in the highest dispersion curve along the Σ branch between the Γ and M points of the BZ, see Figs. 5.5 and 5.6. If the top-most dispersion curve has its maximum energy at other than the Γ-point of the BZ, one might expect to observe evidence for this feature in the second-order Raman spectrum, which should then exhibit an overtone band at an energy greater than twice the energy of either the $E_{2g}^{(2)}$ mode (1581 cm^{-1}) or the E_{1u} mode (1588 cm^{-1}). Note in particular the sharp feature at 3248 cm^{-1} in the second-order Raman spectrum of graphite [5.47] (Fig. 5.8). This band, which is upshifted by fully 72 cm^{-1} from the overtone of the E_{1u} mode, is clear evidence that the top-most phonon dispersion curve of graphite rises 36 cm^{-1} above the maximal Γ-point energy and gives strong support for lattice dynamical models which predict this anomalous feature ([5.42, 48], Chap. 4 and references therein).

5.3.4 Finite-Size Effects

As a part of their studies of the second-order Raman spectra of graphite, *Nemanich* and *Solin* [5.3] also examined the dependence of both the first- and second-order spectra on crystallite size. The in-plane structural correlation length L_a was determined from the known linear variation [5.49] of the relative intensity of the first-order bands at $1355\,cm^{-1}$ and $1581\,cm^{-1}$ with crystal size. As can be seen from *Nemanich* and *Solin*'s [5.3] results shown in Fig. 5.9, the reduction in L_a to molecular dimensions including length scales appropriate to the short range correlations in glassy carbon has a dramatic effect on the Raman spectra. Note particularly the remarkable growth of the first-order band at $1355\,cm^{-1}$ and the development of the doublet at $1581 - 1620\,cm^{-1}$. Equally pronounced is the growth of the $2700\,cm^{-1}$ band with decreasing crystallite size in the second-order spectrum. Also note that both members of the $1581 - 1620\,cm^{-1}$ doublet shift to higher energy by as much as $10\,cm^{-1}$ with decreasing crystallite size.

Nemanich and *Solin* [5.3] noted that both the first- and second-order Raman spectra of microcrystalline graphite can be qualitatively understood in terms of the breakdown of the wave-vector conservation requirement in crystallites which are too small to support Bloch waves. For instance, the $1620\,cm^{-1}$ band was ascribed to phonons along the topmost dispersion curves which are activated by the selection rule relaxation induced by finite crystal size. Similarly, the upshift of the $1581 - 1620\,cm^{-1}$ doublet was attributed to phonons along the rising Σ branch away from the center of the BZ that now participate in the scattering process.

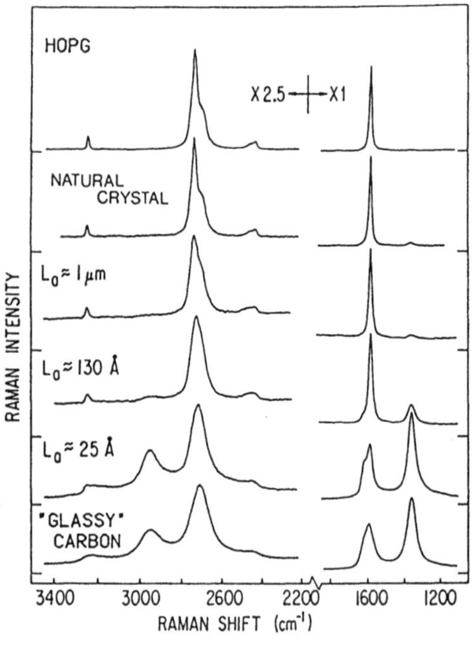

Fig. 5.9. First- and second-order Raman spectra of several different forms of graphite with a range of in-plane correlation lengths, L_a (see text). (From [5.3])

The first-order band at 1355 cm^{-1} was associated with a zone-boundary phonon which would be unobservable if the momentum selection rules obtained. Key features in the second-order spectra were identified by *Nemanich* and *Solin* [5.3] as overtones of the above-cited first-order excitations. For instance, the 3248 cm^{-1} high-frequency mode and the strong band at $\sim$ 2710 cm^{-1} were identified as overtones of the 1620 cm^{-1} and 1355 cm^{-1} modes, respectively. Clearly, as a result of the theoretical and experimental contributions discussed above, the first- and second-order optical spectra not only of single crystal graphite but of various other microcrystalline morphologies, are now reasonably well understood.

5.4 Binary Graphite Intercalation Compounds

The overwhelming majority of optical studies of the lattice dynamical properties of GICs have dealt with binary compounds. Therefore this section will naturally constitute the bulk of the current review.

5.4.1 Lattice Dynamics and First-Order Optical Spectra

(a) General Theoretical Considerations

As in the case of pristine graphite discussed above, prerequisite to a quantitative understanding of the Raman and IR spectra of BGICs is a detailed knowledge of the phonon dispersion curves for each of the stages of a given intercalate under consideration. Fortunately, in the case of the BGICs, model calculations of the stage-dependence of the phonon dispersion curves have been carried out. These have been reviewed thoroughly in the previous chapter. The most accurate of these models is that of *Al-Jishi* and *Dresselhaus* [5.50] who not only calculated the phonon dispersion curves but also the resultant first- and second-order Raman spectra as well. Although their calculations were carried out explicitly for alkali-donor GICs, the principles established in their model and in prior calculations by *Leung* et al. [5.51], who addressed both donor and acceptor BGICs, have broad applicability. For instance, with the proper choice of force constants, these models should work equally well for a variety of TGICs, including homogeneous and heterogeneous forms [5.12]. The versatility of these models results from simplifications associated with the neglect of the in-plane structure of the intercalate layers and the insensitivity of the dispersion curves to the explicit lateral arrangement (e.g., staggered or eclipsed) of the carbon layers. While such a simplification precludes a proper description of the stage-1 alkali BGICs, which have a 3d superlattice structure [5.45], the dispersion curves of most other GICs can be adequately treated.

Not surprisingly, the phonon dispersion curves of the intercalated compounds are closely related to those of pristine graphite. For the reader's convenience

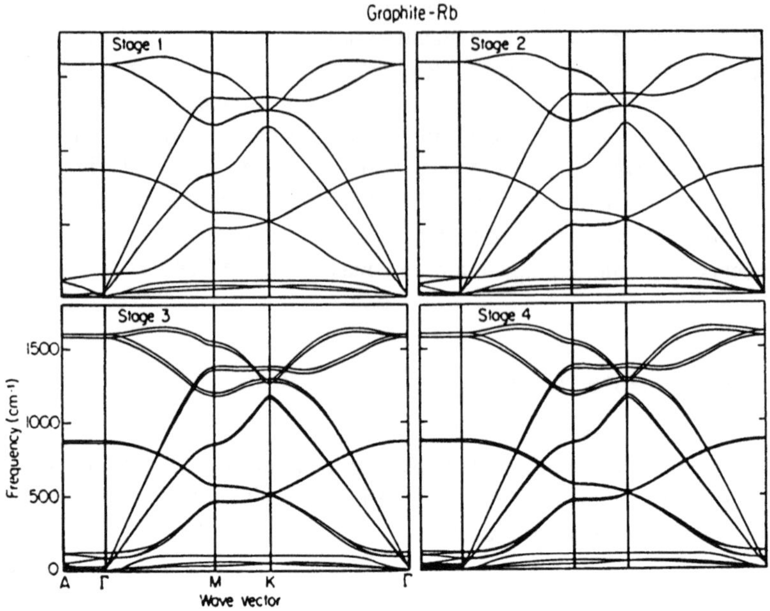

Fig. 5.10. The calculated phonon dispersion curves of rubidium-graphite. (From [5.50])

and to facilitate the following discussion, we reproduce in Fig. 5.10 the *Al-Jishi* and *Dresselhaus* calculation of the phonon dispersion curves for stage 1–4 Rb graphite [5.50]. The essential feature of the *Al-Jishi* and *Dresselhaus* model which influences the first-order Raman and IR spectra of GICs is that the real space expansion of the c-axis unit cell associated with stage formation gives rise to a contraction of the reciprocal lattice along the c^* direction by a factor of $(n+1)/2$. In the limit of weak interlayer interactions, this contraction results in the well-known effect of zone-folding [5.5] of the graphite phonon dispersion curves so that non-zone-center phonons of graphite are mapped back to the zone center in GICs. In addition to the graphitic zone-folded modes, there may also be phonon branches associated with external motions of the intercalant species [5.52].

For higher stage materials, the mapping of the graphitic modes onto the zone-center produces a plethora of optically active modes. However, the weak interlayer interaction results in very little dispersion along the c^* direction in the BZ (Fig. 5.6) so that many of the folded phonons are quasi degenerate and/or unresolvable in optical studies. This is particularly true for the high-frequency modes in the region of the $E_{2g}^{(2)}$ mode of pristine graphite ($\sim 1580\,\mathrm{cm}^{-1}$). Note that on the scale of the ordinates in Fig. 5.10, the slight ($\sim 6\,\mathrm{cm}^{-1}$) splitting of the $E_{2g}^{(2)}$ and E_{1u} modes is not resolvable. But for $n \geq 3$ this unresolved doublet splits into a pair of unresolved doublets. Each member of this pair contains Raman- and IR-active components which should manifest themselves as a doublet in the corresponding optical spectra for $n \geq 3$. In contrast, the first-order Raman and

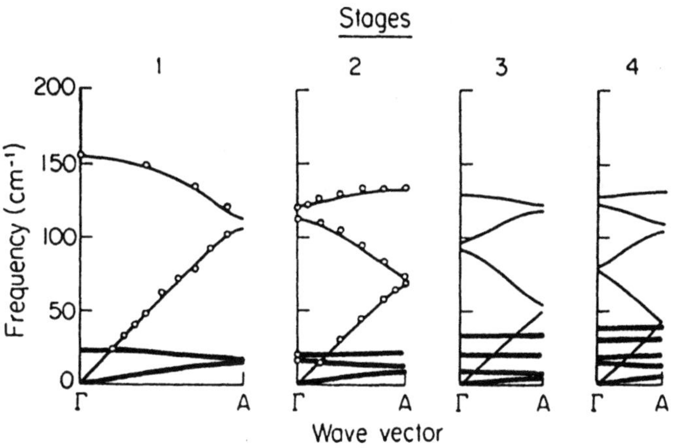

Fig. 5.11. Phonon-dispersion curves along the c^* direction in stage 1, 2, 3 and 4 Rb-GICs. The open circles represent the experimental points (see Chap. 4 and references therein). Heavy lines (light lines) represent calculated dispersions of the rigid-layer shear (beating) modes. (From [5.50])

IR spectra of stage-1 and stage-2 compounds should exhibit only a single mode in the high frequency region.

The low-frequency external modes which involve coupled motions of the host and guest layers are much more sensitive to the mass of the intercalated species and to the guest–host interactions than are the high-frequency modes. The stage-dependence of the c-axis low-frequency dispersion curves of Rb-graphite calculated by *Al-Jishi* and *Dresselhaus* [5.50] is shown in Fig. 5.11. These curves correspond to rigid layer motions of the shear type (heavy lines in Fig. 5.10) or the longitudinal (beating) type (light lines in Fig. 5.10) and involve displacements in which whole layers move as a single entity. The latter are derived from the (optically) silent $B_{1g}^{(1)}$ longitudinal mode of pristine graphite while the former derive from the $E_{2g}^{(1)}$ rigid layer mode. Note the dramatic effect of zone-folding on these branches, which are more easily resolved than some of the high-frequency branches. In particular, a stage-n GIC exhibits two sets of $n+1$ modes, one set associated with beating modes and the other with shear modes. Within each set, one of the modes is acoustic and is thus not observable in first-order Raman or IR measurements. The n remaining optic modes in each set may be Raman and IR allowed, depending on their particular symmetry properties, as discussed in Sect. 5.1.

Before the phonon dispersion curves of GICs were measured or calculated, the above-described stage-dependence of the first-order high-frequency optical spectra of the $E_{2g}^{(2)} - E_{1u}$ region of BGICs was first understood qualitatively using a simple physical picture [5.53–55], which is referred to by *Solin* and co-workers as the nearest layer model (NLM) [5.53, 54]. Since much physical insight into the behavior of GICs can be gained from an understanding of the NLM, we briefly review its salient features here.

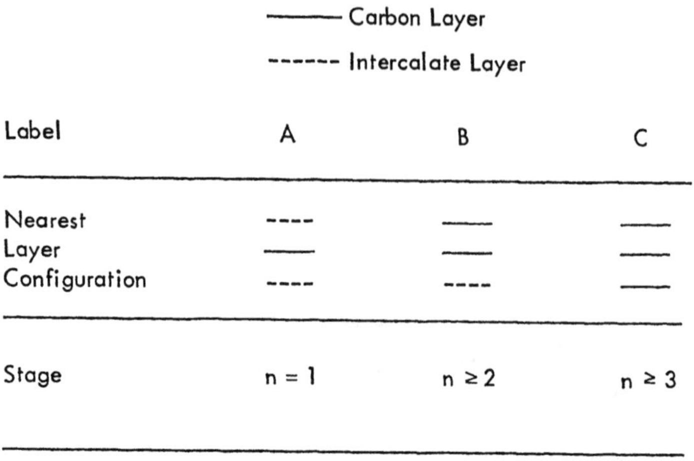

Fig. 5.12. Nearest-layer environments of graphite layers in graphite intercalation compounds

Consider the possible nearest layer environments of a graphite layer in a GIC independent of the in-plane structure of the intercalant layers themselves. There are only three possible nearest-layer environments and these are labeled respectively as types A, B, and C as shown in Fig. 5.12. For a stage-1 compound every carbon plane is flanked on either side by an intercalate layer (type A in Fig. 5.12). These carbon layers are called bounding layers and are the only type which exist in the compound. If one assumes that the intercalate merely perturbs the high-frequency intralayer phonons then one would expect to observe a single (energy shifted) $E_{2g}^{(2)}$-like mode in the first-order Raman spectrum of either a donor or acceptor stage-1 GIC and correspondingly a single E_{1u}-like mode in the IR spectrum. For a stage-2 GIC, there is again only one type of nearest-layer environment, in which every graphite plane is flanked on one side by an intercalate layer and on the other side by another graphite layer (type B in Fig. 5.12). (Although different from the type A layers, type B carbon layers are also called bounding layers.) Again using a perturbation approach, one also expects a single intralayer mode to occur in the Raman and IR spectra of stage-2 GICs but at an energy shifted from the corresponding modes of pristine graphite by an amount different from that observed for the stage-1 compound. Finally for GICs with $n \geq 3$, exactly two types of nearest layer environment exist: the type B environment characteristic of the stage-2 compound and in addition one in which the carbon layer is flanked on both sides by another carbon layer (type C in Fig. 5.12). Therefore, the first-order Raman (IR) spectra of GICs with $n \geq 3$ should exhibit a doublet of $E_{2g}^{(2)}(E_{1u})$ modes, one labeled $E_{2g}^{(2)i}(E_{1u}^{i})$ derived from the interior carbon layers and one labeled $E_{2g}^{(2)b}(E_{1u}^{b})$ derived from the bounding carbon layers.

In the context of the NLM, the intensity of the member of the Raman or IR doublet associated with the interior layers should grow with increasing stage for $n \geq 3$ while the intensity of the bounding layer mode should decrease. The stage dependence of the relative intensities of the Raman doublet, R, has been computed by *Solin* [5.53] by simply "counting" the numbers of type C interior layers and type B bounding layers and assuming that σ_i/σ_b, the ratio of the Raman cross sections for scattering from $E_{2g}^{(2)i}$ and $E_{2g}^{(2)b}$ modes, was a stage-independent constant. He found that for $N \geq 2$

$$R = \frac{I[E_{2g}^{(2)i}]}{I[E_{2g}^{(2)b}]} = \frac{\sigma_i}{\sigma_b}\frac{n-2}{2} . \tag{5.22}$$

(b) Comparison of Theory with Spectroscopic Measurements

The stage-1 alkali donor BGICs exhibit a unique set of optical spectra, which require additional theoretical treatment beyond the lattice dynamics approach taken above. We will now discuss the theoretical and experimental aspects of those compounds and then address the properties of donor BGICs with $n \geq 2$ and of acceptor BGICs with $n \geq 1$.

Stage-1 Donors. The first-order Raman spectra of all stage-1 donor BGICs studied to date [5.6–11, 56–58] are shown in Fig. 5.13. With the exception of LiC_6 [5.56], these spectra are unique relative to other donor and acceptor GICs.

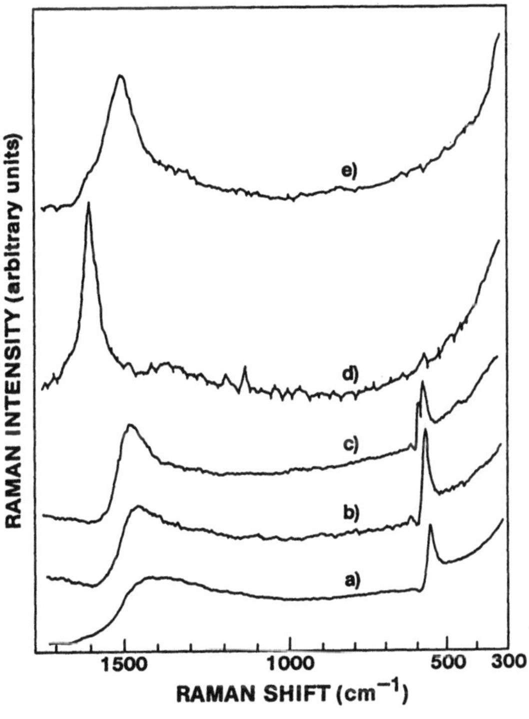

Fig. 5.13. The first-order Raman spectra of stage-1 donor binary graphite intercalation compounds. (*a*) K [5.8], (*b*) Rb [5.8], (*c*) Cs [5.8], (*d*) Li [5.58], (*e*) Eu [5.58]

The uniqueness is associated with the high-frequency asymmetrically broadened band and, for the heavy alkalis K, Rb, and Cs [5.6, 8], the set of modes in the 500–600 cm^{-1} region. It is now well known that the primary source of the unique spectral properties of the stage-1 donors is their crystal structure [5.6, 8]. The intercalated ions in all of these compounds form, at room temperature, a 3d superlattice which is, by definition, commensurate with the host layers [5.5, 45]. In contrast, higher stage donors and almost all acceptors exhibit intercalate layer structures which are incommensurate or disordered [5.5, 45] at room temperature. Therefore, since in-plane order is one of the central structural features of stage-1 donor BGICs, models such as the NLM or phonon dispersion calculations which are based solely on the c-axis structure are inadequate. But they have been appropriately modified to account well for the spectra in Fig. 5.13, as we now show.

Consider first LiC_6 [5.56–58]. It exhibits a broad singlet which is readily associated with the $E_{2g}^{(2)b}$ mode, as expected from the phonon dispersion calculations by *Al-Jishi* and *Dresselhaus* [5.50] and the NLM [5.53]. (In fact, this band is weakly asymmetric for reasons which will soon become clear.) In contrast, the MC_8 compounds [5.8] show, in the bounding mode region, only the highly asymmetric feature noted above. This feature was first identified by *Nemanich* et al. [5.54] as a Fano-Breit-Wigner (FBW) line [5.59], which results from the symmetry-allowed interaction of a discrete (phonon) state with a continuum background. The FBW line shape which is often observed in Raman scattering from semiconductors [5.60] is given by the expression

$$\frac{d^2\sigma}{d\Omega d\omega} = K\left[1+\left(\frac{\omega_S-\omega_0}{Q\Gamma}\right)^2\right]\left[1+\left(\frac{\omega_S-\omega_0}{\Gamma}\right)^2\right]^{-1} , \tag{5.23}$$

where K is a normalization constant, ω_0 is the bare (unrenormalized) frequency of the discrete phonon, Γ is a linewidth and q is the inverse of the FBW coupling coefficient. The renormalized frequency ω_{max} at which the observed intensity maximum occurs is given by

$$\omega_{max} = \omega_0 + \Gamma/Q . \tag{5.24}$$

Note that in the limit $1/Q \to 0$, which corresponds to decoupling between the discrete state and the continuum, the FBW line shape reduces to a Lorentzian form. Equation (5.23) provides an excellent fit to the Raman spectra of the MC_8 alkali GICs in the 1400–1700 cm^{-1} spectral region [5.54, 61]. The resultant FBW parameters have been tabulated by *Dresselhaus* and *Dresselhaus* [5.6]. The EuC_6 spectrum is also characterized by a FBW line shape [5.57, 58]. So too is the LiC_6 spectrum [5.56], although the coupling between the bounding layer phonon and the continuum is much weaker than in the case of the other stage-1 donors, so that the line is much more Lorentzian in that compound.

The origin of the continuum which participates in the FBW interaction in stage-1 donor BGICs has been somewhat controversial. *Nemanich* et al. [5.54]

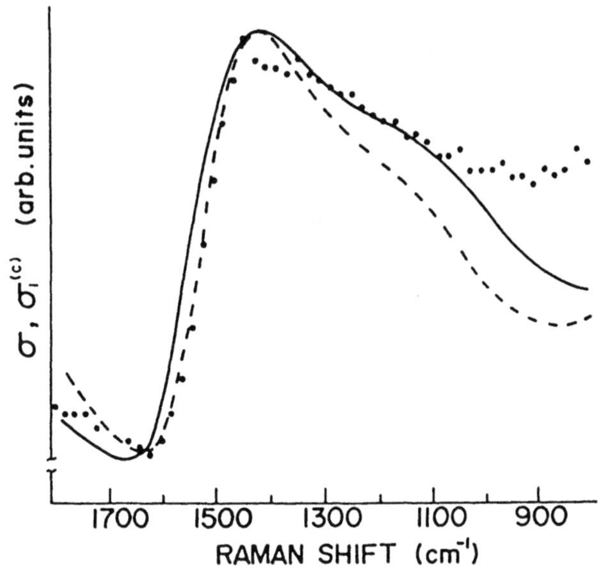

Fig. 5.14. A comparison between the experimental data for first-order Raman scattering (dots) from KC_8 [5.54] with a calculation based on an electronic continuum background. The solid line represents the total contribution to the scattering, σ, which is normalized to fit the observed height at the maximum and minimum, while the dashed line represents the Fano-Breight-Wigner contribution, $\sigma_1^{(c)}$, only. (From [5.62])

ascribed the continuum to electronic interband and intraband transitions in the metallic sample. This suggestion was put on a sound quantitative footing by *Miyazaki* et al. [5.62] who successfully calculated the observed spectrum of KC_8 using a model in which zone-center phonons at $858\,\mathrm{cm}^{-1}$, $1280\,\mathrm{cm}^{-1}$ and $1580\,\mathrm{cm}^{-1}$ were coupled to interband electronic transitions near the Fermi level. The result of their calculation with an electron–phonon coupling parameter of $\sim 0.08\,\mathrm{eV}$ is shown in Fig. 5.14. *Eklund* et al. [5.61] attributed the continuum to frequency-dependent multiphonon scattering on the basis of an extended fit to the CsC_8 data over a wide range of Raman shift up to $3200\,\mathrm{cm}^{-1}$. This fit is shown in Fig. 5.15, which, in the case of the frequency-dependent continuum, yields a *bare* phonon energy close to the $E_{2g}^{(2)}$ energy of pristine graphite. Note however that an electronic continuum could, in principle, produce the same frequency dependence which *Eklund* et al. [5.61] attributed to multiphonon processes. In yet another model, the continuum in EuC_6 has been attributed to disorder-induced one-phonon scattering activated by stacking faults [5.57, 58]. No line shape calculations have yet been forthcoming for that model.

We now address the low frequency excitations in the 550–$650\,\mathrm{cm}^{-1}$ region of Fig. 5.13. This region consists of a triplet of modes, as can be seen from the expanded and polarized spectra for CsC_8 [5.63] shown in Fig. 5.16. Note that the interior member of the triplet is clearly a FBW line, as can be seen from its characteristic shape. These modes have been attributed to M-point intralayer graphitic

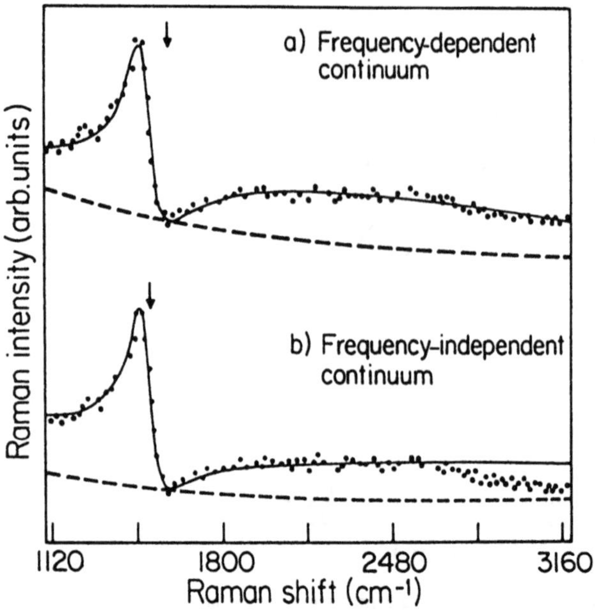

Fig. 5.15. Theoretical fits to the observed room temperature Raman spectrum for stage-1 C_8Cs in the broad frequency range $1100 \leq \omega \leq 3200\,\mathrm{cm}^{-1}$ using a coupling of the $E_{2g}^{(2)b}$ bounding layer mode to a multiphonon continuum which is (*a*) frequency dependent and (*b*) frequency independent. The arrows in (*a*) and (*b*) indicate the frequency of the bare phonon. The dashed lines represent the noninterfering continuum in each case. (From [5.61])

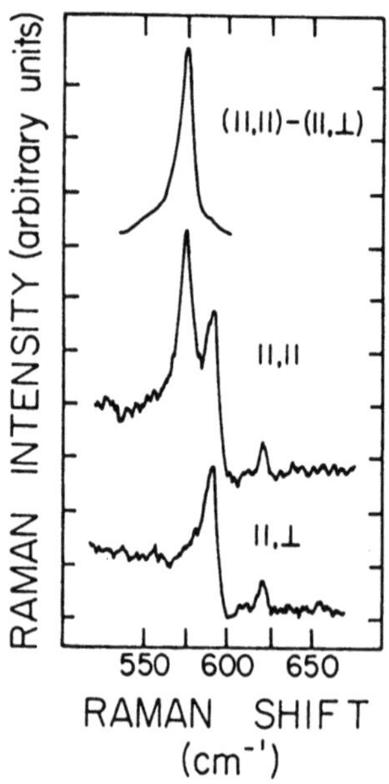

Fig. 5.16. Room-temperature Raman spectrum of the *M*-point phonons in C_8Cs. (From [5.63])

Table 5.1. Frequencies of the stage-1 M-point triplet Raman bands (from [5.8])

Sample	Space Group of ordered structure	Observed energy[a] $[cm^{-1}]$	Calculated energy[b] $[cm^{-1}]$	Calculated and observed symmetry
C_8K	D_{2h}^{24}	554	582	A_g
		557	583	B_{1g}
		-	592	$A_g + B_{1g}$
C_8Rb	D_{2h}^{24}	566	586	A_g
		573	588	B_{1g}
		617	600	$A_g + B_{1g}$
C_8Cs	D_6^4 or D_6^5	574	-	A_1
		585	-	E_2
		620	-	R_2

[a]These energies are accurate to $\pm 1\,cm^{-1}$.
[b]Calculations from [5.64].

phonons [5.4–6, 63] made Raman active by the guest species. This attribution is confirmed by the polarization properties shown in Fig. 5.16 and summarized for all three MC_8 compounds in Table 5.1. In this table, the observed polarization properties are compared with predictions based on the phonon dispersion curves calculated by *Horie* et al. [5.64]. The more recent calculations of *Al-Jishi* and *Dresselhaus* [5.50] (Fig. 5.10) would yield the same polarization properties as those of *Horie* et al. [5.64].

Two mechanisms have been advanced to explain the activation of the M-point phonons. In the in-plane zone-folding mechanism [5.5, 6], the commensurate superposition of the intercalate layers onto the graphitic host gives rise to an expansion of the real space in-plane unit cell of the latter and consequently a contraction of the corresponding cell in reciprocal space. For the $(2 \times 2)R0°$ structure of the MC_8 heavy alkali BGICs [5.45], this contraction results in a mapping of the M point of the cell of pristine graphite onto the zone-center Γ point of the GIC. On the other hand, both LiC_6 and EuC_6 possess a $(\sqrt{3} \times \sqrt{3})R30°$ in-plane superlattice structure [5.45], which would result in a mapping of the K point onto the zone center. *Dresselhaus* and *Dresselhaus* have pointed out [5.6] that the density of states of K-point phonons is much lower than that of those at the M point and so the zone-folding model offers a natural explanation for the absence of the low-frequency Raman bands in the spectra of MC_6 BGICs. However, in pressure-dependent Raman studies of alkali BGICs [5.13] (see below), no M-point modes were detectable even in compounds with pressured-induced $(2 \times 2)R0°$ superlattices.

An alternative mechanism for activation of the M-point phonons is disorder-induced scattering [5.1, 63]. This mechanism was favored by *Caswell* and *Solin* [5.63] because the strong M-point Raman modes persisted for CsC_8 at temperatures well above the melting temperature of the in-plane superlattice (see below). The disorder-induced mechanism does not, however, provide a natural explanation of the Raman spectra of the MC_6 compounds. There has been a tendency to view the two mechanisms described above as mutually exclusive. In fact it is

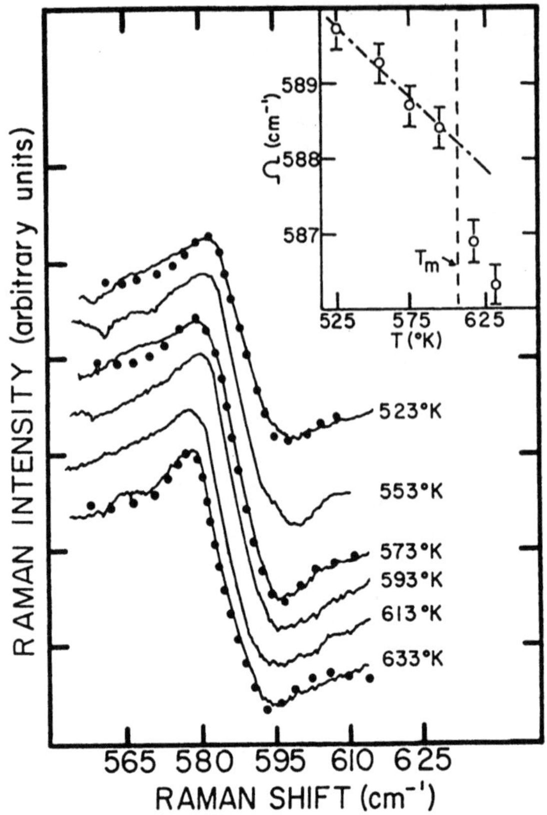

Fig. 5.17. Temperature dependence of the 580 cm^{-1} Fano line of CsC_8. (•) least-squares FBW fit; (o) renormalized energy; (– · –·) least-squares fit; T_m = 608 K is the melting temperature. (From [5.63])

quite possible that both the disorder-induced and zone-folding explanations are simultaneously operative.

The stage 1 alkali BGICs show no evidence of any IR signal associated with vibrational excitations. We defer discussion of this observation to until we address the IR properties of stage-1 donors and acceptors later in this section.

Although the Raman spectra of several stage-1 donor BGICs have been measured at a few different temperatures [5.6], the most systematic study to date is that of CsC_8 near the melting transition. In Fig. 5.17 are shown the data of *Caswell* and *Solin* [5.63] for the temperature dependence of the FBW line in the M-point triplet of CsC_8. As can be seen in the inset, the renormalized phonon energy deduced from the FBW fits to the data (circles in Fig. 5.17) shows a discontinuous drop at the melting temperature T_M = 608 K deduced from X-ray diffraction studies [5.65]. (However, neutron scattering studies [5.66] indicate a higher melting temperature of 806 ± 2 K. This discrepancy is not yet resolved.) From a comparison of the observed Stokes–anti-Stokes ratio with that calculated on the basis of a multiphonon process, *Caswell* and *Solin* [5.63] identified the

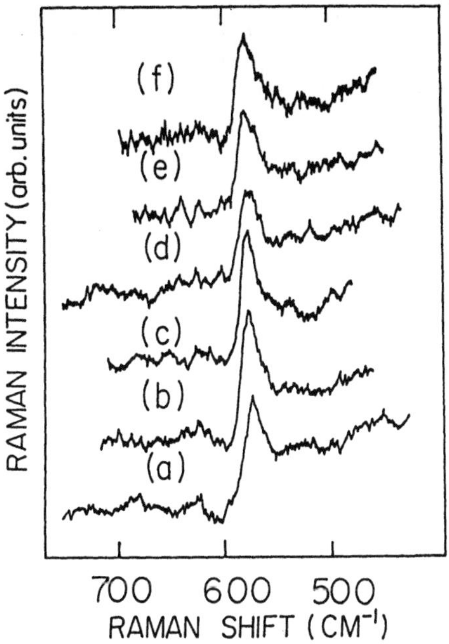

Fig. 5.18. Raman spectra of the 580 cm^{-1} FBW line in RbC_8, taken at (*a*) 0, (*b*) 2.3, (*c*) 4.3, (*d*) 6.8, (*e*) 8.3, and (*f*) 9.6 kbar. (From [5.13])

background contribution to the FBW M-point line as a first-order *single phonon* continuum for Raman shifts $< 600\,\mathrm{cm}^{-1}$.

Wada [5.13] has studied the pressure dependence of the M-point phonons of RbC_8 up to pressures of 10 kbar and his results are shown in Fig. 5.18. (The $E_{2g}^{(2)}$ region was not studied because it was masked by the intense 1332 cm^{-1} Raman scattering from the diamond anvils in the high pressure cell.) As can be seen from Fig. 5.18, there is a slight frequency upshift and increase in background with increasing pressure but there are no dramatic changes in the spectral properties. This is consistent with X-ray studies [5.13, 67], which show no structural phase transitions for RbC_8 in the pressure range up to 10 kbar.

Stage-n, $n \geq 2$, Donors. *Solin* and co-workers [5.4, 8–11] have carried out extensive investigations of the stage dependence of the first-order room temperature Raman spectra of the heavy alkali donor BGICs MC_{12n}, M =K, Rb, Cs, $n = 2$ to 6. Their results for the potassium-graphite system are shown in Fig. 5.19. Qualitatively similar results are obtained with the other heavy alkali metals. Notice that, with the exception of the slight background that increases with decreasing Raman shift for the stage-2 and stage-3 compounds, the only notable features in these spectra are the strong bands at $\sim 1600\,\mathrm{cm}^{-1}$. As expected on the basis of the lattice dynamics and the NL [5.53, 54] models discussed above, the spectrum of the stage-2 compound is a singlet while the spectra of all higher stages exhibit a doublet in the $E_{2g}^{(2)}$ region. The corresponding spectra for RbC_{12n} [5.53] are in quantitative agreement with the phonon dispersion curves reported by *Al-Jishi* and *Dresselhaus* [5.50], see Fig. 5.10. Polarization studies of the Raman spectra

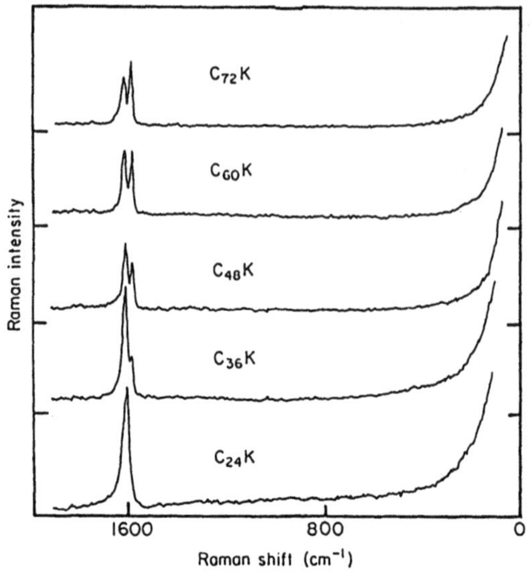

Fig. 5.19. The stage dependence of the first-order Raman spectra of KC_{12n}, $n = 2$–6. (From [5.8])

of the MC_{12n} BGICs are consistent with the assignment of the higher-energy and lower-energy doublet members as an $E_{2g}^{(2)b}$ mode and $E_{2g}^{(2)i}$ mode, respectively.

Solin and *Caswell* have fit the doublet region of all of the first-order Raman spectra of the MC_{12n} BGICs using a pair of Lorentzian lines whose amplitude, Raman shift and width were treated as adjustable parameters in a least-squares fitting routine [5.8]. The results of their analysis are summarized in Table 5.2. From these results, *Solin* and *Caswell* have computed the stage dependence of the integrated intensities $I[E_{2g}^{(2)i}]$ and $I[E_{2g}^{(2)b}]$ of each band in the doublet and plotted their stage-dependent ratio, R, against stage number as a test of the validity of (5.22). The results of this plot are reproduced in Fig. 5.20, together with a linear least squares fit to the data. It is evident from Fig. 5.20 that, to within experimental error, R varies linearly with n and intercepts the abscissa at $n = 2$ in complete agreement with (5.22). From the slope of the straight line fit to the data of Fig. 5.20, *Solin* and *Caswell* [5.8] obtained a value of $\sigma_i/\sigma_b = 0.25 \pm 0.05$.

Recall from Sect. 5.1, that the Raman cross sections σ_i and σ_b depend on the induced polarizations of the interior and bounding layers, respectively. But these polarizations depend on the charge exchange to or from the layer; the higher the electronic charge transfer to the layer, the higher is its induced polarization and Raman cross section for $E_{2g}^{(2)}$-like intralayer vibrations. The above results from Fig. 5.20 indicate that charge transferred from the donor intercalant in heavy alkali BGICs is delocalized along the c-axis with considerably more residing on the bounding layer than the interior layers. Moreover, the charge transfer and its distribution are apparently independent of stage for $n \geq 2$ and are the same, to within experimental error, for all three of the heavy alkali intercalates. These observations are completely consistent with other measurements of the charge transfer properties of these donor compounds [5.68].

Table 5.2. Parameters of the Lorentzian fits to the Raman spectra of stage n, $n \geq 2$, alkali GICs[a] (from [5.8])

Sample	ω_i	Y_i	Γ_i	ω_b	Y_b	Γ_b	R
$C_{24}K$	–	–	–	1610	22.3	26.0	0
$C_{36}K$	1584	5.7	6.9	1610	20.4	18.6	0.10
$C_{48}K$	1584	4.5	7.0	1612	12.6	18.7	0.28
$C_{60}K$	1584	11.6	7.4	1612	11.3	17.8	0.43
$C_{72}K$	1583	22.6	6.7	1613	16.2	13.8	0.68
$C_{24}Rb$	–	–	–	1603	17.2	22.0	0
$C_{36}Rb$	1582	3.8	8.3	1606	13.7	14.9	0.15
$C_{48}Rb$	1582	6.0	8.3	1609	13.8	15.9	0.22
$C_{60}Rb$	1581	8.3	12.1	1608	11.3	18.2	0.49
$C_{72}Rb$	1582	8.7	10.6	1609	8.4	18.4	0.60
$C_{108?}Rb$[b]	1582	10.8	8.0	1611	6.6	11.6	1.14
$C_{24}Cs$	–	–	–	1603	5.8	22.0	0
$C_{36}Cs$	1580	2	7.3	1605	8.3	15.2	0.12
$C_{48}Cs$	1580	4.9	8.0	1606	6.9	15.3	0.37
$C_{60}Cs$[c]	1582	4.6	9.1	1610	3.0	16.2	0.86
	1582	5.2	9.2	1607	6.6	15.8	0.47
	1582	7.1	9.9	1607	6.6	22.0	0.48
$C_{72}Cs$[c]	1581	5.7	10.5	1610	4.0	15.4	–
	1581	5.3	10.0	1610	3.1	14.1	–
	1580	5.2	10.3	1607	3.2	21.6	–
HOPG	1581	14.6	13.8	–	–	–	∞

[a] All energy values are in cm^{-1}. The amplitude, Y, is given in arbitrary units. The Raman shifts, ω, and linewidths, Γ, are accurate to $\pm 0.5\,cm^{-1}$, respectively. The linewidth Γ has been corrected for the spectral slit width. $R = I[E_{2g}^{(2)i}]/I[E_{2g}^{(2)b}]$ as defined in (5.22) of the text.

[b] The composition and stage of this sample is uncertain, as indicated by the "?".

[c] Data given for several samples which exhibited the same X-ray spectra but different Raman spectra.

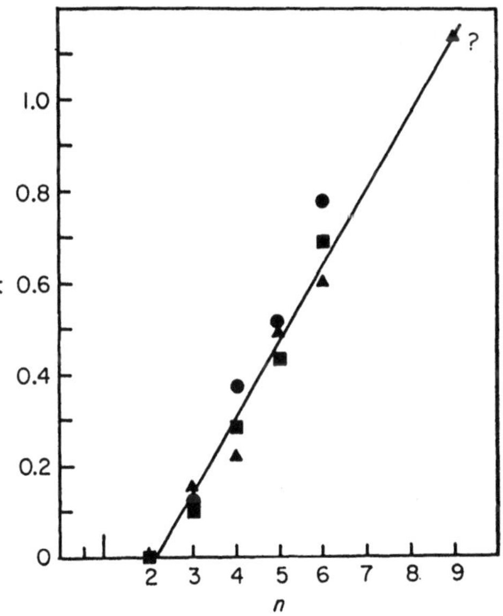

Fig. 5.20. Linear least squares fit to the data for MC_{12n} compounds from Table 5.2 using (5.22) (■) K, (▲) Rb, (●) Cs. (From [5.8])

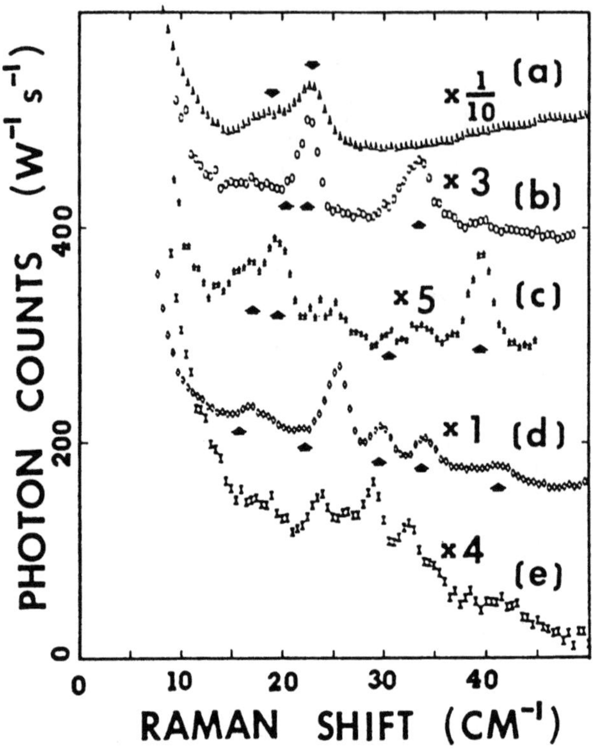

Fig. 5.21. The low-frequency Raman spectra of KC_{12n}, n = 2–6. The arrows indicate the calculated values of the phonon frequencies (see text). (From [5.69])

By overcoming the significant experimental difficulties of acquiring low-frequency phonon Raman spectra in the presence of the very strong parasitic scattering produced by GICs, the surfaces of which are of poor optical quality, *Wada* et al. [5.69] were able to obtain the Raman spectra of the rigid-layer shear modes in stage-2–6 potassium-graphite. Unlike the purely graphitic modes discussed above, the rigid layer modes involve motions of both the carbon layers and the intercalate layers. The rigid layer mode measurements reported by *Wada* et al. are reproduced in Fig. 5.21. The arrows in the figure correspond to the phonon energies calculated on the basis of a linear chain model with nearest neighbor force constants as discussed above. Recall that from that model and from the *Al-Jishi* and *Dresselhaus* phonon dispersion model [5.50] one expects a stage-n GIC to exhibit exactly n zone-center rigid-layer optic shear modes. This expectation is clearly justified for stages 2–5 in Fig. 5.21 and is consistent with the data for stage 6. In addition, the observed phonon frequencies are in good agreement with both models.

The very low frequency Raman modes in stage-2 RbC_{24} were also studied by *Eklund* et al. [5.70] and by *Giergiel* et al. [5.71]. The former observed only two modes at 15 cm^{-1} and 20 cm^{-1} in the room-temperature spectra, very close

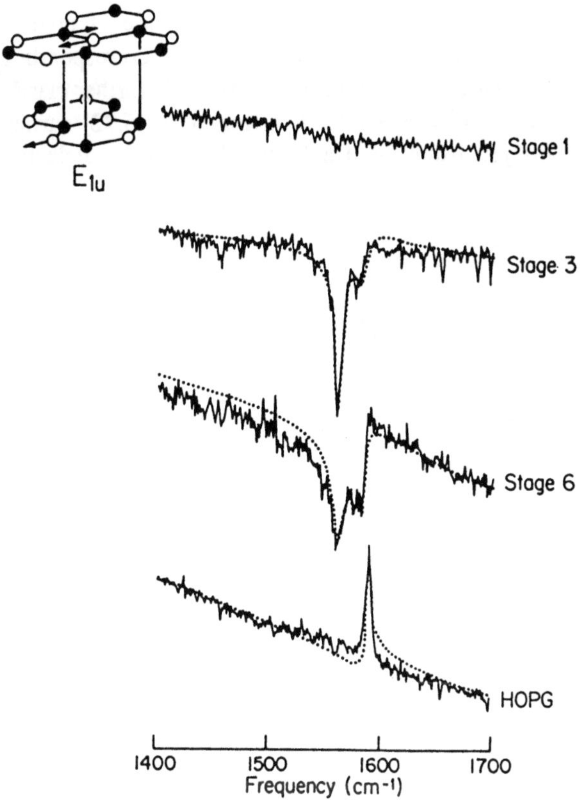

Fig. 5.22. Room-temperature c-face infrared reflectivity spectra in the energy range $1400 < \omega \leq 1700\,\mathrm{cm}^{-1}$ for stages $n = 1, 3$ and 6 Rb-graphite compounds, and for pristine graphite. The fit to experimental data (see text) using parameters given in Table 5.3 is shown by the dotted curve. (From [5.72])

in energy to the rigid-layer shear modes seen by *Wada* et al. [5.69] for KC_{24}. In subsequent studies of the temperature dependence (see below) of the low-frequency modes in RbC_{24}, *Giergiel* et al. [5.71] reported the shear modes to be at $18\,\mathrm{cm}^{-1}$ and $22\,\mathrm{cm}^{-1}$ but identified two additional features, an out-of-plane mode at $115\,\mathrm{cm}^{-1}$ and a mode at $35\,\mathrm{cm}^{-1}$ which only appeared at low temperature and was therefore associated with an in-plane order–disorder transition.

To date, the stage dependence of the first-order IR spectra of donor BGICs has only been studied for $n \geq 3$ and in this case the rubidium-graphite system was probed. The results of these studies, which were reported by *Leung* et al. [5.72], are shown in Fig. 5.22 for the spectral region of the E_{1u} mode. They identified the following notable features of these spectra and of other donor spectra not shown in the figure: (1) The stage-1 and stage-2 heavy alkali BGICs with K, Rb, or Cs show no evidence of IR active lattice modes. (2) The observed spectra for stages $n \geq 3$ appear to contain only two bands, which exchange intensity (oscillator strength) with increasing stage, but careful analysis reveals the presence of several additional bands. (3) While the E_{1u} mode of pristine graphite (HOPG form) appears as a reflectivity maximum, the corresponding spectra of stages with $n \geq 3$ appear as a reflectivity minimum. To further clarify the spectral features of the data shown in Fig. 5.22, *Leung* et al. [5.72] carried out a Lorentzian line shape analysis using the usual form [5.73] of the dielectric function given by

$$\varepsilon(\omega) = \varepsilon_\infty + \varepsilon_{\text{Drude}} + \sum_j F_j \frac{\omega_j^2}{\omega_j^2 - \omega^2 - \mathrm{i}\omega\Gamma_j}, \tag{5.25}$$

where F_j, ω_j and Γ_j are the oscillator strength, mode frequency and linewidth of the jth oscillator and the Drude term represents the free carrier contribution [5.73]. The results of their analysis are summarized in Table 5.3. Note that the minimum number of oscillators required to fit (dotted lines in Fig. 5.22) the stage-3 and stage-6 data were 3 and 4, respectively.

Table 5.3. Infrared lattice mode parameters for graphite–alkali metal intercalation compounds[a] (from [5.72])

Intercalant	Stage (n)	Mode frequency (ω_j) [cm^{-1}]	Width (Γ_j) [cm^{-1}]	Strength (F_j)
Rb	3	1563.5	6	0.100
		1569.0	10	0.040
		1581.0	11	0.065
Rb	6	1562.5	10	0.038
		1569.0	5	0.006
		1575.0	6	0.013
		1583.0	8	0.030
HOPG	∞	1587.5	2	0.008

[a] Although not explicitly shown, there is no observable IR-active graphtic optic mode in stage-1 and 2 intercalation compounds

On the basis of their frequencies and stage-dependent oscillator strengths, the highest and lowest frequency oscillators in the multiplets listed in Table 5.3. were identified by *Leung* et al. [5.72] with the E_{1u}^{b} and E_{1u}^{i} modes in quantitative agreement with the *Al-Jishi* and *Dresselhaus* phonon dispersion curves [5.50] (stage 3, see Fig. 5.10) and in qualitative agreement with the NL model [5.53, 54]. But the interior oscillators have remained unexplained to date. One suspects that they may be an artifact of the fitting procedure itself and specifically related to the inadequacy of the Drude approximation. *Leung* et al. [5.72] pointed out that the absence of a first-order IR spectrum for stage-1 GICs is to be expected on the basis of symmetry selection rules, since the zone-center intralayer vibration of a single graphite layer is not IR active. The lack of an IR lattice response from the stage-2 alkali BGICs was however more problematic, since stage-2 IR activity is symmetry allowed and is observed for acceptor BGICs (see below). *Leung* et al. attributed the loss of a stage-2 donor IR lattice response to screening of the dipole

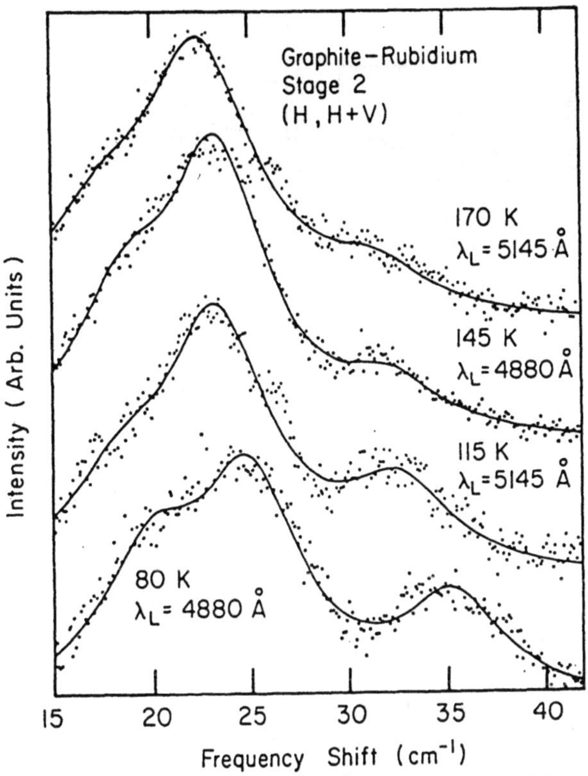

Fig. 5.23. Raman spectra of stage-2 graphite-Rb in the (10–45) cm^{-1} region taken at temperatures T = 170, 145, 115 and 80 K. Spectra were taken with different laser wavelengths and should not be compared for changes in absolute integrated intensities. Incident light was polarized ||, and the scattered light was unanalyzed. Solid lines are the results of the Lorentzian line shape analysis after the subtraction of a linear background. (From [5.71])

moment by high density highly mobile charge in the bounding layers [5.72]. This interpretation has not yet been quantitatively verified and must therefore still be considered speculative. Finally, the change from a maximum in reflectivity for HOPG to a minimum for the intercalated material was attributed to both an increase in the free carrier contribution to $\varepsilon(\omega)$ and an increase in the background reflectivity.

The temperature dependence of the first-order Raman spectra of higher stage donor GICs has been studied by several groups [5.74, 75] in an effort to observe variations of the spectral properties associated with well-known structural phase transitions [5.45] (see also Chap. 2). Unfortunately, to date the observed spectra have been unremarkable vis-à-vis these phase transitions and merely exhibit predictable changes in the linewidth and Raman shift of the intralayer graphitic phonon modes. On the other hand, studies of the very low frequency modes of RbC_{24} reveal a strongly temperature dependent mode at 35 cm^{-1} as can be seen from the work of *Giergiel* et al. [5.71] shown in Fig. 5.23. This mode was

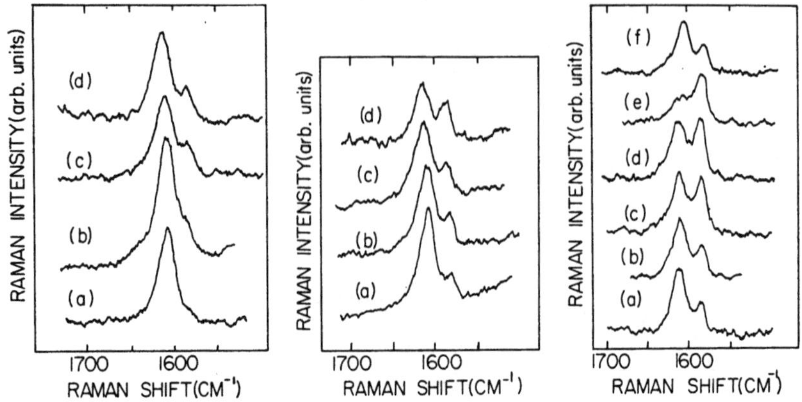

Fig. 5.24. *Left panel*: Raman spectra of RbC_{24} taken at (*a*) 0.0, (*b*) 3.8, (*c*) 5.9, and (*d*) 8.4 kbar. *Center panel*: Raman spectra of RbC_{36} taken at (*a*) 0.0, (*b*) 3.0, (*c*) 4.8, and (*d*) 7.6 kbar. *Right panel*: Raman spectra of RbC_{48} taken at (*a*) 2.2, (*b*) 3.3, (*c*) 5.3, (*d*) 8.8, (*e*) 9.8, and (*f*) 0.0 kbar. All spectra in a given panel were taken consecutively. (From [5.13])

attributed to an in-plane structural ordering which occurs for $T \lesssim T_u = 170\,K$ [5.45] and leads to a zone-folding of $35\,cm^{-1}$ zone-edge phonons to the Γ point, where they are Raman active. *Giergiel* et al. [5.71] also studied the temperature-dependent frequency shift of this mode, of the rigid layer shear modes and of the out-of-plane mode, all of which were discussed above. Each of these modes exhibited the usual upshift with decreasing temperature.

In contrast to the temperature-dependent studies cited immediately above, *Wada* [5.13] has carried out very fruitful studies of both staging and in-plane structural transitions in stage 2–4 rubidium-graphite and in stage-2 CsC_{24}. He used the diamond anvil technique [5.38] to obtain hydrostatic pressures in the range 0–10 kbar at room temperature. His results are shown in Fig. 5.24 for the Rb-BGICs.

In order to address these results, we must first briefly describe the results of pressure-dependent X-ray studies of the structural properties of the alkali BGICs [5.76]. All the stage-2 BGICs MC_{24}, M = K, Rb, and Cs, undergo a pressure-induced structural transition in which there is a staging transition from stage 2 to stage 3, accompanied by an in-plane densification from a disordered intercalate layer with MC_{12} layer stoichiometry to a commensurate $(2 \times 2)R0°$ superlattice structure. This transition is completely reversible. With increasing pressure, the stage 3 phase grows at the expense of stage 2 but both coexist until sufficient pressure is reached to achieve a pure stage-3 material. The transition occurs at constant stoichiometry. If we let s be the layer stoichiometry in a stage-n compound $MC_{n\times s}$, then the pressure-induced transition in MC_{24} can be represented by the equation

$$MC_{2\times12} \xrightarrow{P\uparrow} MC_{3\times8}\ . \tag{5.26}$$

The higher stage alkali BGICs with $n \geq 3$ also show transitions to a *higher* stage with pressure [5.13, 67, 76, 77] although these transitions are not fully reversible and pure stages are not obtained even at pressures up to 30 kbar [5.77]. Pressure-induced transitions to fractional stages have also been reported for some compounds [5.78].

The staging phase transitions noted above can be clearly observed in *Wada*'s pressure-dependent Raman spectra of the E_{2g} doublet region shown in Fig. 5.24. He noted that the ambient pressure stage-2 singlet of RbC_{24} [Fig. 5.24, left panel, trace (a)] evolves continuously with pressure to the doublet spectrum of a stage-3 compound [Fig. 5.24, left panel, trace (d)]. Qualitatively similar behavior was observed for CsC_{24} [5.13]. The transition to higher stage with increasing pressure is also quite apparent in the pressure-dependent spectra of RbC_{36} and RbC_{48}, as can be seen from the center and right panels of Fig. 5.24, respectively. In describing these spectra, *Wada* noted the evolution from a pure stage at ambient pressure to an admixture of a higher lying stage or stages as evidenced by the growth of the intensity of the bounding layer mode of the doublet relative to that of the interior layer mode. He also pointed out that the pressure-induced changes in the higher stage compounds were not completely reversible. This can be seen by noting the differences in the ambient pressure spectra of RbC_{48} taken before and after pressurization [Fig. 5.24, right panel, traces (a) and (f), respectively]. From linear fits of the pressure dependence of the Raman shifts shown in Fig. 5.24, *Wada* [5.13] computed the rate of Raman shift with pressure, $\partial\omega_0/\partial P$, for pristine graphite, CsC_{24}, and CsC_{12n}, $n = 2$–4. His results are summarized in Table 5.4.

Table 5.4. The zero-pressure Raman frequency (ω_0) and the rates of Raman frequency shifts with respect to pressure, $\partial\omega_0/\partial P$, of graphite, CsC_{24}, and RbC_{12n} ($n = 2, 3, 4$).

	Mode	ω_0 [cm^{-1}]	$\partial\omega_0/\partial P$ [cm^{-1} $kbar^{-1}$]
Graphite	$E_{2g}^{(2)}$	1584	0.50 ± 0.04
CsC_{24}	$E_{2g}^{(2)b}$	1603	0.70 ± 0.12
RbC_{24}	$E_{2g}^{(2)b}$	1608	0.51 ± 0.09
RbC_{36}	$E_{2g}^{(2)i}$	1582	0.78 ± 0.08
	$E_{2g}^{(2)b}$	1609	0.44 ± 0.05
RbC_{48}	$E_{2g}^{(2)i}$	1584	0.32 ± 0.12
	$E_{2g}^{(2)b}$	1609	0.54 ± 0.08

Stage-n, $n \geq 1$, Acceptors. The Raman spectra of a number of acceptor BGICs have been studied over the past few years [5.6]. The most thorough of these studies has been carried out by *Eklund* et al. [5.79] on $SbCl_5$ graphite, the compounds of which are reportedly the most chemically stable (in vacuum

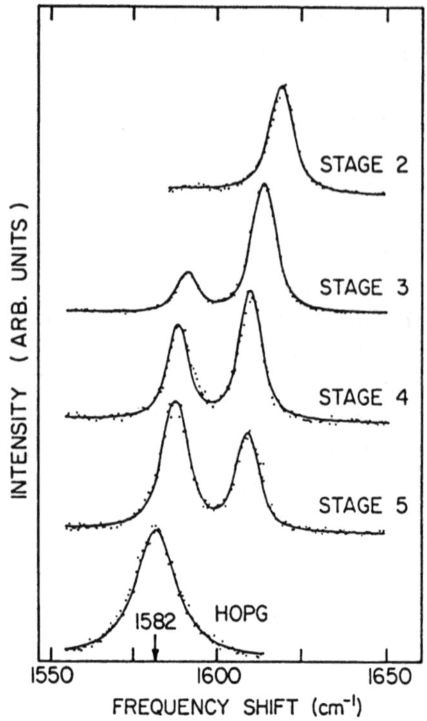

Fig. 5.25. Raman spectra of the zone-center phonons of HOPG and stage2–5 $SbCl_5$ graphite. The solid lines represent the results of a Lorentzian line shape analysis. (From [5.79])

and in air) of any GIC yet synthesized [5.80]. In the following discussion, we shall focus on that system, realizing that its high-frequency Raman spectra are typical of all acceptor GICs. The stage dependence of the first-order Raman spectra of $SbCl_5$ graphite reported by *Eklund* et al. is shown in Fig. 5.25 together with Lorentzian fits to the data. (Although stage-1 spectra were not reported, such spectra of all acceptor GICs studied to date show a single band slightly upshifted in energy from the corresponding stage-2 singlet [5.4, 6, 7].) These spectra exhibit the same qualitative behavior as do the corresponding spectra of the donor compounds (Fig. 5.19), the high-frequency bounding layer mode losing intensity to the downshifted interior layer mode with increasing stage, and they are in excellent agreement with the qualitative predictions of the NL model [5.53, 54].

We have used the data of Fig. 5.25 and that of the other two acceptor BGICs [5.81, 82] for which stage-dependent first-order Raman spectra have been reported to test the applicability of (5.22) to acceptor compounds. Figure 5.26 shows plots of R against stage for each of these intercalation systems together with a linear least sqaures fit to the data for each guest species. The fits are reasonable and intersect the abscissa at approximately $n = 2$, as predicted. Using (5.22) and these linear fits, we find $\sigma_b/\sigma_i = 0.89 \pm 0.15$, 1.03 ± 0.11, and 1.37 ± 0.19, for the $SbCl_5$, $FeCl_3$, and $AlCl_3$ acceptor BGICs, respectively. These Raman cross-section ratios are about a factor of three higher than those exhibited by the donor

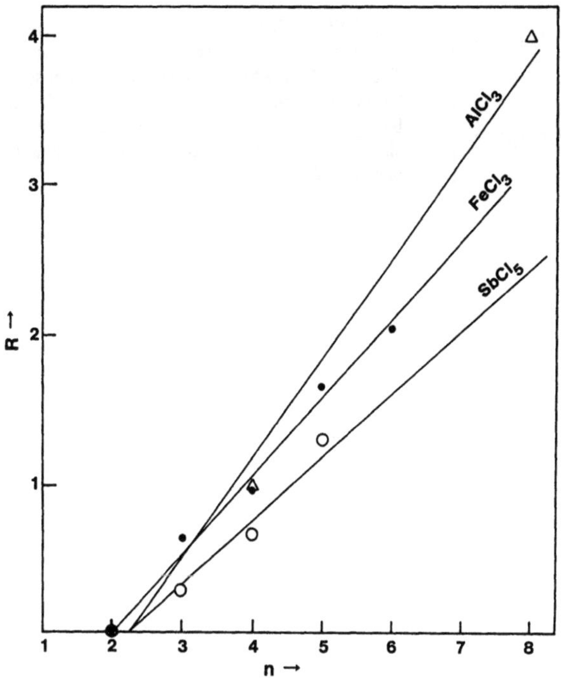

Fig. 5.26. Linear least squares fits to the stage dependence of the relative Raman intensities of bounding and interior layers in acceptor binary graphite intercalation compounds. (o) $SbCl_5$ [5.79], (△) $AlCl_3$ [5.82], (•) $FeCl_3$ [5.81]

compounds (Fig. 5.20) and are, to within experimental error, slightly greater than but of the order of 1. This indicates that the interior and bounding layers in the acceptor BGICs have about equal charge densities and that the charge exchange in these compounds is much lower than that in donors, a fact which is well known [5.6]. Also, the fact that $\sigma_i > \sigma_b$ shows, not surprisingly, that electronic charge is *extracted* from the bounding layers in acceptor BGICs.

Stage-dependent first-order IR spectra have been reported for $FeCl_3$ [5.81], $AlCl_3$ [5.82], and $SbCl_5$ [5.79] acceptor GICs, all of which exhibit qualitatively similar properties. An example of such spectra is shown in Fig. 5.27, which is from the $SbCl_5$ graphite studies by *Eklund* et al. [5.79]. None of the acceptor compounds studied to date exhibit any first-order reflectance signature of lattice modes in the E_{1u} region. This is consistent with group theoretical selection rules as discussed above in the case of donors. However, unlike the donor compounds, all of the acceptor materials do show E_{1u} reflectance features in the stage-2 compounds as well as higher stages. An oscillator analysis identical to that used with (5.25) for the donors has been carried out for $AlCl_3$ [5.82] and $SbCl_5$ [5.79] acceptors. The results obtained by *Eklund* et al. [5.79] from the data of Fig. 5.27 are summarized in Table 5.5. The highest energy oscillator for stages 3–5 has been attributed to the E^{i}_{1u} vibration [5.79], while the downshifted modes have been associated with pairs of carbon layers which contain a bounding layer as one

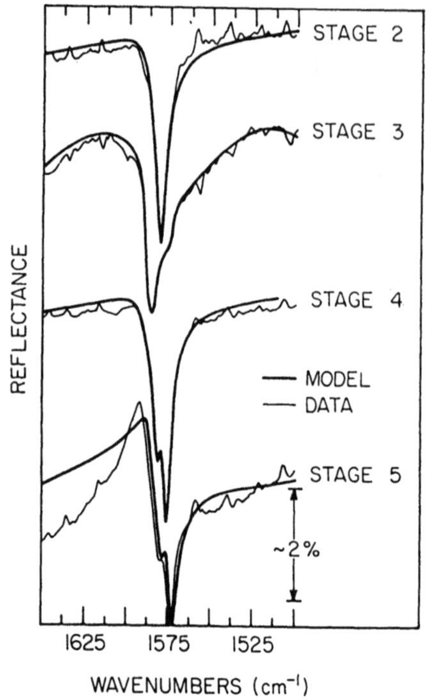

Fig. 5.27. Reflectance spectra of the infrared-active phonons of stage 2–5 $SbCl_5$ graphite. The stage-3 data have a superimposed sinusoidal background which is instrument related. This background was included in the stage-3 analysis. The solid lines are from an oscillator analysis based on (5.25). (From [5.79])

Table 5.5. Oscillator parameters (5.25) used to fit the infrared-active phonon features in the reflectance spectra of stage 2–5 $SbCl_5$ graphite shown in Fig. 5.27 (from [5.79])

Stage, n	Mode frequency, ω_j [cm^{-1}]	Width, Γ_j (FWHM) [cm^{-1}]	Strength, F_j
2	1580.4	3	0.16
	1585	5	0.12
3	1577	4	0.03
	1584.5	4	0.065
	1588.5	3	0.088
4	1576.3	2.5	0.08
	1582.3	3.5	0.14
	1588.5	4	0.035
5	1574.8	2.5	0.035
	1581.3	4	0.062
	1586	2.5	0.02

member [5.82]. The pair of oscillators for the stage-2 compound must clearly be associated with purely bounding layers but their presence in view of the singlet expected from the NL model has not been addressed and requires further analysis.

Since species which form acceptor BGICs are always molecular entities, one would expect to be able to observe their internal molecular vibrations as bands in the Raman and IR spectra of the intercalated compound just as is the case with any other "molecular solid" [5.83]. To date, reports of the observation by optical methods of internal modes in GICs have been very rare and involve only the following acceptors: Br_2 [5.29, 84], ICl [5.37], IBr [5.37], $FeCl_3$ [5.85], and AsF_5 [5.16]. But only in the cases of Br_2 [5.29, 84] and AsF_5 [5.16] are the spectroscopic changes observed for the intercalated molecule sufficiently distinct from the corresponding spectral properties of the free molecule to unambiguously conclude that internal modes of the former have indeed been detected. The case of graphite-Br_2 has been extensively discussed in previous reviews [5.5, 6] and it is now well known that the Raman scattering from the internal molecular modes of that material is resonance enhanced [5.86]. (We will address resonance Raman scattering in the next subsection.) This led to the speculation [5.4, 6] that resonance enhancement was a prerequisite to the observation of internal molecular modes in GICs. Since for the large majority of acceptor species, such enhancement required excitation with generally unavailable deep UV laser sources, it seemed that there was little hope of observing internal modes in other acceptor GICs.

Recently, in a series of very clever experiments, *Ohana* and *Yacoby* [5.16] showed that by increasing the penetration depth of the incident laser radiation in a Raman experiment one could obtain nonresonance enhanced Raman spectra of the intercalant internal modes. Attempts to enhance the penetration depth by tuning the incident photon energy to the window between the plasma and adsorption edges in c-face measurements proved unsuccessful [5.16]. However, by carrying out a-face studies with incident photons polarized parallel to the c-axis, a sufficient increase in penetration depth was achieved to acquire internal mode spectra from AsF_5 graphite. This was possible, as noted by *Ohana* and *Yacoby* [5.16], because the spatial separation between carbon layers in acceptor GICs and basal plane reflection symmetry preclude direct optical transitions between π bands for c-polarized light. Also, absorption due to phonon assisted transitions is minimal in this configuration.

The results of *Ohana* and *Yacoby*'s nonresonance enhanced first-order Raman scattering studies of stage-1 C-AsF_5 and of the pure AsF_5 vapor in which it was immersed are shown in Fig. 5.28 [5.16]. The singlet at $1639\,cm^{-1}$ is the usual band associated with the $E_{2g}^{(2)b}$ graphitic mode. Although selection rule forbidden in the polarization configuration used, it was nevertheless observed because of imperfect alignment of the graphite layers in the a-face. The strong band at $735\,cm^{-1}$ [left panel of Fig. 5.28, (a) and (b)] is due to the symmetric breathing mode of the free molecule but the other three labeled lines in the spectrum of the GIC were all identified as symmetric breathing modes of various molecular entities that are known to be present in C-AsF_5 [5.87]. Specifically, the $683\,cm^{-1}$, $698\,cm^{-1}$, and $753\,cm^{-1}$ modes have been identified by *Ohana* and *Yacoby* [5.16] with AsF_6^-, AsF_3, and AsF_5, respectively.

Yacoby and *Ohana* [5.88] have also developed methods for minimizing the reflectivity of GICs to IR photons, to produce a theoretical 70-fold increase in the sensitivity for observing IR-active vibrational modes, but this method has not yet been demonstrated for internal molecular modes. With the proven ability to

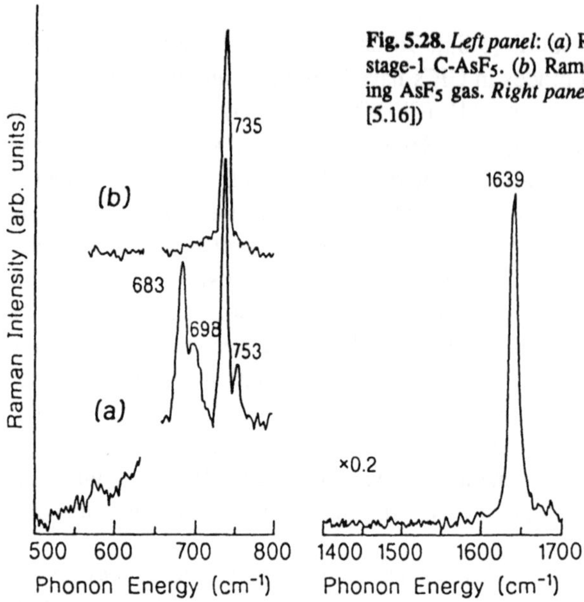

Fig. 5.28. *Left panel*: (*a*) Raman scattering on the *a*-face of stage-1 C-AsF_5. (*b*) Raman scattering from the surrounding AsF_5 gas. *Right panel*: High-frequency region. (From [5.16])

observe such modes with nonresonance Raman scattering and the potential for IR observation, one expects that a better understanding of the effect of guest–host interactions on intercalate modes in GICs will be forthcoming.

As noted above, the work of *Eklund* et al. [5.86] and of *Ohana* and *Yacoby* [5.16] indicates that the detection of the intralayer $E_{2g}^{(2)}$ graphitic modes of GICs by first-order Raman scattering is facilitated by resonance enhancement. Such enhancement occurs when either the incident or scattered photon is resonant with an allowed electronic transition in the host material [5.1]. In GICs, these transitions are between π bands. Although resonance Raman scattering in solids is well understood, its theoretical analysis is based on a quantum mechanical microscopic model, in contrast to the semiclassical macroscopic model described in Sect. 5.1. Therefore we only address the salient theoretical aspects of resonant scattering which are germane to GICs and refer the reader to texts [5.1] and review articles [5.20] for additional details.

In the microscopic model as applied to acceptor GICs by *Eklund* et al. [5.86], the differential scattering cross section is given by

$$\frac{d^2\sigma}{d\omega_I d\Omega} = r_0^2 \left(\frac{\omega_S}{\omega_I}\right) |M|^2 , \tag{5.27}$$

where r_0 is the classical electron radius, all other parameters except M have been previously defined and

$$M = M_0 \int_{E_t}^{E_0} dE_{vc} E_{vc}^2 \{[(E_{vc} - \omega_I + i\Gamma)(E_{vc} - \omega_S - i\Gamma)]^{-1} + [(E_{vc} + \omega_I - i\Gamma)(E_{vc} + \omega_S - i\Gamma)]^{-1}\} . \tag{5.28}$$

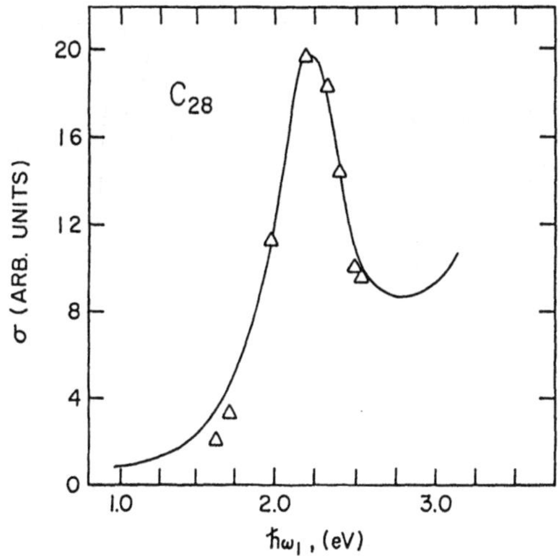

Fig. 5.29. A comparison of the calculated (solid line) and experimental (Δ) values for the Raman cross section of the stage-1 $C_{28} + HSO_4 - (H_2SO_4)_x$. The calculated values are from (5.27) and (5.28) using $E_t = 2.15\,eV$, $E_0 = 4.5\,eV$, and $\Gamma = 0.22\,eV$. (From [5.86])

Here $E_{vc} = E_c(\boldsymbol{q}) - E_v(\boldsymbol{q})$ is the energy for vertical transitions between the valence and conduction bands, E_t and E_0 are interband threshold and cutoff energies and Γ is a damping factor which simulates the lifetime of the excited electrons. The resonance effects are embodied in the square bracketed terms in (5.28).

Using (5.27) and (5.28) *Eklund* et al. [5.86] calculated the dependence of the Raman cross section of the $E_{2g}^{(2)b}$ mode of stage-1 C-H_2SO_4 on incident photon energy. The results of this calculation are compared with their experimental measurements in Fig. 5.29. As is shown by the figure, the calculation agrees well with experiment and accounts for the pronounced resonance enhancement at 2.15 eV.

A careful study of the resonance Raman scattering from the $E_{2g}^{(2)b}$ phonon in stage-1 C-AsF_5 has recently been reported by *Yacoby* and *Ohana* [5.88]. Figure 5.30 shows the Raman spectra they obtained at several different incident photon energies. The solid lines are fits to the data using the FBW line shape analysis [5.59], which *Yacoby* and *Ohana* [5.88] interpreted as indicating that the phonon was interacting with an *electronic* continuum. Notice from the scale factors displayed in Fig. 5.30 that there is about a factor of 80 variation in the scattering signal. This dramatic resonance effect has been accounted for quantitatively with a two-band model [5.88] which yielded the solid line fit to the photon energy dependence of the cross-section data shown in Fig. 5.31. As in the case for stage-1 C-H_2SO_4 [5.86] discussed above, the microscopic theory of resonance-enhanced first-order Raman scattering accounts well for the experimental results.

To date, only in the cases of AsF_5 [5.89] and Br_2 [5.90] species have studies of the temperature dependence of the optically active vibrational excitations of an

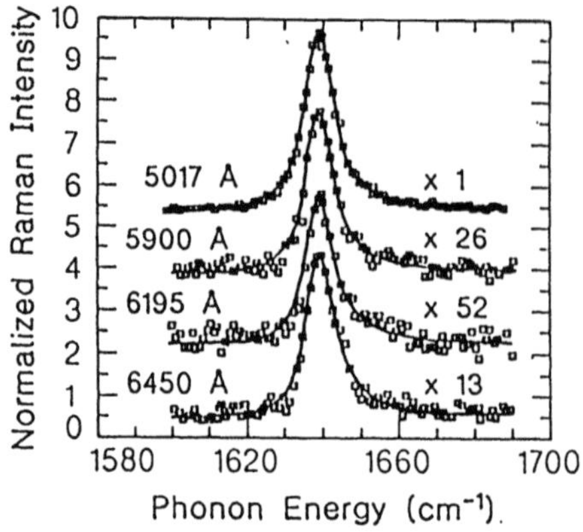

Fig. 5.30. Spectra of the $1640\,cm^{-1}$ Raman line in stage-1 C-AsF_5 for various laser wavelengths. The squares represent experimental points; the solid lines are FBW fits. (From [5.88])

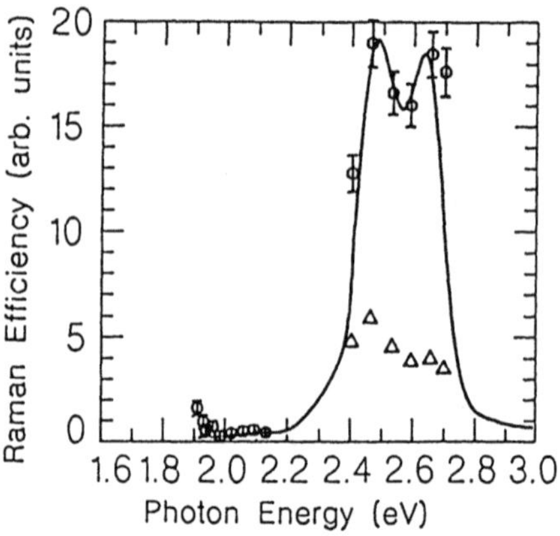

Fig. 5.31. The Raman efficiency of the $1640\,cm^{-1}$ line in stage-1 C-AsF_5 as a function of laser photon energy. The triangles represent raw data, the circles, the data after renormalization and the solid line was calculated using the two band model discussed in the text. (From [5.88])

acceptor GIC been carried out. Because of its thoroughness, we focus on the Br_2 study here. The incentive for this study was to use Raman scattering to monitor changes in both the intercalate internal modes and in the graphite intralayer modes which result from temperature-induced structural phase transformations in the intercalate layers [5.91]. Below $\sim 340\,K$, the Br layers form a 2d crystalline structure which is commensurate with the graphite host layers and which is similar to the structure found in pure solid Br molecular crystals [5.91]. At $\sim 380\,K$, the Br layers undergo an order–disorder-type melting transition [5.92]. In all cases, the Br_2 molecules are arranged with their axes parallel to the graphite layers.

The temperature-dependent Raman spectra of the symmetric stretching mode of the Br_2 molecules in stage-3 bromine-graphite obtained by *Erbil* et al. [5.90] are shown in Fig. 5.32. These spectra were excited with the 4880 Å Ar^+ ion laser line and are resonance enhanced [5.84]. The solid lines in the figure correspond to best fits to the data and were obtained using a Lorentzian form for $T < 335\,K$ and a Gaussian form at higher temperatures. The temperature dependence of the integrated intensities of the Br_2 stretching mode and of the $E_{2g}^{(2)b}$ mode are shown in Fig. 5.33, while that of the corresponding mode frequencies are shown in Figs. 5.34 and 5.35 respectively [5.90].

The data in Figs. 5.32–35 have been interpreted by *Erbil* et al. [5.90] as follows: Below $T_0 = 335 \pm 10\,K$ the Br_2 molecules each "see" the same environment since their layer structure is commensurate with the bounding graphite layers. The Raman spectra of their stretching mode therefore exhibits a natural line shape which is best fit by a Lorentzian form. At temperatures above T_0 but below the melting temperature $T_M = 379 \pm 10\,K$, the Br_2 molecules see a distribution of environments which result in a more inhomogeneous broadening of the Raman

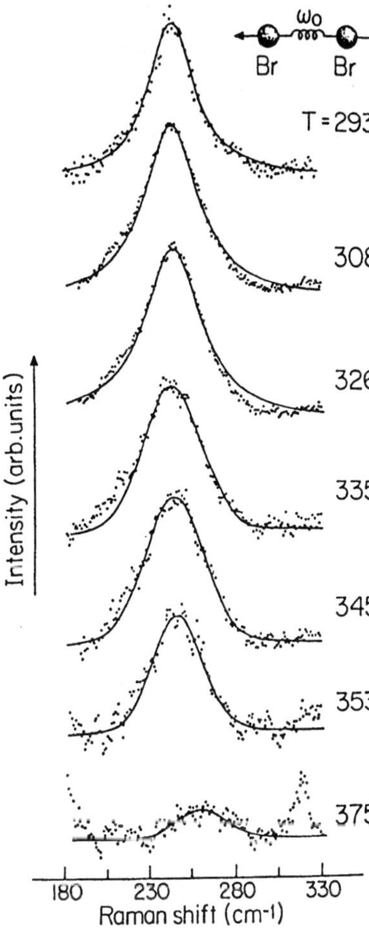

Fig. 5.32. Raman spectra for the bromine symmetric stretch mode in a stage-3 graphite-bromine compound shown for several temperatures in the range $290 < T < 380$ K. For each temperature, the experimental data points are fitted to a Gaussian line shape above 335 K and a Lorentzian line shape below 335 K (solid curves). The Raman line at $320\,cm^{-1}$ in the trace at 375 K is due to free bromine gas. The Raman spectra at different temperatures are shown using different intensity scales for clarity. (From [5.90])

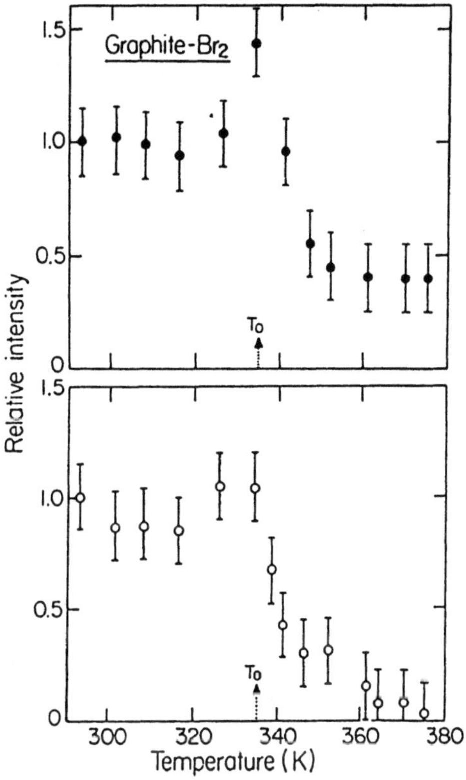

Fig. 5.33. Integrated intensities of the Stokes peaks (using 4880 Å laser excitation) for the bromine stretch mode at 242 cm^{-1} $I(\omega_0)$ (•) and the graphite bounding layer mode (o) as a function of temperature for a stage-3 graphite-bromine compound in which the intensities are normalized to their values at 290 K. The temperature T_0 where the discontinuity in the intensities occurs is identified with a commensurate–incommensurate order–disorder transition in the intercalate layer. (From [5.90])

line and a Gaussian fit to that line. The discontinuous drop in the intensities of the intercalate internal mode and of the graphitic intralayer mode as seen in Fig. 5.33 are induced by the commensurate-incommensurate structural transition through a perturbation of the electronic structure of the molecule. Moreover, the disappearance of the symmetric stretching mode at $\sim T_M$ results from the complete melting of the intercalate layer. As the layer melts, the intermolecular bonding

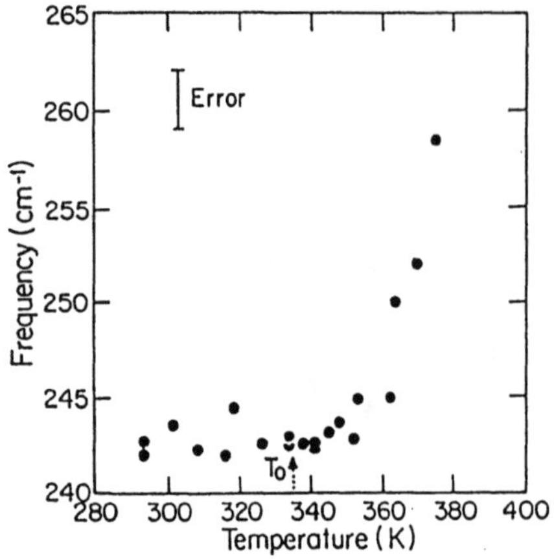

Fig. 5.34. Temperature dependence of the frequency of the intercalate bromine stretch mode for a stage-3 graphite-bromine compound. (From [5.90])

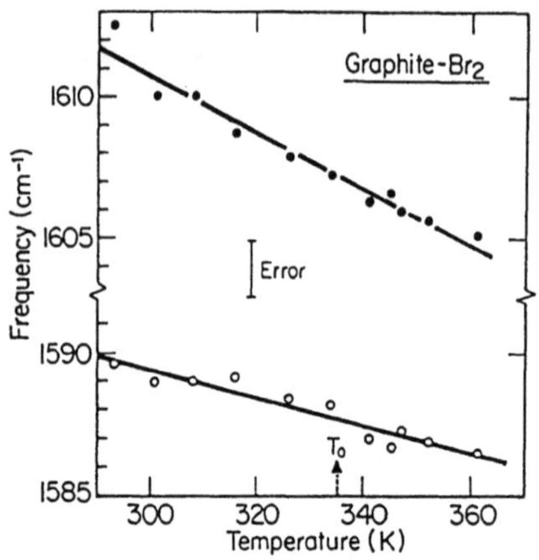

Fig. 5.35. Temperature dependence of the frequencies of the graphite interior layer (o) and the graphite bounding layer (o) modes over the temperature range $290 < T < 360$ K for a stage-3 graphite-bromine compound. (From [5.90])

decreases. This results in the upshift of the symmetric stretch frequency exhibited in Fig. 5.34. Finally, the downshift of the graphitic interior and bounding layer modes (Fig. 5.35) is, by comparison with similar effects in pristine graphite [5.89] and AsF_5 graphite [5.89], attributed to anharmonic changes in the vibrational potential caused by thermal expansion of the graphite layer lattice.

The analysis cited above appears to be valid with but one minor exception. Since we now know that the intercalate internal modes [5.84] *and* the graphitic modes [5.86, 88] are resonance enhanced, it is likely that the discontinuities in intensity shown in Fig. 5.33 both result from changes in the electronic band structure of the intercalation compound itself, rather than from changes in only the molecular levels of the intercalate [5.90].

5.4.2 Lattice Dynamics and Second-Order Raman Scattering

(a) General Theoretical Considerations

In Sect. 5.1, we pointed out that second-order optical processes provide access to non-zone-center phonons, since pairs of phonons whose net wave vector is ~ 0 participate in the Raman or IR interaction. Although there have, to date, been no reported studies of second-order IR measurements of any GIC, second-order Raman scattering from both donor and acceptor BGICs has been reported. Moreover, detailed calculations of the second-order spectra of donor compounds have been carried out by *Al-Jishi* and *Dresselhaus* [5.50]. These authors calculated the Raman cross section for overtone Stokes scattering involving two phonons of equal and opposite wave vector lying on the same branch of the dispersion curves. They employed a microscopic model which, like the macroscopic model discussed in Sect. 5.1, yields the second-order cross section as a weighted density of states. Specifically, *Al-Jishi* and *Dresselhaus* find [5.50] that

$$\frac{d^2\sigma}{d\Omega d\omega} = \sum_{\sigma}\sum_{q} \frac{G(\sigma, q)}{4\{[(\omega_I - \omega_S)/2 - \omega_{\sigma q}]^2 + \Gamma^2\}}, \tag{5.29}$$

where as before σ and q are, respectively, the phonon branch index and wave vector, the delta functions in (5.14) and (5.16) have been replaced by Lorentzian distributions of fixed width Γ, and the weight function, $G(\sigma, q)$ is proportional to the square of the susceptibility derivative given in (5.15) with $\sigma = \sigma'$. Using (5.29) with $\Gamma = 6\,\mathrm{cm}^{-1}$, together with the phonon dispersion curves shown in Fig. 5.10, *Al-Jishi* and *Dresselhaus* [5.50] calculated the second-order overtone Raman spectra for stage 3 and 4 Rb-graphite. These calculated spectra are shown in Fig. 5.36.

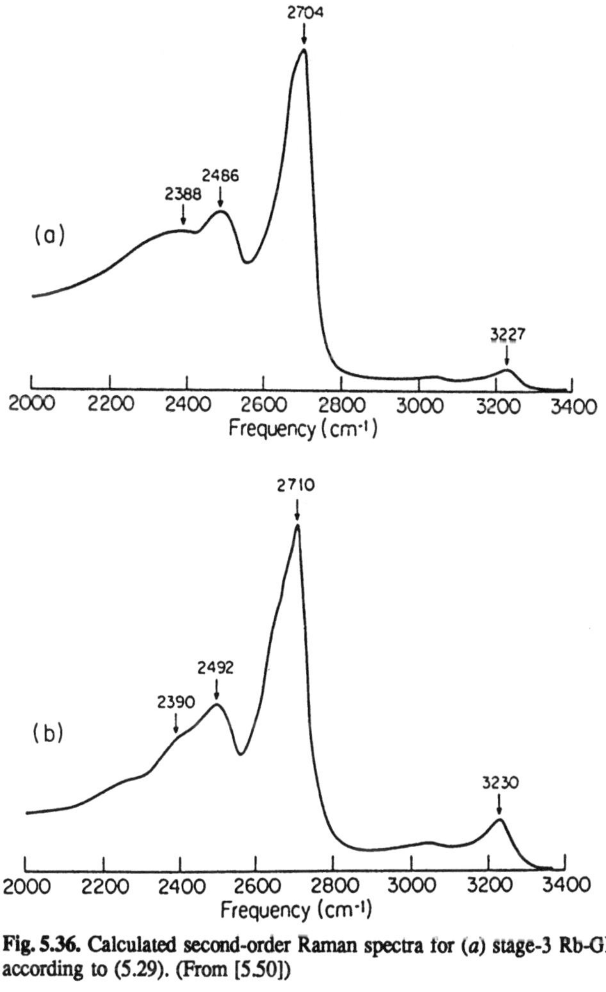

Fig. 5.36. Calculated second-order Raman spectra for (*a*) stage-3 Rb-GICs and (*b*) stage-4 Rb-GICs according to (5.29). (From [5.50])

Al-Jishi and *Dresselhaus* [5.50] also pointed out that the second-order spectra of stage 1 and 2 donor compounds may be featureless. Such compounds contain only bounding layers whose electron density is much larger than that of the interior layers or of the layers in pristine graphite. They argue that this increased electron density will produce negligible damping of the zone-center phonons but very large damping of the non-zone-center phonons. Structural disorder in the intercalate layer of the stage-2 compounds could also contribute to the damping of bounding layer phonons. Thus the zone-center modes are readily observed in first-order light scattering from all stages of donor BGICs, whereas these modes and the non-zone-center modes are washed out in the second-order spectra of stages which contain no interior layers. The theoretical implications of the above discussion will be compared with the corresponding experimental results which we now address.

(b) Comparison of Theory with Spectroscopic Measurement

Low Stage Donors, $n \leq 2$. Although Raman spectra of the low stage ($n \leq 2$) donor BGICs have been measured in the frequency region where second-order scattering might occur [5.70, 71], no second-order features have yet been observed for these materials, as expected on the basis of the analysis presented by *Al-Jishi* and *Dresselhaus* [5.50].

High Stage Donors, $n \geq 3$. The second-order Raman spectra of stage 3 and 4 Rb-graphite have been measured by *Eklund* et al. [5.70]. Their results are shown

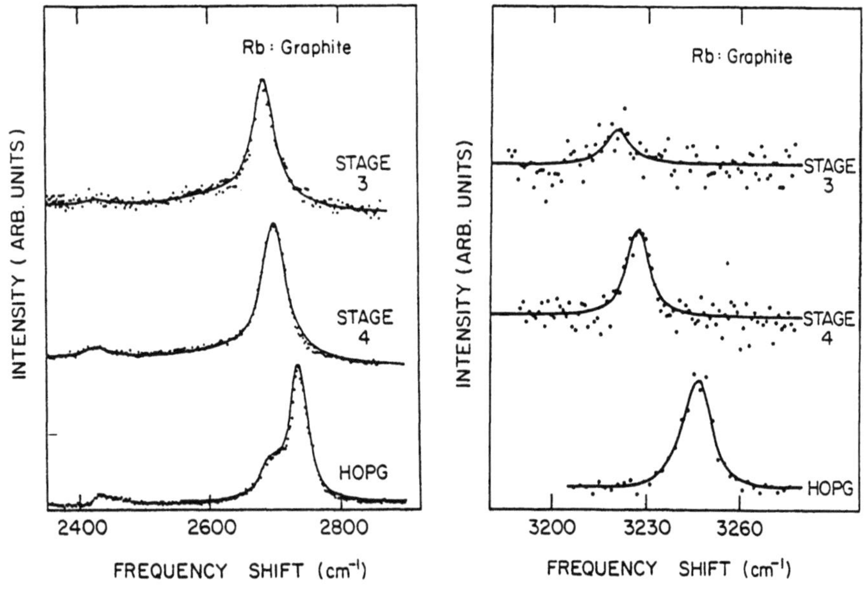

Fig. 5.37. Second-order Raman spectra of HOPG, HOPG-$SbCl_5$ and Rb-HOPG. The solid lines represent the results of Lorentzian line shape analyses. (From [5.70])

in Fig. 5.37, which also includes, for comparison, the corresponding spectrum of pristine graphite. These spectra resemble qualitatively the calculated spectra shown in Fig. 5.36. In particular, for the stage-3 compound the dominant feature is a strong band at 2685 cm^{-1}, but notable peaks also occur at 2420 cm^{-1} and 3220 cm^{-1}. The corresponding features in the theoretical spectra (Fig. 5.36) occur at 2704, 2437, and 3227 cm^{-1}, respectively. However, there is a significant difference in the intensity distributions of the theoretical and experimental spectra. The $\sim$ 2700 cm^{-1} feature has been identified [5.50] as an overtone of high-lying M-point phonons (Fig. 5.10) while that at $\sim$ 3220 cm^{-1} is an overtone of the maximum energy phonons on the Σ line between Γ and K (Figs. 5.6 and 10). Finally, the peak at $\sim$ 2430 cm^{-1} has been associated [5.70] with the mid-zone phonons on the T' branch (Figs. 5.6 and 10).

Stage-n Acceptors, $n \geq 2$. To date, the only second-order Raman studies on acceptor BGICs involve stage-2–5 $SbCl_5$ graphite [5.70], the spectra of which are very similar to that of pristine graphite and of stage 3 and 4 Rb-graphite. No theoretical calculations exist for the second-order acceptor spectra. However, since the charge transfer is much smaller in the acceptor compounds than in the donors, the electronically induced damping of the non-zone-center phonons should also be significantly reduced. This may be why the stage-2 $SbCl_5$ compound exhibits sharp second-order Raman features [5.70].

5.4.3 Charge Transfer and Phonon Frequencies

(a) Theoretical Considerations

It has been known since the work of *Nixon* and *Parry* [5.93] that the in-plane C-C bond distance in donor BGICs *increases* with increasing stage. In contrast, diffraction studies of electrochemical charging [5.18, 40] and of the stage dependence of the C-C distance in acceptor compounds reveal a *decrease* in that distance with increasing stage. Several models have been advanced to explain these observations but the most successful is that due to *Kamitakahara* et al. [5.94]. The above behavior can be qualitatively described as follows. In a donor (acceptor) GIC, charge is transferred into (out of) an antibonding π band. This weakens (stiffens) the intralayer C-C bonds, which results in an expansion (contraction) of the bounding layers. The charge transfer to the interior layers affects them similarly but they experience less relative change than do the bounding layers.

The stiffening of the C-C bonds in acceptors can be expected to result in a stepwise or discrete *increase* in the energies of the bounding and interior layer modes with increasing stage and vice versa for the donor compounds. Using electrochemical methods [5.18, 40], one can vary the charge transfer continuously in some acceptor BGICs and thus monitor the continuous change in phonon frequency. Explicit calculations by *Leung* et al. [5.95] show that the phonon energies should vary linearly with reciprocal stage.

(b) Comparison of Theory with Spectroscopic Measurements

Discrete Charge Transfer. In their review article, *Dresselhaus* and *Dresselhaus* [5.6] have plotted the stage dependence of the $E_{2g}^{(2)}$ interior and bounding layer modes against reciprocal stage for a large number of donor and acceptor BGICs [Ref. 5.6, Fig. 2.14]. All four plots show a linear variation with reciprocal stage, as expected from the discussion above. *Eklund* [5.7] has augmented this data with more recent results, including those derived from second-order Raman scattering, and has determined the best linear fits using a least squares analysis. His results are shown in Fig. 5.38. Note that the interior layer modes of both donors and acceptors extrapolate at $n = \infty$ to the frequency of the $E_{2g}^{(2)}$ mode of pristine graphite. According to *Leung* et al. [5.95], the frequency shifts seen in Fig. 5.38 are dominated by two effects, namely the guest–host interaction, which is responsible for the splitting between the interior and bounding layer modes, and a stage dependence in the host layer force constants. These two effects add in acceptors and produce the large shift in the bounding layer mode. In contrast, these effects tend to cancel in donors, which results in a much smaller shift in the bounding layer mode for those compounds.

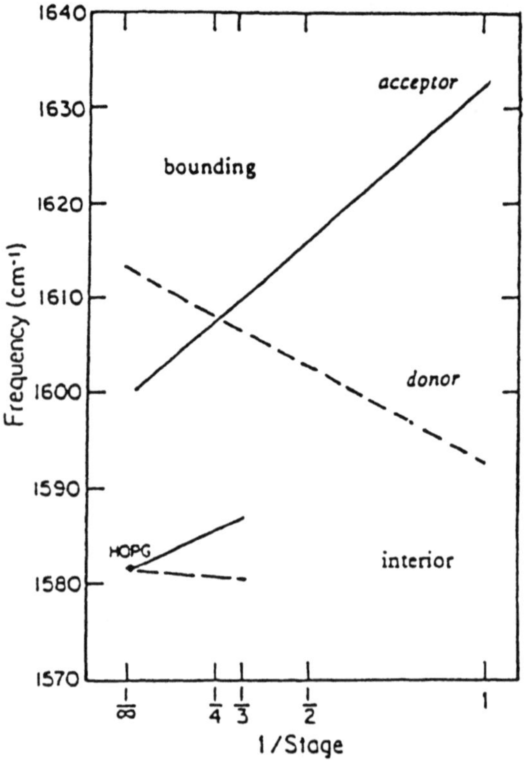

Fig. 5.38. $1/n$ dependence of the Raman active interior and bounding layer mode frequencies for donor (- - -) and acceptor (—) binary graphite intercalation compounds. (From [5.7])

Continuous Charge Transfer. The electrochemical method of intercalation is particularly attractive for the study of charge transfer effects because the charge transfer can be continuously varied and quantitatively controlled and because this method can be used in situ with a wide variety of probes, e.g., neutron scattering, optical reflectivity, and transport [5.40]. In a very nice set of experiments, *Eklund* and co-workers [5.18, 40 and references therein] have carried out a number of in situ investigations including Raman and IR studies of the electrochemical intercalation of H_2SO_4. The compound formed electrochemically has the formula $C_p^+HSO_4^-(H_2SO_4)_x$ where $p = 1/f_c$ is the oxidation number and f_c is the charge transferred to the guest species. (Because this compound contains a microscopic admixture of charged and neutral species, it is in fact a TGIC [5.12]. However, for reasons discussed below in Sect. 5.5.4, we address it here under BGICs.) By controlling the electrochemical current, one can accurately control the charge transfer. Graphite–sulfuric acid evolves with increasing charge transfer through a series of decreasing stages, exhibiting a staircase variation of EMF with Q/M, the charge transferred per unit mass [5.40]. This plot shows plateaus in regions corresponding to the transformation between stages n and $n-1$ and a linear increase in regions corresponding to the overcharging of pure stages. In the overcharging process, charge transfer converts neutral H_2SO_4 into HSO_4^- but at fixed stage, thus providing the ideal opportunity for examining the response of the phonon modes to charge transfer *at constant stage.*

Eklund et al. [5.40] have studied the variation of the frequency of the $E_{2g}^{(2)b}$ mode of graphite-sulfuric acid with charging, using in situ first-order Raman scattering. Their results are shown in Fig. 5.39. Note that the bounding layer mode shifts linearly with charge transfer in the overcharging regions corresponding to constant stage, as evidenced by the straight lines L_1 and L_2 for stages 1

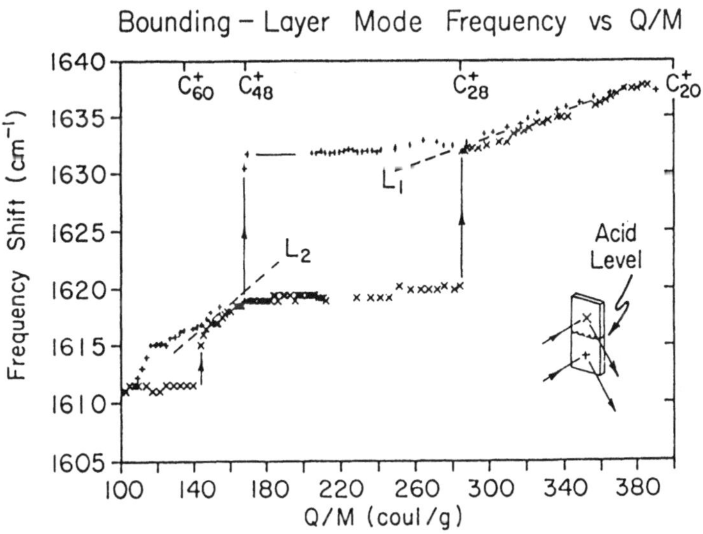

Fig. 5.39. Graphitic bounding layer mode frequency vs charge transfer Q/M. (From [5.40])

and 2, respectively. In contrast, in the region of stage transformation, the mode frequency is constant for each of the two coexisting stages. *Eklund* et al. have compared the data of Fig. 5.39 with the variation in the intralayer C-C distance [5.96] and concluded that interlayer forces play only a very minor role in determining the shift in mode frequency. The dependence of E_{1u} modes on charge transfer in graphite-sulfuric acid has also been studied using in situ IR reflection spectroscopy [5.97] and is compatible with the companion Raman studies [5.40]. However, a detailed interpretation of the IR data still awaits a line shape analysis of the spectral profiles.

5.5 Ternary Graphite Intercalation Compounds

Ternary graphite intercalation compounds (TGICs) contain two distinct guest species in the graphite galleries. Interest in the physics and chemistry of TGICs has accelerated during the past few years, culminating in two reviews. One of these, by *Solin* and *Zabel* [5.12], emphasizes the physics of these systems while the other, by *Setton*, constitutes Chap. 9 of this volume and emphasizes the chemistry and synthesis of TGICs. Although the literature on optical studies of the lattice dynamical properties of TGICs is quite limited relative to that on BGICs, it is sufficient to warrant a brief discussion in this review. Moreover, with the exception of the dual alkali donor TGICs, optical studies of these materials have not been previously reviewed. *Solin* and *Zabel* [5.12] have classified TGICs into four main structural groups, namely, homogeneous, heterogeneous, trilayer and localized [Ref. 5.12, Fig. 2]. Additionally, they are characterized as donor–donor, donor–acceptor, or acceptor–acceptor compounds according to the electronic nature of the guest species. We shall use these labels in the following discussion.

5.5.1 Homogeneous TGICs

(a) Donor–Donor

The stage-1 compounds $K_{1-x}Rb_xC_8$ with $0 \leq x \leq 1$ are in-plane substitutional homogeneous TGICs in which the alkali atoms randomly decorate the lattice sites corresponding to the hexagonal pockets in the bounding graphite layers [5.98]. They are homogeneous because the two guest ions are microscopically intermixed in the same gallery and so all galleries in the compound are statistically identical. Dual-alkali TGICs of this type exhibit composition-dependent anomalies in several of their properties [5.12]. Of particular interest are the force constant anomalies [5.99], reviewed in Chap. 4 of this volume. For K-Rb-graphite, the compressional elastic constant $C_{33}(x)$ shows an anomalous softening at $x \sim 0.66$ [5.99].

In an effort to observe similar anomalies in the optical phonons, *Solin* et al. [5.100] studied the composition dependence of the first-order Raman spectra of

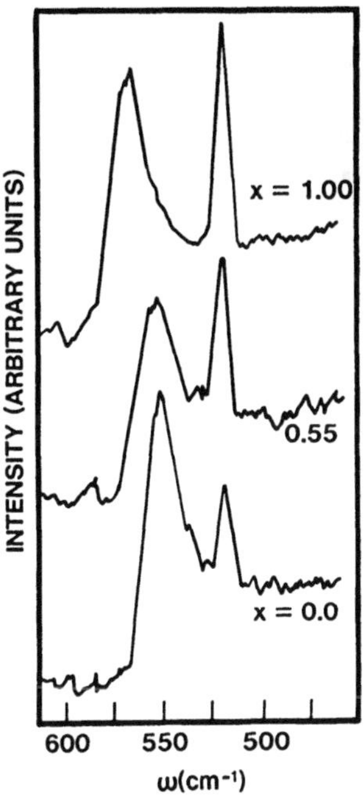

Fig. 5.40. Raman spectra of $K_{1-x}Rb_xC_8$ recorded in the ||, || configuration. The sharp feature at 517.37 cm^{-1} corresponds to the 5286.08 Å Ar^+ emission line. (From [5.100])

HOPG-derived $K_{1-x}Rb_xC_8$ with special emphasis on the sharp M-point FBW line. (The full spectrum of each compound with fixed x was qualitatively indistinguishable [5.100] from that of the stage-1 MC_8 BGICs shown in Fig. 5.13.) Some of the M-point spectra obtained by *Solin* et al. are shown in Fig. 5.40, which reveals an upshift in $\omega_{max}(x)$ [see (5.24) and associated discussion] with increasing Rb content. Moreover, a plot of $\omega_{max}(x)$ vs x revealed a sharp dip at $x = 0.67$ [5.100]. *Medjahed* et al. [5.101] repeated these measurements using specimens prepared from both HOPG and single crystals. Their HOPG results for $\omega_{max}(x)$ agree with those of *Solin* et al. [5.100] but the sharp drop in $\omega_{max}(x)$ at $x = 0.67$ is absent in the single crystal samples, which instead exhibit a sigmoidal response with the maximum slope at $x \sim 0.6$. *Medjahed* et al. [5.101] also carried out an FBW line shape analysis of their data to obtain the composition dependence of the bare phonon frequency $\omega_0(x)$. Their results for both HOPG and single crystal samples are shown in Fig. 5.41. On the basis of Fig. 5.41, *Medjahed* et al. [5.101] dismissed the "anomaly" at $x = 0.67$ as an artifact of the HOPG specimens, the results of which they attributed to strain inhomogeneities. This argument has been challenged [5.12] on the basis of X-ray data. Furthermore, notwithstanding the solid line "guide to the eye" which *Medjahed* et al. superposed on their data, the solid dots in Fig. 5.41 do show a marked plateau, if not a dip, in the region of $x = 0.67$, especially if one includes the point of maximal Raman shift at $x \sim 0.95$ [5.101]. Moreover, even the solid line in Fig. 5.41 shows a sharp rise at $x \gtrsim 0.6$. Thus, whether the data in Fig. 5.41 represent "anomalous" behavior or not may be a matter of semantics. We will leave that judgement up to the reader.

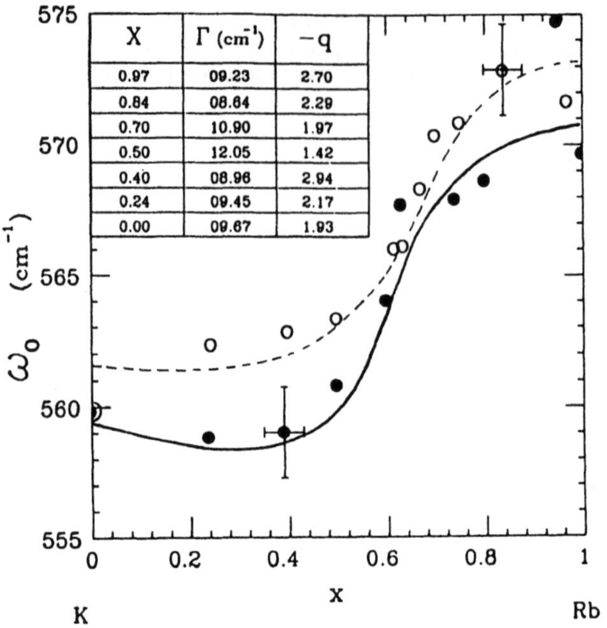

Fig. 5.41. Composition dependence of the bare phonon frequency ω_0 for HOPG samples (o) and single-crystal samples (•). *Inset:* Table of Breit-Wigner parameters for HOPG samples obtained from fits to (5.23) and (5.24). (From [5.101])

(b) Donor–Acceptor

Several donor–acceptor TGICs that are formed from the sequential intercalation of an acceptor molecule into an alkali BGIC such as K-graphite have been examined with first-order Raman scattering [5.102]. Acceptor molecules of interest include NH_3, furan, or tetrahydrofuran (THF). The resultant alkali-molecular TGICs have the following relevant properties as determined from X-ray studies [5.12], see also Chap. 9. When an MC_8 compound is intercalated with an acceptor molecule, the resultant TGIC also has a stage-1 structure but some alkali atoms may be desorbed in the process. When a stage-n MC_{12n} compound with $n \geq 2$ is sequentially intercalated with an acceptor, either a pure stage-$(n-1)$ homogeneous TGIC or a mixed stage $(n-1)+(n-2)+\ldots$ TGIC results. This generally occurs without any loss of the alkali species.

The Raman spectrum of a stage-1 TGIC $K(THF)_xC_{24}$ which was formed from stage 2 KC_{24} and structurally analyzed with X-rays has been measured by *Vora* et al. [5.102] and is shown in Fig. 5.42. The corresponding spectra for the NH_3 and furan molecules are qualitatively similar to that in Fig. 5.42 [5.102]. The singlet at $\sim 1606\,\mathrm{cm}^{-1}$ has an FBW line shape and is considerably broader than the singlet observed in stage-1 acceptor BGICs. Moreover, it is superposed on an intense continuum. Thus the spectrum in Fig. 5.42 is characteristic of a stage-1 MC_6 donor GBIC (see, for example, Fig. 5.13) with weak coupling between the

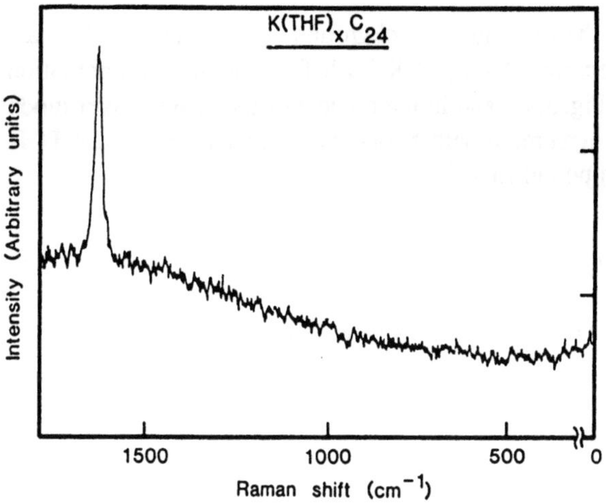

Fig. 5.42. The Raman spectrum of a stage-1 ternary K(THF)$_x$KC$_{24}$ compound. The abscissa is linear in wavelength rather than wave number. (From [5.102])

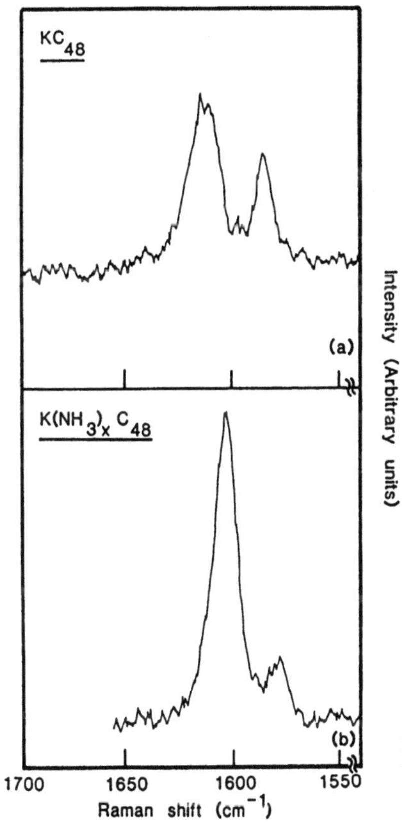

Fig. 5.43. Raman spectra showing the transformation of a stage-4 KC$_{48}$ binary GIC to a stage 2 + 3 K(NH$_3$)$_x$C$_{48}$ ternary GIC. (From [5.102])

bounding layer mode and the continuum background. *Vora* et al. [5.102] have also studied the Raman spectrum of stage-4 KC_{48} before and after ammoniation. Their results are shown in Fig. 5.43, and in the context of the nearest layer model [5.53, 54] (Sect. 5.1) are in agreement with X-ray studies that show that the TGIC is a mixed stage 2 + 3 compound [5.12].

5.5.2 Heterogeneous TGICs

To date, there are no reported Raman studies of any heterogeneous TGIC.

5.5.3 Trilayer TGICs

(a) Donor–Acceptor

Trilayer TGICs contain three distinct layers of two guest species within every host gallery, all galleries being identical [5.12]. For instance, all of the galleries in the stage-1 compound $KHgC_4$ contain a three-layer intercalate sandwich, consisting of a mercury layer bounded by two potassium layers in a 3d commensurate superlattice structure [5.103]. This compound is of considerable interest because of its superconductivity [5.104] and because it exhibits unique first- and second-order Raman spectra relative to that of all other GICs studied to date [5.105].

The Raman spectra of the potassium-amalgam-graphite trilayer TGICs have been studied by *Timp* et al. [5.105], who synthesized not only the stage-2 and stage-3 compounds but also two types of stage-1 material. The "pink" stage-1 has a *c*-axis repeat distance of $I_c = 10.13 \pm 0.05$ Å and in-plane superlattice regions with both $(2\times2)R0°$ and $(\sqrt{3}\times2)R(0°, 30°)$ structures. The "red" stage-1 compound yields $I_c = 10.22\pm0.05$ Å and both $(2\times2)R0°$ and $(\sqrt{3}\times\sqrt{3})R30°$ in-plane superlattice structures [5.105]. The first-order Raman spectra of these two stage-1 TGICs are quite different, as can be seen from the measurements by *Timp* et al., which are shown in Fig. 5.44. The presence of a number of peaks in the 830–1700 cm^{-1} region is unique, since such features have never been observed before for any other GIC. *Timp* et al. [5.105] attributed these peaks to modes which were Raman activated by in-plane zone-folding due to the superlattice structures cited above. They ascribed the spectral differences between the pink and red phases to the different superlattice structures which they possess, since the K and M points fold to the zone-center in the red stage-1 compound, while the M and $\Lambda_{3/4}$ points are folded in the pink stage-1 compound. The band positions and symmetries (polarization properties) which they deduced using this approach were in fair agreement with experiment [5.105]. The first-order Raman spectra of the stage-2 and stage-3 potassium amalgam TGICs measured by *Timp* et al. [5.105] were unremarkable and similar to those reported for other BGICs and TGICs.

Timp et al. also reported unique second-order Raman spectra of the pink stage-1 specimen. Whereas the second-order spectra of all other stage-1 compounds

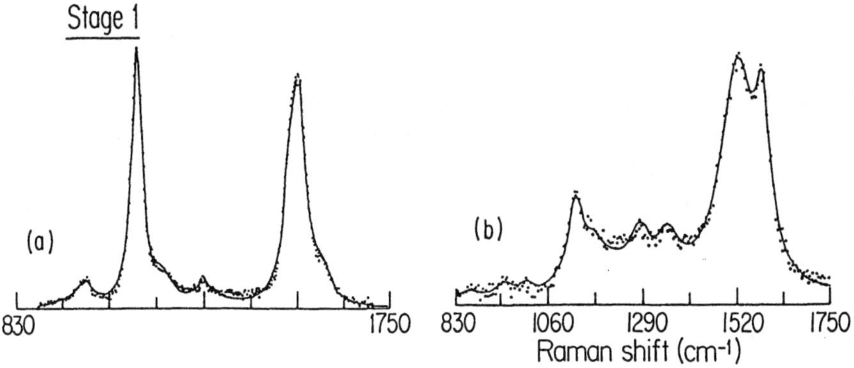

Fig. 5.44. First-order Raman spectra of stage-1 potassium-amalgam GIC, (**a**) for a pink sample and (**b**) for a red sample, both based on a HOPG host material. The unpolarized spectra show no significant changes in peak intensity or line width for ||, || and ||, ⊥ polarizations. (From [5.105])

studied to date are featureless, those of the pink K-Hg-C compound exhibit several prominent overtone and combination bands [5.105]. The unique quality of this second-order spectrum may result from the fact that intercalate layers are ordered, yet the charge transfer is sufficiently reduced, relative to donor BGICs, to minimize the electronic damping of the phonons, see Sect. 5.4.2.

The first-order Raman spectra of the K-H-C trilayer TGICs have also been studied [5.106]. The spectrum of the stage-1 compound KH_xC_4, $x \sim 0.8$, is similar to that of LiC_6 [5.56–58] and is indicative of weak coupling between the intralayer $E_{2g}^{(2)b}$ phonon and the continuum background. This observation is consistent with the continuum being electronic, since the reduced charge transfer in the KH_xC_4 compound relative to that of KC_8 would reduce the continuum and weaken the coupling. The stage-2 K-H-C TGICs yield singlet spectra with a strong band at $1598 \pm 3\,cm^{-1}$ [5.106] and are thus qualitatively similar to the corresponding spectra of other TGICs and BGICs.

5.5.4 Localized TGICs

Localized TGICs contain two microscopically separated *distinct chemical species* of the same or similar molecules within the same gallery such that all galleries are statistically identical [5.12]. A prime example is graphite–nitric acid, which contains, within the same gallery, neutral "spacer" molecules and charged molecules [5.40]. Other examples include acceptor species such as AsF_5 which disproportionate into several distinct entities on intercalation [5.87]. Although they are in fact localized TGICs, GICs formed from such species have been traditionally treated as binary compounds. Therefore, in deference to this history and to the usage employed in the other articles in this volume, we have discussed localized TGICs in the previous sections on BGICs.

5.6 Applications of Optical Phonon Spectroscopy

First-order and in some case second-order Raman spectroscopy has been extensively employed to study the insertion of chemical species in various morphological manifestations of graphite, including various types of graphite fibers ([5.37] and references therein), carbon-based composite materials [5.107], single crystals, HOPG, etc., using several methods such as ion implantation [5.108], electrochemistry [5.15], two bulb vapor transport [5.109], etc. The pristine forms of the above cited materials have also been studied. In several cases, the micro-Raman technique has been used to include spatially resolved spectra, which have proved to be especially useful in studies of fibers [5.37].

In the typical application of Raman spectroscopy, two types of information are acquired: (1) By studying the relative magnitudes and spectral widths of the $1582\,cm^{-1}$ $E_{2g}^{(2)}$ band and the $1355\,cm^{-1}$ disorder or finite-size-induced band, the degree of structural disorder in the pristine or intercalated host material can be semiquantitatively ascertained. (2) By studying the relative intensities of the interior and bounding layer $E_{2g}^{(2)}$ modes, an approximate value of the stage index can be obtained. The use of Raman spectroscopy to characterize graphitic materials in this way has been reviewed by *Dresselhaus* and *Dresselhaus* [5.6] and more recently by *Dresselhaus* [5.37] and by *Doll* et al. [5.107]. The reader is referred to this literature for additional information. Most recently, time resolved and spatially resolved first-order Raman scattering has been used to study the staging process in HOPG electrochemically intercalated with H_2SO_4 [5.15]. We use this work to illustrate the power of the Raman method as a characterization tool.

In their experiment, *Nishitani* et al. [5.15] mounted a thin rectangular piece of HOPG with its long edge perpendicular to and just touching the surface of liquid H_2SO_4 in an electrochemical cell. With an optical multichannel analzyer equipped with a silicon intensified target, they were able to record in real time the first-order spectrum of the $E_{2g}^{(2)}$ doublet as a function of position from the liquid surface. Note that at the measurement temperature used (300 K), the vapor pressure of sulfuric acid is sufficiently small that vapor phase intercalation of the specimen can be ignored. The results of *Nishitani* et al. [5.15] are shown in Fig. 5.45, in which the abscissa represents increasing distance from the liquid surface and the ordinate represents increasing time after initiation of the electrochemical reaction. From the relative intensities of the doublet components, *Nishitani* et al. [5.15] established the time vs position map of the staging kinetics shown in Fig. 5.46. This shows rather dramatically the "movement of the phase boundary between different well staged domains towards the interior of the crystal with time" [5.15]. *Nishitani* et al. [5.15] concluded that the stage transformation process resembles the first-order phase transition of the growth of a crystal from a melt and that their model of staging is most compatible with these results.

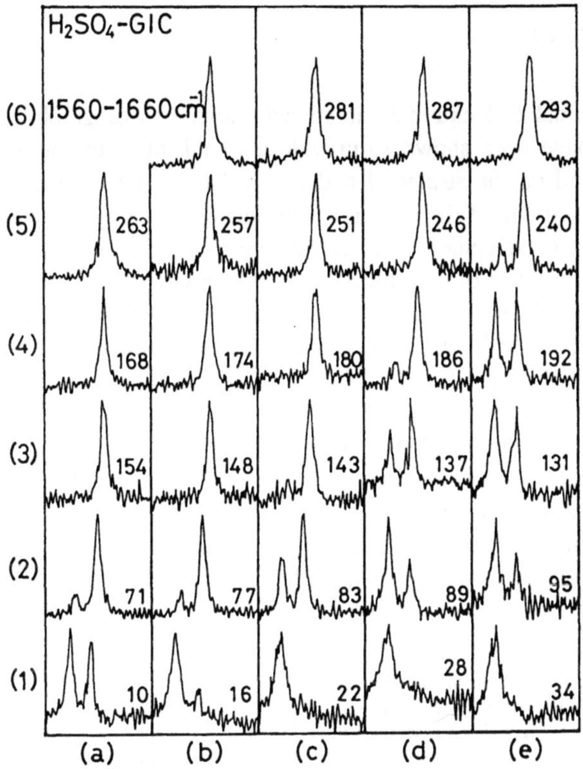

Fig. 5.45. The time- and position-dependent Raman spectra of graphite–sulfuric acid for the intralayer modes in the region 1560–1660 cm^{-1}, during intercalation from graphite through stage 2. The spectra in columns (*a–e*) correspond to those observed at distances of 3, 4, 5, 6, 7 mm, respectively, from the sample edge (liquid surface). The numbers in the figure refer to the time [min] after the onset of the electrochemical reaction. (From [5.15])

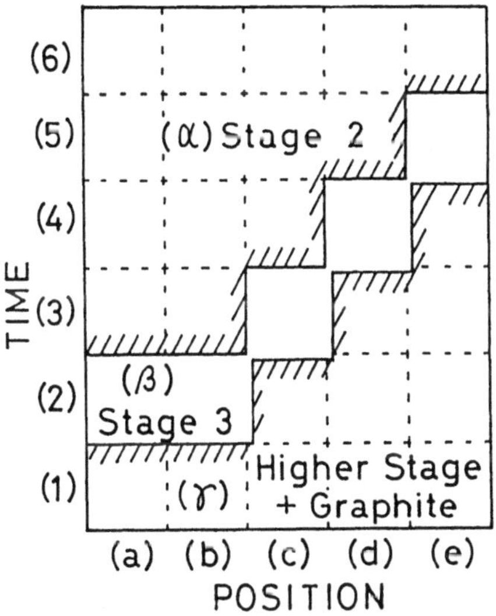

Fig. 5.46. The stage indices in time-position space where each stage index in a frame corresponds to that of each spectrum in Fig. 5.45. (From [5.15])

5.7 Concluding Remarks

Optical studies of the internal and external vibrational excitations of graphite intercalation compounds have been shown to provide a wealth of information on the lattice dynamics, and on the structural and physical properties of these materials. Theoretical calculations of the optical spectra using lattice dynamical models are in semiquantitative agreement with experiments and provide a good basis for understanding the physics of the optical processes in GICs. The sound theoretical base and wealth of current experimental data enable optical studies to be used as an effective method for nondestructive in situ characterization of all graphitic materials, including not only GICs but also pristine substances. Microsampling techniques and real-time measurements provide added versatility in the case of Raman scattering. The adaptation of IR methods, perhaps with laser sources, to microsampling is a potentially important future technological goal. Of fundamental interest for future work would be the use of Brillouin scattering to augment neutron results in the very low frequency region of the phonon spectrum and the optical study of magnetic excitations, which has proved to be so fruitful with other solid systems.

Acknowledgements. It is a pleasure to acknowledge the major contributions of my students and colleagues to the material cited in this chapter under my coauthorship. Their intellectual contributions and insight during the past few years have been essential to the development of our understanding of the optical properties of GICs. Explicit thanks are due to N. Caswell, R. Clarke, C. Horie, H. Miyazaki, R.J. Nemanich, N. Wada, and H. Zabel. Naturally, any errors or omissions in this review are solely my responsibility. I am grateful to the NSF for supporting my research on intercalated solids, most recently under grants DMR 85-17223 and DMR MRG 85-14154.

References

5.1 W. Hayes, R. Loudon: *Scattering of Light by Crystals* (Wiley, New York 1978)
5.2 S.W. Lovesey: *Theory of Neutron Scattering from Condensed Matter* (Oxford University Press, Oxford 1984)
5.3 R.J. Nemanich, S.A. Solin: Phys. Rev. B **20**, 392 (1979)
5.4 S.A. Solin: Physica **99B**, 443 (1980)
5.5 M.S. Dresselhaus, G. Dresselhaus: Adv. Phys. **30**, 139 (1981)
5.6 M.S. Dresselhaus, G. Dresselhaus: In *Light Scattering in Solids III*, ed. by M. Cardona, G. Guntherodt, Topics Appl. Phys., Vol. 51 (Springer, Berlin, Heidelberg 1982) p. 3
5.7 P.C. Eklund: In *Intercalation in Layered Solids*, ed. by M.S. Dresselhaus (Plenum, New York 1986) p. 323
5.8 S.A. Solin, N. Caswell: J. Raman Spectrosc. **10**, 129 (1981)
5.9 S.A. Solin: J. de Phys. **42**, C6–283 (1981)
5.10 S.A.Solin: J. Phys. Soc. Jpn. **49**, 903 (1980)
5.11 S.A. Solin: In *Intercalation in Layered Solids*, ed. by M.S. Dresselhaus (Plenum, New York 1986) p. 313
5.12 S.A. Solin, H. Zabel: Adv. Phys. **37**, 87 (1988)
5.13 N. Wada, Phys. Rev. B **24**, 1065–1078 (1981)
5.14 L. McNeil, J. Steinbeck, L. Salamanca-Riba, G. Dresselhaus: Phys. Rev. B **31**, 2451 (1985)

5.15 R. Nishitani, Y. Sasaki, Y. Nishina: J. Phys. Soc. Jpn. **56**, 1051 (1987)
5.16 I. Ohana, Y. Yacoby: Phys. Rev. Lett. **57**, 2572 (1986)
5.17 B.S. Elman, M.S. Dresselhaus, G. Dresselhaus, E.W. Maby, H. Mazurek: Phys. Rev. B **24**, 1027 (1981)
5.18 P.C. Eklund: In *Intercalation in Layered Solids*, ed. by M.S. Dresselhaus (Plenum, New York 1986) p. 337
5.19 Y.J. Chabal: Surf. Sci. Rep. **8**, 211 (1988)
5.20 M. Cardona (ed.): *Light Scattering in Solids*, 2nd ed., Topics Appl. Phys., Vol. 8 (Springer, Berlin, Heidelberg 1983)
5.21 K.D. Muller, W.G. Rothschild: *Far-Infrared Spectroscopy* (Wiley, New York 1971)
5.22 J.D. Jackson: *Classical Electrodynamics* (Wiley, New York 1962)
5.23 J.D. Landau, E.M. Lifshitz: *Statistical Physics* (Pergamon, Oxford 1969)
5.24 V. Heine: *Group Theory in Quantum Mechanics* (Pergamon, Oxford 1960)
5.25 R. Zallen: *The Physics of Amorphous Solids* (Wiley, New York 1983)
5.26 R.J. Nemanich, S.A. Solin, R.M. Martin: Phys. Rev. B **23**, 6348 (1981)
5.27 J.L. Birman: Phys. Rev. **131**, 1489 (1963)
5.28 P.C. Eklund, G.D. Mahan, J.G. Spolar, E.T. Arakawa, J.M. Zhang, D.M. Hoffman: Solid State Commun. **57**, 567 (1986)
5.29 J.J. Song, D.D.L. Chung, P.C. Eklund, M.S. Dresselhaus: Solid State Commun. **20**, 1111 (1976)
5.30 N. Caswell, S.A. Solin: Solid State Commun. **27**, 961 (1978)
5.31 R.J. Nemanich, G. Lucovsky, S.A. Solin: In Proc. of the Int. Conf. on Lattice Dynamics, ed. by M. Balkanski (Flammarion, Paris 1978) p. 619
5.32 M. Zanini, J.E. Fischer: Mater. Sci. Eng. **31**, 169 (1977)
5.33 M. Zanini, D. Grubisic, J.E. Fischer: Phys. Status Solidi **90**, 151 (1978)
5.34 I. Ohana, D. Schmeltzer, Y. Yacoby: In Extended Abstracts of the 1984 MRS Symposium on Graphite Intercalation Compounds, ed. by P.C. Eklund, M.S. Dresselhaus, G. Dresselhaus (Materials Research Society, Pittsburgh, PA 1984) p. 121
5.35 J.M. Zhang, P.C.Eklund: J. Mater. Res. **2**, 858 (1987)
5.36 P. Kwizera, M.S. Dresselhaus, G. Dresselhaus: Carbon, submitted
5.37 M.S. Dresselhaus: In *Intercalation in Layered Solids*, ed. by M.S. Dresselhaus (Plenum, New York 1986) p. 461
5.38 R.M. Hazen, L.W. Finger: *Comparative Crystal Chemistry* (Wiley, New York 1982)
5.39 N. Caswell, S.A. Solin: Phys. Rev. B **20**, 2251 (1979)
5.40 P.C. Eklund, E.T. Arakawa, J.L. Zaretsky, W.A. Kamitakahara, G.D. Mahan: Synth. Met. **12**, 97 (1985)
5.41 R.W.G. Wycoff: *Crystal Structures*, Vol. 1 (Oxford University Press, Oxford 1962) p. 26
5.42 R. Al-Jishi, G. Dresselhaus: Phys. Rev. B **26**, 5414 (1981)
5.43 S.A. Solin: Physica **99B**, 443 (1980) and references therein
5.44 L.J. Brillson, E. Burstein, A.A. Maradudin, T. Stark: In *Physics of Semimetals and Narrow Gap Semiconductors*, ed. by D.L. Cater, R.T. Bate (Pergamon, Oxford 1971) p. 187
5.45 S.A. Solin: Adv. Chem. Phys. **49**, 455 (1982)
5.46 S.A. Solin, A.K. Ramdas: Phys. Rev. B **1**, 1687 (1970)
5.47 R.J. Nemanich, S.A. Solin: Solid State Commun. **23**, 417 (1977)
5.48 M. Maeda, Y. Kuramoto, C. Horie: H. Phys. Soc. Jpn. **47**, 337 (1979)
5.49 F. Tuinstra, J.F. Koenig: J. Chem. Phys. **53**, 1126 (1970)
5.50 R. Al-Jishi, G. Dresselhaus: Phys. Rev. B **26**, 4523 (1982)
5.51 F.Y. Leung, G. Dresselhaus, M.S. Dresselhaus: Phys. Rev. B **24**, 6083 (1981)
5.52 W.A. Kamitakahara, N. Wada, S.A. Solin: Solid State Commun. **44**, 297 (1982)
5.53 S.A. Solin: Mater. Sci. Eng. **31**, 153 (1977)
5.54 R.J. Nemanich, S.A. Solin, D. Guerard: Phys. Rev. B **16**, 2965 (1977)
5.55 J.J. Song, D.D.L. Chung, P.C. Eklund, M.S. Dresselhaus: Solid State Commun. **20**, 1111 (1976)
5.56 P.C. Eklund, G. Dresselhaus, M.S. Dresselhaus, J.E. Fischer: Phys. Rev. B **21**, 4705 (1980)
5.57 D.M. Hwang, S.A. Solin, D. Guerard: In *Physics of Intercalation Compounds*, ed. by L. Pietronero, E. Tosatti, Springer Ser. Solid-State Sci. Vol. 38 (Springer, Berlin, Heidelberg 1981) p. 187
5.58 D.M. Hwang, D. Guerard: Solid State Commun. **40**, 759 (1981)
5.59 U. Fano: Phys. Rev. **124**, 1866 (1961)

5.60 M.V. Klein: In [5.20], Chap. 4
5.61 P.C. Eklund, K.R. Subbaswamy: Phys. Rev. B **20**, 5157 (1979)
5.62 H. Miyazaki, T. Hatano, G. Kusunoki, T. Watanabe, C. Horie: Physica **105B**, 381 (1981)
5.63 N. Caswell, S.A. Solin: Phys. Rev. B **20**, 2551 (1979)
5.64 C. Horie, M. Maeda, Y. Kuramoto: Physica **99B**, 430 (1980)
5.65 R. Clarke, N. Caswell, S.A. Solin: Phys. Rev. Lett **42**, 61 (1979)
5.66 D.S. Robinson, J.L. Zaretsky: Bull. Am. Phys. Soc. **31**, 587 (1986)
5.67 S.A. Solin: In *Physics of Intercalation Compounds*, ed. by L. Pietronero, E. Tosatti, Springer Ser. Solid-State Sci., Vol. 38 (Springer, Berlin, Heidelberg 1981) p. 214
5.68 P.C. Eklund: In *Intercalation in Layered Solids*, ed. by M.S. Dresselhaus (Plenum, New York 1986) and references therein
5.69 N. Wada, M.V. Klein, H. Zabel: J. de Phys. **42**, C6–350 (1981)
5.70 P.C. Eklund, J. Giergiel, P. Boolchand: In *Physics of Intercalation Compounds*, ed. by L. Pietronero, E. Tosatti, Springer Ser. Solid-State Sci., Vol. 38 (Springer, Berlin, Heidelberg 1981) p. 168
5.71 J. Giergiel, P.C. Eklund, R. Al-Jishi, G. Dresselhaus: Phys. Rev. B **26**, 6881 (1982)
5.72 S.Y. Leung, C. Underhill, G. Dresselhaus, M.S. Dresselhaus: Solid State Commun. **33**, 285 (1980)
5.73 P.C. Eklund, M.M. Yang, G.L. Doll: In *Intercalation in Layered Solids*, ed. by M.S. Dresselhaus (Plenum, New York 1986) p. 2573
5.74 P.C. Eklund, G. Dresselhaus, M.S. Dresselhaus, J.E. Fischer: Phys. Rev. B **16**, 3330 (1977)
5.75 M.S. Dresselhaus, G. Dresselhaus: In *Light Scattering in Solids III*, ed. by M. Cardona, G. Guntherodt, Topics Appl. Phys., Vol. 51 (Springer, Berlin, Heidelberg 1982) and references therein
5.76 For a thorough review see R. Clarke, C. Uher: Adv. Phys. **33**, 469 (1984)
5.77 J.E. Fischer, H.J. Kim: Synth. Met. **12**, 137 (1985)
5.78 C.D. Fuerst, J.E. Fischer, J.D. Axe, J.B. Hastings, D.B. McWhan: Phys. Rev. Lett **50**, 780 (1983)
5.79 P.C. Eklund, D.S. Smith, V.R.K. Murthy, S.Y. Leung: Synth. Met. **2**, 99 (1980)
5.80 P.C. Eklund: In *Intercalation in Layered Solids*, ed. by M.S. Dresselhaus (Plenum, New York 1986) p. 163
5.81 C. Underhill, S.Y. Leung, G. Dresselhaus, M.S. Dresselhaus: Solid State Commun. **29**, 769 (1979)
5.82 G.M. Gualberto, C. Underhill, S.Y. Leung, G. Dresselhaus: Phys. Rev. B **21**, 862 (1980)
5.83 See for example S.A. Solin, A.K. Ramedas: Phys. Rev. **174**, 1069 (1968)
5.84 P.C. Eklund, N. Kambe, G. Dresselhaus, M.S. Dresselhaus: Phys. Rev. B **18**, 7069 (1978)
5.85 N. Caswell, S.A. Solin: Solid State Commun. **27**, 961 (1978)
5.86 P.C. Eklund, G.D. Mahan. J.G. Spolar, E.T. Arakawa, J.M. Zhang, D.M. Hoffman: Solid State Commun. **57**, 567 (1986)
5.87 N.R. Bartlett, R.N. Biagioni, E.M. McCarron, B.W. McQuillan, F.T. Tanzela: In *Molecular Metals*, ed. by W.E. Hatfield (Plenum, New York 1985) p. 356
5.88 Y. Yacoby, I. Ohana: Synth. Met. **23**, 3321 (1988)
5.89 P.C. Eklund, E.R. Falardeau, J.E. Fischer: Solid State Commun. **32**, 631 (1979)
5.90 A. Erbil, G. Dresselhaus, M.S. Dresselhaus: Phys. Rev. B **25**, 5451 (1982)
5.91 K.K. Bardhan, J.C. Wu, D.D.L. Chung: Synth. Met. **2**, 109 (1980)
5.92 D.D.L. Chung, G. Dresselhaus, M.S. Dresselhaus: Mater. Sci. Eng. **37**, 213 (1979)
5.93 D.E. Nixon, G.S. Parry: J. Phys. C **2**, 1732 (1969)
5.94 C.T. Chan, W.A. Kamitakahara, K.M. Ho, P.C. Eklund: Phys. Rev. Lett **58**, 1528 (1987) C.T. Chan, W.A. Kamitakahara, K.M. Ho: Synth. Met. **23**, 327 (1988)
5.95 S.Y. Leung, G. Dresselhaus, M.S. Dresselhaus: Solid State Commun. **38**, 175–178 (1981); Phys. Rev. B **24**, 6083 (1981)
5.96 W.A. Kamitakahara, J.L. Zaretsky, P.C. Eklund: Synth. Meth. **12**, 301 (1985)
5.97 A. Rao, J.M. Zhang, G.W. Lehman, P.C. Eklund: *Graphite Intercalation Compounds: Extended Abstracts*, ed. by M.S. Dresselhaus, G. Dresselhaus, S.A. Solin (Materials Research Society, Pittsburgh, PA 1986) p. 108
5.98 G. Furdin, B. Carton, A. Hérold, C. Zeller: C.R. Acad. Sci. B **280**, 653 (1975)
5.99 D.A. Neumann, H. Zabel, J.J. Rush, N. Berk: Phys. Rev. Lett **53**, 56 (1984) ref5.100S.A. Solin, P. Chow, H. Zabel: Phys. Rev. Lett **53**, 1927 (1984)
5.101 D. Medjahed, R. Merlin, R. Clarke: Phys. Rev. B **36**, 9345 (1987)

5.102 P. Vora, B.R. York, S.A. Solin: Synth. Met. **1**, 355 (1983)
5.103 P. Lagrange, M. El Mahrini, A. Hérold: Rev. Chim. Miner. **20**, 229 (1983)
5.104 G. Dresselhaus, A. Chaiken: In *Intercalation in Layered Solids*, ed. by M.S. Dresselhaus (Plenum, New York 1986) p. 387
5.105 G. Timp, B.S. Elman, R. Al-Jishi, G. Dresselhaus: Solid State Commun. **44**, 987 (1982)
5.106 N.C. Yeh, T. Enoki, L.E. McNeil, G. Roth, L. Salamanca-Riba, M. Endo, G. Dresselhaus: In *Extended Abstracts for the Symposium on Graphite Intercalation Compounds*, ed. by P.C. Eklund, M.S. Dresselhaus, G. Dresselhaus (Materials Research Society, Pittsburgh, PA 1984) p. 246
5.107 G.L. Doll, R.M. Sakya, J.T. Nichols, J.S. Speck, M.S. Dresselhaus: Synth. Met. **23**, 481 (1988)
5.108 B.S. Elman, M.S. Dresselhaus, G. Dresselhaus, E.W. Maby, H. Mazurek: Phys. Rev. B **24**, 1027 (1981)
5.109 A. Hérold: Bull. Soc. Chim. Fr. 999 (1955)

6. Intercalate Diffusion

Andreas Magerl

With 21 Figures

From the ease and the rapidity of the intercalation process in graphite it is evident that particle diffusion in GICs must be quite fast. A diffusion constant on the order of 10^{-5}cm^2/s is estimated from a large HOPG sample with linear dimensions of 1 cm and an intercalation time of a few hours. Often, the intercalation process proceeds via a well-defined intercalation front which can be observed readily, e.g., from a change in the sample thickness or color. A quantitative evaluation gives valuable information about the chemical diffusion constant D^* related to the macroscopic transport of mass in the sample. Simple visual observation reveals that intercalation always proceeds coplanar with the graphene sheets, showing the two-dimensional (2d) character of the diffusion process.

The flow of mass into a graphite sample is a complex phenomenon. In the first place, room has to be made available to accommodate the intercalate. The van-der-Waals gaps between graphene sheets have to be opened up at the edge of the sample, which seems to require a certain threshold energy. More material will then follow readily into the once widened galleries and the diffusion is driven mainly by the gradient in the chemical potential related to the local concentration of the intercalate. Similar phenomena will control the diffusion across the numerous inner surfaces of large samples, rendering D^* very sensitive to sample properties, including previous intercalations.

Microscopic techniques such as nuclear magnetic resonance (NMR) and quasi-elastic neutron scattering (QENS) access directly the self-diffusion constant D related to the particle mobility in thermodynamic equilibrium, i.e., without a macroscopic flow of mass. The values of D^* and D can be very different. For example, for highly concentrated TaH_x, it has been shown by a comparison of QENS and NMR with Gorsky effect measurement that D^* exceeds D by more than one order of magnitude [6.1], whereas the opposite behavior is found at the critical point in NbH_x [6.2]. Therefore, care has to be taken when diffusion constants obtained with different methods are compared.

Two important parameters concern the activation energy E_a and the prefactor D_0, which can be obtained from an Arrhenius plot of the temperature dependence of the measured diffusion constant.

In this chapter we are mainly concerned with the self-diffusion process of intercalates in GICs. Among the large number of different techniques applied, we shall focus on QENS in this chapter; however, we will also include other methods where relevant. The unique advantage of QENS is in its ability to access on an atomic scale the time constants involved in the diffusion (from the value of the energy broadening of a monochromatic neutron beam) and simultaneously the geometry of the diffusive path (from the wave vector dependence of the energy broadening). Therefore – if applicable – QENS is able to provide very detailed information about the microscopic mechanisms involved in particle diffusion.

In the following section, we give a brief and illustrative presentation of QENS as it is needed to follow the later discussion. This is by no means intended to serve as a textbook introduction to this technique. For further details the reader should refer to the original literature [6.3–7]. In Sect. 6.2, we review the diffusion data on alkali metal-GICs. These are relatively simple systems because the intercalate has no internal degrees of freedom and only long range or translational diffusion modes are expected. In molecular intercalates or in ternary compounds (Sect. 6.3) additional local diffusion, also called rotational modes, has to be considered. These modes imply that the center of gravity of the diffusing unit is not displaced with time. In general, at elevated temperatures both diffusion modes will be present simultaneously. Generally, the time constants involved in the two processes differ greatly, a fact that can be used to distinguish these modes.

To date, most of the measurements have been performed on time-of-flight instruments on steady-state sources, which are very well suited to quasi-elastic studies on powder samples or HOPG material. In this technique, the continuous and white neutron beam from the reactor is prepared in the primary part of the spectrometer to give a short pulse (duty cycle $< 1\,\%$) of monochromatic neutrons which then impinge onto the sample. Diffusive processes lead to a relatively small energy loss or energy gain of the neutron. This increase of the energy width of the neutron beam is called quasi-elastic scattering. In favorable cases it can be detected by monitoring the arrival times of the neutrons at the detectors after they have traveled a flight path of a few meters. In general, neutrons scattered at the sample into a large solid angle are counted simultaneously in a multi-detector array which provides a good yield of the incident beam. In addition, time-of-flight spectrometers served by primary beams from different thermal sources offer a wide range of energy resolutions, from $10\,\mu$eV to several meV. The diffusion constants accessible on modern spectrometers, including backscattering equipment, range from $D \sim 10^{-4}\text{cm}^2/\text{s}$ to $D \sim 10^{-8}\text{cm}^2/\text{s}$.

6.1 Quasi-Elastic Neutron Scattering

As outlined in Chap. 2 of this book, GICs show a rich variety of crystallographic structures. At high temperatures, disordered phases dominated by short range correlations are generally found. They are commonly called liquid phases. However, one should be careful with this a priori assignment, keeping in mind that, from a purely structural point of view, a liquid and an amorphous state are rather similar. Their main difference concerns the excitation spectrum, including the diffusion dynamics. In a liquid the atoms or molecules change position rapidly and diffusion constants as large as $D \simeq 10^{-4}\text{cm}^2/\text{s}$ are observed, whereas the largely limited particle mobility in the amorphous state reduces D by many orders of magnitude. A proper distinction between an amorphous material and a liquid can be based on a study of the diffusion dynamics. QENS is a particularly promising technique because it goes well beyond the determination of the time constants involved. At the same time, it provides very detailed insights into the diffusive path and therefore into short range structural properties.

6.1.1 Scattering Cross Sections

Thermal and near-thermal neutrons have long wavelengths $\lambda(\sim 1\text{–}10\,\text{Å})$ compared to the range of the neutron–nucleus interaction ($\sim 10^{-5}\,\text{Å}$). Therefore, the neutron flux scattered from an isolated nucleus is isotropic and the strength of the interaction has a constant value, called the scattering length b. The neutron itself and also often the scattering nuclei carry a spin with a spin-dependent interaction strength. Because of the random orientation of nuclear spins and the neutron spin (under usual experimental conditions), only the mean value of the spin-dependent scattering lengths can lead to constructive interference, while the deviations from this mean value give spatially isotropic scattering, also called incoherent scattering. In addition, most elements consist of a natural mixture of isotopes with individual scattering lengths. Again, their random distribution constitutes another origin of incoherent scattering. As a consequence, the total cross section for the neutron–nucleus interaction σ^{tot} can be divided into two parts:

$$\begin{aligned} \sigma^{\text{tot}} &= \sigma^{\text{coh}} + \sigma^{\text{inc}} \quad \text{with} \\ \sigma^{\text{coh}} &= 4\pi \langle b \rangle^2 \quad \text{and} \\ \sigma^{\text{inc}} &= 4\pi \langle (b - \langle b \rangle)^2 \rangle \,, \end{aligned} \tag{6.1}$$

where $\langle \cdot \rangle$ indicates the ensemble average for both spin- and isotope-incoherent scattering. The term σ^{coh} reflects collective features within the scattering object leading to a scattered intensity which is structured in space (e.g. Bragg peaks from a solid or diffuse peaks in a liquid). In contrast, σ^{inc} is associated with the single particle behavior whose scattered intensity is spatially isotropic. It must be emphasized that both coherent and incoherent scattering can be either partially or totally elastic, quasi-elastic or inelastic.

Table 6.1 gives cross sections for selected elements relevant to this chapter. Note that in some cases values for individual isotopes are listed. Most cross sections are on the order of a few barns $[10^{-24}\text{cm}^2]$. The heavy alkali and carbon nuclei are essentially coherent scatterers. For an ordered structure, e. g. the stage-1 compounds at low temperature, the scattered intensity is entirely contained in the Bragg peaks (in this chapter we set the Debye-Waller factor to unity, which means that weak scattering contributions, for example from phonons, will be neglected). In contrast, ^{1}H has an extremely large incoherent cross section. In general it overwhelms all other scattering contributions between the Bragg peaks and in this case only information about the hydrogen motions can be obtained.

Table 6.1. Coherent and incoherent cross sections for selected elements (from [6.8])

Element	$\sigma^{coh}[10^{-24}cm^2]$	$\sigma^{inc}[10^{-24}cm^2]$
^1H	1.76	79.91
^2H	5.60	2.04
Li	0.45	0.91
^7Li	0.62	0.78
C	5.55	< 0.01
O	4.24	< 0.01
N	11.01	0.49
S	1.02	< 0.01
K	1.73	0.25
Rb	6.30	0.3
Cs	3.69	0.21

6.1.2 Scattering Laws and Correlation Functions

Let a neutron beam with energy $\hbar\omega_i$ and momentum $\boldsymbol{k}_i$ (Fourier coordinates) impinge on a sample (Fig. 6.1). The intensity scattered with energy $\hbar\omega_f$ (energy transfer $\hbar\omega$) and with momentum $\boldsymbol{k}_f$ (momentum transfer $\boldsymbol{Q}$) into a solid angle element $d\Omega$ is given by

$$\frac{d^2\sigma}{d\Omega d(\hbar\omega_f)} = \frac{k_f}{k_i}\left(\frac{\sigma^{coh}}{4\pi}S^{coh}(\boldsymbol{Q},\omega) + \frac{\sigma^{inc}}{4\pi}S^{inc}(\boldsymbol{Q},\omega)\right) . \tag{6.2}$$

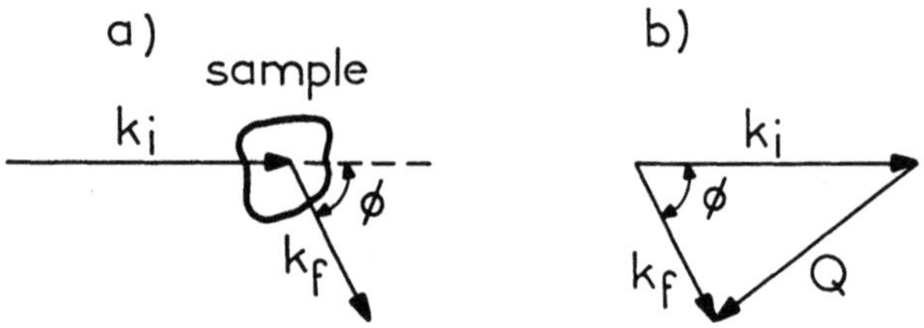

Fig. 6.1. Scattering geometry in real space (**a**) and in Fourier space (**b**). Note that k_i and k_f have different lengths for inelastic scattering

There are two terms in (6.2) arising from the two different cross sections as emphasized above. They are associated with the scattering laws $S^{coh}(\boldsymbol{Q},\omega)$ and $S^{inc}(\boldsymbol{Q},\omega)$ which depend only on the properties of the samples. They are readily interpreted as the time and space Fourier transforms of the real-space correlation functions $G^s(\boldsymbol{r},t)$ and $G^p(\boldsymbol{r},t)$ [6.9]:

$$\begin{aligned} S^{inc}(\boldsymbol{Q},\omega) &= \frac{1}{2\pi}\int G^s(\boldsymbol{r},t)e^{i(\boldsymbol{Q}\cdot\boldsymbol{r}-\omega t)}d\boldsymbol{r}\,dt , \\ S^{coh}(\boldsymbol{Q},\omega) &= \frac{1}{2\pi}\int \left[G^s(\boldsymbol{r},t) + G^p(\boldsymbol{r},t)\right] e^{i(\boldsymbol{Q}\cdot\boldsymbol{r}-\omega t)}d\boldsymbol{r}\,dt . \end{aligned} \tag{6.3}$$

The self-correlation function $G^s(\boldsymbol{r},t)$ gives the evolution with time of the trajectory of a particular atom, whereas the pair correlation function $G^p(\boldsymbol{r},t)$ describes the time evolution of the relative spatial coordinates of different atoms. Note that the dominant contribution to $S^{coh}(\boldsymbol{Q},\omega)$ is from $G^p(\boldsymbol{r},t)$.

For diffusive motions, the particle is displaced in space in a random fashion (periodic motions lead to inelastic scattering like phonons, which we will neglect in this chapter). At time zero, it can be assumed that the atom of interest is at a particular site with a probability of 1. With time, this probability will either decrease to 0, if the atom can diffuse to infinite distance (translational diffusion), or it will decrease to a finite value if the particle remains in a restricted volume (local diffusion or rotational diffusion). Accordingly, the self-correlation function can be broken up into two terms:

$$G^s(\boldsymbol{r},t) = G^s(\boldsymbol{r}) + \tilde{G}^s(\boldsymbol{r},t) \,, \tag{6.4}$$

where $G^s(\boldsymbol{r})$ gives the probability of finding the considered atom at infinite time at a distance $\boldsymbol{r}$ from its starting point. It is a time-independent function and its Fourier transform

$$F(Q) = \int G^s(\boldsymbol{r}) e^{i\boldsymbol{Q}\cdot\boldsymbol{r}} d\boldsymbol{r}\, \delta(\omega) \tag{6.5}$$

is strictly elastic in energy. This Fourier transform $F(\boldsymbol{Q})$ is also called the elastic incoherent structure factor (EISF). It provides very important information because it is directly related to the geometry of the diffusive path.

In a liquid, for instance, there is an infinitely large volume available for a diffusive particle. The probability of finding it at any particular position at infinite time is zero. Thus $G^s(\boldsymbol{r}) \equiv 0$ and the scattered neutron intensity is entirely quasi-elastic, i.e., $F(Q) \equiv 0$. As another example we consider an atom performing random jumps in a double minimum potential where the minima are separated by $\boldsymbol{a}$. Then $G^s(\boldsymbol{r}) = 1/2$ for $\boldsymbol{r} = \pm\boldsymbol{a}/2$ and $F(Q) = (1+\cos \boldsymbol{Q}\cdot\boldsymbol{a})/2$. For $\boldsymbol{Q}\cdot\boldsymbol{a} = n2\pi$, the scattered intensity will be entirely elastic, and for $\boldsymbol{Q}\cdot\boldsymbol{a} = (n+1/2)2\pi$ it will be entirely quasi-elastic. For any other value of $\boldsymbol{Q}$ the spectrum will be composed of an elastic and a quasi-elastic component. For a powder sample a proper average has to be taken and the EISF becomes $F(Q) = [1 + \sin(\boldsymbol{Q}\cdot\boldsymbol{a})/\boldsymbol{Q}\cdot\boldsymbol{a}]/2$.

With respect to the time-dependent part of the correlation function, it is assumed that it decays exponentially towards its limiting value $G^s(\boldsymbol{r})$. Then

$$S^{inc}(\boldsymbol{Q},\omega) = F(Q)\delta(\omega) + [1 - F(Q)]\frac{\Gamma/\pi}{\Gamma^2+\omega^2} \,, \tag{6.6}$$

where the width of the Lorentzian function Γ relates to the decay constant. There are numerous systems, such as molecular crystals, with more than one reorientational motion. In this case, a corresponding number of Lorentzian lines with different widths and individual weight factors have to be considered [6.10]. Still, the energy-integrated intesity of $S^{inc}(\boldsymbol{Q},\omega)$ has to be unity:

$$\int S^{inc}(\boldsymbol{Q},\omega)d\omega = 1 \,. \tag{6.7}$$

Indeed, this condition is very useful to verify that no fast components in the spectrum have been overlooked, perhaps due to the limited energy range of a spectrometer. So far, the simple scattering law given by (6.6) has been adequate to describe the measurements on GICs.

For rotational jump diffusion, the neutron will experience the same time constant regardless of the wavelength with which the system is probed. Consequently, the linewidth Γ in (6.6) is Q independent, while the fraction of the quasi-elastic broadened intensity varies with Q (EISF).

For translational diffusion, the decay time of the self-correlation function depends on the length scale on which it is probed, consequently the linewidth Γ becomes Q dependent. For small Q values corresponding to long distances in real space, the relation

$$\Gamma = \hbar D Q^2 \tag{6.8}$$

is always valid.

In an ideal liquid there are no preferred displacement vectors and (6.8) holds over a wide Q range [6.11]. In contrast, the linewidth Γ becomes a periodic function in Q if the particle diffuses on a regular substrate, as for a lattice gas [6.12], where only well-defined jump vectors exist. Then

$$\Gamma = \frac{1}{n\tau} \sum_{i=1}^{n} [1 - \exp(-\mathrm{i}\boldsymbol{Q} \cdot \boldsymbol{r}_i)] , \tag{6.9}$$

where τ is the mean residence time of the atom at a given site and $\boldsymbol{r}_i$ are the jump vectors leading to the n neighbor sites.

Translational and rotational diffusion can coexist at sufficiently high temperatures. It can be assumed that these two modes are decoupled if the rotational diffusion is much faster than the translational diffusion. The scattering law is still composed of two components: the original EISF of the rotation becomes broadened, reflecting directly the long-range diffusion and the width for the local motion will increase slightly according to the translational component. A schematic representation of various scattering laws is given in Fig. 6.2.

Coherent scattering in principle involves the interference of all particles (6.3). Therefore, the interpretation of a spectrum with respect to particle diffusion is less straightforward than for incoherent scattering. The structure factor $S(\boldsymbol{Q})$ with

$$S(\boldsymbol{Q}) = \int S^{\mathrm{coh}}(\boldsymbol{Q}, \omega) d\omega \tag{6.10}$$

can be regarded as a direct measure of how many atoms do indeed contribute to the coherent scattering. $S(\boldsymbol{Q})$ can be obtained either from a proper energy integration of a neutron spectrum or more directly from an X-ray diffractogram.

In a liquid, $S(\boldsymbol{Q})$ will approach unity for high enough Q (e.g. $Q > 5\,\text{Å}^{-1}$) indicating that a single-particle approach is justified. Coherency effects have to be taken into account for lower Q in approximative ways. Following *Sköld* [6.13], the coherent scattering law can be related to the incoherent one by

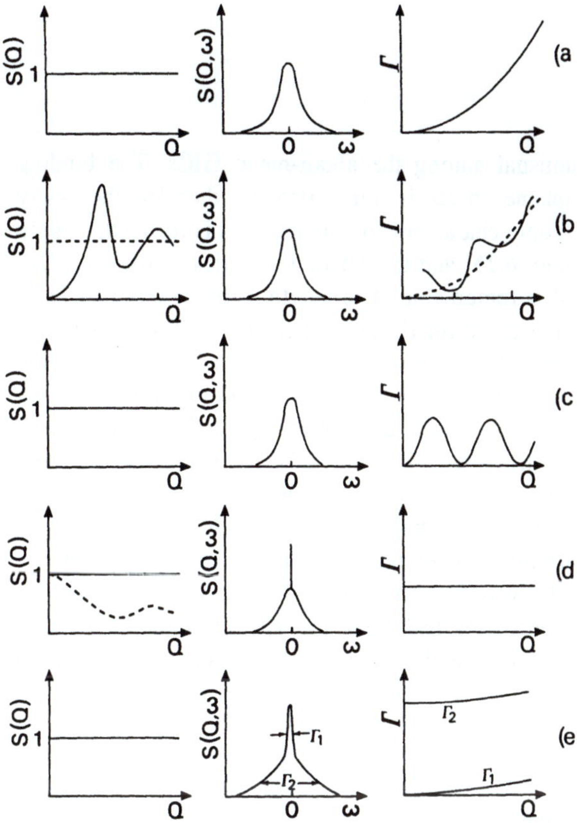

Fig. 6.2. Schematic representation of structure factors $S(Q)$, scattering laws $S(Q,\omega)$ and the Q dependence of the quasi-elastic linewidth Γ for an incoherent scattering liquid (a), a coherent scattering liquid (b) (regions of de Gennes narrowing are marked by ticks, the incoherent case is reproduced by dashed lines), and incoherent scattering lattice gas (c), a rotational diffusion (d) [the dashed line in $S(Q)$ represents the EISF], and a simultaneous rotational (Γ_2) and translational diffusion (Γ_1) (e)

$$S^{coh}(Q,\omega) = S(Q) S^{inc}\left(\frac{Q}{\sqrt{S(Q)}},\omega\right) . \qquad (6.11)$$

Following this argument, the scattering law for quasi-elastic coherent scattering again is a Lorentzian function where the energy width Γ^{coh} is related to the incoherent width Γ by

$$\Gamma^{coh} S(Q) = \Gamma . \qquad (6.12)$$

Thus, the coherent width is smaller than the incoherent width if $S(Q)$ is larger than 1, and vice versa, where the correction itself is given by $S(Q)$. This leads to an additional structure in Γ^{coh} also known as *de Gennes oscillations* [6.14].

6.2 Alkali GICs

6.2.1 Stage-1 Compounds

Stage-1 Li-GIC (LiC_6) is unusual among the alkali-metal GICs. The binding energy of the Li to the graphene sheets is much stronger than for the heavy alkali metals. Thus the anisotropic character of the compound is reduced, as seen for example from the electronic [6.15] and the elastic properties [6.16, 17]. Also, the small size of the unit cell is remarkable. The length of the c-axis is 3.71 Å, which is only 11 % larger than the layer distance in graphite. The in-plane Li structure is $(\sqrt{3}\times\sqrt{3})$R30° with three equivalent sites (Fig. 6.3). However, only one subset is occupied in the ordered phase below $T_c = 715$ K, yielding an $A\alpha A\alpha$ stacking sequence (A and α for graphene and alkali layers, respectively). Above T_c, a disordered lattice liquid structure is formed [6.17, 18] where the α, β and γ sites are occupied with equal probability. In addition to its unique physical properties, LiC_6 deserves particular attention because its technical applications, such as Li-based batteries, are more developed than those of most other GICs.

The diffusion of Li in LiC_6 was first studied on powder samples with conventional NMR by the line narrowing technique [6.19]. The quoted activation energy of 0.2 eV is likely to be too small due to the experimental conditions as discussed in [6.20].

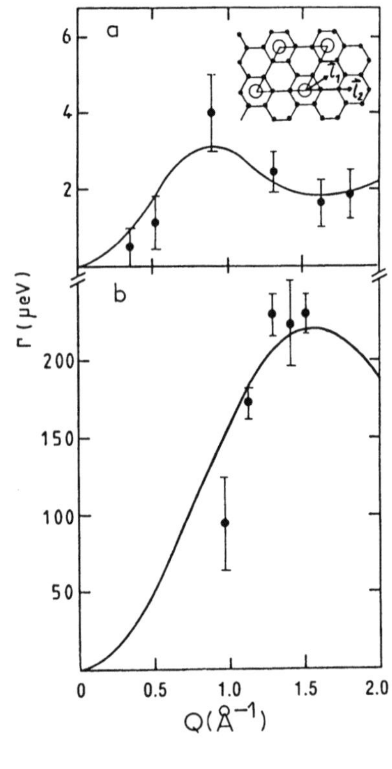

Fig. 6.3. Q dependence of the Lorentzian linewidth Γ from translational diffusion of Li in LiC_6 at 660 K (a) and at 720 K (b). The solid lines are model curves assuming jump vectors l_2 and l_1 for panels (a) and (b), respectively. Inset in panel (a) gives the $(\sqrt{3}\times\sqrt{3})$ Li structure below the melting temperature and the nearest neighbor and next nearest neighbor jump vectors l_1 and l_2 [6.22]

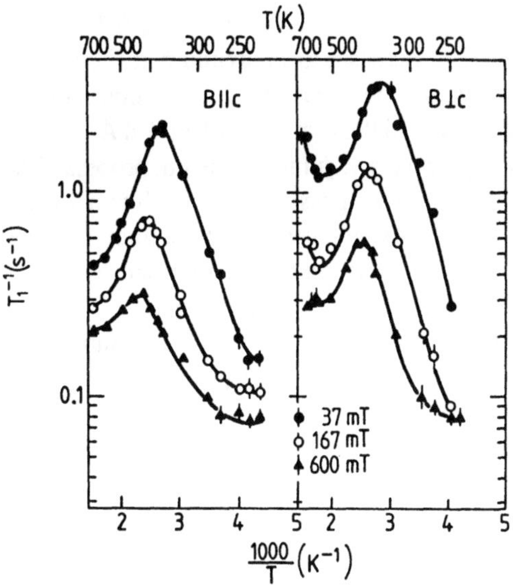

Fig. 6.4. Temperature dependence of the spin lattice relaxation rate of ^{8}Li in LiC_6 at various magnetic fields $\boldsymbol{B}$ parallel and perpendicular to the c-axis [6.20]

In βNMR, a polarized ^{8}Li nucleus is produced via the capture of a polarized neutron by ^{7}Li. The ^{8}Li polarization and its decay constant via the spin lattice relaxation rate are determined through the asymmetry of the β-radiation with respect to a magnetic field [6.21]. Figure 6.4 gives the relaxation rate in the ordered phase of LiC_6 between 250 K and 700 K. An activation energy of (0.6 ± 0.2) eV has been determined for the low-temperature regime, i.e. below the maximum of the relaxation rate. In addition, the low-dimensional character of the diffusion has been emphasized for the high-temperature regime. However, it has not been possible to distinguish between 2d diffusion parallel to the graphene sheets and 1d diffusion along the hexagonal axis, as might be expected considering the $A\alpha A\alpha$ stacking.

In particular, when LiC_6 is made from isotope enriched ^{7}Li it is a favorable material for neutron scattering. It is the only alkali with an appreciable incoherent scattering cross section of 0.8 b (barn) which is about equal to its coherent cross section. (Only Na has a larger incoherent cross section, of 1.6 b. However, Na cannot be intercalated.)

The coherent cross section is entirely concentrated in the Bragg peaks in the ordered phase below T_c and a quasi-elastic spectrum will stem entirely from the incoherent cross section. For the lattice gas/lattice liquid structure above T_c, it has been estimated that at most 25 % of the neutron intensity in between the Bragg peaks is due to coherent scattering [6.22]. Thus, the Li diffusion can be studied by QENS in both the ordered and the disordered phases.

Measurements have been performed between room temperature and 720 K [6.22]. No quasi-elastic broadening of the scattering function was observed when the out-of-plane motion was probed up to 675 K. Therefore, no diffusion along the channels in the hexagonal direction was observed within the resolution of the instruments ($D < 10^{-8}$ cm^2/s). In contrast, in-plane diffusion became observable above 630 K. The Q dependence of the linewidth Γ is shown in Fig. 6.3 for data sets taken below and above T_c. At 660 K Γ increases to a maximum at $Q \simeq 1\,\text{Å}^{-1}$ followed by a decrease for higher Q values, whereas at 720 K Γ continues to increase beyond $Q = 1\,\text{Å}^{-1}$ and reaches a plateau around $Q \simeq 1.4\,\text{Å}^{-1}$. This oscillatory behavior of Γ immediately reveals a jump diffusion process. The crystallographic structure of LiC_6 suggests two possibilities, as shown by the insert in Fig. 6.3: either jumps to the nearest hexagon centers with a jump distance $l_1 = 2.46$ Å or jumps to the next nearest hexagon centers as imposed by the Li in-plane structure with a distance $l_2 = 4.26$ Å. The Q dependence of Γ taking into account the 2d powder nature of the pyrolytic graphite (PG) samples has been calculated in [6.22] and it is presented by the solid lines in Fig. 6.3 for the distances l_1 and l_2 in the disordered and the ordered phase, respectively. This shows that there is a jump diffusion mechanism in LiC_6 both below and above T_c. However, the jump vector changes at the melting temperature.

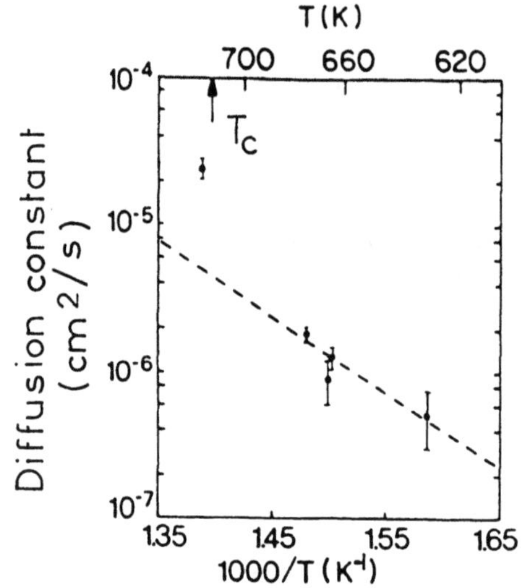

Fig. 6.5. Diffusion constant for LiC_6. The dashed line corresponds to an activation energy $E_a = 1$ eV according to the data in the ordered phase. The arrow marks the melting temperature T_c [6.22]

The diffusion constants shown in an Arrhenius plot in Fig. 6.5 have been derived from the absolute values of the line broadening Γ. A large value for the activation energy $E_a = 1$ eV is obtained for the ordered phase. However, the error bar on this number is considerable (about 30 %), which makes these data coherent with the results from βNMR taken at lower temperatures. It should be added that high activation energies for LiC_6 have also been obtained from theoretical arguments. In a density-functional approach, *DiVincenzo* and *Mele* [6.23] calculated $E_a = 1.3$ eV, in good agreement with the experiments.

As already emphasized, no quasi-elastic measurements are possible in the ordered phases of intercalated heavy alkali-metal atoms due to their negligibly small incoherent cross sections, see Table 6.1. A GIC made from a mixture of different alkali metals with different coherent cross sections would show incoherent scattering well suited for QENS. But this has not yet been attempted.

To date, measurements have only been performed on KC_8 [6.24]. As expected, no quasi-elastic scattering could be observed below T_c. However, a rapid K diffusion is found above T_c. The Q dependence of the linewidth Γ as shown in Fig. 6.6 reveals a pronounced structure. First, Γ reaches a maximum around $1.15\,\text{Å}^{-1}$ followed by a drop to a minimum at $1.5\,\text{Å}^{-1}$. This position corresponds to the first peak in the structure factor $S(Q)$ for KC_8, as known from X-ray scattering or as obtained from an energy integration of the Lorentzian component. Therefore, the minimum in Γ at $Q = 1.5\,\text{Å}^{-1}$ is identified as a de Gennes narrowing effect.

The absolute value of $S(Q)$ has to be known to correct for coherency effects (6.12) and to extract a diffusion constant from Γ^{coh}. However, this is not well known, in part due to the limited Q-range of the measurements. On the other hand, $S(Q)$ must pass the value of 1 before Γ starts to drop towards its minimum. From this argument and considering the general shape of $S(Q)$ we assume that $S(Q) = 1$ for $Q = 1.1\,\text{Å}^{-1}$, from which diffusion constants $D = 5.6 \times 10^{-5}\text{cm}^2/\text{s}$ and $D = 8.1 \times 10^{-5}\text{cm}^2/\text{s}$ are obtained for 623 K and 698 K, respectively. This yields an activation energy $E_a = 0.18\,\text{eV}$, which is markedly smaller than for LiC_6, and a prefactor $D_0 = 16 \times 10^{-4}\text{cm}^2/\text{s}$. All values for E_a and D_0 are also summarized in Table 6.2. This reduced value for E_a has also been predicted by *DiVincenzo* and *Mele* [6.23]. According to their calculation, it originates from the increased c-axis lattice constant so that the K atoms diffuse in a host charge density with a much smaller in-plane corrugation amplitude.

Table 6.2. Activation energies and prefactors for particle diffusion in graphite

Compound	E_a [eV]	D_0 [cm^2s^{-1}]	Comments	Ref.
Stage-1 AGICs				
LiC_6	0.6		By βNMR	[6.20]
	1.0	51	By QENS, ordered phase	[6.22]
KC_8	0.18	16×10^{-4}	Liquid phase	[6.24]
Stage-2 AGICs				
KC_{24}	0.126	1.2×10^{-4}		[6.25]
RbC_{24}	0.063	2.5×10^{-4}		[6.25]
CsC_{24}	0.077	4.3×10^{-4}		[6.25]
Molecular Intercalates and Ternary Compounds				
$C_{10}HNO_3$	0.075		Semi-incommensurate phase	[6.26]
	0.10		Liquid phase (additional translation not Arrhenius like)	[6.26]
H_2SO_4 Stage 1	0.18 – 0.21		Depending on acid concentration	[6.27]
$K(NH_3)_{4.3}C_{24}$	0.005		Rotational diffusion for $T < 180\,\text{K}$	[6.28]
	0.10		Rotational diffusion for $T > 180\,\text{K}$	[6.28]

AGIC: Alkali graphite intercalation compound

6.2.2 Stage-2 Compounds

With the exception of LiC_{12}, all stage-2 compounds have a smaller in-plane intercalate density than their stage-1 relatives. This includes LiC_{18}, which is a second, low density composition of a stage-2 Li-GIC. Both the intercalate–graphene and the intercalate–intercalate interactions are reduced. This is manifest from the lower order–disorder transition temperatures T_c (10 K for LiC_{18}, 123 K for KC_{24} and about 165 K for both RbC_{24} and CsC_{24}) and from the rich pattern of crystallographic structures at low temperatures, as elucidated in Chap. 2 by *Moss* and *Moret*. Naturally, these differences will also be reflected in the dynamical properties of these compounds.

No measurement on either stage-2 Li compound has been reported to date, in spite of the obvious interest in comparing the diffusion of the two species with each other and with their stage-1 relative.

Among the heavy alkali-metal intercalates, RbC_{24} has been measured over the widest temperature range [6.29–32]. Quasi-elastic scattering becomes observable above T_c = 165 K. Both the energy width Γ and the intensity contained in the Lorentzian component increase continuously until about 350 K. At higher temperatures, Γ continues to increase, whereas the intensity remains constant as shown in Figs. 6.7, 8. (The slight decrease of $S(Q)$ between 400 K and 600 K is probably due to the Debye-Waller factor, which we neglect in this chapter.) Simultaneously, the intensity of the δ-function decreases between T_c and 350 K to a small value and it remains constant above this temperature. These data reveal the coexistence of solid- and liquid-like dynamical behavior above T_c. The number of Rb atoms in the liquid state increases gradually over a wide temperature range at the expense of Rb which are dynamically blocked. *Zabel* et al. argue [6.32] that this particular dynamical behavior is a fundamental property of the melting of a 2d discommensuration domain lattice on a periodic substrate. The melting transition is smeared out in temperature due to inequivalent sites occupied by the intercalate. First diffusion becomes possible for loosely bound atoms close to the domain walls and it is only at higher temperatures that the particles in the domain center become mobile as well. This aspect of a gradual melting in 2d is also reflected in other properties such as specific heat [6.33], layer shear elastic constants ([6.32, 34] and Chap. 4 of this volume) or in the spin lattice relaxation [6.35].

To discuss the Q dependence of the quasi-elastic linewidth, we need to know the alkali structure factor $S(Q)$ to correct for coherency effects as given by (6.12). The structure factor $S(Q)$ can be obtained from an energy integration of the intensity in the Lorentzian line, as shown for RbC_{24} in Fig. 6.9 or directly from X-ray scattering, also shown in Fig. 6.9 for KC_{24} [6.36] and CsC_{24} [6.37]. For both techniques, broad diffuse peaks at $Q \simeq 1.2\,\text{Å}^{-1}$ and $Q \simeq 2.4\,\text{Å}^{-1}$ characteristic for the liquid-like short range correlations are observed. Indeed, the similarity of the structure factors in Fig. 6.9 provides the most direct evidence that the quasi-elastic scattering stems exclusively from the alkali-metal diffusion.

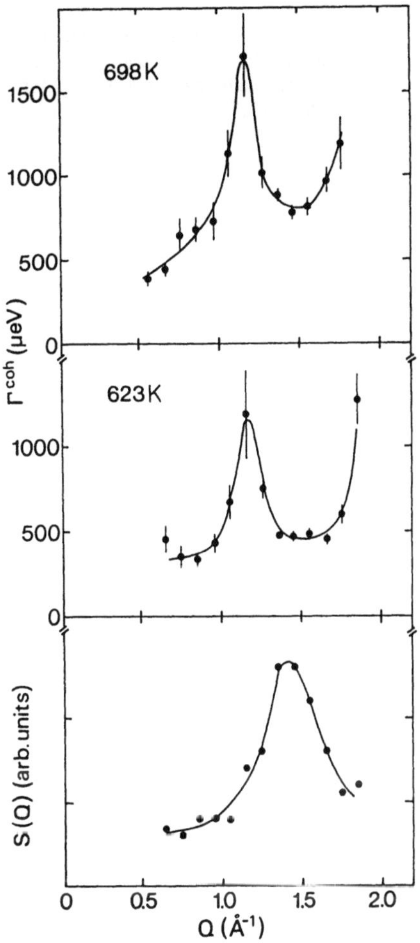

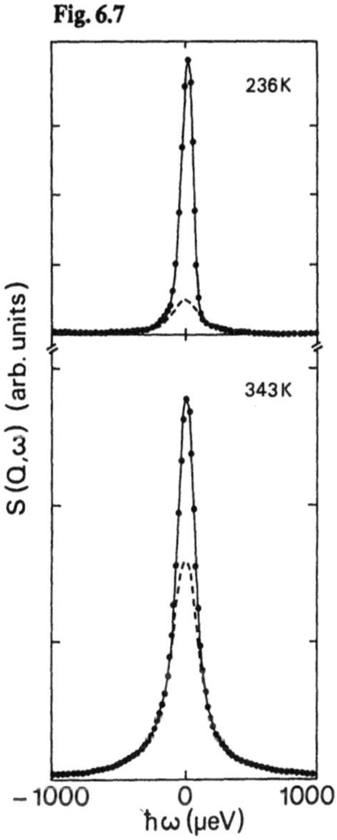

Fig. 6.6. Q dependence of the Lorentzian linewidth Γ for KC_8 at 623 K and 698 K and of the structure factor $S(Q)$ as obtained from an energy integration of the quasi-elastic spectra at 623 K. Solid lines are guides to the eyes [6.24]

Fig. 6.7. Neutron spectra of RbC_{24} at $Q = 1.25\,\text{Å}^{-1}$ measured at 236 K and 343 K. The solid lines show a fit with a δ-function and a Lorentzian line both convoluted with the instrumental resolution. The dashed line gives the Lorentzian component only. Note that both the width and the relative intensity of the Lorentzian component have increased at the higher temperature

Of special interest are the X-ray experiments on CsC_{24}. They have been measured out to large Q so that $S(Q)$ can be determined on an absolute scale. It is seen that $S(Q) = 1$ for the first time at $Q \simeq 1.0\,\text{Å}^{-1}$. This scaling has also been applied to the Rb data. Following the above arguments, for a stage-1 GIC the diffusion constant D can now be calculated directly from Γ at $Q = 1.0\,\text{Å}^{-1}$.

Figure 6.10 shows D for K-, Rb- and Cs-GICs in an Arrhenius plot [6.25]. It is noted that the data are well described by an activation energy. In particular there is no break, e.g. for RbC_{24} at 350 K, as might be inferred from Fig. 6.8.

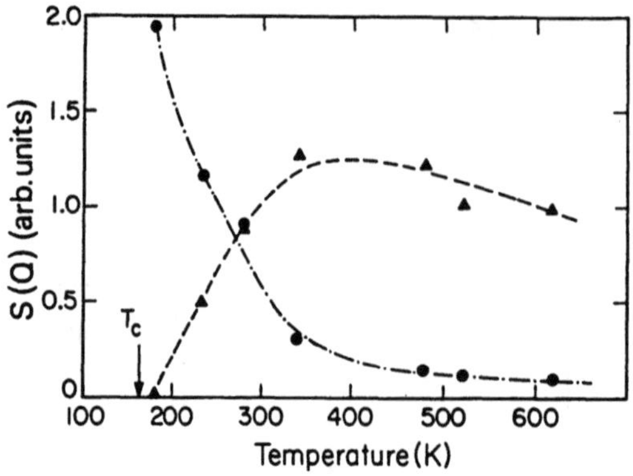

Fig. 6.8. Temperature dependence of the elastic scattering (dash-dotted line) and of the energy-integrated quasi-elastic scattering (dashed line) for RbC_{24} [6.32]

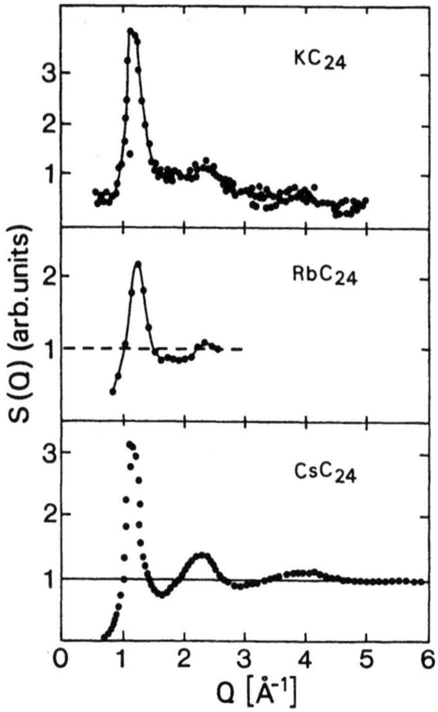

Fig. 6.9. Alkali structure factors $S(Q)$ from X-ray scattering (KC_{24}, CsC_{24}) and from an energy integration of quasi-elastic neutron scattering (RbC_{24}) [6.30]. See text for details

This shows again that the melting transition is smooth over a wide temperature range. The activation energies are of equal magnitude (126, 63 and 77 meV for K-, Rb- and Cs-GICs, respectively) and for KC_{24} this is about half the value for stage-1 KC_8. This is not at all unexpected in view of the similar alkali–graphene interactions and the reduced in-plane alkali density of stage 2 as compared to stage 1. It is remarkable that the heavy Cs diffuses faster, in particular at low temperatures, than the lighter K and Rb atoms.

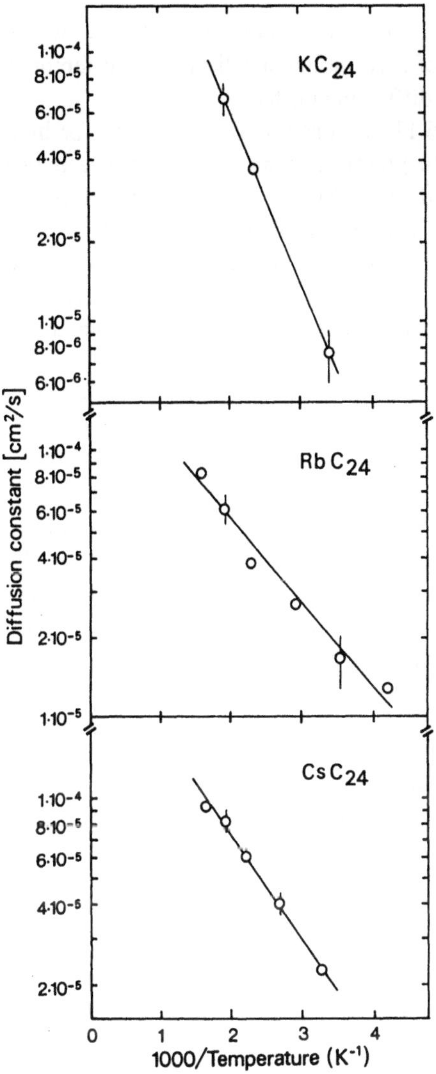

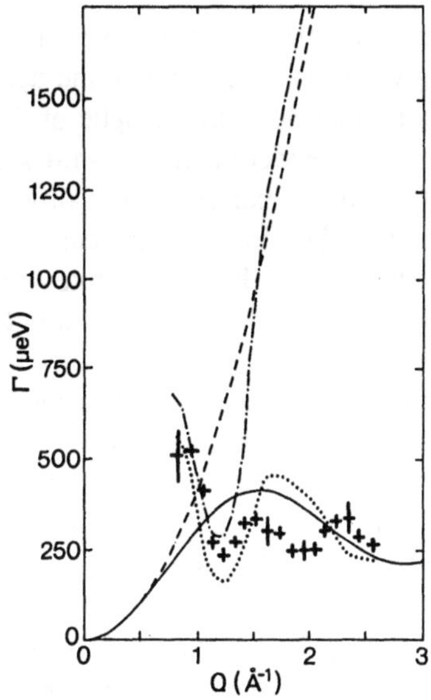

Fig. 6.11. Measured quasi-elastic linewidth Γ for RbC_{24} at 523 K (+) compared with the expected linewidth for a normal incoherent (- - -) and coherent (-·-·-) scattering liquid and an incoherent (—) and coherent (· · ·) scattering lattice liquid

◀ **Fig. 6.10.** Arrhenius plots of the diffusion constants for stage-2 heavy alkali-metal GICs [6.25]

The Q dependence of Γ for RbC_{24} at 523 K is compared in Fig. 6.11 with the expected quasi-elastic linewidth for a normal 3d liquid with either an incoherent or coherent cross section. In all cases, $D = 6.1 \times 10^{-5} \mathrm{cm}^2/\mathrm{s}$. In the incoherent case, Γ should increase proportional to Q^2 so that Γ at $2\,\text{Å}^{-1}$ is four times its value at $Q = 1\,\text{Å}^{-1}$. The structure factor $S(Q)$ from Fig. 6.9 has been used to obtain Γ^{coh}. It oscillates with decreasing amplitude about the incoherent linewidth. The first minimum in the observed linewidth at $Q \simeq 1.2\,\text{Å}^{-1}$ is consistent with the expected Q dependence for Γ^{coh}. This de Gennes narrowing should only be considered as a correction for short range coherency effects and has little implication for the discussion of the diffusive path. There is an obvious disagreement between experimental data and model curves at higher Q, which gives clear evidence that the dynamics of a 2d molten alkali layer is different to a normal 3d liquid in spite of the large diffusion constant.

The experimental Γ values in Fig. 6.11 approach a constant value for high Q. This is expected for a diffusion with preferred jump vectors as in a lattice liquid. The jump vector can be imposed either by the internal structure of the intercalate or by the graphite substrate. The expected quasi-elastic width for a 2d jump diffusion is also shown in Fig. 6.11. A jump distance of 2.46 Å according to the hexagon–hexagon distance has been assumed and a proper powder averaging for the graphite sample has been considered. Again, the linewidths for incoherent and coherent scatterers are shown.

The incoherent width has a maximum at $Q = 1.5\,\text{Å}^{-1}$ (compare LiC_6 above T_c, Sect. 6.2.1) and a shallow minimum at $Q \simeq 2.8\,\text{Å}^{-1}$. The coherent linewidth shown for the experimental Q-range reveals a characteristic structure with a pronounced minimum at $Q \simeq 1.2\,\text{Å}^{-1}$ and an almost constant value for $Q \geq 2.0\,\text{Å}^{-1}$. This particular Q dependence is also observed experimentally, providing strong evidence for a lattice-liquid-like structure of the alkali layers. However, the amplitudes of the maximum and minimum values of Γ are less pronounced than predicted by the model. This might be due to an overestimation of the coherency correction with $S(Q)$, which is not known accurately enough, or alternatively it is conceivable that the jump length of 2.46 Å is rather a mean value with a certain distribution. This results in a damping of the oscillations of Γ as observed in the data. A more accurate determination of the jump length and the width of the distribution of the jump vector seems to be called for. However, this will require more precise data measured over an extended Q-range.

6.3 Molecular Intercalates and Ternary Compounds

In contrast to alkali GICs which show only translational diffusion, molecular intercalates and ternary compounds also show local diffusion. In general, this rich pattern of diffusion modes can be unraveled because the time constants involved differ greatly. As outlined above, QENS is a particularly powerful tool for identifying various diffusion mechanisms because both the characteristic times and individual structure factors are determined.

From the increasing number of compounds studied recently we shall focus in the following on two binary (HNO_3 and H_2SO_4) and two ternary compounds [alkali-NH_3 and alkali-tetrahydrofuran (THF)] that have revealed the most detailed information on intercalate diffusion. Further interesting compounds like the trilayer metal-hydrogen-GIC $C_8KH_{2/3}$ have been studied as well. However, diffusion seems to be too slow as to be directly measurable and only a rough estimate of D has been given [6.38, 39].

For the following, it should be noted that all the molecules contain H so that the self-diffusion is measured directly by QENS. However, it should always be kept in mind that the motion of the molecule is only monitored via the H diffusion.

6.3.1 Stage-2 HNO_3 GIC ($C_{10}HNO_3$)

The structure and dynamics [6.40, 41] of HNO_3 GIC have been attracting scientific interest for a number of years. *Touzain* [6.42] and later *Pinto* et al. [6.43] have proposed an upright orientation of the NO_3 unit with respect to the graphene sheet, not excluding a slight tilt angle. Recently, the structure of stage-2 HNO_3 GIC has been studied by *Samuelson* et al. [6.44] over an extended temperature range. At low temperatures, an ordered commensurate structure with large unit cell dimensions was found. Above 210 K, a stripe-domain phase is identified: one unit cell orientation remains in registry with the graphite whereas the second in-plane orientation becomes incommensurate, with an increasing number of domain walls at higher temperatures. Above 250 K, only liquid-like short range order is observed.

These three regimes are also distinguished in a quasi-elastic study [6.26, 45, 46]. In the ordered phase, only elastic scattering is observed, even with the best energy resolution available. For $210\,K \leq T \leq 257\,K$ the spectra are composed of an elastic component and a Lorentzian profile with Q-dependent intensity. This immediately identifies a local diffusion. Above 257 K, the central peak also becomes quasi-elastic, revealing simultaneous rotational and translational diffusion in the liquid phase. These drastic changes in the quasi-elastic spectra, which are determined by scattering from the H only, and the structural phase transitions as determined by X-ray diffraction on the NO_3 unit, have identical temperature patterns. Therefore, it can be concluded that the H is strongly associated with the NO_3 unit in the entire temperature range and that the H dynamics reflects the diffusion properties of the entire HNO_3 molecule.

Figure 6.12 shows neutron data at 310 K for two scattering geometries probing the in-plane and out-of-plane diffusion. In the first case, the spectrum shows a broad Lorentzian line with an intrinsic width of $\Gamma = 220\,\mu eV$ HWHM and on top of it a much narrower Lorentzian component with an intrinsic width of $\Gamma = 13.5\,\mu eV$ HWHM, corresponding to rotation and translation, respectively. In the second case, the scattering is dominated by only one elastic peak with no intrinsic broadening. Thus diffusive motions even in the liquid phase are confined to H displacements parallel to the graphene sheets. In the following, we shall first consider the rotational diffusion and later the translational mode.

Rotational diffusion becomes observable already in the stripe domain phase starting at about 210 K. The fraction of the quasi-elastic broadened intensity increases with temperature, showing that the fraction of molecules participating in the rotation is getting larger (Fig. 6.13) [6.26, 47]. A similar unusual behavior has already been observed for the stage-2 alkali GIC above the melting transition (Sect. 6.2.2) and an analogous interpretation seems appropriate, invoking nonequivalent molecules which are either weakly bound and mobile at the dis-

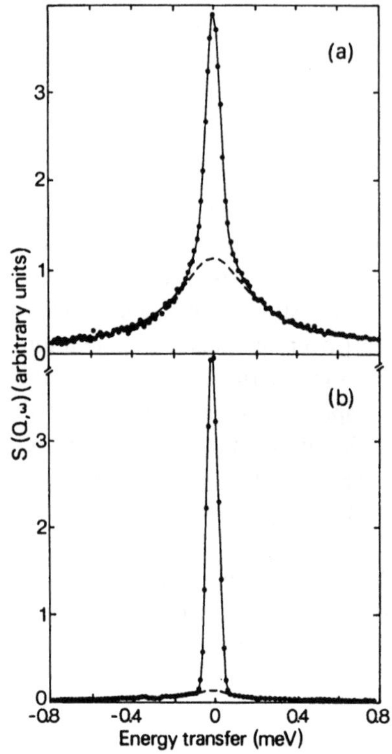

Fig. 6.12. Neutron spectra of stage-2 HNO_3 GIC at 310 K to measure in-plane (*a*) and out-of-plane (*b*) diffusion. The solid line shows a fit of the data with the sum of a narrow and a wide Lorentzian component. The dashed lines show just the wide Lorentzian component. The residual quasi-elastic component in (*b*) is probably due to the sample mosaicity and the relaxed *Q* resolution of the time-of-flight spectrometer used [6.44]

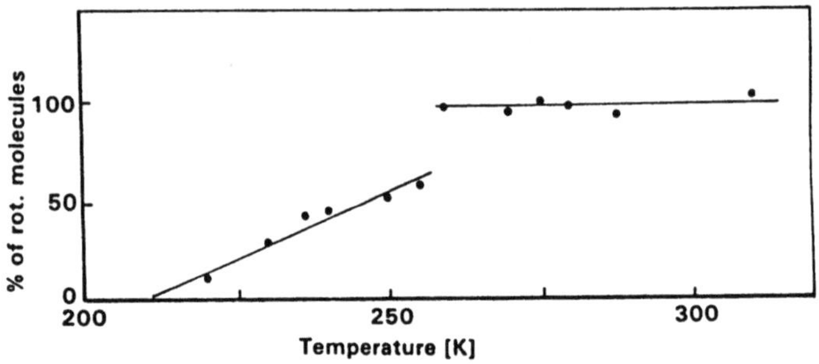

Fig. 6.13. Fraction of HNO_3 molecules participating in rotational diffusion in a stage-2 compound [6.26]

location lines or firmly embedded and static in the compression zones within the intercalate islands. With an increase in temperature, the incommensurability [6.44] and consequently the number of discommensuration lines increases and more and more HNO_3 molecules become rotationally unlocked. At a high enough dislocation density, the translational long range order breaks down and a structure analysis reveals a disordered liquid-like state. At the same time, all intercalated molecules become dynamically mobile.

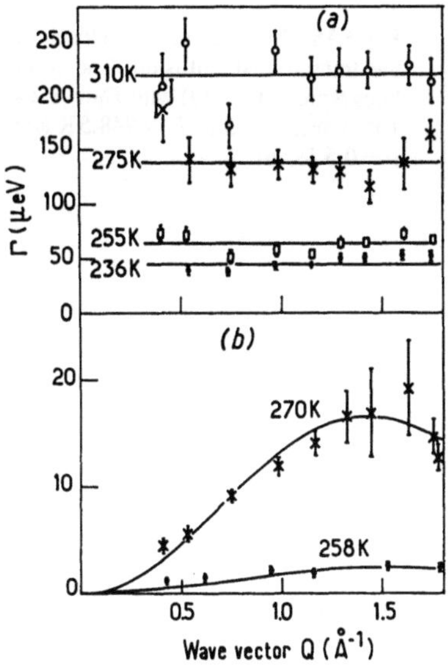

Fig. 6.14. Q dependence of Lorentzian components for the rotational (a) and the translational (b) diffusion modes. The solid lines in (b) are fits to a Chudley Elliott model [6.44]

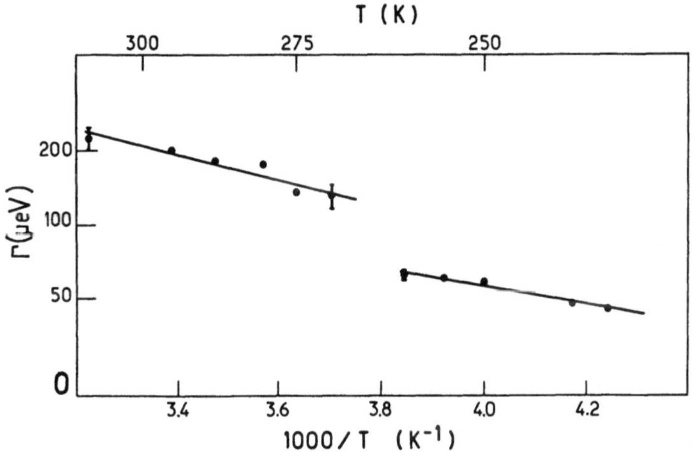

Fig. 6.15. Arrhenius plot of the Lorentzian linewidth for the rotational diffusion mode [6.44]

The Q dependence of the broad Lorentzian linewidth as shown in Fig. 6.14 is constant within experimental errors, demonstrating the rotation–diffusion character of the motion involved. An Arrhenius plot of the linewidth Γ is given in Fig. 6.15. The phase transition at about 255 K is clearly marked by a drastic change in the absolute value of Γ, whereas the activation energies of 75 meV and 100 meV for the semi-incommensurate phase and liquid phase, respectively, are only slightly different. Similar activation energies were found in earlier NMR work [6.40, 41]. However, they have been attributed to different diffusion modes.

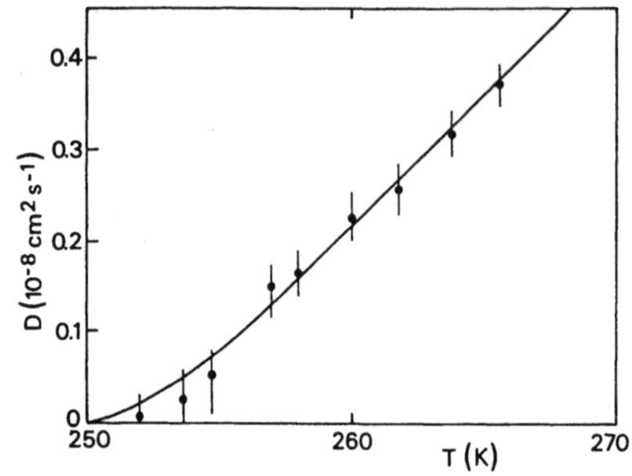

Fig. 6.16. Temperature dependence of the translational diffusion. The solid lines represents (6.19) with $D_0 = 8.2 \times 10^{-7} cm^2/s$, $b = 5.5$, $T_c = 248.5$ K and $\nu = 0.5$ [6.46]

A second quasi-elastic component with a considerably smaller width becomes observable in the liquid phase. The Q dependence of the width shown in Fig. 6.14 uniquely identifies a translational diffusion. The maximum of the linewidth at $Q = 1.5 \, Å^{-1}$ reveals a jump diffusion on a characteristic length scale of 2.5 Å.

The temperature dependence of the diffusion constant D as given in Fig. 6.16 does not follow an Arrhenius law. *Rosenman* et al. show [6.46] that the melting is continuous within the experimental energy resolution, in contrast to the melting of normal 3d solids, which reveals a discontinuity of the diffusion constant at the melting temperature.

X-ray measurements [6.44] in the same temperature regime show that the HNO_3 intercalate is still orientationally correlated, whereas the long range translational order is already lost. These structural features together with the continuous onset of molecular diffusion have been regarded as the hallmark of the hexatic phase in the 2d dislocation-driven melting theory [6.48, 49]. From the calculations of *Zippelius* et al. [6.50], it follows that the diffusion is controlled by the number of free dislocations, yielding

$$D = D_0 \exp[-2b/(T - T_c)^\nu] \qquad (6.13)$$

with a critical exponent $\nu = 0.5$ [6.51]. A fit of the temperature dependence of D as given in Fig. 6.16 yields $D_0 = 8.2 \times 10^{-7} cm^2/s$ and $T_c = 248.5$ K. The agreement of T_c with the value of the melting temperature from diffraction (247 K) confirms the description of the melting of HNO_3 GIC within the picture of a hexatic phase.

The intercalation of HNO_3 into graphite has been followed recently by means of the time variation of the ESR line shape [6.52]. It has been found that the intercalation front progresses proportional to the square root of the exposure time with a macroscopic diffusion constant $D^* = 7.3 \times 10^{-9} cm^2/s$ at room temperature. This is about two orders of magnitude smaller than the value of D for translational diffusion from QENS, again emphasizing that different aspects of particle diffusion are probed by macroscopic and microscopic techniques.

6.3.2 Stage-1 H_2SO_4 GICs

QENS investigations on H_2SO_4 GICs have concentrated on stage-1 compounds with particular emphasis on the influence of the residual water solvant on the intercalate dynamics [6.27]. Measurements on HOPG samples intercalated with 96 % and 99 % pure sulfuric acid do not show any quasi-elastic broadening below 250 K, and only one Lorentzian line is observed at higher temperatures. The energy width Γ is Q independent, which proves the rotational character of this mode. No translational diffusion has been observed in the measured temperature regime up to 300 K.

The intensity contained in the Lorentzian component increases continuously with temperature for $T > 250$ K (Fig. 6.17), showing that more and more molecules become mobile. These dynamical features are quite analogous to the diffusion properties of stage-2 HNO_3 GIC (Sect. 6.3.1) in the stripe domain phase. Hence, it can be speculated that basic structural features involving a temperature-dependent discommensurability with a variable density of discommensuration lines are similar in these two compounds, although a direct confirmation by diffraction techniques is still lacking.

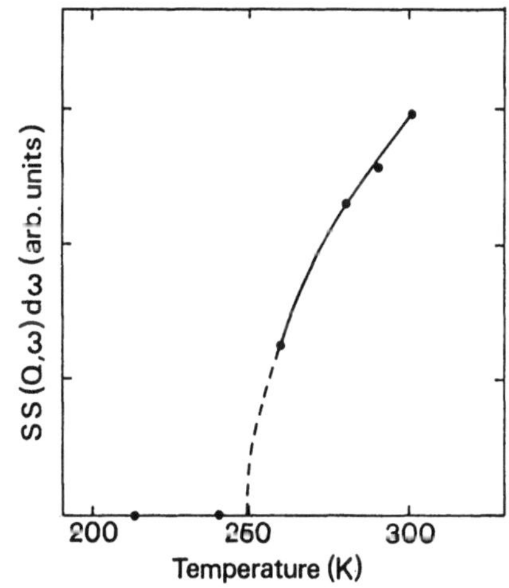

Fig. 6.17. Temperature dependence of the energy-integrated intensity contained in the Lorentzian component of a stage-1 H_2SO_4 GIC [6.27]

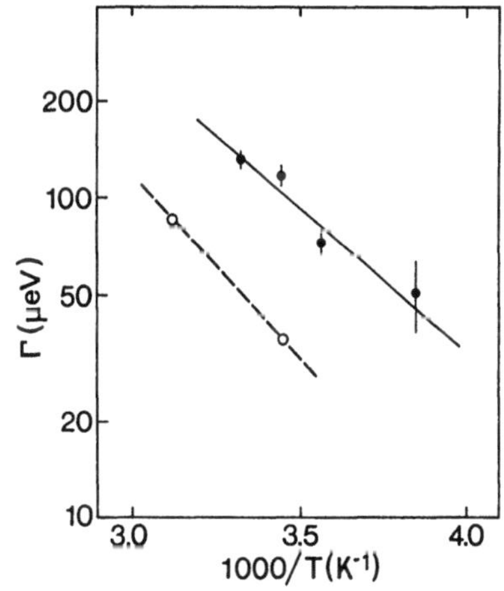

Fig. 6.18. Arrhenius plot of Lorentzian linewidth Γ for 99 % H_2SO_4 (•) and 96 % H_2SO_4 (o) GICs [6.27]

From the temperature dependence of the Lorentzian linewidth as depicted in Fig. 6.18, it is seen that the 99 % H_2SO_4 GIC shows faster diffusion than the 96 % H_2SO_4 GIC. In addition, the activation energy for the higher concentration acid (180 meV) seems to be slightly lower than that of the 96 % H_2SO_4 GIC (210 meV). This demonstrates that even a small amount of water leads to covalent bonds which drastically reduce the mobility of the intercalate.

6.3.3 Ternary $Cs(THF)_x$ Compounds

Tetrahydrofuran (THF) is added to alkali GICs to form ternary compounds. The alkali–THF complex depends on the intercalate concentration and on the polarizing power of the metal. In contrast to binary alkali GICs, the layer thickness decreases from the Li-based to the Cs-based ternary compounds.

The dynamics of THF in the Cs-based GIC has been studied recently with particular emphasis on a comparison between stage-1 $Cs(THF)_{1.7}C_{24}$ and stage-2 $Cs(THF)_{1.05}C_{24}$ [6.53].

For the stage-2 compound, no diffusion was observable up to 318 K and only weak quasi-elastic scattering was found at higher temperatures. In contrast, a stage-1 sample shows pronounced quasi-elastic scattering above 245 K. This can be seen from the change of the elastic scattering with temperature as given in Fig. 6.19. There is a well-defined break at this temperature, which is attributed to a gradual freezing of the molecules [6.53]. The value of the EISF at 300 K shows that only a fraction of the THF molecules are diffusing. This has been confirmed by a partial deintercalation from stage 1 to stage 2, where all THF molecules are immobile. This finding is also in agreement with proton NMR measurements at 295 K [6.54]. These results are consistent with a description in which 50 % of the THF are immobile and 50 % are rotating freely around the hexagonal axis.

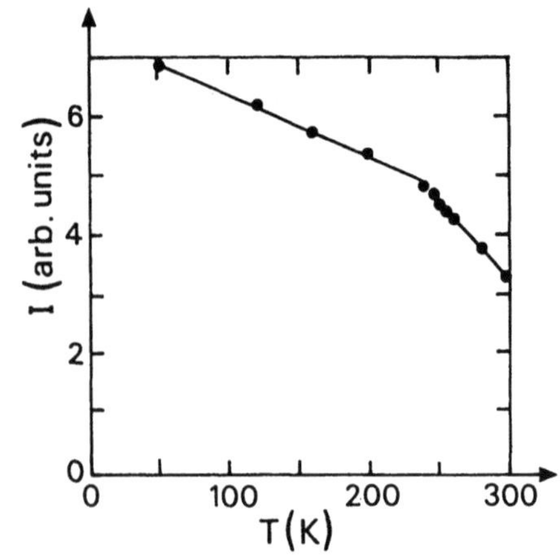

Fig. 6.19. Temperature dependence of the elastic intensity I for stage-1 $Cs(THF)_{1.7}C_{24}$ [6.53]

6.3.4 Stage-1 $K(NH_3)_{4.3}C_{24}$

In stage-1 $K(NH_3)_{4.3}C_{24}$, the major part of the NH_3 molecules are strongly associated with the alkali, forming a planar $K(NH_3)_4$ complex, and the remaining NH_3 are loosely bound in between these molecular structures [6.55, 56].

A survey of the diffusion dynamics has been conducted [6.28, 57, 58] and the main results are reproduced in Fig. 6.20. At a low temperature of 78 K, the neutron spectra show a Lorentzian component of weak intensity with Γ = 400 μeV,

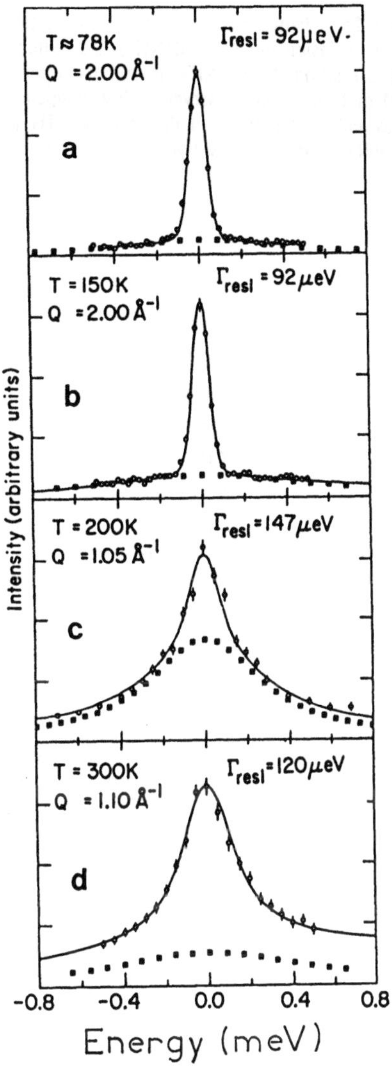

Fig. 6.20. Quasi-elastic spectra of in-plane diffusion of stage-1 $K(NH_3)_{4.3}C_{24}$. The solid line is a fit to the experimental data (circles) by a δ function and a Lorentzian line (*a–c*) or by two Lorentzian lines (*d*) convoluted with the instrumental resolution. The squares show the scattering due to the wide Lorentzian line [6.55, 56]. Note the different instrumental resolutions Γ_{resl}

which is tentatively attributed to a rotational motion. At 150 K, Γ has increased to 650 μeV, indicating a surprisingly low activation energy of 5 meV. Above the solid–liquid phase transition at 185 K, this first motion becomes very rapid and is no longer measurable under the conditions of the experiment. However, a second rotational mode with a considerably higher activation energy of $E_a = 100$ meV is observed in the measured temperature regime up to 300 K. Simultaneously, in this liquid phase, a translational diffusion is observed in analogy to the other molecular intercalates.

The Q dependence of the quasi-elastic scattering has not yet been determined and therefore the diffusive paths cannot be derived directly from the experimental data. Nevertheless, it is proposed [6.28, 57, 58] that the low-temperature diffusion

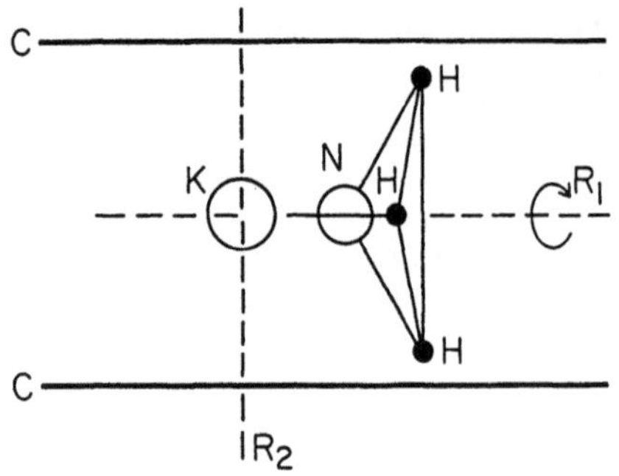

Fig. 6.21. Schematic diagram of proposed rotational diffusion modes in $K(NH_3)_4C_{24}$ featuring a rotation R_1 of NH_3 around the threefold molecular axis observed at low temperatures and a rotation R_2 of the entire $K(NH_3)_4$ complex observed in the liquid phase [6.57]

reflects a rotation of the NH_3 around the threefold molecular axis (Fig. 6.21) and that the high-temperature quasi-elastic scattering originates from a local motion of the NH_3 around the central K atom. Furthermore, the translational mode is either a displacement of entire $K(NH_3)_4$ complexes or motions of individual NH_3 migrating between complexes. A direct confirmation of this picture could come from a careful structure factor measurement for the individual diffusion modes.

6.4 Conclusion

In this chapter, a survey of studies of atomic/molecular diffusion in GICs has been presented with particular emphasis on QENS. To date, only a few systems have been investigated. Nevertheless, this limited number of examples reveals a rich variety of unusual diffusion patterns which seem to be directly related to the low-dimensional nature of these compounds. A great deal of work still remains to be done to explore the stage dependence of various diffusion modes, for example, and to relate these dynamical aspects to the rich structural patterns of GICs.

Acknowledgements. A large number of the experiments referred to were performed at the ILL with the help and advice of my colleagues A. Dianoux, I.S. Anderson, H. Blank and G. Kearly. The enjoyable collaboration with many visitors during the experiments (in particular H. Zabel, J.J. Rush, J.M. Rowe, I. Rosenman, Ch. Simon, F. Batallon and M. Bindl) and the data processing (M. Misenheimer) are gratefully acknowledged. I thank my wife Angelika for her patience and the hospitality she showed towards our guests, which kept our spirits high at all times.

References

6.1 U. Potzl, J. Völkl, H. Wipf, A. Magerl: Phys. Status Solidi B **123**, 85 (1984)
6.2 J. Völkl: Ber. Bunsenges. Phys. Chem. **76**, 797 (1972)
6.3 G.L. Squires: *Introduction to the Theory of Thermal Neutron Scattering* (Cambridge University Press, Cambridge 1978)
6.4 W. Marshall, S.W. Lovesey: *Theory of Thermal Neutron Scattering* (Clarendon, Oxford 1971)

6.5 T. Springer: *Quasielastic Neutron Scattering for the Investigation of Diffusive Motions in Solids and Liquids*, Springer Tracts Mod. Phys., Vol. 64 (Springer, Berlin, Heidelberg 1972)
6.6 T. Springer: In *Dynamics of Solids and Liquids by Neutron Scattering*, ed. by S.W. Lovesey, T. Springer, Topics Curr. Phys., Vol. 3 (Springer, Berlin, Heidelberg 1977) pp. 255-300
6.7 H. Zabel: Quasi-elastic Neutron Scattering: "A Powerful Tool for Investigating Diffusion in Solids", in *Nontraditional Methods in Diffusion*, ed. by G.E. Murch, H.K. Birnbaum, J.R. Cost, Conf. Proc. of the Metallurgical Society of AIME (AIME, Pittsburgh, PA 1983)
6.8 V.F. Sears: Thermal-Neutron Scattering Lengths and Cross Sections for Condensed Matter Research; AECL-8490, Chalk River, Ontario (1984)
6.9 L. Van Hove: Phys. Rev. **95**, 249 (1954)
6.10 W. Press: *Single-Particle Rotations in Molecular Crystals*, Springer Tracts Mod. Phys., Vol. 92 (Springer, Berlin, Heidelberg 1981)
6.11 K. Sköld, J.M. Rowe, G. Ostrowski, P.D. Randolph: Phys. Rev. A **6**, 1107 (1972)
6.12 C.T. Chudley, R.J. Elliott: Proc. Phys. Soc. **77**, 353 (1961)
6.13 K. Sköld: Phys. Rev. Lett. **19**, 1023 (1967)
6.14 P.D. de Gennes: Physica **25**, 825 (1959)
6.15 N.A.W. Holzwarth, S. Rabii, L. Girifalco: Phys. Rev. B **18**, 5190 (1978)
6.16 H. Zabel, A. Magerl, J.J. Rush: Phys. Rev. B **27**, 3930 (1983)
6.17 J. Rossat-Mignod, D. Fruchart, M.J. Moran, J.W. Milliken, J.E. Fischer: Synth. Met. **2**, 143 (1980)
6.18 D.S. Robinson, M.B. Salamon: Phys. Rev. Lett. **48**, 156 (1982)
6.19 P. Lauginie, M. Letellier, H. Estrade, J. Conard, D. Guerard: In Proc. of the Vth London Int. Carbon and Graphite Conf. 1977 (Imperial College, London 1978) p. 645
6.20 P. Freilander, P. Heitjans, H. Ackermann, B. Bader, G. Kiese, A. Schirmer, H.-J. Stöckmann, C. Van der Marel, A. Magerl, H. Zabel: Z. Phys. Chem. NF **151**, 93 (1987)
6.21 P. Heitjans: Solid State Ionics **18/19**, 50 (1986)
6.22 A. Magerl, H. Zabel, I.S. Anderson: Phys. Rev. Lett. **55**, 222 (1985)
6.23 D.P. DiVincenzo, E.J. Mele: Phys. Rev. Lett. **53**, 52 (1984)
6.24 A. Magerl, H. Zabel: Unpublished
6.25 A. Magerl, H. Zabel, M. Misenheimer: Unpublished
6.26 F. Batallan, I. Rosenman, A. Magerl, H. Fuzellier: Phys. Rev. B **32**, 4810 (1985)
6.27 M. Bindl: Thesis, TU München (1986); and to be published
6.28 D.A. Neumann, H. Zabel, J.J. Rush, Y.B. Fan, S.A. Solin: J. Phys. C **20**, L761 (1987)
6.29 H. Zabel, A. Magerl, A.J. Dianoux, J.J. Rush: Phys. Rev. Lett. **50**, 2094 (1983)
6.30 A. Magerl, H. Zabel, J.J. Rush, A.J. Dianoux: Synth. Met. **7**, 227 (1983)
6.31 H. Zabel, M. Suzuki, D.A. Neumann, S.E. Hardcastle, A. Magerl, W.A. Kamitakahara: Synth. Met. **12**, 105 (1985)
6.32 H. Zabel, S.E. Hardcastle, D.A. Neumann, M. Suzuki, A. Magerl: Phys. Rev. Lett. **57**, 2041 (1986)
6.33 H. Suematsu, M. Suzuki, H. Ikeda: J. Phys. Soc. Jpn. **49**, L835 (1980)
6.34 H. Zabel, W.A. Kamitakahara, R.M. Nicklow: Phys. Rev. B **26**, 5919 (1982)
6.35 F. Borsa, M. Corti, A. Rigamonti, S. Torre: Phys. Rev. Lett. **53**, 2102 (1984)
6.36 H. Zabel, Y.M. Jan, S.C. Moss: Physica **99B**, 453 (1980)
6.37 R. Clarke, N. Caswell, S.A. Solin: Phys. Rev. Lett. **42**, 61 (1979)
6.38 J.P. Beaufils, T. Trewern, R.B. Thomas, J.W. White: J. Chem. Soc., Faraday Trans. I, **78**, 2387 (1982)
6.39 T. Trewern, R.K. Thomas, J.W. White: J. Chem. Soc., Faraday Trans. I, **78**, 2399 (1982)
6.40 A. Avogadro, M. Villa: J. Chem. Phys. **66**, 2359 (1977)
6.41 A. Avogadro, M. Villa: J. Chem. Phys. **70**, 109 (1979)
6.42 P. Touzain: Synth. Met. **1**, 3 (1979/1980)
6.43 H. Pinto, M. Melamud, O. Shahal, R. Moreh, H. Shaked: Physica **121B**, 121 (1983)
6.44 E.J. Samuelson, R. Moret, H. Fuzellier, M. Klatt, M. Lelaurain, A. Hérold: Phys. Rev. B **32**, 417 (1985)
6.45 I. Rosenman, F. Batallan, A. Magerl, H. Fuzellier: Synth. Met. **12**, 117 (1985)
6.46 I. Rosenman, Ch. Simon, F. Batallan, A. Magerl: Europhys. Lett. **3**, 1013 (1987)
6.47 A. Magerl: Europhys. News **16**, 16 (1985)
6.48 D.R. Nelson: Phys. Rev. B **18**, 2318 (1978)
6.49 D.R. Nelson, B.I. Halperin: Phys. Rev. B **19**, 2457 (1979)
6.50 A. Zippelius, B.I. Halperin, D.R. Nelson: Phys. Rev. B **22**, 2514 (1980)

6.51 S. Ostlund, B.J. Halperin: Phys. Rev. B **23**, 335 (1981)
6.52 I. Palchan, D. Davidov, V. Zevin, G. Polatsek, H. Selig: Synth. Met. **16**, 215 (1986)
6.53 M. Goldmann, A.J. Dianoux, B. Gonzalez, F. Beguin, H. Estrade Szwarckopf, J. Conard: Synth. Met. **23**, 55 (1988)
6.54 F. Beguin, C. Laroche, B. Gonzalez, M.F. Quintin, M. Goldmann: Synth. Met. **23**, 155 (1988)
6.55 X.W. Qian, D.R. Stump, B.R. York, S.A. Solin: Phys. Rev. Lett. **54**, 1271 (1985)
6.56 X.W. Qian, D.R. Stump, S.A. Solin: Phys. Rev. B **33**, 5756 (1986)
6.57 H. Zabel, D.A. Neumann: Can. J. Chem. **66**, 666 (1988)
6.58 H. Zabel: Synth. Met. **23**, 37 (1988)

7. Microscopic Properties of Graphite Intercalation Compounds

Dah-Min David Hwang

With 23 Figures

In this chapter we review the subject of microstructural properties in the graphite intercalation compounds (GICs). GICs, obtained by intercalating various atomic or molecular species into the van der Waals galleries between the host graphite crystalline layers, are not expected to exist with perfect single-crystalline structures. Microstructure in GICs is generally unavoidable for the following reasons: (1) the finite lateral extent of the intercalant layers associated with the staging phenomenon, (2) the multiphase nature of the in-plane structures due to quenching from the elevated intercalation temperature (even if thermal equilibrium has been reached at the intercalation temperature) to the observation temperature, and (3) the inherent defects from the host graphite crystals and the defects resulting from "incomplete" annealing. Other factors, such as intercalant disproportionation, stoichiometric variation, incomplete intercalation, and partial desorption, further complicate the microstructure.

If a stage-n GIC had a perfect crystalline structure, i.e., the crystal were made of alternate stacking of a planar layer of intercalant and n planar layers of graphite sheets extending laterally over the entire crystal, then any staging transformation due to intercalation (or deintercalation) would involve a complete removal of some macroscopic sheets of intercalant from the graphite interlayer galleries and reinsertion of some of them into other galleries. Such a process would be very time consuming and contradict the fact that a high-stage compound can evolve continuously into a compound of lower and lower stage upon further intercalation. To explain the continuous evolution of staging, *Daumas* and *Hérold* proposed [7.1] the model shown in Fig. 7.1. According to the Daumas-Hérold

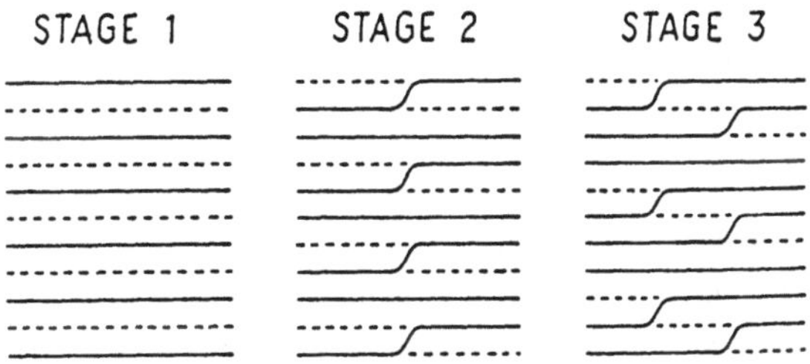

Fig. 7.1. The Daumas-Hérold model of staging [7.1]. Graphite layers are represented by solid lines while intercalant layers are represented by dashed lines. For stage $n \geq 2$ compounds, intercalant exists in every graphite interlayer gallery but condenses into a two-dimensional solid or liquid which occupies only a fraction of the total lateral area

model, the intercalant in a high-stage ($n \geq 2$) GIC is present in every graphite interlayer gallery but condenses into domains of a two-dimensional liquid or solid which occupy only a fraction of the total lateral area. Daumas-Hérold domains belonging to different graphite galleries repel each other and result in a stacking sequence which is locally well defined (except near the domain boundaries where the graphite layers are deformed). The lateral dimensions of Daumas-Hérold domains and their correlation length along the stacking direction (the c-direction) are larger than a few tens of nanometers for "well-staged" GICs since the broadening of the diffraction peaks due to the finite size of Daumas-Hérold domains is usually not resolved in X-ray diffraction. The upper bound of Daumas-Hérold domain size is probably limited by the sluggishness of the in-plane diffusion and the pinning of the domain boundaries by the host crystalline defects.

Although the Daumas-Hérold domain model has been well accepted by the scientific community, only a handful of experimental justifications have been reported. Knowledge of the physical dimensions of the Daumas-Hérold domains is still fragmentary. Staging phenomena and intercalation kinetics are reviewed in Chap. 3 by *Kirczenow*. In Sect. 7.1 of this chapter, we review direct experimental observations of staging phenomena using microscopic probes and try to establish a general description of the physical extent of Daumas-Hérold domains.

Except for a few acceptor compounds (e.g. HNO_3, H_2SO_4, and Br_2), the diffusion rate of the intercalant in the graphite galleries is so slow at room temperature that the intercalation process is usually carried out at elevated temperatures. Typically, a piece of graphite host material [single crystal graphite, highly oriented pyrolytic graphite (HOPG), or powdered graphite] is exposed to a fixed chemical potential of intercalant at an elevated temperature for between a few hours and a few weeks until a well-staged homogeneous compound has been obtained. Assuming thermal equilibrium has been reached at the intercalation temperature, each Daumas-Hérold domain would then have a single in-plane structure with a uniform in-plane density. Stage transformation is not usually observed when the specimen is quenched to the ambient or other observation temperature, because stage transformation involves macroscopic coherent rearrangement of the intercalant, which is a slow process.

If a GIC has an ordered solid-like in-plane structure at the intercalation temperature (e.g. $FeCl_3$ graphite and stage-1 alkali-metal graphite), the in-plane density is a constant over a range of intercalant chemical potentials and temperatures. The ordered in-plane structure may persist to lower temperatures without any phase transition. On the other hand, if a GIC exhibits disordered in-plane structure at the intercalation temperature, the in-plane density has no discrete value and is a continuous function of both the intercalant chemical potential and the intercalation temperature. Upon cooling, the disordered in-plane structure may transform into other structural phases. If any in-plane phase transition occurs and produces a new phase with an in-plane density different from the original phase, then in order to maintain the original average in-plane density, at least one additional structural phase with a different in-plane density must be formed, resulting in the coexistence of multiple structural phases in the same Daumas-Hérold domain. Furthermore, nucleation of commensurate or incommensurate in-plane superlattices may occur simultaneously at several equivalent graphite lattice sites and result in domains of translational and/or rotational degeneracy. Some intercalants have been found to disproportionate into several chemical species, which may segregate into domains of different chemistry.

GICs are usually studied in their metastable states, quenched from the intercalation temperature. The Daumas-Hérold domain structure may preserve its morphology at the intercalation temperature, but the intercalant in each Daumas-Hérold domain may segregate into subdomains of different in-plane structures or chemical species. The size distribution of domains depends not only on the nature of the intercalant, but also on extrinsic parameters such as intercalant in-plane density, host crystal defect structures, and the thermal history of the specimens.

GICs have a rich variety of structural phases and complicated phase transitions which have been studied extensively using X-ray and neutron diffraction. A tremendous amount of information on the structural properties of GICs has been accumulated and several excellent review articles have been written by *Dresselhaus* and *Dresselhaus* [7.2], *Solin* [7.3], *Clarke* and *Uher* [7.4], and *Moss* and *Moret* (Chap. 2 of this volume). However, both X-ray and neutron diffraction have limitations regarding the study of the aforementioned multiphase multidomain microstructure of GICs. The in-plane dimensions of these microdomains and their extent in the *c*-direction are therefore discussed very little in those reviews. Section 7.2 of this chapter complements those reviews and summarizes the information on in-plane structures obtained by microscopic probes.

X-ray diffraction typically uses a probe beam of millimeter diameter and neutron diffraction uses a probe beam of centimeter diameter. The diffraction pattern obtained from a GIC with several coexisting structural phases is a superposition of diffraction from all the structural phases and orientations in the probed volume. The superposed pattern may become too complicated to deconvolute. At the same time, the broadening of the diffraction peaks due to the finite domain size ($\geq$ a few tens of nanometers) may be too subtle to be detected readily.

Optical probes, such as transmission, reflection, and Raman spectroscopies, can reveal the structural properties indirectly. The probed volume, equal to the probe beam size multiplied by the penetration depth (or the escape depth), also inevitably contains many microstructural domains in GICs.

This chapter reviews the detection and analysis of the microstructure in GICs using microscopic probes, mainly transmission electron microscopy (TEM), scanning transmission electron microscopy (STEM), and scanning ion microscopy (SIM). The capabilities and limitations of TEM and STEM are briefly described here. A brief description of SIM is given in Sect. 7.1.2.

Both TEM and STEM utilize electron probe beams with a typical energy of 100 keV or higher which can penetrate solid thin-film specimens a few tens of nanometers thick. In TEM, the transmitted (or scattered) electrons are collected by an objective lens and projected onto a fluorescent screen or a photographic plate to produce a bright-field (or dark-field) image. In STEM, the probe beam is focused and scans across the specimen. The total intensity of the transmitted or scattered electrons is detected and used to modulate a video display that is synchronized with the scanning signal. Modern TEM or STEM instruments can resolve spatial features as small as 0.2 nm.

Electron diffraction patterns are readily obtained from the probed area. If a focused electron beam is used, the diameter of the probed area can be as small as a few nanometers for TEM or a few tenths of a nanometer for STEM. However, the resulting diffraction pattern exhibits disks instead of spots, since the probe beam is highly convergent. Interpretation of diffraction patterns with overlapped disks is nontrivial. In TEM, it is possible to use a near-parallel illuminating beam and select the area contributing to the diffraction pattern by an aperture located at the back image plane of the objective lens. A selected area diffraction (SAD) pattern with well-defined spots is thus obtained from an area as small as a fraction of a micrometer. In both TEM and STEM, characteristic X-ray fluorescences induced by the electron probe beam can be analyzed with an energy dispersive spectrometer (EDS) or a wavelength dispersive spectrometer (WDS). Elementary composition from the probed volume can be quantified with a sensitivity up to $\sim$ 0.1 at.%. Therefore, TEM and STEM are powerful tools for identifying structural and chemical properties of microscopic domains.

There are several difficulties associated with using TEM and STEM to study GICs. TEM and STEM specimens must be thin enough for electron transparency. Due to the anisotropic nature of graphite and GICs, thin flakes with the thin direction along the c-axis are readily made by repeated cleaving. In-plane domain morphology and its diffraction are studied in this orientation with the electron probe beam looking along the c-axis. However, the probe beam must be normal to the c-axis in order to record images of the layer stacking arrangement (to be referred to as c-axis images or c-axis fringes). Thin sections with the thin direction perpendicular to the c-direction are very difficult to prepare from graphite and GICs. To date, most c-axis images showing the layer structures of graphite or GICs are obtained from finely ground powders or from the edges of cleaved thin platelets. Presumably, some of the edges of the cleaved platelets are curled up and fractions of the specimens may have their c-axis perpendicular to the electron beam direction. In the strict sense of high-resolution electron microscopy, these images are "noninterpretable". This is discussed further in Sect. 7.1.1b.

Another requirement for electron microscopy is that the specimen must be stable in vacuum. This automatically excludes in situ studies of GICs under thermal equilibrium. Even GICs regarded as "stable" in the macroscopic sense may have partially deintercalated surfaces after the external intercalant source has been removed. Surface reaction and contamination may occur during transfer of the specimen from the intercalation chamber to the microscope. All these possible sources of artifacts must be carefully evaluated when interpreting the data.

Several apparent discrepancies between the results of electron microscopy analysis and X-ray diffraction studies have been reported. Section 7.3 discusses some of the possible causes and gives an overall summary.

7.1 Layer Structure and Staging

7.1.1 High-Resolution Transmission Electron Microscopy Studies

X-ray studies of GICs have revealed that the characteristic c-axis repeat distance, d_n, of a stage-n GIC can be given [7.5] by the classical formula $d_n = (n-1)d_1 + d_2$, where d_1 is the distance between adjacent carbon layers in the pristine graphite and d_2 is the package thickness of intercalant layer, equal to the separation of the two carbon layers that sandwich the intercalant layer. The observation of this discrete c-axis repeat distance results in the phenomenological or "classical" model of staging, in which a stage-n GIC consists of alternate stacking of a planar layer of intercalant and n planar layers of graphite sheets. To facilitate the interpretation of the continuous evolution of stage numbers in GICs, *Daumas* and *Hérold* proposed [7.1] a domain-structural model as shown in Fig. 7.1, in which, excepting the stage-1 compounds, the intercalant forms domains with finite lateral extent. As can be seen in Fig. 7.1, further intercalation of the stage-2 compound results in the expansion of the Daumas-Hérold domains and a part of the specimen is converted into stage 1, while deintercalation of the stage-2 compound results in a reduction of the fraction of the graphite gallery occupied by the Daumas-Hérold domains. The formation of stage-3 regions is achieved through some rearrangement of the intercalant in the same graphite gallery over a distance less than the Daumas-Hérold domain in-plane dimensions. Details of the staging kinetics are reviewed by *Kirczenow* in Chap. 3. Here we present only key features of the Daumas-Hérold model observed experimentally.

Although the "classical" staging model and the Daumas-Hérold domain model have been in existence for many years, direct imaging of the layer arrangement in GICs has only recently become possible, thanks to advances in high-resolution TEM. First we present some typical lattice images showing the layer arrangement in GICs, followed by discussions on the faithfulness of those images to the real structures.

(a) Experimental Observations

High-resolution c-axis lattice images of $FeCl_3$ graphite have been studied by *Thomas* and co-workers [7.6–9]. Representative results are shown in Figs. 7.2 and 3. Figure 7.2 reveals [7.8] the statistical distribution of the intercalant layers. Note that the planar intercalant layers and graphite layers are extended across the field of view of $\sim$ 60 nm. Although the specimen is predominantly stage 4 according to X-ray diffraction [7.8], the "stage numbers" shown in Fig. 7.2 run, from top to bottom, 5, 5, 16, 8, 6, 8, 10, and 8. Figure 7.3 was taken from a different region of the specimen, showing the kinking of the graphite layers and the edges of the intercalant layers, the characteristic features of the Daumas-Hérold domain model.

The above results were obtained from specimens of finely ground powders. The c-axis layer images were obtained by searching for small regions at the edges

Fig. 7.2. High resolution c-axis lattice image [7.8] from $FeCl_3$ graphite showing the statistical distribution of intercalant layers. The specimen is predominantly stage 4 as determined by X-ray diffraction

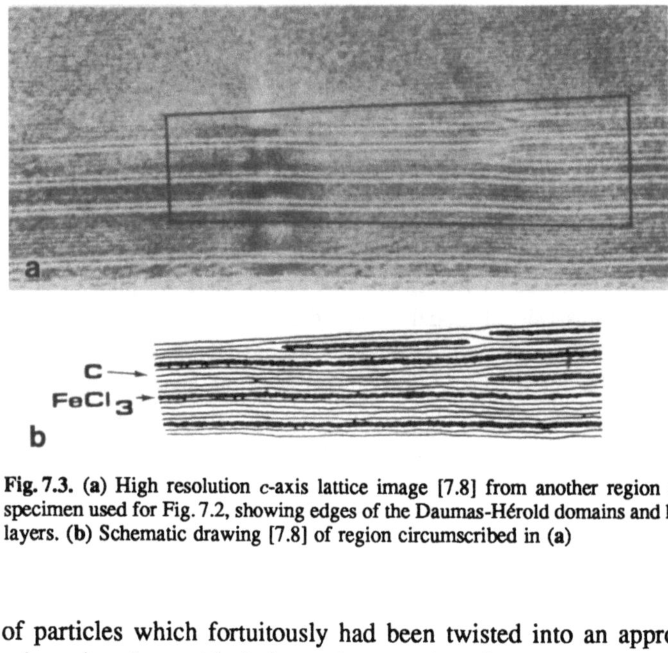

Fig. 7.3. (**a**) High resolution c-axis lattice image [7.8] from another region of the $FeCl_3$ graphite specimen used for Fig. 7.2, showing edges of the Daumas-Hérold domains and kinking of the graphite layers. (**b**) Schematic drawing [7.8] of region circumscribed in (**a**)

of particles which fortuitously had been twisted into an appropriate "edge-on" orientation, i.e., with their c-axis normal to the electron beam [7.9]. *Armstrong* et al. [7.10] prepared edge-on TEM specimens of $FeCl_3$ graphite using ion beam thinning of flakes embedded in epoxy resin. Their results are not significantly different from those shown in Figs. 7.2 and 3.

Millward et al. [7.9] and *Thomas* [7.11] also reported c-axis layer images for a nominal stage-2-$FeCl_2$ graphite specimen. A region revealing mixed stage-2 and stage-3 stacking is shown in Fig. 7.4. Images showing regions of regular stage-1 stacking, mixing of high-stage stacking, and termination of $FeCl_2$ layers have also been obtained from the same specimen. A rare and unexpected observation was that of two contiguous sheets of $FeCl_2$ intercalated into a single graphite gallery [7.9]. $FeCl_3$ graphite was found to be less stable than $FeCl_2$ graphite,

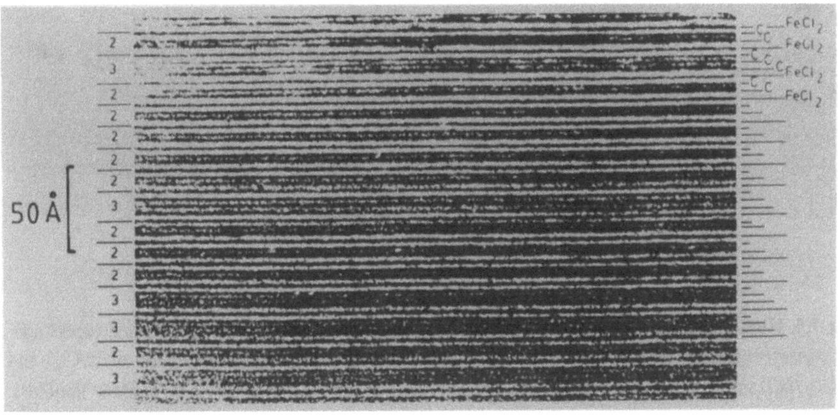

Fig. 7.4. High resolution c-axis lattice image [7.9] from a nominal stage-2 $FeCl_2$ graphite specimen showing mixed stage-2 and stage-3 stacking

as many deintercalated regions were observed in the c-axis layer images of a nominal stage-1 $FeCl_3$ graphite specimen [7.9].

Chieu et al. [7.12] reported the observation of c-axis lattice fringes from a stage-1 $FeCl_3$-intercalated benzene-derived fiber (T_{HT} = 2900°C). The lattice fringes exhibit regular stage-1 regions extending up to 5 nm × 40 nm. They concluded that graphite fibers are easier to intercalate than HOPG and single crystals, and are more air and vacuum stable.

Timp et al. [7.13] obtained c-axis lattice images from a stage-2 $SbCl_5$ graphite specimen. The images exhibit the 1.3 nm lattice fringes of stage-2 $SbCl_5$ throughout the field of view 200 nm in width. No stacking disorder or intercalant layer discontinuity is observed. A relatively small objective aperture, encompassing the [000] and [002] superlattice reflections, was used and limited the spatial resolution.

A c-axis lattice image of stage-2 $SbCl_5$ graphite of higher resolution was reported by *Salamanca-Riba* et al. [7.14] and is shown in Fig. 7.5 along with the computer simulated images obtained for the models in which the intercalant layers consist of a mixture of either $SbCl_6^-$ and $SbCl_3$ or $SbCl_6^-$ and $SbCl_5$. The TEM image appears to be consistent with either simulation, see Sects. 7.1.1b and 7.2.1b.

$SbCl_5$ graphite does not necessarily exhibit better-ordered staging than $FeCl_3$ graphite. *Hwang* [7.15] reported c-axis lattice images obtained from a nominal stage-3 $SbCl_5$ graphite specimen, revealing a stacking disorder and intercalant layer discontinuity similar to that shown in Figs. 7.2 and 3.

Hauw et al. [7.16] obtained layer fringes from a nominal stage-2 $CuCl_2$ graphite. They identified the statistical distribution of stage-1, stage-2, and stage-3 regions as well as the discontinuity of $CuCl_2$ layers. They also found a region showing the insertion of a pure six-layer-thick $CuCl_2$ crystalline slab into a graphite gallery.

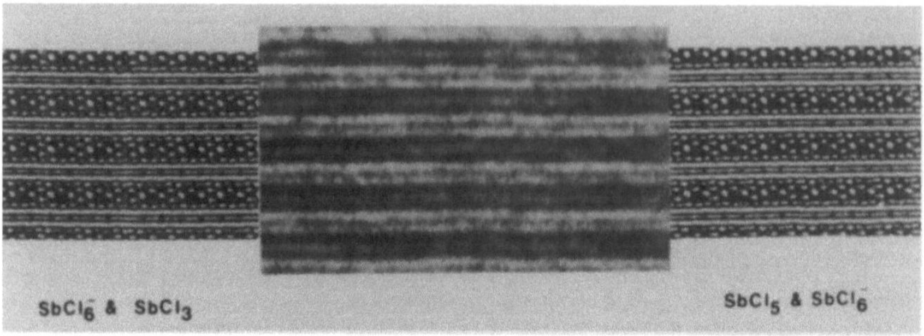

Fig. 7.5. High-resolution *c*-axis lattice images [7.14] for stage-2 $SbCl_5$ graphite obtained experimentally *(center)* and by computer simulation using the multislice method for a mixture of $SbCl_6^-$ and $SbCl_3$ *(left)* and a mixture of $SbCl_6^-$ and $SbCl_5$ *(right)* in the $(\sqrt{7} \times \sqrt{7})R19.1°$ in-plane structure

High-resolution TEM analysis was carried out on a nominal stage-2 $NiCl_2$ graphite specimen by *Dorignac* et al. [7.17]. Similar to the aforementioned results, they observed regions of regular stage-1, mixing of various stages with random statistical distribution, and termination of intercalant layers in graphite galleries. They observed contrast reversing of the same intercalant layers across the field of view, attributed to the change in specimen height or thickness. Simulated images were calculated [7.17] for a stage-6 compound, which will be discussed in Sect. 7.1.1b.

With a magnification high enough to resolve the *c*-axis layer images, the field of view in the lateral direction is limited to a few hundred nanometers, which is comparable to or smaller than the Daumas-Hérold domain size expected for most GICs. Therefore, in those GICs, termination of intercalant layers in the graphite matrix is expected to be observed only infrequently, as confirmed in the aforementioned experiments. However, a very different layer stacking and domain arrangement was obtained by *Matsuura* et al. [7.18] from a nominal stage-2 $CoCl_2$ graphite specimen, as shown in Fig. 7.6. The layer image [7.18] shown reveals $CoCl_2$ islands of typical size only 15–17 nm, consistent with the analysis of in-plane Moiré patterns (to be discussed in Sect. 7.2), which suggests island-like clusters of diameter $\sim$ 10 nm.

$MnCl_2$ graphite appears to have an in-plane structure with island-like clusters similar to that of $CoCl_2$ graphite. An electron diffraction pattern and a high-resolution TEM image obtained from a nominal stage-1 $MnCl_2$ graphite [7.19] are shown in Fig. 7.7. The wavy appearance of the layers probably reflects the small intercalant islands. Cross fringes of spacing 0.29 nm are seen in the TEM image, indicating stacking correlations of the intercalant layers.

Evans and *Thomas* [7.6] reported the lattice image shown in Fig. 7.8 from a graphite powder specimen that had been saturated with potassium. Figure 7.8 reveals many regions of pristine graphite spacing, indicating that the specimen had been considerably vacuum deintercalated.

c-axis lattice fringes have been obtained from an intercalated graphite fiber prepared by chemical absorption of hydrogen into a stage-1 potassium-intercalated

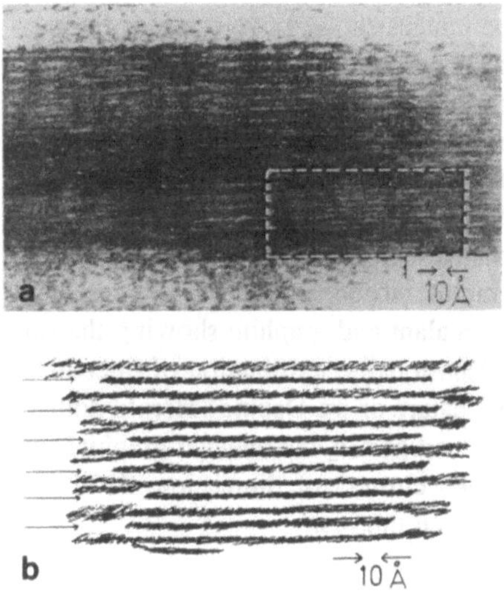

Fig. 7.6. High-resolution c-axis lattice image [7.18] from a nominal stage-2 $CoCl_2$ graphite specimen showing small $CoCl_2$ islands of lateral width 15–17 nm

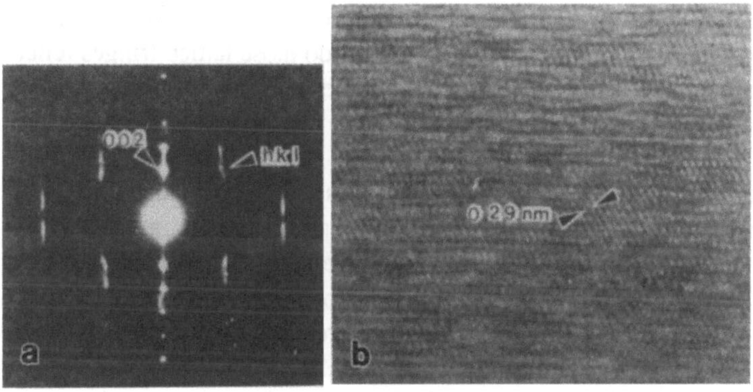

Fig. 7.7. (**a**) Selected area electron diffraction pattern and (**b**) high-resolution TEM image [7.19] of stage-1 $MnCl_2$ graphite viewed normal to the c-axis. The wavy layer image in (**b**) probably reflects the small island-like intercalant clusters of dimension 10–50 nm. The cross fringes suggest the existence of an inter-intercalant layer stacking correlation

Fig. 7.8. High-resolution c-axis lattice image [7.6] of K-graphite. The specimen had been saturated with potassium and the presence of pristine graphite spacings indicates considerable vacuum desorption

fiber by *Salamanca-Riba* et al. [7.20]. The images showed two types of regions: one with a fringe spacing of 1.18 nm corresponding to stage-2 KH_x-graphite; the other with a spacing of 0.87 nm corresponding to stage-2 K-graphite.

In summary, the c-axis lattice imaging studies of layer stacking arrangement in GICs yield the following results: Most GICs, including $FeCl_3$ graphite, $SbCl_5$ graphite, and K-graphite, have planar intercalant layers extending laterally over hundreds of nanometers. There are well-staged regions showing ordered alternate stacking of an intercalant layer and n graphite sheets, as well as regions with disordered stacking of planar layers of intercalant and graphite showing the statistical distribution of intercalant layers. Occasionally, termination of intercalant layers in graphite galleries, a characteristic feature of the Daumas-Hérold domain model, was detected. In contrast, for $CoCl_2$ and $MnCl_2$ intercalated graphite, the intercalant appears to form small island-like clusters of dimension 10–20 nm. Wavy c-layers are observed in the c-axis lattice images.

(b) Image Simulation and Interpretation

Two concerns usually arise regarding the validity of the high-resolution lattice fringes as shown in Figs. 7.2–8: (1) How accurately do those lattice fringes reflect the real microstructure? (2) How representative are the regions showing lattice fringes for the majority of the specimens? These two concerns are addressed below.

Interpretation of high-resolution lattice images is a very complicated exercise. Establishing the unit cell structures or the defect arrangements in crystals has only recently become possible in certain favorable cases. This remains a field of active research where, unfortunately, simple procedures and rules for image interpretation cannot always be given. Interested readers should consult books on high-resolution electron microscopy, such as [7.21, 22].

The appearance of a high-resolution lattice image depends strongly on the characteristics of the microscope and the operating conditions. The latter include the specimen thickness and the defocus setting of the objective lens; both are difficult to determine experimentally for less-than-ideal specimens. Usually, a high-resolution image has to be compared to a series of computer simulations with various specimen thicknesses and objective defocusing in order to confirm the proposed structure. Only under strict idealized conditions can a high-resolution image be approximated by a projection of the crystal structure convolved with the finite instrumental resolution.

An example of computer image simulation for GICs is shown in Fig. 7.9, from a study of $NiCl_2$ graphite by *Dorignac* et al. [7.17]. The simulation was done for the repeating superlattice cell, as shown at the top of Fig. 7.9, which includes one $NiCl_2$ layer surrounded by three graphite layers on each side and sufficiently extended laterally for adequate analysis of the experimental images. The free metal-halide structure has been assumed for the $NiCl_2$ layer. As can be seen from Fig. 7.9, the calculated images not only depend on the microscope characteristics (e.g. the 200 keV thermionic gun versus the 120 keV field emission gun), but are

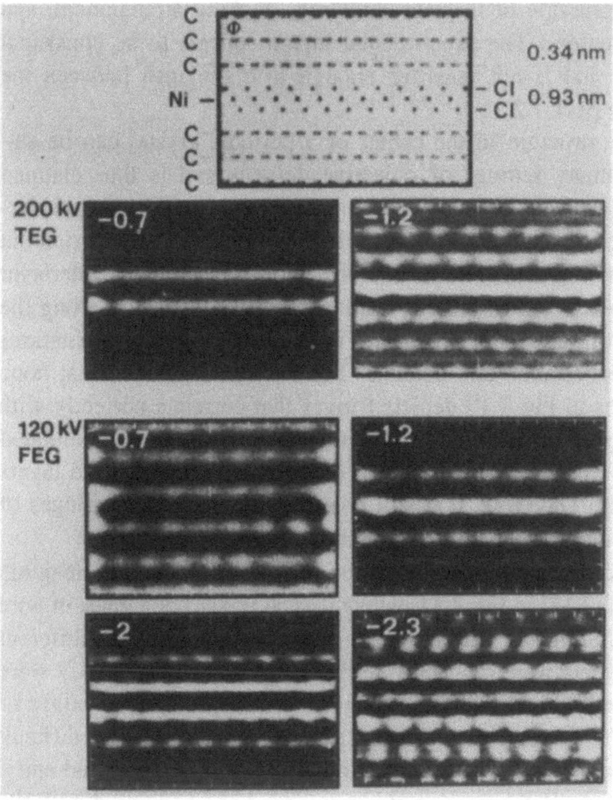

Fig. 7.9. Projected potential (ϕ) and focal series of computer simlated images [7.17] for an extended cell (130 × 90 pixels) including one $NiCl_2$ layer intercalated between 6 graphite planes, viewed normal to the c-axis. Focus is indicated in reduced units and a crystal thickness of 10 nm is used. (*top*) Beam voltage 200 keV and thermionic gun; (*bottom*) 120 keV and field emission gun. Note that the contrast of $NiCl_2$ layers and carbon layers reverses with focus variation

also a sensitive function of the defocus setting. The calculated images change rapidly with defocus, the contrast of the $NiCl_2$ layer and graphite layers being simultaneously reversed. In general, the intercalant layers can be distinguished easily from the graphite layers and the stage number can be deduced. However, none of the simulated images resembles a smeared view of the projected structure shown in the top of Fig. 7.9. High resolution lattice images also change rapidly with the specimen thickness [7.16, 21] the contrast being reversed with every change of a few nanometers of thickness. A specimen thickness of 10 nm was used in calculating Fig. 7.9.

Salamanca-Riba et al. [7.14] presented two simulated images for stage-2 $SbCl_5$ graphite, shown with the experimental TEM image in Fig. 7.5. The simulation was done for a mixture of $SbCl_6^-$ and $SbCl_3$ in an $A\alpha BA\delta BA\gamma B\ldots$ stacking sequence and for a mixture of $SbCl_6^-$ and $SbCl_5$ in an $A\alpha BA\alpha BA\alpha B\ldots$ stacking sequence. The in-plane structure was assumed to be $(\sqrt{7}\times\sqrt{7})R19.1°$.

(See Sect. 7.2 for a discussion of in-plane structure, stacking arrangement, and chemical disproportionation.) The experimental image appears to be consistent with either simulation and is not sensitive enough to distinguish between the intercalant species, see Sect. 7.2.1b.

Although the layer structure in the center of a periodic crystal can be satisfactorily imaged at many settings of objective defocus and is thus claimed to be "resolved", this may not be true if the interlayer spacings vary considerably. The first computer simulation of nonperiodic layer imaging was carried out by *Millward* and *Thomas* [7.23]. The result for a model structure of interlayer spacings varying from 0.34 nm to 1.0 nm, with a thickness of 0.5 nm along the 100 keV electron beam direction, is shown in Fig. 7.10 for several combinations of objective spherical aberration and defocus. The authors concluded that none of the simulated images in Fig. 7.10 depicts fringes that correlate correctly with the positions of the layers in the model. Nevertheless, in most of the simulated images shown in Fig. 7.10 it is not difficult to detect the presence of the layers with abnormal spacings. Therefore, it is still possible to associate the fringes of abnormal spacing or contrast in Figs. 7.2–8 with the intercalant layers.

Defects in a periodic structure can easily be overlooked in lattice imaging. This point is illustrated [7.12] in Fig. 7.11, where optical images of a piece of wire gauze with "defects" illuminated with laser light were recorded at three different focus settings. The defects were seen only in the exact focus image; they were lost in the first- and second-order Fourier images recorded at defocus values of $\Delta f_0 \equiv 2d^2/\lambda$ and $2\Delta f_0$, respectively. The exact focus setting in TEM is difficult to determine unless there exists an amorphous region in the neighborhood and a series of images at various defocus settings are recorded and compared with the computer simulation. In practice, a lattice image may be recorded with a severe defocus to maximize the contrast of the fringes of the main period. In some cases, a small objective aperture is used to deliberately block the diffractions of high spatial frequencies and to obtain an image which appears to have regularly spaced fringes. Therefore, without carefully evaluating the experimental conditions, one cannot rule out the presence of defects based on lattice images showing perfect periodicity.

Termination of intercalant layers in graphite galleries has been observed in many GICs, as was illustrated in Figs. 7.3 and 6. Although the core structure of this type of edge dislocation cannot be established without image simulation, the presence of the terminating layer can be assured by examining the regular periodic structure around the core. Intuitively, when a domain is viewed edge-on, one would expect to see the domain boundary only when the boundary has a finite lateral displacement through the specimen thickness, typically of tens of nanometers. This leads to difficulty in interpreting the island-like clusters of diameter $\sim$ 10 nm shown in Fig. 7.6. Probably, images with wavy layers as shown in Fig. 7.7 are more acceptable.

With the exception of one study [7.10], in which the edge-on TEM specimens were prepared by ion beam thinning of flakes embedded in epoxy resin, all reported TEM investigations on GICs were obtained from finely ground powders

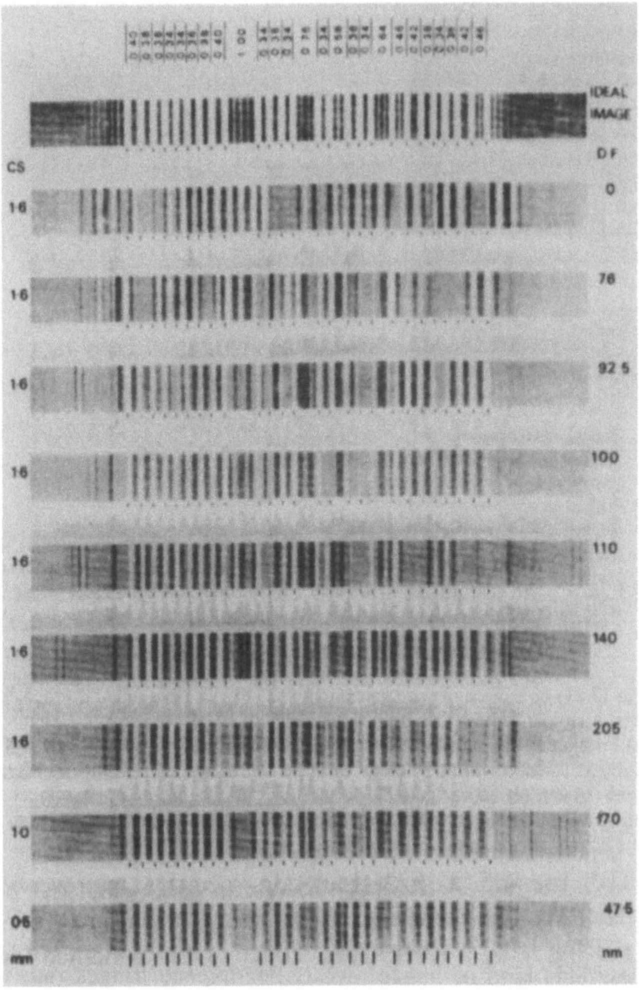

Fig. 7.10. Computer simulated images [7.23] representing high-resolution edge-on views of a model layer structure of various interlayer spacings as indicated at the top of the figure. A specimen thickness of 0.5 nm was used. Objective spherical aberration and underfocus are indicated in the margin. For comparison, an "ideal" image is shown, calculated by assuming a constant phase-contrast transfer value of 1, up to a spatial frequency of 5 nm^{-1}

or repeatedly cleaved thin flakes. Presumably, *c*-axis layer images and electron diffractions are obtained from the regions near the edges of flakes that have fortuitously been twisted into the appropriate "edge-on" orientation. The exact morphology of such a GIC specimen is not known and a means of estimating the thickness of the region of proper orientation and considering the effect of misoriented material above and below it poses a challenge to microscopists.

It is not unusual for lattice fringes to be obtained from specimens of poorly defined morphology. Lattice images of spacing 0.34 nm from partially graphitized

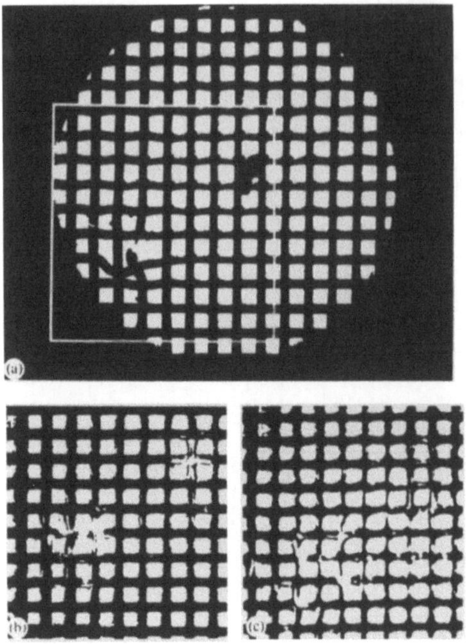

Fig. 7.11a-c. Optical images [7.21] of a piece of wire gauze illuminated with laser light recorded at three different focus settings. A crystal "defect" has been simulated by piercing the gauze with a pin. (**a**) Exact focus image. (**b**) First-order Fourier image from the enclosed region of (**a**) recorded at a defocus value of $\Delta f_0 \equiv 2d^2/\lambda$. The second-order Fourier image recorded at $2\Delta f_0$ is shown in (**c**). Note that the Fourier images (**b**) and (**c**) appear periodic so that the effect of the defect is masked. An intermediate focus setting can be found at which the contrast of the defect alone is maximum. Similar effects are seen in electron microscope images of defects in crystals

carbon are routinely recorded for resolution checks and calibrations of high-resolution microscopes. This kind of image reveals interference fringes rather than structure images. Similarly, although cross sections of graphite fibers have a typical scanning electron microscopy (SEM) image [7.24] as shown in Fig. 7.12, it is possible to obtain lattice fringe images exhibiting extended "dislocation-free" regions from graphite fibers [7.25] and stage-1 $FeCl_3$ intercalated graphite fibers [7.12]. This kind of image still provides indispensable information. For example, intergrowth of stage-2 KH_x-graphite and stage-2 K-graphite was detected from graphite fibers prepared by chemical absorption of hydrogen into stage-1 potassium intercalated fiber [7.20].

Now let us consider the representativeness of the regions showing lattice fringes for the majority of specimens. To obtain faithful high-resolution lattice images, the specimen thickness is limited to 10–50 nm. If an edge-on specimen of ideal morphology is prepared (by ion thinning or grinding), such a specimen is prone to deintercalation in air or in the TEM vacuum chamber, since all intercalant is only tens of nanometers away from the edges of graphite galleries. Therefore, it is not surprising to find many c-axis lattice images obtained from

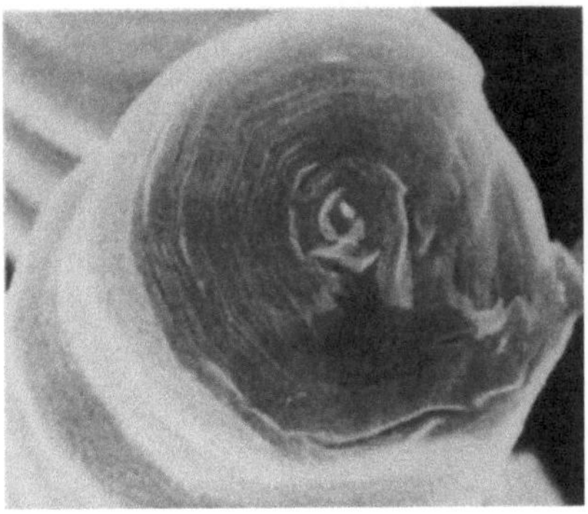

Fig. 7.12. SEM image [7.24] of the cross section of an $\sim 5\,\mu m$ benzene-derived graphite fiber

specimens characterized as well-staged by X-ray diffraction revealing regions of higher-stage spacing. Lattice images shown in Figs. 7.2–4 and 8 may represent partially vacuum-desorbed regions and deviate from the bulk majority.

7.1.2 High-Resolution Scanning Ion Probe Studies

Viewing a thin flake of GIC along the *c*-axis through the entire thickness would not produce a contrast of Daumas-Hérold domains, since the domains are staggered along the *c*-axis. Planar view of Daumas-Hérold domains may be obtained if the probe is surface sensitive with a probed depth less than the *c*-axis separation of the intercalant layers.

Freshly cleaved surfaces of graphite and stage-2 and stage-4 $SbCl_5$ graphite have been studied by *Levi-Setti* et al. [7.26] using a newly developed high-resolution scanning ion microprobe (SIM) [7.27]. This instrument focuses a 40-keV beam of Ga^+ ions from a liquid-metal ion source to a spot size as small as 20 nm on the surface of a bulk specimen. The ion-induced secondary electrons (ISEs) can be collected to produce images showing topographic and primary ion-channeling contrast by roster scanning the ion beam. The ion-induced secondary ions can be mass filtered by an RF quadrupole secondary ion mass spectrometer (SIMS) and yield elemental maps. Due to the short escape depth of the secondary ions (~ 0.5 nm), the SIMS maps represent the elemental distributions on the top first or second atomic layers of the probed specimen with a lateral resolution of the probe beam size.

ISE images of a freshly cleaved surface of HOPG and a freshly cleaved surface of stage-4 $SbCl_5$ graphite are shown in Fig. 7.13. Since the critical angle for axial channeling of 40-keV Ga^+ ions in graphite is $\sim 2°$, the pleated appearance of the HOPG surface shown in Fig. 7.13a may be due to primary-ion-channeling effects, reflecting the mosaic spread of crystallite orientations in HOPG. Some pleats exhibit faint, nearly parallel lines, which are identified (see below) as surface steps resulting from cleaving.

The ISE image of stage-4 $SbCl_5$ graphite shown in Fig. 7.13b exhibits bead-like domains and networks of lines. The $^{35}Cl^-$ SIMS maps shown in Fig. 7.14

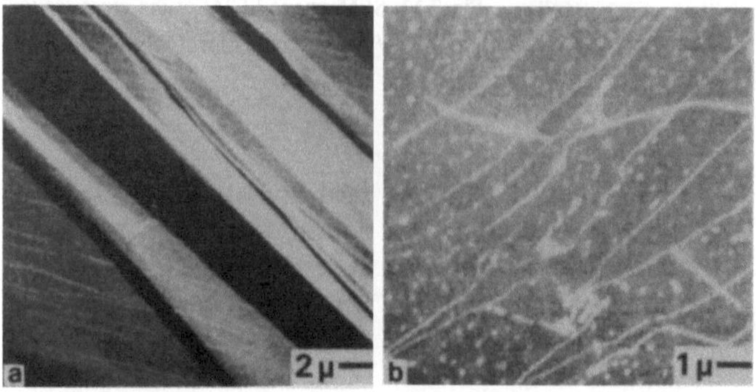

Fig. 7.13. Ion-induced secondary electron (ISE) images of freshly cleaved surfaces of (*a*) HOPG and (*b*) stage-4 $SbCl_5$ graphite obtained by a high-resolution scanning ion microprobe [7.26]

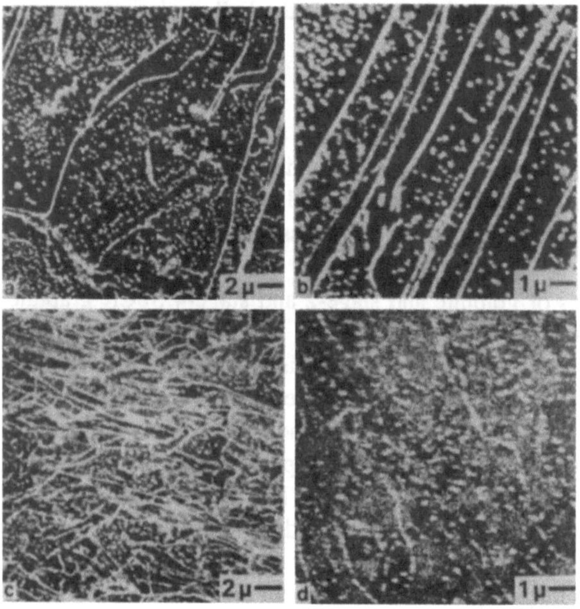

Fig. 7.14. $^{35}Cl^-$ SIMS maps of freshly cleaved surfaces of (*a, b*) stage-4 and (*c, d*) stage-2 $SbCl_5$ graphite obtained by a high-resolution scanning ion microprobe [7.26]

for freshly cleaved surfaces of stage-4 and stage-2 $SbCl_5$ graphite identified those bead-like domains and networks of lines to be rich in chlorine (and presumably intercalant). The peak-to-valley intensity ratio is $\sim$ 10 in Fig. 7.14a and b for the stage-4 compound and $\sim$ 3 in Fig. 7.14c and d for the stage-2 compound [7.26]. The bead-like domains are distributed at random over most of the surface area with a typical dimension of $\sim$ 200 nm. The pattern of the line networks is strongly sample dependent.

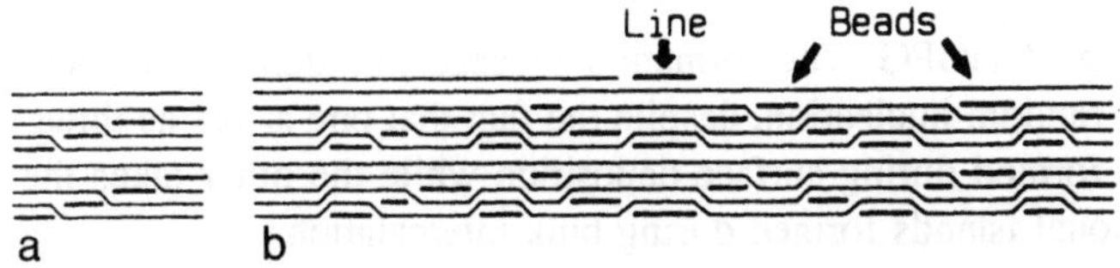

Fig. 7.15. (*a*) Daumas-Hérold domain model [7.1] for a stage-4 GIC. The thin lines represent the graphite layers, while the thick line segments represent the intercalant domains. (*b*) Elaborated version [7.26] of the Daumas-Hérold domain model which allows for the random nature of the domain size and lateral distribution and shows a crystal surface step. Note the different vertical and horizontal scales used. The graphite interlayer spacing is 0.34 nm. The typical Daumas-Hérold domain size was found [7.26] to be $\sim$ 200 nm for stage-4 $SbCl_5$ graphite

An elaborated version [7.26] of the Daumas-Hérold domain model is shown in Fig. 7.15b for a stage-4 GIC, representing a freshly cleaved surface. This model allows a random distribution of the domain size and the evaporation of the intercalant from the exposed surface. Figure 7.15b also indicates a crystal step where the intercalant domains on the plateau side diffuse toward the step and accumulate there. In view of the short escape depth of secondary ions ($\sim$ 0.5 nm) and the shallow erosion depth ($\sim$ 0.15 nm) involved in obtaining each SIMS map shown in Fig. 7.14, the bead-like domains in the SIMS maps are attributed [7.26] to the Daumas-Hérold domains one mono-graphite layer underneath the surface as pointed out in Fig. 7.15b. The line networks [7.26] represent the accumulation of the intercalant along the surface steps.

Bretz and *Clarke* [7.28] pointed out that Fig. 7.14b appears very similar to the images obtained with scanning electron microscopy (SEM) from HOPG surfaces decorated with vapor deposited gold of 0.9 nm average coverage ($\sim$ 3 atomic layers). They proposed another interpretation; that the patterns shown in Figs. 7.13, 14 represent the intercalant relaxed on the exposed surface. They argued that the surface-exposed intercalant had sufficient time, after the room-temperature cleavage, to migrate and to decorate the HOPG surface and therefore was not related to the Daumas-Hérold domains. *Hwang* and *Levi-Setti* [7.29] responded that, if this were the case, then it would be difficult to understand how a 1/8 monolayer of intercalant on the HOPG surface would entirely condense into two-dimensional islands occupying $\sim 1/4$ of the surface area (which would have an in-plane density of about half of that of the intercalant domains inside the bulk) and showing no sign of desorption or aggregation over a period of weeks under 10^{-9} torr vacuum. It also cannot explain why the bead-like domains are depleted only on one side of the surface-step lines as delineated in Fig. 7.14a and b and other high-resolution SIMS maps [7.30]. The sputtering yield of the intercalant is at least two orders of magnitude higher than that of graphite. The fact that the images of the beads can survive a few high-resolution SIMS mappings indicates that the beads are not adsorbed on the surface.

The similarity between the SEM images of gold-decorated HOPG surfaces and the SIMS maps of cleaved surfaces of $SbCl_5$ graphite is not accidental. In both cases, the line patterns reflect the decoration of the surface steps resulting from random cleaving of HOPG. The formation of bead-like domains in both systems results from random nucleation, despite the fact that one arises as three-dimensional clusters formed during surface deposition while the other takes the form of two-dimensional islands formed during bulk intercalation.

7.2 In-Plane Structure

GICs have a rich variety of in-plane structural phases. This section reviews the contribution of TEM studies to our understanding of the multiphase multidomain in-plane structures. Many in-plane structures are ordered superstructures commensurate with the graphite honeycomb layers. For the convenience of the following discussion, a schematic sketch of the honeycomb structure of a graphite layer is shown in Fig. 7.16. Carbon atoms are located at the intersections of line segments. The numbers in Fig. 7.16 represent the commensurate distances, i.e., the distances from the centers of the hexagons marked with the numbers to the center of the hexagon marked with 0, in units of the graphite in-plane lattice constant a_0 = 0.246 nm. An ordered intercalant in-plane lattice is denoted as $(a \times b)R(\theta, \phi)$ where a and b are the lengths of the in-plane translation vectors of the intercalant layers in units of a_0, and θ and ϕ are the angles between the two intercalant translation vectors and the two graphite translation vectors. For hexagonal systems, the notation $(a \times a)R\theta$ can be used where θ is the relative rotation between the intercalant lattice and the graphite lattice. $R(\theta, \phi)$ and $R(\theta)$ are used when the lengths of the translation vectors are implicit. The three possible relative stacking positions of the graphite layers are denoted by A, B and C. The possible equivalent sites of the intercalant commensurate superlattice are denoted by α, β, γ...

Selected area diffraction (SAD) in TEM can be obtained from an area with a diameter of a fraction of a micrometer. For convergent-beam electron diffrac-

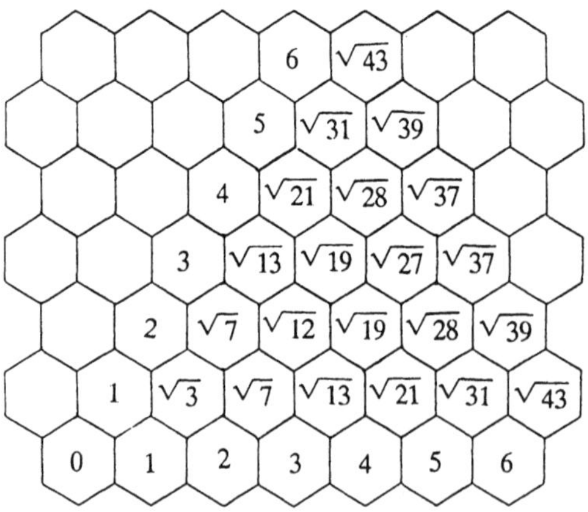

Fig. 7.16. A schematic sketch of a graphite honeycomb layer. Carbon atoms are located at the intersections of line segments. The numbers represent the commensurate distances, i.e., the distances from the centers of hexagons marked with the numbers to the center of the hexagon marked with 0, in units of the graphite in-plane lattice constant, 0.246 nm

tion, the diffraction area is defined by the probe size, which can be less than a nanometer. The electron beam penetrates through the specimen, which usually has a thickness of tens to hundreds of nanometers. If a single-phase diffraction pattern is recorded, it implies that the in-plane structure not only extends over the probed area but also correlates through the specimen thickness.

Electrons interact strongly with solid thin films and multiple diffraction is difficult to avoid. Furthermore, electron diffraction patterns are usually recorded on photographic plates whose response to the electron intensity is nonlinear and difficult to calibrate. Therefore, unlike X-ray diffraction, electron diffraction is not a quantitative technique for measuring the scattering intensity, and diffuse scattering can easily be overlooked. The scattering wave vectors are difficult to measure with an accuracy better than 1 %. On the other hand, electron diffraction patterns can be monitored continuously while moving the specimen around and can be recorded within a few seconds. A minor phase with a simple diffraction pattern may attract more attention from the microscopist than do complicated phases.

7.2.1 Acceptor GICs

(a) $FeCl_3$, $CoCl_2$, $MnCl_2$, and $CuCl_2$ Intercalated Graphite

The $FeCl_3$, $CoCl_2$, $MnCl_2$ and $CuCl_2$ compounds are stable in air and vacuum and can be easily cleaved for TEM studies. Their pristine forms exhibit layer structures. Each basic layer is composed of a double layer of chlorine atoms with the metal layer in between. This basic layer structure is preserved upon intercalation and results in a solid-like incommensurate in-plane structure. The intercalant lattices and the graphite lattice have preferred or locked orientational relations which are intercalant and stage dependent. No in-plane phase transition has been observed in these compounds.

The first GIC to have been studied by electron diffraction was $FeCl_3$ [7.31]. *Cowley* and *Ibers* [7.31] obtained an in-plane diffraction pattern from a 55.5 % $FeCl_3$ graphite (presumably stage 2). The pattern would seem, at first glance, to result from diffraction from a mixture of graphite and pristine $FeCl_3$, with the respective *a*- and *b*-axes of the two hexagonal lattices rotated by 30° to one another around the common *c*-axis. However, the appearance of the {100} and {200} $FeCl_3$ diffraction spots, which are forbidden in pristine $FeCl_3$, suggests that the normal stacking sequence of pristine $FeCl_3$ is not present. The intensities of those diffractions are not sensitive to the crystal tilt, indicating a random translational stacking of $FeCl_3$ layers in $FeCl_3$ graphite along the *c*-axis.

The above results have been confirmed in subsequent TEM studies [7.6, 32]. The in-plane structure of $FeCl_3$ graphite is stage independent. The $FeCl_3$ layers have a preferred orientation 30° off the graphite lattice with some misoriented spots scattered around the preferred orientation. Such a structure is denoted by R(30° $\pm\,\delta$), where δ indicates the scattering of diffraction spots around the preferred 30° orientation. The degree of the orientational preference is specimen dependent. The fact that the diffraction patterns exhibit discrete misoriented spots from speciments with a thickness of hundreds of layers indicates an orientational correlation of the intercalant layers along the c-axis, in spite of the disordered translational stacking. Moiré fringes with spacings of 15 nm [7.6] and 18 nm [7.32] are observed, attributed to the slight departure of $FeCl_3$ layers from the preferred 30°-off orientation. These Moiré fringes [7.6, 32] extend over 1 μm, reflecting the lateral range of the orientational correlation.

Matsuura et al. [7.18] studied in-plane electron difraction and imaging from stage-2 $CoCl_2$ graphite as part of their c-axis lattice study which was shown in Fig. 7.6. They found the same in-plane structure of intercalated $CoCl_2$ as of the pristine $CoCl_2$. The in-plane diffraction spots of $CoCl_2$ spread in the direction perpendicular to the scattering wave vectors and form arc-like patterns. Dark-field images from the intercalant spots exhibit rotational Moiré patterns, indicating that the intercalant layers are composed of many island-like clusters with an average diameter of ~ 10 nm. *Hao* et al. [7.33] confirmed the diffraction arcs for the intercalant in-plane structure in $CoCl_2$ graphite. They have obtained, as special cases, an electron selected area diffraction (SAD) pattern from the stage-1 compound which exhibits only a single orientation, R2°, and an SAD pattern from the stage-2 compound with two orientations, R29° and R25°. This indicates a long range of orientational correlation, both laterally and along the c-axis. The lateral correlation is much longer than 15 nm, which is believed to be the typical size of the intercalant islands. *Hao* et al. concluded that stage-1 $CoCl_2$ has preferred orientation around R0°, while stage-2 $CoCl_2$ has preferred orientation around R30°.

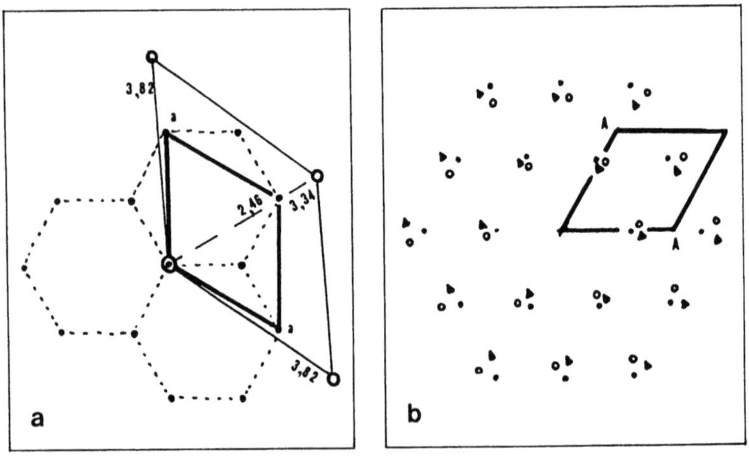

Fig. 7.17. (a) In-plane lattices [7.16] of graphite (•) and $CuCl_2$ (o) and their orientational relation for stage-2 $CuCl_2$ graphite. (b) Schematic representation [7.16] of the in-plane electron diffraction of stage-2 $CuCl_2$ graphite. The 3-fold splitting of the $CuCl_2$ diffractions is due to the 3 equivalent orientations of the monoclinic $CuCl_2$ layers with respect to the hexagonal graphite crystal

Speck et al. [7.19] studied the in-plane electron diffraction of stage-1 $MnCl_2$ graphite. They found that the layer structure of $MnCl_2$ also persists after intercalation with a preferred orientation $R(30° \pm \alpha)$, where $0° < \alpha < 4°$. Occasionally, a $R(0° \pm \beta)$ phase was observed with $\beta \sim \alpha$. This result suggests that the free energy curve has minima near both 0° and 30°. Moiré fringes were observed in the $[110]_{\text{intercalant}}$ dark-field images. The average coherence length of the fringes was found to be ~ 20 nm, which should be related to the average island size.

Hauw et al. [7.16] reported electron diffraction studies of stage-2 $CuCl_2$ graphite. As shown in Fig. 7.17a, the in-plane structure of $CuCl_2$ in $CuCl_2$ graphite is monoclinic, slightly distorted from hexagonal, preserving the layer structure of pristine $CuCl_2$. Due to the three equivalent orientations of the monoclinic layers with respect to the graphite, each $CuCl_2$ spot splits into three spots in the diffraction pattern shown in Fig. 7.17b. *Hao* et al. [7.33] confirmed the result [7.16] of Fig. 7.17 that the orientational locking angle for the stage-2 compound is $\sim 0°$. They found the orientational locking angle for the stage-1 compound to be $\sim 30°$. This stage-dependent orientational locking is just the opposite of that of $CoCl_2$ graphite, in which the preferred orientation is 0° for the stage-1 compound and 30° for the stage-2 compounds.

(b) $SbCl_5$ Intercalated Graphite

$SbCl_5$ graphite has complicated in-plane structures and phase transitions. The in-plane structure is disordered at temperatures above $\sim$ 180°C. Most $SbCl_5$ graphite samples are intercalated at temperatures above 180°C and therefore their in-plane densities may vary over a range depending on the exact intercalation temperature and intercalant chemical potential. As discussed previously, any in-plane transition occurring upon cooling will result in a coexistence of multiple phases in order to preserve the average in-plane density. The structure of $SbCl_5$ graphite is further complicated by the disproportionation of $SbCl_5$. Molecular species $SbCl_3$, $SbCl_4^-$, $SbCl_5$, and $SbCl_6^-$ have been identified in $SbCl_5$ graphite by *Boolchand* et al. [7.34] using Mössbauer spectroscopy.

At room temperature, $SbCl_5$ graphite usually consists of at least an ordered phase and a disordered phase. The most commonly observed ordered phase is the $(\sqrt{7} \times \sqrt{7})R \pm 19.1°$ commensurate superlattice. Another commensurate superlattice, $(\sqrt{39} \times \sqrt{39})R \pm 16.1°$, has been identified by *Timp* et al. [7.13] and isolated by *Hwang* [7.35] using electron diffraction. X-ray studies by *Clarke* and co-workers [7.36–38] found that the $(\sqrt{39} \times \sqrt{39})R \pm 16.1°$ phase appears to grow out of the diffuse phase at temperatures below 230 K and the spacing is approximately 2% compressed relative to the commensurate $\sqrt{39}$ periodicity. Regularly spaced ($\sim$ 60 nm) fringes oriented approximately parallel to the graphite [110] direction are observed [7.36] in low-temperature TEM micrographs. They proposed [7.38] that the structure is not a uniformly compressed lattice but rather a modulated in-plane structure. Other high-order commensurate superlattices, $(14 \times 14)R0°$ and $(28 \times 28)R0°$, have been reported [7.38] in some specimens following X-ray diffraction studies.

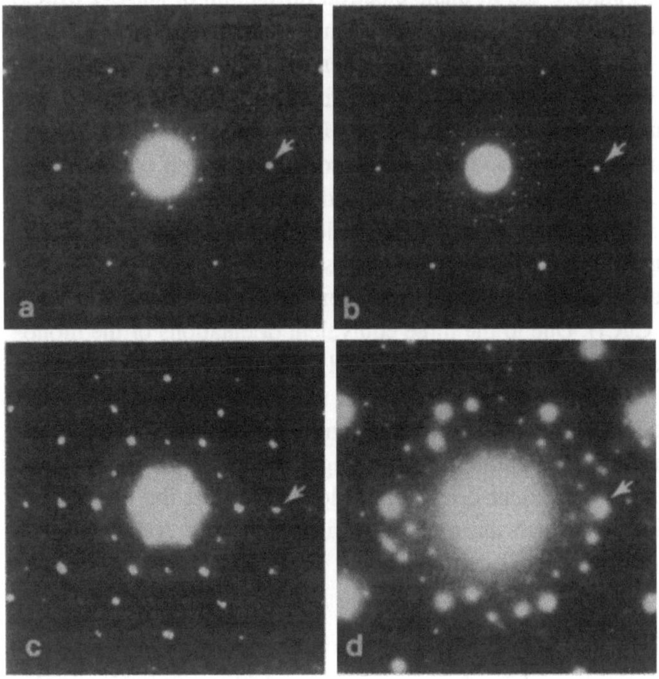

Fig. 7.18. In-plane diffraction patterns [7.55] of a nominal stage-3 $SbCl_5$ graphite specimen from 0.47 μm selected areas. The arrows indicate the graphite [100] in-plane diffractions. The in-plane superlattices are (*a*) $(\sqrt{7}\times\sqrt{7})R-19.1°$ with traces of $(\sqrt{7}\times\sqrt{7})R+19.1°$ and $(\sqrt{39}\times\sqrt{39})R\pm16.1°$, (*b*) $(\sqrt{39}\times\sqrt{39})R\pm16.1°$, (*c*) $(3\times3)R0°$, and (*d*) $(\sqrt{13}\times\sqrt{13})R+13.9°$. Note the lack of rotational degeneracy in patterns (*a*) and (*d*)

Some room-temperature in-plane electron patterns obtained by *Hwang* [7.35] from a nominal stage-3 $SbCl_5$ graphite recorded with a 0.47 μm selected area aperture are shown in Fig. 7.18. Figure 7.18a is a $(\sqrt{7}\times\sqrt{7})R-19.1°$ pattern. Note that the other rotational-degenerate $(\sqrt{7}\times\sqrt{7})R+19.1°$ pattern is almost invisible. Figure 7.18b is the $(\sqrt{39}\times\sqrt{39})R\pm16.1°$ pattern in which, unlike in the previous studies, no trace of $(\sqrt{7}\times\sqrt{7})R\pm19.1°$ spots are detected. Other commensurate in-plane phases were also observed. Single-phase diffraction patterns of $(3\times3)R0°$ and $(\sqrt{13}\times\sqrt{13})R+13.9°$ are shown in Fig. 7.18c,d. The thin TEM specimens may be partially vacuum deintercalated and therefore have lower in-plane densities. The presence of various in-plane structures may be governed by the local in-plane density. The fact that it is possible to record a single oriented diffraction pattern from an in-plane structure with rotational degeneracy, indicates a strong orientational correlation of the in-plane structures. As deduced from Fig. 7.18, the in-plane structures of $SbCl_5$ graphite exhibit a correlation range of at least 470 nm (the selected area aperture diameter) in the lateral directions and at least 50 nm (the specimen thickness) along the c-axis.

More commensurate in-plane phases were observed by *Mittlemen* et al. [7.39] using a STEM with a focused probe beam of diameter $\sim$ 10 nm. They identified several new phases: $(\sqrt{19}\times\sqrt{19})R\pm$23.4°, $(\sqrt{31}\times\sqrt{31})R\pm$8.95°, and possibly $(\sqrt{76}\times\sqrt{76})R\pm$23.4°. They did not observe any phase transition between 77 K and 295 K.

Electron diffraction studies by *Timp* et al. [7.13], *Salamanca-Riba* et al. [7.39] and *Suzuki* et al. [7.41] reported the observation that, at temperatures below 185 K, the $(\sqrt{7}\times\sqrt{7})$R $\pm$ 19.1° phase converts into a disordered glassy phase which persists until 230 K upon warming up. Later on, *Salamanca-Riba* et al. [7.14] discovered that this disordered phase is induced by the bombardment of high-energy probing electrons. The threshold dosage for the formation of the glassy state is $\sim$ 54 000 electrons/nm^2 at $\sim$ 50 K for 200 keV electrons [7.14]. This disordered state is "annealed" back to the ordered state at temperatures above 230 K. This electron damaged glassy state has not been observed in other GICs.

Clarke et al. observed [7.36], in room-temperature TEM micrographs for stage-2 $SbCl_5$ graphite, small ($\sim$ 5 nm) irregular "bubbles" of darker contrast, occupying perhaps 15 % of the total area. They proposed [7.36] that these bubbles indicate the presence of a segregated minority phase, possibly disordered islands of neutral $SbCl_3$. Fitting the (001) X-ray intensities, *Homma* and *Clarke* [7.38] concluded that the major phase consists of $SbCl_6^-$.

Hwang et al. [7.30, 42, 43] studied stage-4 $SbCl_5$ graphite using an STEM equipped with an energy dispersive X-ray spectrometer (EDS) for elemental analysis. Bright-field and dark-field scanning images are shown in Fig. 7.19a and b, respectively. High density islands of lateral dimension 50–100 nm are found randomly distributed throughout the specimen, occupying (11 $\pm$ 2) % of the total imaged area. EDS analysis revealed that the chemical composition in the island region is $SbCl_{3.1\pm0.3}$, while that of the background region is $SbCl_{7.0\pm2.0}$. Figure 7.19c is the relative antimony L-line and chlorine K-line X-ray signals as a function of position acquired with the electron beam scanning across an island. Note that chlorine yields a constant signal and is thus uniformly distributed over the line of scan, whereas the antimony signal doubles in the island region compared to the background region. Substantial differences were observed [7.42] in the electron energy-loss spectra (EELS) from the island and the background regions in the low-energy range (0–40 eV), whereas the EELS in the range of the carbon $1S$ core excitation ($\sim$ 280 eV) was found [7.42] to be spatially invariant. The latter observation indicates that the charge is laterally delocalized.

Microdiffraction studies by *Hwang* and *Cowley* [7.44] showed that the islands are disordered and yield diffuse rings, whereas the background exhibits a pattern which can be associated with a $(\sqrt{7}\times\sqrt{7})$R $\pm$ 19.1° structure and some, as yet unindexed, additional spots. These results [7.30, 42–44] indicate that the chemistry and structure of the island and background regions are correlated through the thickness of the specimen ($\sim$ 50 nm thick) with the islands adopting an irregularly shaped columnar structure.

Based on the consideration of incomplete disproportionation [7.34] and charge delocalization, *Hwang* et al. [7.42] proposed that the islands are com-

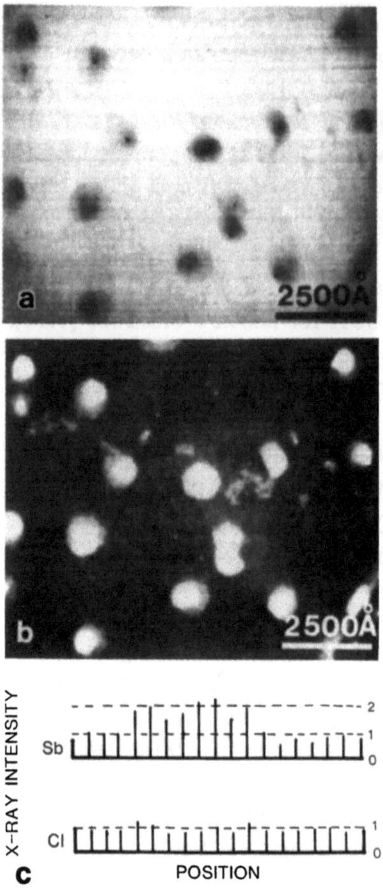

Fig. 7.19. In-plane (**a**) bright-field and (**b**) dark-field scanning-images [7.43] of stage-4 $SbCl_5$ graphite obtained with a scanning transmission electron microscope (STEM). High density islands were observed throughout the specimen. (**c**) Relative X-ray intensities of the antimony L-line and the chlorine K-line as a function of the electron beam position scanned across an island, indicating that the islands are antimony rich

posed of Sb-rich $SbCl_3/SbCl_4^-$-like material whereas the background consists of Sb-deficient $SbCl_5/SBCl_6^-$-like material. Given the island area occupancy of 11% and the uniform chlorine distribution, a unique intercalant macroscopic chemical composition is derived: $(SbCl_3)_{1/8}(SbCl_4^-)_{1/32}(SbCl_5)_{9/16}(SbCl_6^-)_{9/32}$, i.e., the molecular ratio of $SbCl_3$, $SbCl_4^-$, $SbCl_5$, and $SbCl_6^-$ in the intercalant layer is 4:1:18:9.

Salamanca-Riba et al. [7.14] reported X-ray diffraction and TEM analysis of a stage-2 $SbCl_5$ graphite specimen which exhibits only the $(\sqrt{7}\times\sqrt{7})R\pm 19.1°$ phase from room temperature down to 18 K. In-plane lattice images were obtained with an objective aperture that encompassed superlattice reflections up to $\{100\}$ of $(\sqrt{7}\times\sqrt{7})R19.1°$. As expected, hexagonal cross fringes corresponding to the $(\sqrt{7}\times\sqrt{7})R19.1°$ superlattice were obtained. The lattice images were compared to simulated images calculated by assuming the intercalant species to be pure $SbCl_3$, $SbCl_5$, or $SbCl_6^-$, or a mixture of $SbCl_5$ and $SbCl_6^-$ or $SbCl_3$ and $SbCl_6^-$. The calculation was carried out with various stacking sequences for the $(\sqrt{7}\times\sqrt{7})R19.1°$ structure. It was found that either a mixture of $SbCl_3$ and

$SbCl_6^-$ with $AB\alpha AB\delta AB\gamma\dots$ stacking or a mixture of $SbCl_5$ and $SbCl_6^-$ with $AB\alpha AB\alpha AB\alpha\dots$ stacking yields an equally "very good agreement" with the TEM images [7.14].

(c) Br_2 Intercalated Graphite

Br_2-graphite also has complicated in-plane structures and phase transitions. A recent detailed review is given by *Chung* [7.45]. Br_2-graphite has a high vapor pressure of Br_2 and well-staged samples cannot be analyzed in instruments with vacuum specimen chambers. However, a metastable residue compound is obtained when a Br_2-graphite specimen is allowed to desorb in air or vacuum for weeks. The residue compound can be studied in a TEM and appears to preserve some in-plane structure of well-staged Br_2. Here, we present only a TEM study of the residue compound by *Ghosh* and *Chung* [7.46] to illustrate the use of microstructure probes for the study of multidomain materials.

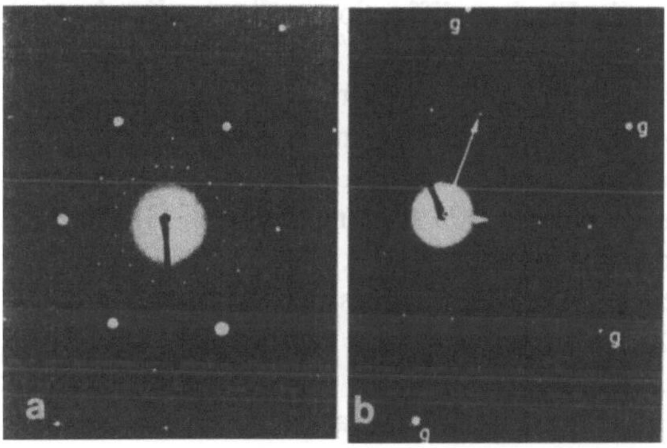

Fig. 7.20. In-plane electron diffraction patterns [7.46] from a residue compound of Br_2-graphite recorded with a 600 nm selected area aperture showing (**a**) threefold twinning of the monoclinic $(\sqrt{3}\times\sqrt{13})R(-30°, 13.9°)$ superlattice and (**b**) a single oriented domain of $(\sqrt{3}\times\sqrt{13})R(-30°, 13.9°)$

Figure 7.20a is an in-plane electron diffraction pattern from a Br_2-graphite residue compound showing 6-fold symmetry. With a 600 nm selected area aperture, it is possible to obtain a single domain diffraction pattern as shown in Fig. 7.20b. From Fig. 7.20b, it is easy to deduce the in-plane unit cell to be primitive monoclinic $(\sqrt{3}\times\sqrt{13})R(-30°, 13.9°)$ [or centered rectangular $(\sqrt{3}\times 7)R(-30°, 0°)$]. The pattern in Fig. 7.20a is a superposition of the pattern in Fig. 7.20b with three equivalent orientations with respect to the graphite lattice.

7.2.2 Donor GICs

Alkali-metal GICs contain atomic or ionic intercalant species and have simple inter-intercalant interaction. Stage-1 alkali GICs have simple primitive in-plane commensurate superlattices: $(\sqrt{3} \times \sqrt{3})$R30° for Li and (2×2)R0° for K, Rb, and Cs. This ordered structure is established at the intercalation temperature and therefore possesses a discrete composition LiC_6 and MC_8(M = K, Rb, and Cs). No phase transition has been observed from 4 K up to the in-plane melting temperature around 700 K.

X-ray diffraction studies indicate that high-stage ($n \geq 2$) heavy-alkali GICs exhibit disordered in-plane structures at the intercalation temperatures. Upon cooling to below room temperature, many complicated phase transitions have been observed. This is a field of extensive current research and is the subject of many reviews [7.2 – 4] (see also Chap. 2). In this section, we present only TEM results and try to understand some apparent conflicts between TEM and X-ray investigations.

(a) Li Intercalated Graphite

Lithium has a smaller ion radius and forms a more compact in-plane structure than that of the heavy alkali metals. Electron diffraction studies by *Kambe* et al. [7.47] on stage-1 Li-graphite confirmed the $(\sqrt{3} \times \sqrt{3})$R30° in-plane structure of the intercalant layers as established from X-ray diffraction. In the temperature range 220 K $< T <$ 300 K, $(\sqrt{3} \times \sqrt{3})$R30° diffraction spots were observed, indicating that the stacking sequence is not ordered $\alpha\beta\gamma$. [Note that, from structure factor considerations, a completely ordered $\alpha\beta\gamma$ stacking of $(\sqrt{3} \times \sqrt{3})$R30°, and similarly a completely ordered $\alpha\beta\gamma\delta$ stacking of (2×2)R0°, would lead to the vanishing of in-plane superlattice spots.] This is consistent with the $\alpha\alpha$ stacking sequence established by X-ray diffraction. In the temperature range 100 K $< T <$ 220 K, the simple graphite diffraction pattern was observed, suggesting an ordered stacking sequence of $\alpha\beta\gamma$. As the temperature was increased above room temperature, an irreversible transition was observed and the intercalant diffraction spots changed into incommensurate diffraction rings which can be associated with randomly oriented fcc crystals of an as yet unidentified nature.

(b) K Intercalated Graphite

The first TEM study of alkali-metal graphite was carried out by *Evans* and *Thomas* [7.6] on K-graphite. Surprisingly, an ordered $(\sqrt{3} \times \sqrt{3})$R30° diffraction pattern was observed. They found that stage-1 K-graphite is unstable in vacuum. The c-axis lattice image shown in Fig. 7.8 of powdered graphite that had been saturated with potassium indicates considerable deintercalation.

Chung et al. [7.48] observed the (2×2)R0° in-plane electron diffraction pattern for stage-1 K-graphite and stage-1 Cs-graphite at temperatures between 77 and 500 K. They found [7.48] that, after partial "desorption in a nitrogen atmosphere" at room temperature for a few minutes or longer, the (2×2)R0° pattern of MC_8 (M = K, Rb, Cs) was gradually replaced by complicated and as yet unidentified patterns. Completely "nitrogen desorbed" compounds yielded diffuse diffraction rings at room temperature and unidentified spot patterns at 77 K

[7.48]. When stage-1 K-graphite and Cs-graphite were "vacuum desorbed", the in-plane diffraction pattern [7.48] became $(\sqrt{3}\times\sqrt{3})$R30° which was attributed to a surface epitaxial layer of K or Cs. *Hwang* et al. [7.49] found that when a stage-1 K-graphite specimen had been kept at room temperature and in $\sim 3 \times 10^{-10}$ torr vacuum for a month, its diffraction pattern changed from the (2×2)R0° pattern to an incommensurate hexagonal pattern associated with changes in the electron energy loss spectra.

Berker et al. [7.50, 51] studied stage-2 K-graphite of nominal composition KC_{21}. Below 86 K, they inferred the coexistence of a disordered phase and a $(\sqrt{3}\times\sqrt{3})$R30° commensurate ordered phase of $\alpha\beta\gamma$ stacking, since no in-plane superlattice spots were detected. In rare instances, $(\sqrt{7}\times\sqrt{7})$R19.1° diffraction spots were observed. Between 86 and 130 K, sharp ring patterns are obtained by electron diffraction. The ring patterns were attributed to ordered islands of dilated $\sqrt{3}\times\sqrt{3}$ and contracted $\sqrt{7}\times\sqrt{7}$ in-plane structures with dimensions ≥ 5 nm and orientations unlocked from the graphite lattice and thus also unlocked from each other. At temperatures above 130 K, the small randomly oriented islands coalesced into larger islands and became orientationally locked with the graphite lattice without an observable change in wave vectors. TEM bright-field imaging and X-ray fluorescence analysis revealed the island morphology and the potassium distribution. The $\sqrt{3}\times\sqrt{3}$ rings were found to come from potassium-rich dark regions while the $\sqrt{7}\times\sqrt{7}$ rings come from potassium-deficient light regions.

(c) Rb Intercalated Graphite

Kambe et al. [7.51, 52] found well-annealed stage-1 Rb-graphite exhibiting the simple graphite in-plane electron diffraction pattern without superlattice spots in the temperature range $100 < T < 300$ K, implying an ordered $\alpha\beta\gamma\delta$ stacking of (2×2)R0° structure. For well-annealed RbC_8 at $300 < T < 330 \pm 10$ K and unannealed as-grown RbC_8 at $110 < T < 320$ K, (2×2)R0° in-plane diffraction patterns were observed, indicating a non-$\alpha\beta\gamma\delta$ stacking. As the temperature was increased above 300 K, desorption became significant after several minutes, as revealed by irreversible changes of the diffraction pattern.

For the higher-stage Rb-graphite, an in-plane electron diffraction pattern corresponding to the $(\sqrt{7}\times\sqrt{7})$R $\pm$ 19.1° structure was observed [7.52] below a stage-dependent temperature denoted as T_u(T_u = 170 K for stage 2 and T_u = 140 K for stage 3). This pattern has been observed in stage 2, 3, 4 and 7 Rb-graphite, as well as in high-stage K-graphite and Cs-graphite, at low temperatures [7.52].

In a temperature range of 20–30 K above T_u, a reversible structure transition occurs and the $(\sqrt{7}\times\sqrt{7})$R19.1° pattern changes into a distinctly different ordered pattern schematically represented in Fig. 7.21a. This pattern appears above T_u for stage $2 \leq n \leq 7$ Rb-graphite as well as for high-stage ($n \geq 2$) Cs-graphite. TEM bright-field images obtained for stage-2 Rb-graphite exhibit small dark clusters of $\sim$ 4 nm diameter at temperatures below 170 K, whereas larger clusters of $\sim$ 18 nm diameter and channels are found above 170 K.

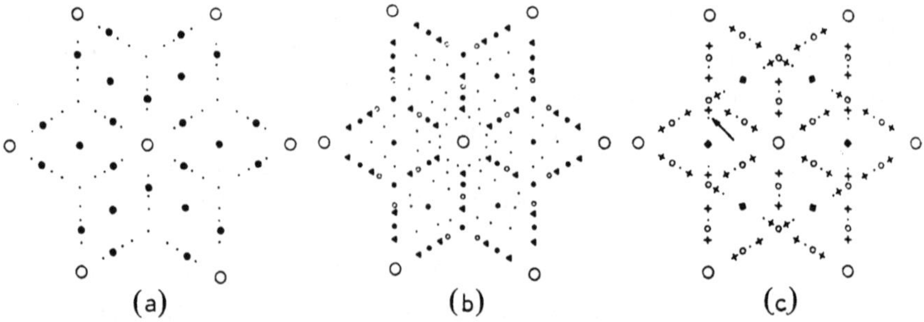

Fig. 7.21. (*a*) Schematic representation of the in-plane diffraction for high-stage Rb-graphite at and around room temperature observed by *Kambe* et al. [7.52]. (*b*) The superposition of diffraction patterns of the $(\sqrt{12} \times \sqrt{12})R30°$ (large dots) and the $(\sqrt{39} \times \sqrt{39})R(16.1°, 43.9°)$ (triangles and small dots) superlattices constructed by *Kambe* et al. [7.52]. The small dots indicate the extra spots in (*b*) not observed in (*a*) whereas the small circles indicate the extra spots in (*a*) not matched by (*b*). (*c*) The superposition of diffractions from the $(\sqrt{3} \times \sqrt{7})R(30°, 40.9°)$ (small dots), $(\sqrt{3} \times 3)R(-30°, 0°)$ (small circles), and $(\sqrt{3} \times 4)R(-30°, 0°)$ (crosses) superlattices, each with three equivalent orientations, proposed by *Hwang* [7.53] to match (*a*). The pair of spots indicated by the arrow in (*c*) and their equivalent are the only extra spots not observed in (*a*)

Kambe et al. suggested [7.52] that the ordered in-plane superlattice shown in Fig. 7.21a could be related to a superposition of a honeycomb $(\sqrt{12} \times \sqrt{12})R30°$ superlattice and an oblique $(\sqrt{39} \times \sqrt{39})R(16.1°, 43.9°)$ superlattice. The calculated superposed pattern is shown in Fig. 7.21b. It has been pointed out by *Hwang* [7.53] that Fig. 7.21b does not match well with Fig. 7.21a. Many diffraction spots for the $(\sqrt{39} \times \sqrt{39})R(16.1°, 43.9°)$ superlattice do not appear in Fig. 7.21a. Actually, the majority of spots are off in position. In analogy to the highly-anisotropic superlattices proposed [7.54] for high-stage Cs-graphite, *Hwang* [7.53] suggested that the pattern can be related to a superposition of three superlattices: the oblique $(\sqrt{3} \times \sqrt{7})R(30°, 40.9°)$ [or centered rectangular $(\sqrt{3} \times 5)R(-30°, 0°)$], the rectangular $(\sqrt{3} \times 3)R(-30°, 0°)$, and the rectangular $(\sqrt{3} \times 4)R(-30°, 0°)$ as individually shown in Fig. 7.22. The superposition of these three superlattice diffractions, each with three equivalent orientations, is shown in Fig. 7.21c. The agreement between Fig. 7.21a and c is much more reasonable.

Kambe et al. [7.52] reported that the pattern shown in Fig. 7.21a persists up to another transition temperature T_m ($T_m = 620$ K for stage-2 Rb-graphite and $T_m = 625$ K for stage-2 Cs-graphite). Above T_m, the pattern of Fig. 7.21a transforms into another complicated and yet unidentified spotty pattern. The transition is fully reversible with little thermal hysteresis [7.52].

Mittlemen [7.55] used an STEM with a probe size of ~ 20 nm to monitor the deintercalation of Rb-graphite in a 10^{-10} torr chamber at room temperature. A saturated Rb-graphite specimen was transferred from the intercalation chamber through a 10^{-4} torr interlock to the vacuum chamber and quenched to 79 K within 30 minutes. A rectangular $(\sqrt{3} \times 2)R(-30°, 0°)$ electron diffraction pattern was observed intially. The pattern changed, over a period of a few hours, into the common $(2 \times 2)R0°$ pattern, which remained stable when the specimen was kept

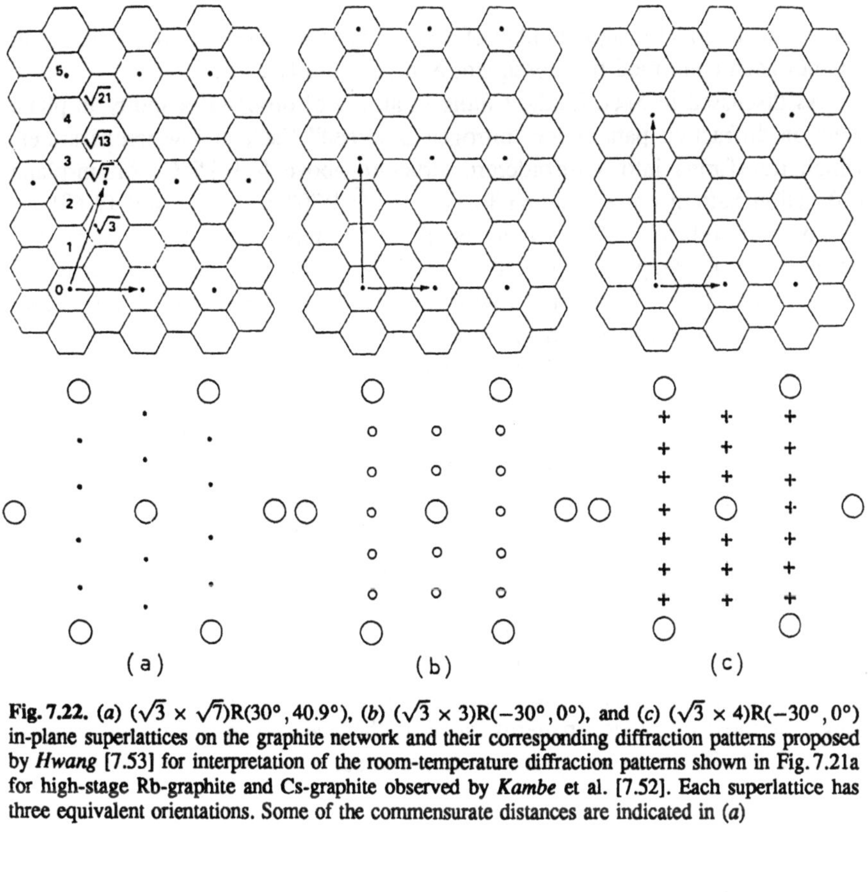

Fig. 7.22. (*a*) $(\sqrt{3} \times \sqrt{7})$R(30°, 40.9°), (*b*) $(\sqrt{3} \times 3)$R(−30°, 0°), and (*c*) $(\sqrt{3} \times 4)$R(−30°, 0°) in-plane superlattices on the graphite network and their corresponding diffraction patterns proposed by *Hwang* [7.53] for interpretation of the room-temperature diffraction patterns shown in Fig. 7.21a for high-stage Rb-graphite and Cs-graphite observed by *Kambe* et al. [7.52]. Each superlattice has three equivalent orientations. Some of the commensurate distances are indicated in (*a*)

at 79 K. The specimen was then vacuum-desorbed by raising the temperature to room temperature. The specimen had been periodically cooled back to 79 K for diffraction studies. A diffuse ring was found to coexist with the (2×2)R0° pattern after a few days of desorption. In addition, commensurate hexagonal patterns $(\sqrt{3} \times \sqrt{3})$R30°, $(\sqrt{12} \times \sqrt{12})$R30°, and (3×3)R0° were observed in some regions. After room-temperature desorption for a week, incommensurate square patterns $(0.46 \times 0.46\,\text{nm})$R(0°, 30°) and $(0.43 \times 0.43\,\text{nm})$R(0°, 30°) appeared and were interpreted as high-density local clusters where inter-intercalant interactions dominate.

(d) Cs Intercalated Graphite

X-ray diffraction studies of high-stage Cs-graphite at liquid nitrogen temperature conducted by *Parry* [7.56] identified the following in-plane superlattices: (2×2)R0° (stage 2 only), oblique $(\sqrt{7} \times 3)$R(−19.1°, 0°) (majority phase for stages 2–4), oblique $(\sqrt{39} \times \sqrt{39})$R(16.1°, 43.9°) (minority phase for stages 2–4), and hexagonal $(\sqrt{7} \times \sqrt{7})$R ± 19.1° (stage 5 only). The $(\sqrt{7} \times 3)$R(−19.1°, 0°) and $(\sqrt{39} \times \sqrt{39})$R(16.1°, 43.9°) phases were characterized by diffraction patterns having an abnormally large number of unobservable reflections [7.56]. This result has not been confirmed by subsequent X-ray studies [7.2–5].

As discussed in Sect. 7.2.2a, *Chung* et al. [7.48] observed a spotty in-plane electron diffraction pattern for "nitrogen desorbed" CsC_8 at low temperatures, which transforms into a disordererd strucutre above $5 \pm 15°C$. An in-plane diffraction pattern corresponding to $(\sqrt{3} \times \sqrt{3})R30°$ was observed [7.48] for "vacuum desorbed" CsC_8. *Kambe* et al. [7.52] reported that high-stage Cs-graphite and Rb-graphite have similar in-plane electron diffraction patterns: the $(\sqrt{7} \times \sqrt{7})R \pm 19.1°$ pattern at low temperatures and the spotty pattern shown in Fig. 7.21a around room temperature.

Hwang et al. [7.54] studied Cs-graphite using a STEM with an electron beam size of $\sim$ 80 nm. They found that the saturated Cs-graphite, transferred into the STEM via a 10^{-6} torr vacuum interlock and quenched to 98 ± 2 K, exhibits uniform bright-field images and the $(2 \times 2)R0°$ in-plane diffraction pattern. After the specimen has been vacuum (10^{-9} torr) desorbed at room temperature for one week and cooled to 98 ± 2 K again, the formation of islands with characteristic dimensions of $\sim 1\,\mu m$ was found by bright-field imaging. Several different types of diffraction patterns, including hexagonal $(2 \times 2)R0°$, rectangular $(\sqrt{3} \times 2)R(-30°, 0°)$, and oblique $(\sqrt{3} \times \sqrt{13})R(-30°, 13.9°)$ [or centered rectangular $(\sqrt{3} \times 7)R(-30°, 0°)$] have been observed by moving the 80 nm electron probe beam a few micrometers across the specimen. Single-phase single-oriented diffraction patterns, such as the one shown in Fig. 7.23 for the $(\sqrt{3} \times 2)R(-30°, 0°)$ superlattice, were recorded.

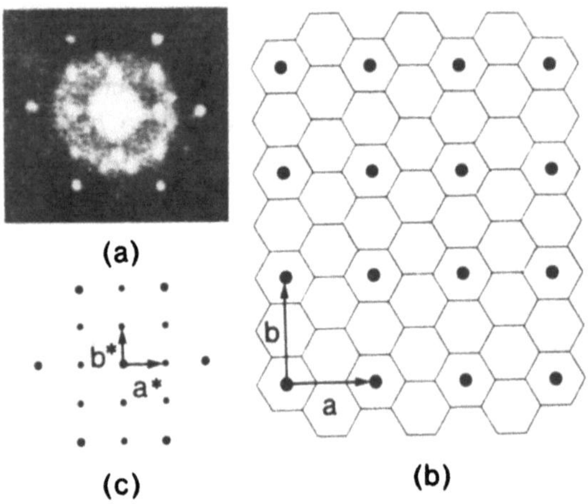

Fig. 7.23. (*a*) One of the electron diffraction patterns [7.54] observed from an unsaturated Cs-graphite at 98 ± 2 K. (*b*) The primitive $(\sqrt{3} \times 2)R(-30°, 0°)$ in-plane superlattice. (*c*) The reciprocal lattice of (*b*). The large dots are due to the graphite lattice. The small dots result from the in-plane superlattice

(e) KHg Intercalated Graphite

The in-plane structure of stage-1 KHg-graphite was studied by *Timp* et al. [7.57]. They found that the compound prepared with a zero temperature difference between the graphite target and the KHg source, i.e. $\Delta T = 0°C$, exhibited predominantly $(2 \times 2)R0°$ and $(\sqrt{3} \times 2)R(-30°, 0°)$ phases. The introduction of a temperature difference $\Delta T = 5°C$ enhanced the occurrence of the $(\sqrt{3} \times \sqrt{3})R30°$ superlattice. These in-plane superlattices have macroscopic correlation ranges extending over a few hundred nanometers as revealed in in-plane images with regular lattice fringes.

7.3 Summary and Concluding Remarks

Microscopic properties of GICs have been studied with various microscopic probes which require the specimens to be subjected to a vacuum environment. The effect of vacuum-desorption must be considered in evaluating the data. Computer simulation [7.14, 17, 23] of high-resolution TEM images indicated that the intercalant layers can be easily identified in *c*-axis lattice images. Attempts at using high-resolution TEM to establish the intercalant species and defect structures have not been successful.

TEM lattice images showing the layer stacking arrangement along the *c*-direction have been obtained from many GICs. Although the specimens have usually been characterized as well-staged by X-ray diffraction, many regions reveal the statistical distribution of intercalant layers in the graphite matrix. The average intercalant layer spacings observed in the lattice images are usually higher than the spacings indicated by X-ray diffraction, suggesting that the specimens have been vacuum desorbed. To obtain high resolution *c*-axis lattice images, the specimens must be oriented with their *c*-axis normal to the electron beam and with thicknesses of less than tens of nanometers. Such a specimen is prone to deintercalation, since the intercalant is only tens of nanometers away from the edges of the graphite galleries. Therefore, the portions of the specimen yielding lattice images probably have a higher stage number than that of the bulk.

With a magnification high enough to resolve the *c*-axis lattice images, the field of view in a TEM micrograph is limited to a few hundred nm. Most GICs, including those intercalated with $FeCl_3$ [7.6–10, 12], $FeCl_2$ [7.9, 11], $SbCl_5$ [7.13–15], $CuCl_2$ [7.16], and $NiCl_2$ [7.17] appear to have intercalant domains extending laterally over hundreds of nanometers. In those GICs, termination of intercalant layers in graphite galleries, the characteristic feature of the Daumas-Hérold domain model, was only occasionally observed within the field of view. Some GICs, e.g., $CoCl_2$ graphite [7.18] and $MnCl_2$ graphite [7.19], appear to have island-like intercalant clusters of dimensions 10–20 nm. Their *c*-axis lattice images exhibit wavy fringes.

GICs can be easily cleaved into thin flakes suitable for TEM investigation of the in-plane structures. Specimens hundreds of nanometers thick can be used to produce acceptable electron diffraction patterns. Intercalants in those plane-view specimens may be tens of micrometers away from the edges of the graphite galleries, and are therefore more stable than in the specimens suitable for c-axis lattice imaging.

Many GICs have complicated multiphase multidomain in-plane structures. In many cases, single-phase single-oriented patterns were obtained by selected-area electron diffraction or convergent-beam electron diffraction. This indicates that, in those multiphase systems, the in-plane structures are correlated strongly in the plane (exceeding the selected-area aperture diameter, typically around 0.5 μm, or the probe beam diameter, 10–100 nm) and along the c-axis (exceeding the specimen thickness, typically around 100 nm). Many new in-plane structural phases, not observed by X-ray diffraction, have been identified by electron diffraction.

The layer structures of pristine $FeCl_3$ [7.6, 31, 32], $CoCl_2$ [7.18, 33], $MnCl_2$ [7.19], and $CuCl_2$ [7.16, 33] are preserved in their intercalated graphite compounds. The intercalant layers exhibit preferred orientational relations with respect to the graphite lattice, with a relative rotation of either 0° or 30°. The degree of preference is specimen dependent. No phase segregation or transition has been observed. These results are consistent with the X-ray results.

The in-plane structure of $SbCl_5$ graphite is disordered at intercalation temperatures above 180°C. At room temperature and below, ordered commensurate phases develop from the disordered phase. The commonly observed phases are $(\sqrt{7}\times\sqrt{7})R\pm19.1°$ and $(\sqrt{39}\times\sqrt{39})R\pm16.1°$. Other high-order commensurate phases, $(14\times14)R0°$ and $(28\times28)R0°$ by X-ray diffraction [7.38] and $(3\times3)R0°$, $(\sqrt{13}\times\sqrt{13})R\pm13.9°$ [7.35], $(\sqrt{19}\times\sqrt{19})R\pm23.4°$, $(\sqrt{31}\times\sqrt{31})R\pm8.95°$, and $(\sqrt{76}\times\sqrt{76})R\pm23.4°$ [8.39] by electron diffraction, have been reported to exist in some specimens. The degree of $SbCl_5$ disproportionation into $SbCl_3$, $SbCl_4^-$, and $SbCl_6^-$ is not conclusive. All four of these molecular species have been detected by Mössbauer spectroscopy [7.34]. High density islands were observed by *Clarke* et al. [7.36, 38] and *Hwang* et al. [7.30, 42, 43]. The former attributed the islands to neutral $SbCl_3$ and the background to $SbCl_6^-$, while the latter attributed the islands to a complex of $SbCl_3/SbCl_4^-$ and the background to a complex of $SbCl_5/SbCl_6^-$, suggesting charge homogenization and incomplete chemical disproportionation. *Salamanca-Riba* et al. [7.14] reported that $SbCl_5$ graphite could exist with the $(\sqrt{7}\times\sqrt{7})R\pm19.1°$ single phase and the intercalant was a mixture of $SbCl_3$ and $SbCl_6^-$.

Stage-1 heavy alkali-metal-graphite MC_8(M = K, Rb and Cs) has a simple $(2\times2)R0°$ in-plane structure, observed by both X-ray and electron diffractions. However, controversial results were obtained from the corresponding high-stage ($n\geq2$) compounds. X-ray studies observed a disordered room-temperature phase which developed into complicated incommensurate structures at low temperatures [7.2–4] (see also Chap. 2). On the other hand, electron diffraction studies [7.6, 48, 50–55] identified many ordered commensurate phases, including hexagonal phases $(\sqrt{3}\times\sqrt{3})R30°$, $(2\times2)R0°$, and $(\sqrt{7}\times\sqrt{7})R\pm19.1°$, and a series of parallel-chain structures $(\sqrt{3}\times 2)R(-30°,0°)$, $(\sqrt{3}\times\sqrt{7})R(30°,40.9°)$, $(\sqrt{3}\times 3)R(-30°,0°)$, $(\sqrt{3}\times\sqrt{13})R(-30°,13.9°)$, and $(\sqrt{3}\times 4)R(-30°,0°)$. Several other spotty ordered diffraction patterns were obtained from specimens at room temperature and higher.

The discrepancy between the X-ray and electron diffraction studies of $SbCl_5$ graphite and high-stage heavy-alkali-metal-graphite may be understood from the following considerations: Due to the disordered nature of the intercalant layers at the intercalation temperature, well-staged GICs do not have unique intercalant densities. Although these GICs are regarded as vacuum-stable in the bulk form, there is evidence [7.6, 48, 49, 55] that vacuum-desorption occurs at room temperature on a microscopic scale. The various in-plane structures probably reflect various in-plane densities. Since the stage transformation is expected to be more sluggish than the in-plane structural transformation, vacuum desorption at low temperatures may result in unusual in-plane densities that cannot be achieved with high-temperature intercalation. Due to the low intercalant diffusion rate at low temperatures, it will take a long time to change the composition of a bulk specimen and those phases with unusual in-plane densities can escape detection in X-ray studies. TEM is more sensitive regarding detection of minor phases with ordered structures. It is interesting to note that $SbCl_5$ graphite has a tendency to form hexagonal commensurate superlattices, at least nine of which have been observed, while high-stage heavy-alkali-metal-graphite has a tendency to form parallel-chain-like commensurate structures $(\sqrt{3}\times b)$, with $b=\sqrt{3},2,\sqrt{7},3,\sqrt{13},4$ and 5 being observed.

The microscopic properties of GICs are complicated due to the variation in in-plane densities and the coexistence of multiple structural phases in microscopic domains. Although many experimental results appear to be contradictory, general trends can be established on analyzing them. More research is needed to construct a clear general picture.

Acknowledgements. In preparing this chapter I drew heavily on knowledge gained from helpful discussions with S.A. Solin and other collaborators, including R. Levi-Setti, X.W. Qian, M. Utlaut, N.W. Parker, R. Mittlemen, G. Nicolaides, G. Crow, Y.L. Wang, M.S. Isaacson, and A.V. Crewe. I also benefited greatly from discussions with M.S. Dresselhaus, J.M. Gibson and S.A. Schwarz

References

7.1 N. Daumas, A. Hérold: C.R. Acad. Sci. C **286**, 373 (1969)
7.2 M.S. Dresselhaus, G. Dresselhaus: Adv. Phys. **30**, 139 (1981)
7.3 S.A. Solin: Adv. Chem. Phys. **49**, 455 (1982)
7.4 R. Clarke, E. Uher: Adv. Phys. **33**, 469 (1984)
7.5 W. Rüdroff, E. Schulze: Z. Inorg. Chem. **277**, 156 (1954)
7.6 E.L. Evans, J.M. Thomas: J. Solid State Chem. **14**, 99 (1975)
7.7 J.M. Thomas, G.R. Millward, N.C. Davies, E.L. Evans: J. Chem. Soc., Dalton Trans. **23**, 2443 (1976)
7.8 J.M. Thomas, G.R. Millward, R.F. Schlögl, H.P. Boehm: Mat. Res. Bull. **15**, 671 (1980)

7.9 G.R. Millward, J.M. Thomas, W. Jones, R. Schlögl: In Inst. Phys. Conf. Ser. No. 61: Electron Microscopy and Analysis Group Conference, Cambridge, 1981 (Institute of Physics, London 1982) p. 399
7.10 C.P. Armstrong, J.A. Little, W.M. Stobbs: In Inst. Phys. Conf. Ser. No. 78: Electron Microscopy and Analysis Group Conference, Newcastle upon Tyne, 1985 (Adam Hilger, Bristol 1985) p. 577
7.11 J.M. Thomas: Ultramicroscopy **8**, 13 (1982)
7.12 T.C. Chieu, G. Timp, M.S. Dresselhaus, M. Endo: In Materials Research Society Symposia Proc., Vol. 20: Intercalated Graphite, ed. by M.S. Dresselhaus, G. Dresselhaus, J.E. Fischer, M.J. Moran (Elsevier, New York 1983) p. 51
7.13 G. Timp, M.S. Dresselhaus, L. Salamanca-Riba, A. Erbil, L.W. Hobbs, G. Dresselhaus, P.C. Eklund, Y. Iye: Phys. Rev. **26**, 2323 (1982)
7.14 L. Salamanca-Riba, G. Roth, J.M. Gibson, A.R. Kortan, G. Dresselhaus, R.J. Birgeneau: Phys. Rev. B **33**, 2738 (1986)
7.15 D.M. Hwang: Presented in MRS 1986 Fall Meeting Symposium K: Graphite Intercalation Compounds, unpublished
7.16 C. Hauw, J. Gaultier, S. Flandrois, O. Gonzalez, D. Dorignac, R. Jagut: Synth. Met. **7**, 313 (1983)
7.17 D. Dorignac, M.J. Lahana, R. Jagut, B. Jouffrey, S. Flandrois, C. Hauw: In Materials Research Society Symposia Proc., Vol. 20: Intercalated Graphite, ed. by M.S. Dresselhaus, G. Dresselhaus, J.E. Fischer, M.J. Moran (Elsevier, New York 1983) p. 33
7.18 M. Matsuura, Y. Murakami, K. Takeda, H. Ikeda, M. Suzuki: Synth. Met. **12**, 427 (1985)
7.19 J.S. Speck, J.M. Delgado, S.T. Chen, N.C. Yeh, M.S. Dresselhaus, J.M. Gibson: In Proc. of MRS 1986 Fall Meeting, Symposium K: Graphite Intercalation Compounds, ed. by M.S. Dresselhaus, G. Dresselhaus, S.A. Solin (Materials Research Society, Pittsburgh, PA 1986) p. 44
7.20 L. Salamanca-Riba, N.C. Yeh, T. Enoki, M.S. Dresselhaus, M. Endo: In Proc. of MRS 1984 Fall Meeting, Symposium I: Graphite Intercalation Compounds, ed. by P.C. Eklund, M.S. Dresselhaus, G. Dresselhaus (Materials Research Society, Pittsburgh, PA 1984) p. 249
7.21 J.C.H. Spence: *Experimental High-Resolution Electron Microscopy* (Clarendon, Oxford 1981)
7.22 J.M. Cowley: *Diffraction Physics* (North-Holland, Amsterdam 1975)
7.23 G.R. Millward, J.M. Thomas: Carbon **17**, 1 (1979)
7.24 M.Z. Tahar, O.C. de Hodgins, M.S. Dresselhaus, M. Endo: In Proc. of MRS 1984 Fall Meeting, Symposium I: Graphite Intercalation Compounds, ed. by P.C. Eklund, M.S. Dresselhaus, G. Dresselhaus (Materials Research Society, Pittsburgh, PA 1984) p. 217
7.25 E. Minami, X. Hao, J.S. Speck, M.S. Dresselhaus, M. Endo: In Proc. of MRS 1986 Fall Meeting, Symposium K: Graphite Intercalation Compounds, ed. by M.S. Dresselhaus, G. Dresselhaus, S.A. Solin (Materials Research Society, Pittsburgh, PA 1986) p. 213
7.26 R. Levi-Setti, G. Crow, Y.L. Wang, N.W. Parker, R. Mittlemen, D.M. Hwang: Phys. Rev. Lett. **54**, 2615 (1985)
7.27 R. Levi-Setti, Y.L. Wang, G. Crow: J. de Phys. Colloq. **45**, C9 (1984)
7.28 M. Bretz, R. Clarke: Phys. Rev. Lett. **55**, 2506 (1985)
7.29 D.M. Hwang, R. Levi-Setti: Phys. Rev. Lett. **55**, 2507 (1985)
7.30 D.M. Hwang, R. Levi-Setti, G. Crow, Y.L. Wang, N.W. Parker, R. Mittlemen, X.W. Qian, S.A. Solin: Synth. Met. **12**, 73 (1985)
7.31 J.M. Cowley, J.A. Ibers: Acta Crystallogr. **9**, 421 (1956)
7.32 P.P. Vaishnava, M. José-Yacamán, P.A. Montano: J. Phys. Chem. Solids **47**, 187 (1986)
7.33 X. Hao, J.S. Speck, M.S. Dresselhaus: In Proc. of MRS 1986 Fall Meeting, Symposium K: Graphite Intercalation Compounds, ed. by M.S. Dresselhaus, G. Dresselhaus, S.A. Solin (Materials Research Society, Pittsburgh, PA 1986) p. 47
7.34 P. Boolchand, W.J. Bresser, D. McDaniel, K. Sisson, V. Yeh, P.C. Eklund: Solid State Commun. **49**, 1049 (1984)
7.35 D.M. Hwang: In Proc. of MRS 1986 Fall Meeting, Symposium K: Graphite Intercalation Compounds, ed. by M.S. Dresselhaus, G. Dresselhaus, S.A. Solin (Materials Research Society, Pittsburgh, PA 1986) p. 60
7.36 R. Clarke, M. Elzingga, J.N. Gray, H. Homma, D.T. Morelli, M. J. Winokur, C. Uher: Phys. Rev. B **26**, 2520 (1982)
7.37 R. Clarke, H. Homma: In Materials Research Society Symposia Proc., Vol. 20: Intercalated Graphite, ed. by M.S. Dresselhaus, G. Dresselhaus, J.E. Fischer, M.J. Moran (Elsevier, New York 1983) p. 3

7.38 H. Homma, R. Clarke: Phys. Rev. B **31**, 5865 (1985)
7.39 R.K. Mittlemen, N.W. Parker, A.V. Crewe: Phys. Rev. B **36**, 7590 (1987)
7.40 L. Salamanca-Riba, G. Timp, L.W. Hobbs, M.S. Dresselhaus: In Materials Research Society Symposia Proc., Vol. 20: Intercalated Graphite, ed. by M.S. Dresselhaus, G. Dresselhaus, J.E. Fischer, M.J. Moran (Elsevier, New York 1983) p. 9
7.41 M. Suzuki, R. Inada, H. Ikeda, S. Tanuma, K. Suzuki, M. Ichihara: J. Phys. Soc. Jpn. **53**, 3052 (1984)
7.42 D.M. Hwang, X.W. Qian, S.A. Solin: Phys. Rev. Lett. **53**, 1473 (1984)
7.43 D.M. Hwang, X.W. Qian, S.A. Solin: In Proc. of MRS Fall Meeting, Symposium I: Graphite Intercalation Compounds, ed. by P.C. Eklund, M.S. Dresselhaus, G. Dresselhaus (Materials Research Society, Pittsburgh, PA 1984) p. 155
7.44 D.M. Hwang, J. Cowley: Unpublished
7.45 D.D.L. Chung: Phase Transitions **8**, 35 (1986)
7.46 D. Ghosh, D.D.L. Chung: Mater. Res. Bull. **18**, 727 (1983)
7.47 N. Kambe, M.S. Dresselhaus, G. Dresselhaus, S. Basu, A.R. McGhie, J.E. Fischer: Mater. Sci. Eng. **40**, 1 (1979)
7.48 D.D.L. Chung, G. Dresselhaus, M.S. Dresselhaus: Mater. Sci. Eng. **31**, 107 (1977)
7.49 D.M. Hwang, M. Utlaut, M.S. Isaacson, S.A. Solin: Phys. Rev. Lett. **43**, 882 (1979)
7.50 A.N. Berker, N. Kambe, G. Dresselhaus, M.S. Dresselhaus: Phys. Rev. Lett. **45**, 1452 (1980)
7.51 M.S. Dresselhaus, N. Kambe, A.N. Berker, G. Dresselhaus: Synth. Met. **2**, 121 (1980)
7.52 K. Kambe, G. Dresselhaus, M.S. Dresselhaus: Phys. Rev. B **21**, 3491 (1980)
7.53 D.M. Hwang: Phys. Rev. B **27**, 1119 (1983)
7.54 D.M. Hwang, N.W. Parker, M. Utlaut, A.V. Crewe: Phys. Rev. B **27**, 1458 (1983)
7.55 R.K. Mittlemen: Phys. Rev. B **36**, 6001 (1987)
7.56 G.S. Parry: Mater. Sci. Eng. **31**, 99 (1977)
7.58 G. Timp, B.S. Elman, M.S. Dresselhaus, P. Tedrow: In Materials Research Society Symposia Proc., Vol. 20: Intercalated Graphite, ed. by M.S. Dresselhaus, G. Dresselhaus, J.E. Fischer, M.J. Moran (Elsevier, New York 1983) p. 201

8. New Graphite–Donor Compounds and Synthesis

Philippe Lagrange and Ralph Setton

With 9 Figures

This chapter aims to survey the new combinations of graphite with donor species found since 1980: it will therefore constitute a sequel to the very comprehensive review by *Dresselhaus* and *Dresselhaus* [8.1]. Thanks to the efforts of a number of laboratories, many new species have been synthesized and studied. We have chosen to limit this survey to the synthesis, structure and the most important chemical properties of the compounds, leaving aside the very large number of monographs dealing with the study and determination of physical properties. However, we wish to underline the fact that many of the ternary phases containing an alkali metal and hydrogen or another metal (Hg, Tl, Bi) are superconducting at low temperatures [8.2–5]. Superconductivity had already been found in the MC_8 binary compounds (M = K, Rb, Cs) for temperatures below a few hundred millikelvin [8.6, 7], but the nature of the problem changed with the ternary phases, for which much higher critical temperatures (up to 4 K) have been found.

Section 8.1 presents a brief reminder of the main features of the graphite–donor metal compounds, classified according to their nature and structure. Section 8.2 describes binary phases obtained by the action of pressure and/or temperature on already well-known graphite–metal combinations. Sections 8.3 and 4 are directly concerned with ternary phases, which have been classified according to the structure of the intercalated layer:

- Phases containing different single-layer sheets of alkali metal (known as *heterostructures*).
- Phases with multiple-layer intercalated sheets containing an alkali metal and another element, the latter so far not intercalated by itself in graphite (H, Hg, Tl, Bi). Numerically more important, these phases present a vast field of study to the chemist and the physicist.

8.1 Preliminary Remarks Concerning the Donor–Graphite Intercalation Compounds

The heavy alkali metals are the most electropositive elements of the periodic classification, and the low energy of ionization of their outermost s electron is probably one of the reasons for the ease of their intercalation into graphite. Furthermore, owing to the facility with which higher-stage compounds are formed,

K, Rb and Cs are at the basis of a large family of binary derivatives extensively studied by physicists and which constitute, in a way, "reference compounds" because of their chemical simplicity and relatively simple structure. Lithium and sodium, the other two alkali metals, are not so easily intercalated: the latter only yields higher-stage compounds with a low metal content while the former, which does give a number of derivatives with a high proportion of metal (stages 1, 2 and 3), only reacts under fairly extreme conditions of preparation.

Similar in many ways to lithium, the alkaline-earth metals show a comparable reactivity and can also be intercalated into graphite by similar methods, at least for strontium and barium (and calcium, but only on the surface). Equivalent results have been obtained with some of the lanthanides, namely europium and ytterbium (and samarium, but only on the surface), so that 9–11 metals, all of them donors, can be said to have been directly intercalated into graphite, in spite of widely varying reactivities. As shown in Table 8.1, the metals also differ considerably in their ability to form the higher stages. Nevertheless, all the binary graphite–metal compounds which have so far been prepared by direct reaction between the elements have in common the characteristic that the intercalated species forms a monatomic layer between the sheets of graphene, whatever the value of d_i, the distance separating the midplanes of the bounding graphene layers.

Table 8.1. Binary graphite–metal phases that have been isolated

Metal	Stage, s	Interlamellar distance, d_i [Å]	Formulae	References
K	1–11	5.35	$MC_8 (s = 1)$	[8.8, 9]
Rb	1–8	5.62	$MC_{12 \times s} (s \geq 2)$	[8.10, 11]
Cs	1–8	5.92	$MC_{12 \times s} (s \geq 2)$	[8.10, 11]
Na	4–8	4.52	$NaC_{8 \times s}$	[8.12, 13]
Li	1–3	3.70	$LiC_6 (s = 1)$ $LiC_{6 \times s - 9 \times s} (s \geq 2)$	[8.14]
Sr	1	4.95	MC_6	[8.15, 16]
Ba	1	5.25	MC_6	[8.15, 16]
Eu	1	4.87	MC_6	[8.17]
Yb	1	4.57	MC_6	[8.17]

Some bimetallic ternary phases are like the binary GICs just described, in that the intercalated layer is a one-atom-thick solid solution of one species uniformly dispersed in the intercalated layer of the other species [8.18–20].

Other ternaries, however, are quite different, since the intercalated sheet is, in this case, composed of a number of superposed and distinct layers. The first ternary compound of this type was obtained by hydrogenation of stage-1 KC_8: the reaction produces a pure second stage formulated $KH_{2/3}C_8$. The increase in stage is due to the freeing of half the available interlamellar area by superposition of one half of the intercalated layer of potassium on the other half, with an intermediary layer of hydrogen as binder. Similarly, but with a much decreased rate of reaction, RbC_8 transforms into $RbH_{2/3}C_8$, with a similar structure, unlike CsC_8, which does not react with hydrogen.

The intercalate-free interlamellar spaces in $MH_{2/3}C_8$ accept more alkali metal (K, Rb or Cs, not necessarily identical to the metal already present) to form new ternary or quaternary *heterostructures* (also known as *bi-intercalation compounds*) with alternating single or triple layers of intercalate [8.25, 26]. As far as is known, hydrogenation of the one-atom-thick alkali metal layer is not possible.

Other compounds with a triple intercalated sheet having a metal instead of hydrogen in the median plane will be examined in detail in Sect. 8.4.2. For the sake of completeness we shall mention here a graphite-sodium-barium compound in which the structure of the intercalated layer is Na-Ba-Na. Thus, in all these stage-2 derivatives, we find, along the direction of the *c*-axis, a regular alternation of layers of highly electropositive atoms (of alkali metal) and less electropositive species (carbon, hydrogen or barium), as shown in Table 8.2. In the present case, the barium compound is not as well defined as the other members of this class of GICs, since there seems to be a large amount of substitution both in the median layer of barium and in the flanking layers of sodium.

Table 8.2. Ternary graphite–donor phases with multiple-layer sheets

Alkali metal	2nd Element	Stage	Interlamellar distance d_i[Å]	Formulae	References
K	H	2	8.53	$KH_{2/3}C_8$	[8.21–23]
Rb	H	2	9.03	$RbH_{2/3}C_8$	[8.24]
Na	Ba	≥ 2	7.38	Undefined	[8.27, 28]

8.2 New Binary Phases Obtained at High Pressure and/or High Temperature

Early observations pointed to the presence of purple-colored phases only visible at "high" temperatures (about 300–400°C), whose composition was probably close to KC_{10} and CsC_{10} [8.29–31], i.e., intermediate between those of stage 1 and stage 2, and for which no method of quenching had been found because of the apparent ease of the reaction

$$8MC_{10} \rightarrow 7MC_8 + MC_{24} \, .$$

Although the stoichiometry of these phases is rather badly defined, the compounds are most probably disordered stage 1 without three-dimensional order.

Furthermore, *Clarke* et al. [8.32, 33] showed that pressures greater than 7000 bar reversibly modify KC_{24} to a highly ordered stage-3 compound with identical stoichiometry, in which the intercalated layers are reorganized into the $(2 \times 2)R0°$ octal structure (Fig. 8.1), with good three-dimensional order.

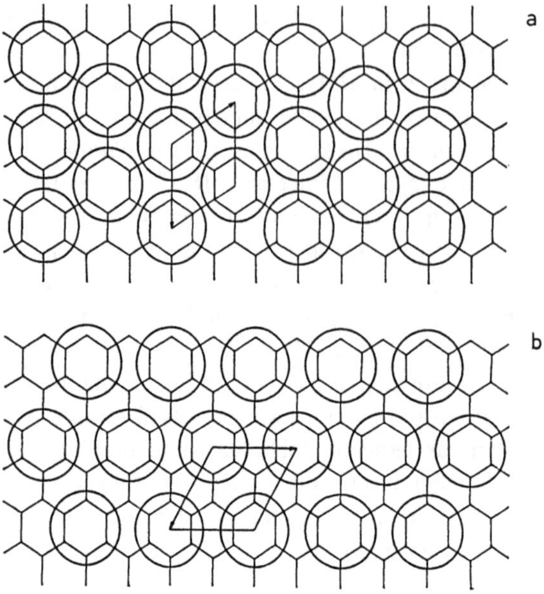

Fig. 8.1. In-plane unit cell for (a) the $(\sqrt{3} \times \sqrt{3})R30°$ structure and (b) the $(2 \times 2)R0°$ structure. The carbon atoms are placed at the corners of the hexagons and the intercalants appear as circles. The intercalate and graphite layers are projected into a single plane

These results are, in a way, forerunners of recent work on the graphite-lithium and graphite-potassium systems. By subjecting known compounds to various temperatures and pressures, new transitions have been discovered, which can be classified into two main groups:

- Crystallographic transitions which occur in the distribution of the two-dimensional intercalated phase, or in the three-dimensional mode of stacking, with only minor modifications of the density.
- Transitions which profoundly affect the density of the intercalated layer or the stage of the compound.

We shall only discuss this latter type of transition in the remainder of this section, having arbitrarily decided that the only "new" compounds are those which have a stoichiometry different from the one usually met, or which have a known stoichiometry but a different stage. This dichotomy is sometimes not easy to resolve as recent results show that, in some cases, the two types of transitions overlap [8.34].

8.2.1 The Graphite–Lithium System

This system was studied systematically by *Fischer* et al. as a function of temperature [8.35, 36]. This led them to an (x, T) phase diagram, in which x is a function of the chemical composition of the binary Li_xC_6 ($x = 6$ Li/C) and T is

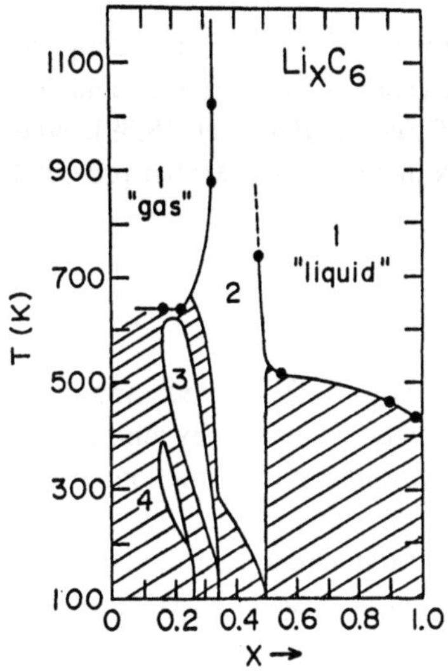

Fig. 8.2. Experimental staging phase diagram for Li_xC_6 derived from X-ray and neutron diffraction. (From [8.35])

the absolute temperature. There are a number of regions in the phase diagram, each corresponding either to a single stage (the clear zones in Fig. 8.2), or to a mixture of stages (shaded zones). Each clear zone includes compounds whose stage number is well defined but which contain an intercalated layer whose density can vary. In general, the limit of the clear zone associated with the larger values of x corresponds to an ordered phase in which the intercalated metallic layers exist as compact $(\sqrt{3} \times \sqrt{3})R30°$ lattices (Fig. 8.1). At sufficiently low temperatures ($T \simeq 100$ K), only this type of structure exists.

A particularly interesting new phenomenon is illustrated in the phase diagram. For sufficiently high temperatures, all the higher-stage compounds become stage-1 derivatives, whatever the value of x. Thus, the compound LiC_{62}($x = 0.097$) changes from stage 4 to stage 1 as T changes from 195 to 557 K [8.37]. The concentration of Li atoms is now identical in all the interlamellar spaces and is about one-tenth that in LiC_6.

The stage-2 region has an unexpected extension in the high-temperature domain, which illustrates the impossibility of transforming a stage-2 into a stage-1 compound by mere heating within that temperature range. The concentration corresponding to $x = 0.33$ is therefore "superstable". At lower temperatures, the intercalated layers, which were initially stage 2, regroup into a $(\sqrt{3} \times \sqrt{3})R30°$ lattice, thereby producing a stage-3 compound (Fig. 8.2). The sequence of carbon and metal layers is then $\ldots ABA\alpha ABA \ldots$. Lastly, one should note that the same process of regrouping in the intercalated layers, with the consequent increase in stage number ($2 \to 3$), can be obtained at ambient temperature by subjecting the GIC to 2.9 kbar.

Miyazaki et al. [8.38] have developed a model for the binary compounds in which the various interaction energies between the metal and the carbon layers are taken into account. Using this model, they were able to show that, under any pressure (and at 0 K), the most stable structure for the Li derivatives belongs to the series $LiC_{6\times s}$, where s is the stage. This corresponds to the results obtained by *Fischer* et al. (Fig. 8.2), but not to those of *Semenenko* et al. [8.39], who claimed that LiC_2 is formed when the pressure is increased to 800 kbar provided sufficient metallic Li is available.

8.2.2 The Graphite–Potassium System

Most of the recent work was performed on the binaries in this system, and many original results have been obtained. The phase transitions induced by temperature variations (modifications in the stacking, order–disorder transformations in the intercalated layer, stage changes, etc.) were thoroughly investigated by *Suematsu* and co-workers, who thus gathered important thermodynamics and kinetics data [8.40–43]. Most new phases were however obtained by *Fischer* and co-workers, using combinations of temperature and pressure on well-known compounds prepared by classical methods.

(a) Metal-Rich Systems (KC_8 Under Pressure)

At ambient temperature and 15 kbar [8.35, 37], KC_8 undergoes a partial condensation of the metallic layers, some of which forego the $(2 \times 2)R0°$ hexagonal lattice for the $(\sqrt{3} \times \sqrt{3})R30°$ lattice (also hexagonal) [8.44]. Since some of the interlayer surface is freed, the rest of the metallic layers now form a new fractional phase whose stage number is no longer 1 but 3/2, with the stacking sequence ...CKCCKCKCCK... formed through the reaction

$$
\begin{array}{lccc}
 & 6\,KC_{8\times1}\ \text{(stage 1)} \rightarrow & 4\,KC_{6\times1}\ \text{(stage 1)} + & K_2C_{8\times3}\ \text{(stage 3/2)}. \\
\text{2d lattice} & (2\times2)R0° & (\sqrt{3}\times\sqrt{3})R30° & (2\times2)R0°
\end{array}
$$

The identity period of $K_2C_{8\times3}$ is the sum of the interlayer distances of the stage-1 and stage-2 compounds at that pressure ($I_c = 13.61$ Å). The compound therefore constitutes the first instance of a fractional phase, with a stage number between 1 and 2.

Between 15 and 19 kbar, the two-phase system again changes into another mixture of two phases, respectively stage 1 and stage 2. In this range of pressure, the fractional-stage GIC is transformed into another "superdense" stage-2 compound, as described by the reaction

$$
\begin{array}{lcc}
 & K_2C_{8\times3}\ \text{(stage 3/2)} \rightarrow & 2\,KC_{6\times2}\ \text{(stage 2)}. \\
\text{2d lattice} & (2\times2)R0° & (\sqrt{3}\times\sqrt{3})R30°
\end{array}
$$

Thus, KC_8 has finally been transformed into KC_6 (stage 1) and KC_{12} (stage 2), both superdense. The same behavior is shown by RbC_8 and CsC_8 at slightly higher pressures.

It is interesting that the stage 4/3 compound with superdense layers is never formed, in spite of the fact that one could expect it to be the result of the reaction

$$3\,KC_{8\times1}\ (\text{stage 1}) \rightarrow K_3C_{6\times4}\ (\text{stage 4/3}).$$
$$\text{2d lattice} \qquad (2\times2)R0^\circ \qquad (\sqrt{3}\times\sqrt{3})R30^\circ$$

It seems that the mixture $2KC_{6\times1}$ (stage 1)+$KC_{6\times2}$ (stage 2) is more stable than the single phase $K_3C_{6\times4}$ (stage 4/3).

(b) Systems with Intermediate Concentrations (KC_{24} and KC_{36})

When submitted to increasing pressure, pure KC_{24} (stage 2) undergoes a continuous modification associated with two transitions at 3.4 and 6.7 kbar [8.34, 45, 46]. The continuous transformation is associated with the nucleation process whereby islands of dense structure gradually form from the disordered and liquid-like intercalated layer. The transition at 6.7 kbar corresponds to the formation of a single compact layer with the 2d hexagonal $(2\times2)R0^\circ$ lattice in $KC_{8\times3}$ (stage 3), but the origin of the transition at 3.4 kbar is unknown.

The stage-3 KC_{36} also shows a gradual nucleation of "solid" islands from the liquid-like intercalated layer as soon as the pressure increases, tending asymptotically to a stage-4 compound. A transition then occurs at 7.5 kbar, leading to a mixture of stage 4 and stage 5 [8.34]. This is not surprising since the stoichiometry 36 for the number of C atoms in the starting material is not a multiple of 8, so that one must have a reaction of the type

$$2KC_{36}(\text{stage 3}) \rightarrow KC_{8\times4}(\text{stage 4}) + KC_{8\times5}(\text{stage 5})\ .$$

(c) Systems with Low Concentrations (Stage ≥ 4)

Systems with a low concentration of potassium all behave similarly [8.34]. As described for the stage-3 KC_{36}, there is a gradual nucleation of islands starting as soon as the pressure is increased, caused by the overall decrease in volume of the system. It necessarily leads to the formation of higher stages and ends at the onset of the sudden transition. The latter must correspond to a critical state for which the only outlet is the formation of an ordered dense 2d $(2\times2)R0^\circ$ lattice and the loss of elastic strain energy associated with the disappearance of the defects in the graphene layers which must exist at the border of each individual island. All these points are well illustrated by the stage vs pressure diagram in Fig. 8.3 showing the evolution of a binary which, initially, was approximately stage 8 and finally stage 13 at 8.5 kbar. Note that with an exact stoichiometry of an ideal stage 8, namely KC_{96}, the high-pressure phase would have been expected to be a pure stage 12.

Miyazaki et al. [8.38] have tried to establish the possible stable compositions for the potassium–graphite system; the results point to stable compositions of $KC_{14\times s}$[1] and $KC_{8\times s}$ for low and high pressures, respectively. The superdense

[1] It is now admitted [8.34] that the high-stage potassium GICs possess in their ground state (at low temperatures) metallic layers with 2d commensurate lattice $(\sqrt{7}\times\sqrt{7})R19.1^\circ$; the corresponding stoichiometry is precisely $KC_{14\times s}$. Consequently, cooling the phase KC_{24} leads to the formation of the 2d lattices $(2\times2)R0^\circ$ and $(\sqrt{7}\times\sqrt{7})R19.1^\circ$.

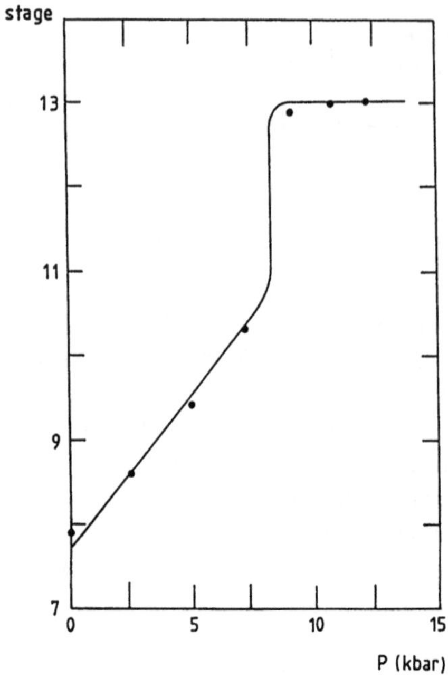

Fig. 8.3. Average stage vs pressure for a potassium graphite intercalation compound of stage 8 at $P = 0$. (From [8.34])

$KC_{6\times s}$ phases described above do not seem to appear in this theoretical evaluation.

8.3 The New Ternary Compounds with Multiple Single Layers

Up until 1983, the intercalated layer in compounds containing two metallic species was a statistically homogeneous one-atom-thick solid solution, best described by the formula $M_x M'_{1-x} C_8$ (with M, M' = Na, K, Rb, Cs) [8.18–20]. New species were then synthesized, in which the same elements are also found, but as individually distinct layers [8.47]. These are the *bi-intercalation compounds*, obtained by successive, rather than simultaneous, intercalation of the two metals.

The action of the second reagent on a second-stage intercalation compound had long since been tried, in the hope that this would result in the formation of a phase with separate different layers occupying alternate interlayer spaces (in the Rüdorff model). This had yielded, for instance, the compounds $MH_{2/3}C_8.M'C_8$, with M =K, Rb, and M' =K, Rb, Cs, with the required structure [8.25, 26]. Nevertheless, this goal is not always easily attained, because of the following possible situations:

a) There may be a reaction between the two chemical species, with the formation of a nonintercalated "inclusion" compound trapped at defects of the graphite matrix.

b) The invading second species can expel and take the place of the first intercalated species.
c) The two species can form a homogeneous solid solution.
d) The two species can react to form a new intercalated single phase.
e) The first species can regroup to form a stage-1 compound and free graphite into which the invading species intercalates, forming another stage-1 compound (formation of two distinct and nonmiscible phases).

8.3.1 Synthesis of Heterostructures with Two Heavy Alkali Metals

For any of the reasons detailed above, the outcome of the reaction of K with CsC_{24} could have been expected to fail. The successful preparation [8.48] of the bi-intercalation compound was performed on CsC_{24} (prepared from HOPG in the standard way) by transferring the platelet to a glass tube containing metallic potassium, sealing under vacuum, and heating to 70°C with the platelets immersed in the molten potassium. At the end of the proper reaction time (12 days), excess potassium was eliminated by spinning the glass tube in a centrifuge maintained at 70°C. As shown by the X-ray diffraction spectra, the product was a mixture of two phases, KC_8 and $KCsC_{16}$, which of course, could not be separated mechanically. The ternary compound was a heterostructure with all its interlayer spaces filled alternately with single-layer sheets of pure potassium or pure cesium according to the sequence ...C K C Cs C K C Cs... along the *c*-axis, with an identity period equal to the sum of the identity periods of KC_8 and CsC_8:

$$I_c = d_i(KC_8) + d_i(CsC_8) = 5.35 + 5.92 = 11.27\,\text{Å}\,.$$

The chemical reaction

$$CsC_{24} + 2K \rightarrow KCsC_{16} + KC_8$$

is therefore a correct representation of the creation of a polyphased mixture.

Originally, *Solin* and his group [8.48] hoped to obtain a single phase formed according to the reaction

$$CsC_{24} + 2K \rightarrow K_2CsC_{24}$$

corresponding to the sequence ...C K C K C Cs C K C K C Cs... and the identity period $I_c = 2d_i(KC_8) + d_i(CsC_8) = 16.62\,\text{Å}$. The formation of this type of ternary has never been observed. On the basis of thermodynamic calculations, *Miyazaki* et al. [8.49] proved that the free energy of such a heterostructure would be much larger than that corresponding to the mixture of separate phases which is actually formed, mainly because of the entropy term, suggesting that this type of structure might be stable at a sufficiently low temperature.

Using similar experimental conditions but a longer time of reaction, *York* et al. obtained $KRbC_{16}$ (in admixture with KC_8), with the identity period $I_c = 5.35 + 5.65 = 11.00\,\text{Å}$, corresponding to the sequence ...C K C Rb C K C Rb... [8.48].

8.3.2 Structures and Reaction Mechanisms

In these heterostructures [8.48, 50], the metal atoms in the intercalated single-layer sheets form epitaxial 2d (2 × 2)R0° lattices on the adjacent layers of graphene. This arrangement gives rise to four possible positions for the intercalated layers with respect to the carbon layers, which form an $\ldots AAA \ldots$ rather than a $\ldots ABA \ldots$ sequence. In KC_8 and RbC_8, the sequence of layers is $\ldots A\alpha A\beta A\gamma A\delta A \ldots$, where α, β, γ and δ identify the nonsuperposed intercalated layers (Fig. 8.4), so that $c = 4d_i$. In CsC_8, the sequence is $\ldots A\alpha A\beta A\gamma A \ldots$ and c is now $3d_i$. It was therefore particularly interesting to find out which of the two possible modes of stacking would be found in the heterostructure. From the hkl reflections, the sequence was unambiguously found to be

$$\begin{array}{ccccccccccc} \ldots & A & \alpha & A & \beta & A & \gamma & A & \delta & A & \ldots \\ \ldots & \mathrm{C} & \mathrm{K} & \mathrm{C} & \mathrm{Cs} & \mathrm{C} & \mathrm{K} & \mathrm{C} & \mathrm{Cs} & \mathrm{C} & \ldots \end{array}$$

i.e., identical to the one in KC_8. The parameters of the 3d orthorhombic lattice are $a = 4.94$ Å, $b = 8.52$ Å, and $c = 22.54$ Å.

Examination of the reaction mixture after 2 hours, 2 days and 12 days revealed that CsC_{24} changes to the final two-phase mixture along the following lines:

- evolution of CsC_{24} towards a three-phase mixture of CsC_{24}, KC_8 and CsC_8;
- evolution to another three-phase system of $KCsC_{16}$, KC_8 and KC_{24};
- evolution to the final two-phase system $KCsC_{16}$ and KC_8.

In the initial stage of the reaction, the invasion of potassium atoms results in a regrouping of Cs atoms to form CsC_8. In the interlamellar space thus freed,

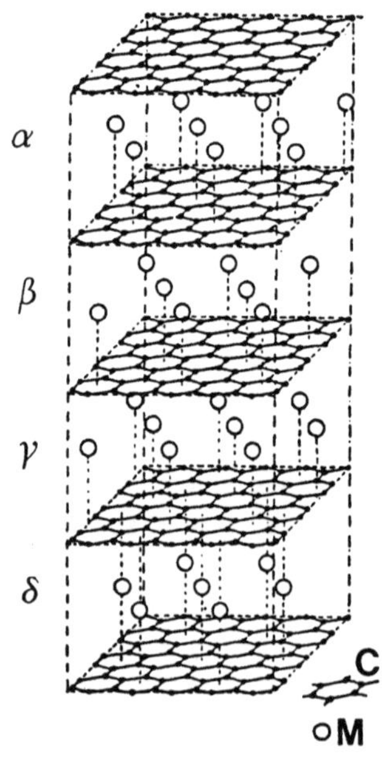

Fig. 8.4. The four equivalent intercalate sites labelled α, β, γ and δ, for the (2 × 2)R0° structure. On a given intercalate layer, only one of these sites is occupied. (From [8.9])

potassium then forms KC_8 and KC_{24}. Although it is tempting to imagine the next step as being merely a reaction between KC_8 and CsC_8 to form $KCsC_{16}$, it is clear that this reaction could not occur without some, as yet unknown, go-between to ensure the formation of the final highly ordered structure, the last step then being the gradual disappearance of KC_{24} to give KC_8 found in the final mixture.

From consideration of the energetics involved, it is likely that a sufficiently prolonged high-temperature annealing of the KC_8+ $KCsC_{16}$ mixture would result in the formation of a single $K_{0.67}Cs_{0.33}C_8$ phase in which the two alkali metals would form a solid solution. To our knowledge, this experiment has not yet been attempted.

8.4 The New Ternary Compounds with Multiple-Layer Intercalated Sheets

Most of the new phases recently synthesized belong to this category, and two lines of development are now evident: on the one hand, the synthesis of hydrogenated derivatives has been considerably extended by using the direct action of alkali metal hydrides on graphite, so that synthesis of sodium derivatives has now been obtained; on the other, the action of binary alloys containing a heavy alkali metal and a second metal of low electropositivity, whose direct intercalation seemed excluded, has also led to a large number of new and interesting phases.

8.4.1 Hydrogen Derivatives

(a) The Graphite–Potassium–Hydrogen Phases

The new method of preparation uses the direct action of a hydride such as KH on graphite [8.51–54]. The first advantage of this method is that the reaction is no longer restricted to powdered graphite, whereas the direct hydrogenation of KC_8 is only successful on powders of the binary [8.21–23]. The second advantage lies in the fact that the method offers direct access to the stage-1 hydrogenation compounds, since C/K is no longer limited to 8 but can be fixed by the initial amounts of reagents.

The reaction is performed in an evacuated and sealed glass tube in which platelets of graphite are buried under a large excess of hydride. Depending on the reaction temperature, a stage-1 or stage-2 compound can be obtained, at 300°–450°C or 250°–300°, respectively. Elimination of the excess hydride is simplified by the fact that it remains as a powder. Obviously, it is necessary to start with stoichiometric mixtures (KH + 4C or KH + 8C) when powdered phases are required. After about a week, the initially gray mixture is transformed into a homogeneous purple to violet or blue phase. Only the stage-1 and stage-2 compounds have been obtained in this way, independently of the type of graphite used. With lower reaction temperatures (200°C) X-ray diffractograms have shown the presence of small amounts of the stage-3 compound.

The compositions of the compounds have been determined by weight-uptake measurements and direct chemical analysis, while examination of the set of 00*l* reflections and *c*-axis dilation measurements yielded values of I_c and interplanar distances. Table 8.3 presents some of the relevant data. Note that the method of preparation affects the hydrogen content, which is 0.67 when the compounds are prepared by direct hydrogenation and 0.8 when the reaction with solid KH is used.

Table 8.3. Graphite-potassium-hydrogen ternary compounds

Stage	Chemical formula	Repeat distance, I_c [Å]	Interlamellar distance, d_i [Å]
1	$KH_{0.8}C_4$	8.53	8.53
2	$KH_{0.8}C_8$	11.88	8.53

The triple-layer sandwich-like structure of the intercalated sheet seems to be accepted quite generally, but there is no agreement yet as to the actual parameters of the cell. These have been given as a = 8.57 Å, b = 12.38 Å [corresponding to a $(2\sqrt{3} \times 5)$ rectangular 2d unit cell], c = 17.06 Å [8.52, 53, 55], or as either $(2 \times 2)R0°$ or $(\sqrt{3} \times \sqrt{3})R30°$ for the 2d unit cell [8.56], according to the preparation temperature.

(b) The Graphite–Sodium–Hydrogen Phases

The method of preparation of these hitherto unknown phases [8.53, 54] is identical to the one described involving KH. Here, too, a large excess of NaH is used, and elevated temperatures lead to metal-rich compounds while lower reaction temperatures yield higher stages.

It has not been found possible to isolate the pure stage-1 compound, which is always formed, at 500°C, as a blue mixture, with the stage-2 compound. The latter can be obtained as a pure phase at 450°C. Other ternaries, all gray but of stages 3–8, are formed at temperatures between 430° and 350°C.

Chemical analysis brings out the important fact that the Na/H ratio is always practically equal to 1, whereas the C/Na ratio is rather ill defined and can vary from $1.8 \times s$ to $4.5 \times s$ (s is the stage), depending on the preparation. These variations may be due to the presence of islands of sodium hydride (NaH) trapped within the structure of the compound, especially in the lower-stage derivatives prepared in the presence of a large excess of hydride. Their presence is demonstrated by X-ray Laüe photographs. The presence of free metal hydride may be one of the causes of the fluctuations in the chemical formula. Thus, d_i varies between 7.30 and 7.60 Å, and there may be a correlation between these fluctuations and the presence of hydride. For these reasons, the exact formula of these compounds is still unknown.

(c) Mechanisms of Formation of the Intercalated Hydrides

The intercalation of potassium hydride is a two-step process [8.53]. Initially, only potassium is intercalated (with perhaps small quantities of dissolved hydrogen), clearly indicating dissociation of the solid hydride. At this point, a binary GIC is formed, which is stage 2 if the temperature is 300°C or stage 1 if it is 400°C.

During the next step in the intercalation, hydrogen and further quantities of potassium are jointly intercalated, in an "iso-stage" process which does not modify the stage but brings the composition of the intercalated species to its final value and forms the triple K-H-K sheet:

$$\text{at } 300°\text{C} \qquad KC_{24} + 1.2H_2 + 2K \rightarrow 3KH_{0.8}C_8 \,,$$

$$\text{at } 400°\text{C} \qquad KC_8 + 0.8H_2 + K \rightarrow 2KH_{0.8}C_4 \,.$$

With sodium hydride, the mechanism is very different, since low-stage GICs of sodium do not exist, and there is simultaneous intercalation of hydrogen and of the metal. One should however remember that the presence of hydrogen gas was found to catalyze the intercalation of sodium [8.57].

(d) Chemical Stability of the Ternary Hydrogen Compounds

In general, the reactivity of the potassium-hydrogen GICs increases with the fineness of the sample, but these compounds are definitely more stable towards air than the corresponding binaries, even if the reaction with water is still quite vigorous. In contrast, the sodium compounds show a very low reactivity. Depending on the stage, these ternary GICs will remain unaltered in air for up to 7 weeks, and in water from a few hours to 3 weeks. Chemical stability [8.54] increases with the stage and is substantially improved by annealing the compounds after their preparation.

8.4.2 Compounds of Mercury, Thallium and Bismuth

In these ternaries [8.58], all the compounds contain a polylayered intercalated sheet in which an otherwise nonintercalatable metal is included. The first member of this series, and also one of the simplest, was synthesized in 1978 [8.59]. Its preparation paved the way for the synthesis of all the other ternaries described in this section.

(a) Preparative Methods

The reactive binary alloy [8.60–62] is obtained under vacuum by mixing the pure liquefied metals, in weighed quantities calculated to yield the required composition. The solid alloy is then powdered in an inert atmosphere and placed in a glass tube containing graphite (single crystal or HOPG). After evacuation, the tube is sealed and heated at a temperature sufficiently high to melt the alloy which will come into contact with the graphite. The quantity of alloy is always sufficiently large to ensure that any preferential intercalation of one of its components will not seriously affect the overall composition. At the end of the preparation, the sample is collected, again in an inert atmosphere, and carefully cleansed of the excess alloy usually stuck to its external surface.

Other methods can also be used with amalgams, since the vapor pressure of the components is sufficiently high that intercalation into graphite can be obtained from the vapor phase. Successive intercalation of the alkali metal and of mercury has also been used. Then, the C/M ratio is always 8 and cannot be changed, also, the amount of mercury distilled onto the binary must be exactly that required to give the correct final formula, with $Hg/M = 1$, and the ternary is necessarily stage 2. The main advantage here is that the compound does not need to be cleansed of excess alloy [8.60].

(b) Nature and Chemical Formulae of the Compounds

Mercury, thallium or bismuth do not form ternary GICs when alloyed with lithium or sodium; these are only obtained from alloys of these metals with the three heavy alkali metals. Only seven of the nine possible binary systems lead to isolatable ternaries with well-defined stages, chemical composition and interlamellar distances (Table 8.4), since the alloys of mercury and thallium with cesium are practically unreactive towards graphite [8.63]. Intercalation is only possible when these two metals are alloyed with the less electropositive potassium or rubidium [8.60, 61, 64]. The bismuth alloys of the three alkali metals all react, even if in somewhat different ways, giving the results shown in Table 8.4 [8.62, 65].

In most cases, two distinct phases, α and β, may be obtained, having the same stage number but slightly different interlamellar distances [8.63, 65, 66]; "β" is the name given to the phase with the larger value of the interlamellar distance. In some cases, only one of the two phases can be isolated, although both are seen in the X-ray diffractograms. In other cases, when both phases can be isolated, chemical and physical analysis of each is possible. Only in the stage-1 α-$CsBi_{0.53}C_4$ and β-$CsBiC_4$ have the two phases been found to have different compositions [8.62 65, 67]. Other cases of dimorphism correspond to different structural arrangements of the components (for instance, α- and β-$RbTl_{1.5}C_4$ [8.64]). Moreover, one of the two phases is usually thermodynamically unstable, but this instability is not systematically associated with the α or the β form (Table 8.4).

The difference in reactivity of the three heavy metals (Hg, Tl, Bi) clearly shows up in the formulae collected in Table 8.4 [8.62, 68]. With potassium, the intercalated multiple layers contain 60 at. % of Tl, 50 at. % of Hg, or 38 at. % of Bi. The heavy metal, much less electropositive than the alkali atoms, may perhaps be in competition with the graphene for a share of the electron donated by the potassium. If that is the case, the electron-acceptor characteristics of the three heavy metals can be looked upon as increasing in the order $Tl < Hg < Bi$. According to this view [8.62, 65], it would be the high electron affinity of bismuth that would permit the formation of the Bi-Cs-GICs (to be compared with the

Table 8.4. Mercury, thallium and bismuth ternary GICs

Alloy	Stage	Form	Interlamellar distance, d_i [Å]	Chemical formula	Color	Stability[a]
K-Hg	1	α	10.16	$KHgC_4$	Pink	+
	1	β*	10.76	-	Yellow	-
	2	α	10.16	$KHgC_8$	Blue	+
	3	α	10.16	-	-	+
Rb-Hg	1	α	10.76	$RbHgC_4$	Pink	+
	2	α	10.76	$RbHgC_8$	Blue	+
K-Tl	1	α	12.10	$KTl_{1.5}C_4$	Silver-gold	+
	2	α	12.10	$KTl_{1.5}C_8$	Pink	+
Rb-Tl	1	α	12.65	$RbTl_{1.5}C_4$	Silver-gold	-
	1	β	13.40	$RbTl_{1.5}C_4$	Silver-gold	+
	2	α	12.65	$RbTl_{1.5}C_8$	Pink	+
K-Bi	2	α	9.94	$KBi_{0.6}C_8$	Blue	+
	2	β	10.86	$KBi_{0.6}C_8$	Blue	-
	3	α	9.90	$KBi_{0.6}C_{12}$	Gray-blue	+
	4	α	9.87	-	Gray	+
	5	α	9.87	-	Gray	+
Rb-Bi	1	α	10.09	$RbBi_{0.6}C_4$	Violet	+
	2	α	10.16	$RbBi_{0.6}C_8$	Blue	+
	2	β*	10.52	-	-	-
	3	α	10.05	-	Gray-blue	+
	3	β*	10.52	-	-	-
	4	α*	10.05	-	Gray	?
	4	β*	10.52	-	Gray	?
	6	α*	10.05	-	-	?
	7	β*	10.52	-	-	?
Cs-Bi	1	α	10.60	$CsBi_{0.53}C_4$	Violet	-
	1	β	11.44	$CsBiC_4$	Blue-green	+
	2	α*	10.70	-	-	-
	2	β	11.50	$CsBiC_8$	Blue-green	+
	3	α*	10.70	-	-	?
	3	β*	11.50	-	-	?
	4	α*	10.70	-	-	?
	4	β*	11.50	-	-	?

* Compound that has not yet been isolated
[a] +: stable; -: metastable; ?: unknown

nonexistence of the Hg-Cs or Tl-Cs homologs), and also the formation of high-stage bismuth ternary compounds (to be compared with the existence of stages 1 and 2 only, with mercury and thallium).

Although this could be true for Tl and Bi, with electron affinities of 0.20 and 0.95 eV, respectively (the electron affinity of Hg is still unknown), there must also exist some as yet unknown additional factor since many other metals have affinities well within this range. Recent work [8.69] on the band-structure of $KHgC_4$ indicates that the bands lying astride the Fermi level are probably Hg 6*p* bands with a considerable admixture of K 4*s* states. Within the ternary compound, the heavy metal–alkali metal combination and the graphite must be competing for the valence electron of the alkali atom, with each carbon atom in the graphene layer being capable of accepting a certain maximum excess electric charge. If the charge on the C atoms is insufficient, there can be no intercalation but, on the other hand, it seems that if the charge is in excess of some threshold, only a binary compound will be formed.

(c) Influence of Alloy Composition and Reaction Temperature on the Nature of the Products Formed

The three liquid heavy metals (mercury, thallium and bismuth) are miscible in all proportions with the three liquid heavy alkali metals. At lower temperatures, the phase diagrams are more complicated, showing the formation of compounds, sometimes with congruent melting points [8.70]. Only alloys which do not contain an excess of either component lead to ternary GICs. When the atomic concentration of the alkali metal is too high, the stage-1 MC_8 binary is formed; when it is too low, no reaction occurs or, at best, a higher stage binary is formed. Figure 8.5 presents useful compositions for the seven binary systems known to react with graphite [8.58, 63, 65]. In most cases, the range of compositions can be divided into subunits, each corresponding to the formation of a particular stage of ternary GIC, as will be seen below. The center of gravity of the range changes from 65.6 to 57 and to 41.5 at.%, respectively, for the Tl, Hg and Bi alloys. This evolution is quite similar to the one found in the formulae of the ternaries, and it may be due to the same causes, see Sect. 8.4.2b.

The nature of the product formed by the reaction of the alloy with graphite depends on the temperature at which the reaction is performed, as well as on

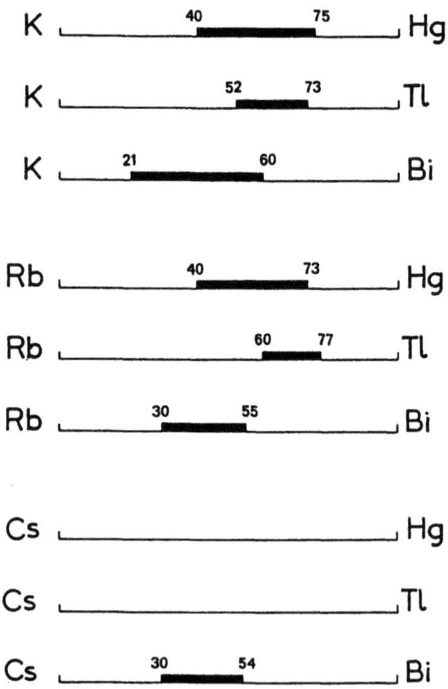

Fig. 8.5. Ranges of composition of the binary alloys leading to ternary phases with graphite. Only seven systems lead to ternary compounds; Cs-Hg and Cs-Tl do not give ternaries. (From [8.58])

the composition of the alloy. The minimum temperature is fixed by the need to liquefy the alloy and, in this respect, small variations in the composition of the alloy can have most important consequences: the melting points of the K-Bi alloys containing 30 and 40 at. % Bi are, respectively, 440° and 625°C [8.70]. Similarly, the temperatures of reaction are strongly dependent on the nature of the heavy metal: 300°C is the maximum permissible temperature in the K-Hg system, while many reactions with the K-Bi alloys are run at 700°C [8.58]. One should also bear in mind that, for most of the alloys, raising the temperature of the reaction results in the formation of higher-stage compounds. This behavior is similar to that met in the preparation of binary compounds, but unlike that needed in the preparation of the alkali-metal–hydrogen GICs.

Some of the compounds listed in Table 8.4 can only be obtained by stopping a preparation at the appropriate point. This is particularly true for the metastable phases which, by their very nature, cannot be the result of an equilibrium process. It is therefore impossible to give details here of the individual preparations, which will be illustrated by just two examples from the K-Hg and the Cs-Bi systems.

The influence of concentration on the stage of the compound obtained is particularly clear in the case of the potassium amalgams [8.58, 71]. As shown in Fig. 8.6, the concentration range 40–75 at. % Hg represents three sub-domains which produce GICs with stage number of respectively 1, 2 and 3. The systems Rb-Hg, K-Tl and Rb-Tl show similar behavior, but K-Bi and Rb-Bi behave differently [8.65] since high concentrations of alkali metal do not lead to the stage-1 compounds, probably because of the high temperature needed to melt the alloy.

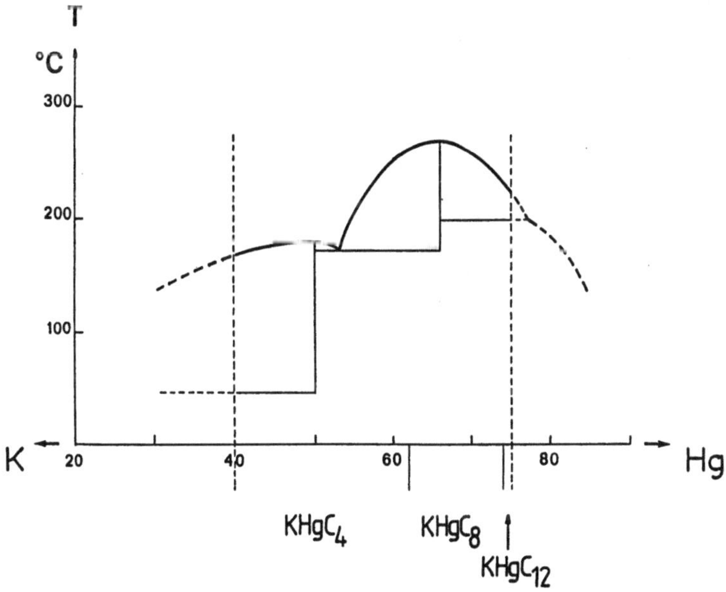

Fig. 8.6. Influence of the composition on the intercalation into graphite of potassium amalgams. (From [8.71])

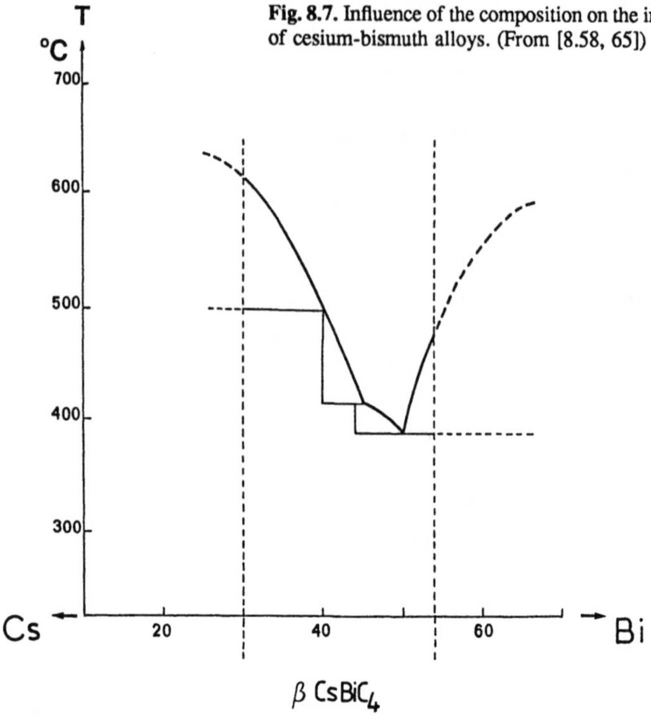

Fig. 8.7. Influence of the composition on the intercalation into graphite of cesium-bismuth alloys. (From [8.58, 65])

The Cs-Bi system is very simple, as there are no subsystems between 30 and 54 at. % Bi, and the stage-1 β-$CsBiC_4$ is always formed at equilibrium [8.65] (Fig. 8.7). Other phases can only be prepared by cutting short the time of reaction or, if higher stages are required, by raising the temperature far above the melting point of the alloy [8.65].

(d) Structures and Reaction Mechanisms

One of the simplest structures in this series of compounds is probably that of the mercurographitides [8.72–74]. In the K and the Rb derivatives, the basic structure of the $KHgC_{4\times s}$ and $RbHgC_{4\times s}$ phases is an intercalated M-Hg-M sheet (M = K or Rb) with the alkali metal layers forming a commensurate $(2 \times 2)R0°$ superstructure on the adjacent graphene planes, and with the K (or Rb) atoms of each layer directly above or below an alkali metal of the other layer. Within each prismatic trigonal site thus created between the two layers, one Hg atom is lodged. These mercury atoms are all disposed in the median plane (Fig. 8.8), or alternately above or below it [8.63]. Recent electron diffraction studies have shown, however, the presence of $(\sqrt{3} \times 2)R(30°, 0°)$ and of $(\sqrt{3} \times \sqrt{3})R30°$ superlattices, the former being particularly associated with the stage-1 K-Hg GIC – form β (Table 8.4) [8.74]. The three-dimensional stacking sequence for $MHgC_4$ has been found to be identical with that of KC_8 or RbC_8, with $c = 4d_i$. The corresponding 3d lattices are therefore orthorhombic, with a = 4.92 Å, b =

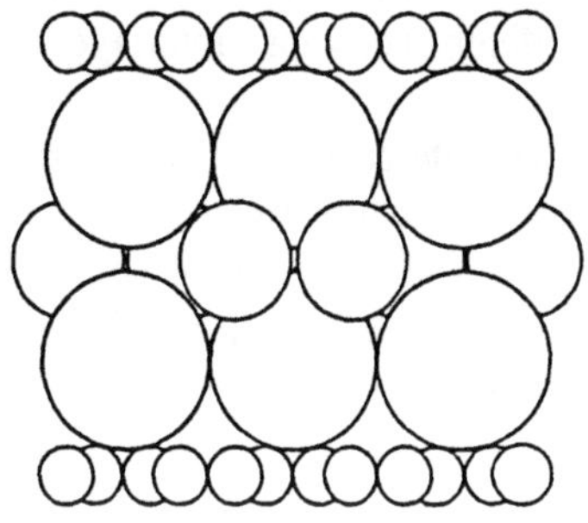

Fig. 8.8. Structure of a triple layer in $KHgC_4$. (From [8.72])

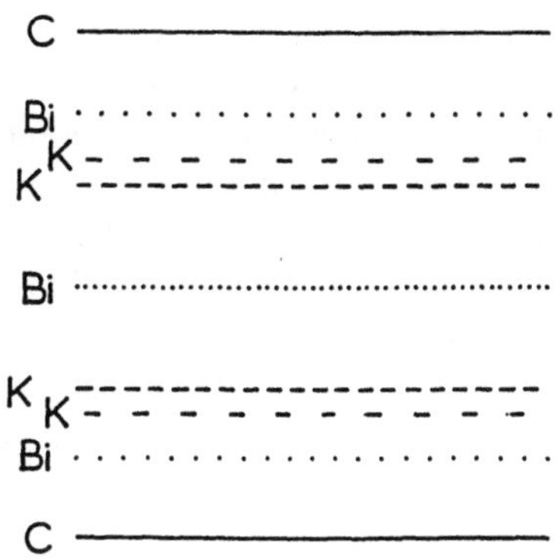

Fig. 8.9. Schematic diagram illustrating the structure of a multiple-layer sheet in a K-Bi graphite intercalation compound. (From [8.65])

8.52 Å and $c = 40.64$ Å or 43.04 Å respectively for $KHgC_4$ or $RbHgC_4$. Because their structures are so similar, the two compounds are miscible in all proportions [8.63]; the c value of the quaternary $K_xRb_{1-x}HgC_4$ increases regularly with x from 40.64 to 43.04 Å with hardly any deviation from Vegard's law.

Little is known of the structure of the thallium compounds, but considering the rather large value of d_i, the structure of the intercalated multiple sheet may well be more complex than that of the mercurographitides.

A little more is known about the bismuthographitides [8.65, 67, 75]. The first structure proposed also involved a triple M-Bi-M layer, but it seems that it is necessary to admit the presence of additional metallic layers. As an example (Fig. 8.9), the intercalated sheet of the stage-2 α-$KBi_{0.6}C_8$ presents a bismuth central layer, four symmetrical potassium layers and, in the neighborhood of the graphene planes, two symmetrical bismuth layers containing a small quantity of metal. On the other hand, the 2d arrangement of the metallic sheets depends largely on the nature of the alkali metal, the stage number and the form (α or β) of the various phases. The 2d unit cell of the metallic sheet can be commensurate or not with respect to that of graphite, either oblique or hexagonal, and is generally large in size. The increase of the stage number seems to promote the commensurability between these two sublattices (metallic and graphitic).

The reaction mechanisms have only been tackled for the compounds richest in intercalated alloy. The reaction occurs in two steps [8.68, 76]. Initially, selective intercalation of the alkali metal yields a binary which is stage 1 (MC_8) with the Hg or Tl alloys, stage 4 (MC_{48}) with Cs-Bi, and stage 5 (MC_{60}) with K-Bi or Rb-Bi. This is followed by the simultaneous intercalation of both components of the alloy; with thallium or mercury, this is an isostage process during which the intercalated single layers are transformed into multiple-layer sheets:

$$MC_8 + 2Hg + M \rightarrow 2MHgC_4 ,$$

$$\text{or} \quad MC_8 + 3Tl + M \rightarrow 2MTl_{1.5}C_4 .$$

This does not happen with Bi, which is intercalated in the course of a polystage process during which a stage-4 or 5 ternary compound is formed, depending on the alkali metal present, after which the empty interlayer spaces are progressively filled up.

Mercury and thallium can therefore only be intercalated when the graphite is already saturated with the alkali metal, while bismuth is taken up long before that point is reached. In other words, the high electron affinity of Bi enables it to compete quite early on with graphite for the electron, but Hg or Tl must wait until the layers of graphene are saturated before they can start capturing some of the available charge.

(e) Chemical Stability

The thallium derivatives seem to be the most stable of these ternary compounds. They are practically inert towards water and air and can therefore be handled without any particular precautions [8.63]. The bismuth compounds are somewhat more fragile. They are slowly oxidized in air and decomposed by water [8.65]. The mercury compounds should not be exposed to air as they are quite easily oxidized and decomposed by water [8.63].

(f) Remarks Concerning the Superconducting Character of the Alkali Metal–Heavy Metal GICs

All researchers have tended to find the same values of critical temperatures for the mercury and thallium ternary compounds. However, several samples of $KHgC_8$, which presents a critical temperature of 1.9 K, became superconducting only at lower temperatures [8.77]. The investigators who observed this phenomenon, proposed a hydrogenation treatment in order to restore the regular critical temperature.

The bismuth compounds present greater difficulties. Several samples of α-$CsBi_{0.53}C_4$ obtained by *Lagrange* et al. [8.5] indeed became superconducting at 4.05 K, the samples prepared by *Chaiken* et al. [8.78] did not show any superconducting transition, and those prepared by *Stang* et al. [8.79] became superconducting (at T_c = 4.7 K) because of the presence in the samples of very small quantities of the binary alloy $CsBi_2$, whose critical temperature is precisely 4.75 K. It is probable that, as in numerous other systems, some weak fluctuations of chemical composition in these bismuth compounds are able to induce very strong variations of critical temperature.

8.5 Conclusion and Vistas

The number of new phases synthesized in the last six or seven years is considerable, probably exceeding the number of compounds known at the beginning of that period of time. The methods by which these new phases were obtained were all original, even if tentative approaches in the same direction had previously been made. The methods involved:

- subjecting binary compounds to extreme conditions of temperature or pressure;
- successive action of two different reagents on graphite, under conditions sufficiently mild to avoid reaction between co-intercalated species;
- action on graphite of alloys or original combinations of metals, one of which would not, by itself, be intercalated but is likely to form new structures with the alkali metal if co-intercalation occurs, hence profoundly modifying both the structure and the physical properties of the old combination.

These methods have not yet been fully exploited; they will undoubtedly be used in the future to obtain many new intercalation compounds. Yet, however interesting and original these methods might appear, it is likely that additional ways of synthesis will also be found, which will yield a crop of new lamellar species in the years to come.

References

8.1 M.S. Dresselhaus, G. Dresselhaus: Adv. Phys. **30**, 139 (1981)
8.2 K. Suzuki, I. Tsujikawa, M. Kobayashi, H. Inokuchi, Y. Oda, A. Sumiyama, H. Nagano, Y. Kimishima: Synth. Met. **12**, 389 (1985)
8.3 Y. Iye, S. Tanuma: Phys Rev. B **32**, 4583 (1982)
8.4 R.A. Wachnik, L.A. Pendrys, F.L. Vogel, P. Lagrange: Solid State Commun. **43**, 5 (1982)
8.5 P. Lagrange, A. Bendriss-Rerhrhaye, J.F. Marêché, E. McRae: Synth. Met. **12**, 201 (1985)
8.6 Y. Koike, S. Tanuma, H. Suematsu, K. Higuchi: J. Phys. Chem. Solids **41**, 1111 (1980)
8.7 M. Kobayashi, I. Tsujikawa: J. Phys. Soc. Jpn. **50**, 3245 (1981)
8.8 K. Fredenhagen, G. Cadenbach: Z. Anorg. Allg. Chem. **158**, 249 (1926)
8.9 W. Rüdorff, E. Schulze: Z. Anorg. Allg. Chem. **277**, 156 (1954)
8.10 A. Hérold: Bull. Soc. Chim. Fr. 999 (1955)
8.11 E. McRae, D. Billaud, J.F. Marêché, A. Hérold: Physica B **99**, 489 (1980)
8.12 R.C. Asher, S.A. Wilson: Nature **181**, 409 (1958)
8.13 A. Métrot, D. Guérard, D. Billaud, A. Hérold: Synth. Met **1**, 363 (1979)
8.14 D. Guérard, A. Hérold: C. R. Acad. Sci. C **275**, 571 (1972)
8.15 D. Guérard, A. Hérold: C. R. Acad. Sci. C **279**, 455 (1974)
8.16 D. Guérard, A. Hérold: C. R. Acad. Sci. C **280**, 729 (1975)
8.17 D. Guérard, A. Hérold: C. R. Acad. Sci. C **281**, 929 (1975)
8.18 D. Billaud, A. Hérold: Bull. Soc. Chim. Fr. 103 (1972)
8.19 D. Billaud, A. Hérold: Bull. Soc. Chim. Fr. 2407 (1974)
8.20 D. Billaud, A. Hérold: Bull. Soc. Chim. Fr. 2715 (1974)
8.21 D. Saehr, A. Hérold: Bull. Soc. Chim. Fr. 3130 (1965)
8.22 M. Colin, A. Hérold: Bull. Soc. Chim. Fr. 1982 (1971)
8.23 P. Lagrange, A. Hérold: Carbon **16**, 235 (1978)
8.24 M. Colin, A. Hérold: C. R. Acad. Sci. C **269**, 1302 (1969)
8.25 P. Lagrange, A. Métrot, A. Hérold: C. R. Acad. Sci. C **278**, 701 (1974)
8.26 P. Lagrange, D. Guérard, A. Hérold, D.G. Onn: Proc. 15th Biennial Conf. on Carbon, Philadelphia, PA (1981) p. 383
8.27 D. Billaud, A. Hérold: Bull. Soc. Chim. Fr. 131 (1978)
8.28 D. Billaud, A. Hérold: C. R. Acad. Sci. C **281**, 305 (1975)
8.29 F.J. Salzano, S. Aronson: J. Chem. Phys. **47**, 2978 (1967)
8.30 B. Carton, A. Hérold: Bull. Soc. Chim. Fr. 1337 (1972)
8.31 B. Carton: Thesis, Nancy (1971)
8.32 R. Clarke, N. Wada, S.A. Solin: Phys. Rev. Lett **44**, 1616 (1980)
8.33 N. Wada, R. Clarke, S.A. Solin: Synth. Met. **2**, 27 (1980)
8.34 J.E. Fischer, H.J. Kim: Phys. Rev. B **35**, 6826 (1987)
8.35 J.E. Fischer, C.D. Fuerst, K.C. Woo: Synth. Met. **7**, 1 (1983)

8.36 K.C. Woo, H. Mertwoy, J.E. Fischer, W.A. Kamitakahara, D.S. Robinson: Phys. Rev. B **27**, 7831 (1983)
8.37 J.E. Fischer, H.J. Kim: Synth. Met. **12**, 137 (1985)
8.38 H. Miyazaki, Y. Kuramoto, C. Horie: J. Phys. Soc. Jpn. **53**, 1380 (1984)
8.39 K.I. Semenenko, V.V. Andrev, V.E. Mordkovitch: Dokl. Akad. Nauk SSSR **271**, 1402 (1983)
8.40 R. Nishitani, Y. Uno, H. Suematsu, Y. Fujii, T. Matsushita: Phys. Rev. Lett. **52**, 1504 (1984)
8.41 H. Minemoto, H. Suematsu: Synth. Met. **12**, 33 (1985)
8.42 H. Suematsu: Ann. Phys. **11**, 75 (1986)
8.43 R. Nishitani, K. Suda, H. Suematsu: J. Phys. Soc. Jpn. **55**, 1601 (1986)
8.44 J.M. Bloch, H. Katz, D. Moses, V.B. Cajipe, J.E. Fischer: Phys. Rev. B **31**, 6785 (1985)
8.45 H.J. Kim, J.E. Fischer: Phys. Rev. B **33**, 1329 (1986)
8.46 B. Sundqvist, J.E. Fischer: Phys. Rev. B **34**, 3532 (1986)
8.47 B.R. York, S.K. Hark, S.A. Solin: Phys. Rev. Lett. **50**, 1470 (1983)
8.48 B.R. York, S.K. Hark, S.A. Solin: Synth. Met. **7**, 25 (1983)
8.49 H. Miyazaki, C. Horie, T. Watanabe: J. Phys. Soc. Jpn. **53**, 3843 (1984)
8.50 S.K. Hark, B.R. York, S.A. Solin: J. Chem. Phys. **82**, 921 (1985)
8.51 D. Guérard, C. Takoudjou, F. Rousseaux: Proc. Conf. Carbone 82, Bordeaux (1982) p. 85
8.52 D. Guérard, C. Takoudjou, F. Rousseaux: Synth. Met. **7**, 43 (1983)
8.53 D. Guérard, N.E. Elalem, C. Takoudjou, F. Rousseaux: Synth. Met. **12**, 195 (1985)
8.54 N.E. Elalem, D. Guérard, F. Rousseaux, G. Loupias, H. Estrade-Szwarckopf, P. Lauginie, J. Conard: Proc. Conf. Carbon '86, Baden-Baden (1986) p. 433
8.55 W.A. Kamitakahara, G. Doll, P.C. Eklund: In *Graphite Intercalation Compounds*, Proc. MRS Fall Meeting, Boston 1986, ed. by M.S. Dresselhaus, G. Dresselhaus, S.A. Solin (MRS, Pittsburgh, PA 1986) p. 69
8.56 L. Salamanca-Riba, N.C. Yeh, M.S. Dresselhaus, M. Endo, T. Enoki: J. Mater. Res. **1**, 177 (1986)
8.57 M.L. Dzurus, G.R. Hennig, G.L. Montet: Proc. 4th Carbon Conf. (Pergamon, Oxford 1960) p. 165
8.58 P. Lagrange: J. Mater. Res. **2**, 839 (1987)
8.59 M. El Makrini, P. Lagrange, D. Guérard, A. Hérold: C. R. Acad. Sci. C **288**, 303 (1979)
8.60 M. El Makrini, P. Lagrange, D. Guérard, A. Hérold: Carbon **18**, 203 (1980)
8.61 M. El Makrini, P. Lagrange, A. Hérold: Carbon **18**, 374 (1980)
8.62 P. Lagrange: In *Graphite Intercalation Compounds*, Proc. MRS Fall Meeting, Boston 1986, ed. by M.S. Dresselhaus, G. Dresselhaus, S.A. Solin (MRS, Pittsburgh, PA 1986) p. 155
8.63 M. El Makrini: Thesis, Nancy (1980)
8.64 P. Lagrange, M. El Makrini, A. Hérold: C. R. Acad. Sci. C **290**, 283 (1980)
8.65 A. Bendriss-Rerhrhaye: Thesis, Nancy (1986)
8.66 P. Lagrange, M. El Makrini, A. Bendriss: Synth. Met. **7**, 33 (1983)
8.67 M.H. Yang, P.C. Eklund: In *Graphite Intercalation Compounds*, Proc. MRS Fall Meeting, Boston 1986, ed. by M.S. Dresselhaus, G. Dresselhaus, S.A. Solin (MRS, Pittsburgh, PA 1986) p. 161
8.68 A. Bendriss-Rerhrhaye, P. Lagrange, A. Hérold: Proc. Conf. Carbon '86, Baden-Baden (1986) p. 436
8.69 L. Senbetu, H. Ikezi, C. Umrigar: In *Graphite Intercalation Compounds*, Proc. MRS Fall Meeting, Boston 1984 (MRS, Pittsburg, PA 1984) p. 74
8.70 M. Hansen: *Constitution of Binary Alloys* Mat. Sci. and Eng. Ser. (McGraw-Hill, New York 1958)
8.71 P. Lagrange, M. El Makrini, D. Guérard, A. Hérold: Physica B **99**, 473 (1980)
8.72 P. Lagrange, M. El Makrini, A. Hérold: Rev. Chim. Min. **20**, 229 (1983)
8.73 P. Lagrange: Carbon **22**, 579 (1984)
8.74 M.H. Yang, P.C. Eklund, W.A. Kamitakahara: In *Graphite Intercalation Compounds*, Proc. MRS Fall Meeting, Boston 1984 (MRS, Pittsburgh, PA 1984) p. 125
8.75 A. Bendriss-Rerhrhaye, P. Lagrange, F. Rousseaux: Synth. Met. **23**, 89 (1988)
8.76 P. Lagrange, M. El Makrini, D. Guérard, A. Hérold: Synth. Met. **2**, 191 (1980)
8.77 G. Roth, A. Chaiken, T. Enoki, N.C. Yeh, G. Dresselhaus, P.M. Tedrow: Phys. Rev. B **32**, 533 (1985)
8.78 A. Chaiken, N.C. Yeh, G. Dresselhaus, P.M. Tedrow: In *Graphite Intercalation Compounds*, Proc. MRS Fall Meeting, Boston 1986, ed. by M.S. Dresselhaus, G. Dresselhaus, S.A. Solin (MRS, Pittsburgh, PA 1986) p. 158
8.79 I. Stang, K. Lüders, V. Geiser, H.J. Güntherodt: Synth. Met. **23**, 371 (1988)

9. Ternary Systems

Ralph Setton

In the ternary intercalation compounds of graphite, two of the three components which form the system are intercalated in the third. In Sects. 9.1–4, the compounds described contain the two components as solid solutions of various types: there are true homogeneous dispersions of component A in B; others contain atoms or molecules of type A in strong interaction – somewhat similar to solvation – with atoms (or molecules) of component B. Even if the interactions are sufficiently strong to form a new intercalated species, the homogeneity of the dispersion legitimizes their classification as a solid solution. In the compounds described in Sect. 9.5, each intercalated component is localized in fairly extensive domains free of atoms (or molecules) of the other kind, thus forming "bi-intercalation" or, if there are more than two intercalated species, "multi-intercalation" compounds. In Sect. 9.6, the compounds described contain one of the species intercalated in a fluorinated carbon lattice formed by the reaction of the second "intercalated" species (namely fluorine) on the graphite matrix. Strictly speaking, this series should not be classified with the intercalation compounds of graphite. Nevertheless, in view of its direct filiation from graphite and of the similarity in the mode of preparation, it was deemed worthwile to retain its description.

The ternary systems containing an alkali metal and hydrogen, or two alkali metals forming a bi-intercalation compound, or the ternary multilayer compounds containing an alkali and mercury, thallium or bismuth are fully described in Chap. 8, and will therefore not be considered here. We will focus here on the synthesis and chemistry of ternary GICs. This chapter is thus complementary to the recent extensive review of the physics of ternary GICs by *Solin* and *Zabel* [9.1].

9.1 Ternary Compounds Containing Only Alkali Metals or Halogens

9.1.1 Solid Solution Ternaries Involving Two Alkali Metals

A number of ways have been found to obtain the simultaneous intercalation of two alkali metals in graphite, and the choice of the method depends mainly on the identity of the pair whose action is sought. Not all of the ten possible binary combinations for alkali metals have been tried. In some cases, the formation, in the alloy, of high melting point intermetallic compounds limits the range of usable compositions, while in others – and this is especially true of the lithium alloys – the high temperatures needed to work with the gaseous or molten alloy might lead to the formation of carbides.

Some of the ten binary combinations (such as Li/Na and Li/K) are known to give pure binary graphite intercalation compounds rather than ternary solid solution intercalates. When the intercalate is a true solid solution, i.e., a homogeneous and uniform distribution of two intercalated species in the interlayer spaces, the composition of the intercalated mixture $A_{1-x}B_x$ is, in general, different from that of the free mixture $A_{1-y}B_y$; the two compositions are connected by the thermodynamic relation [9.2]

$$x/(1-x) = K(T)y/(1-y) \tag{9.1}$$

with

$$\begin{aligned} RT\ell\mathrm{n}K &= \Delta G^0(B\mathrm{C}_8) - \Delta G^0(A\mathrm{C}_8) \\ &= \Delta H^0(B\mathrm{C}_8) - \Delta H^0(A\mathrm{C}_8) - T[\Delta S^0(B\mathrm{C}_8) - \Delta S^0(A\mathrm{C}_8)] \,, \end{aligned}$$

where G, H and S are respectively the free energy, the enthalpy and the entropy, T is the absolute temperature, and where it is assumed that both alkali metals A and B give, when alone, binaries of the form $M\mathrm{C}_8$. Lack of ideality can manifest itself by deviations from *Vegard's* law, anomalies in the magnetic susceptibility or in the phonon spectra when the composition changes.

(a) Ternary Compounds Containing Lithium or Sodium

No ternary compounds containing solid solutions of lithium and of another alkali metal are known. Dilute solutions (50 ppm) of potassium in lithium yield, at 400°C, a mixture of the two stage-1 binaries KC_8 and LiC_6, while the action of K (or Li) vapor on LiC_{12} (or KC_{24}) also yields a mixture of the two stage-1 compounds [9.2, 3].

The liquid Na/K and Na/Cs alloys react with graphite to form $M_x\mathrm{Na}_{1-x}\mathrm{C}_8$, which are true solid solution ternaries with $x > 0.45$ [9.3, 4]. The basal plane spacings show positive deviations from *Vegard's* law, i.e., they are greater than the values calculated by proportional interpolation between the values for the pure sodium and potassium or cesium binaries. The compounds are metastable: prolonged contact with an excess of the liquid phase leads to a gradual replacement of sodium by the heavy alkali metal. The electrical resistivities of the ternary compounds are generally lower than the resistivities of the corresponding pure binaries.

(b) Solid Solution Ternaries Involving Two Heavy Alkali Metals

The three binary alkali metal alloys formed from K, Rb and Cs give stage-1 ternaries $M_xM'_{1-x}C_8$ which are found to exhibit considerable deviations from the ideal behavior: all three I_c vs x relationships are S-shaped rather than straight, as required by *Vegard's* law [9.2]. Similarly, measurements of the magnetic susceptibility show enhancements of the mean paramagnetism for the values $x = 0.3$–0.5 in $Cs_{1-x}K_xC_8$, $x = 0.6$–0.8 in $K_{1-x}Rb_xC_8$ and $x = 0.7$–0.9 in $Cs_{1-x}Rb_xC_8$ [9.5–7][1]: the temperature-independent paramagnetism is multiplied by a factor of 2 or 3 in the first two ternaries for the given values of x. Later work found anomalies in the phonon-Raman spectra of $K_{1-x}Rb_xC_8$, manifested by an asymmetric softening of the graphite phonon modes at the M-point for $x \simeq 0.67$ [9.9], in the value of the elastic constant C_{33} [9.10] and in the characteristics of the line shapes of the diffraction spectra [9.11]. However, other studies on the same ternary – such as new measurements of the magnetic susceptibility, the in-plane C-C distance, or the XPS [9.12], or even new measurements of the interlayer spacing as a function of x [9.13] – failed to reveal anomalies around the "critical" concentration. Equally negative results were obtained by a calculation of the density of states, which gave a monotonic variation as x changed from 0.1 to 0.9 in $Cs_{1-x}K_xC_8$ [9.14] but later work by the same authors revealed a slight splitting at the Fermi level in the DOS for $0.5 < x < 0.7$ in $Cs_{1-x}K_xC_8$ [9.15]. Other authors were able to account for the elastic anomalies in $K_{1-x}Rb_xC_8$ by showing that first-principles band-structure calculations led to the presence of a cusp-like *Van Hove* singularity due to a three-dimensional saddle point in the uppermost energy bands of the two alkali metals near the *Fermi* level [9.16].

The explanation of these discrepancies may lie in the results of a new study on $K_{1-x}Rb_xC_8$ [9.17] in which it was shown that the anomalous softening of the phonon modes and the deviations from *Vegard's* law are only found in samples prepared from HOPG, not in samples obtained with natural single crystals. The anomalies would then be artifacts due to the presence of void defects in HOPG or in multidomain natural crystals, which would preferentially incorporate one of the two metallic species. This process, in which the law of mass-action would favor retention of whichever alkali metal is present in the higher concentration, could explain the sigmoid d_I vs x curves observed, while the combined effects of the presence of voids and the difference in size of the two ions [9.7] would account for the anomalies observed in the magnetic susceptibility and the elastic properties in some – but not in all – of the sample.

If this finding is confirmed, one may legitimately be concerned about the interpretation of a number of previous measurements which probed the electronic properties of samples prepared from HOPG. Obviously, much work is still needed to erase all doubts attached to this question.

---

[1] In accordance with the rules of nomenclature and terminology of graphite intercalation compounds [9.8], the two metals in the ternary are cited in alphabetical order; this is not always the order chosen by the authors of the monographs cited, and this should be taken into account when referring back to the original results

9.1.2 Solid Solution Ternaries Involving Two Halogens Other than Fluorine

The four halogens, from fluorine to iodine, can form six binary systems. In each of them, true interhalogen compounds are found, many of which can be intercalated into graphite. These have all been recently reviewed [9.18] and will therefore not be further described or examined here.

The simultaneous action of Cl_2 and Br_2 on graphite has been found to give stage-2 compounds of mean composition $C_{8.2}Br_{0.525}Cl_{0.475}$, with $I_c = 10.2$ Å. The ratio Br:Cl = 0.53:0.47 in the compound is maintained for values of the ratio in the halogenating mixture which vary between 0.75:0.25 and 0.5:0.5 [9.19–21]. When the difference between the temperature of the solid phase and that of the halogenating mixture was increased to about 60°C, a stage-4 compound $C_{16.3}Br_{0.58}Cl_{0.42}$ was also prepared.

The Cl/I system is slightly more complicated since the stable compounds ICl and ICl_3 exist, of which at least the first is known to give a GIC [9.22]. One should therefore consider three separate ternary systems: (a) the system I_2/ICl, (b) the system ICl/ICl_3, and (c) the system ICl_3/Cl_2 [9.23]; in all cases, the pressure and the composition of the vapor depend only on the temperatures of the solid phase and of the halogenating mixture.

The following list summarizes the results obtained with each system and with the two interhalogen compounds:

graphite/I_2/ICl:	$C_{6.7}Cl_{0.52}I_{0.48}$ (also C_8 compound)
graphite/ICl:	$C_{4-6}Cl_{0.55}I_{0.45}$ (also C_8 compound)
graphite/ICl/ICl_3:	$C_{3.9}Cl_{0.57}I_{0.43}$ (also C_8, C_{12} compounds)
graphite/ICl_3:	$C_{4-5}Cl_{0.60}I_{0.40}$ (also C_{12} compound)
graphite/ICl_3/Cl_2:	$C_{7.4}Cl_{0.78}I_{0.22}$

In all the compositions given above, there is an excess of chlorine with respect to the nearest integral value of the Cl/I ratio. In general, all the compounds in C_4 are stage 1, with $I_c = 7.09$–7.14 Å, those in C_8 are stage 2, with $I_c = 10.24$–10.44 Å, while those in C_{12} are stage 3, with $I_c = 13.8$ Å [9.24].

The only interhalogen compound found in the Br/I system is IBr, so that only two domains need be considered: (a) the system I_2/Br_2 and (b) the system IBr/Br_2. The results obtained [9.25] are quite similar to those described above:

graphite/I_2/IBr:	$C_{11.2}Br_{0.55}I_{0.45}$
graphite/IBr:	$C_{7.9}Br_{0.55}I_{0.45}$
graphite/IBr/Br_2:	the solid phase is gradually enriched in Br and tends towards the graphite-Br_2 binary compound.

In all the compounds, there seems to be a "preferred" nonstoichiometry, which does not, however, systematically favor the more electronegative of the two halogens present, as would normally be expected. The small quantity of excess halogen, which is of the order of 0.1 atom, is probably associated with the necessity of providing a negative charged halide counter-ion (Hal^{-1}) to be opposed to the positively charged layers of graphene. If this is indeed the case, the nature of the excess halogen is then dictated by the value of the free energy associated with the reaction

$$0.5(\mathrm{Hal})_2(\mathrm{g}) + \mathrm{e}^- \rightarrow \mathrm{Hal}^-(\text{condensed state}) \tag{9.2}$$

for which the values of the free energies are, approximately, $\Delta G = -237$, -233 and -228 kJ respectively for $\mathrm{Hal}^{-1} = \mathrm{Br}^-$, Cl^- and I^-. In other words, both binary systems containing bromine as one of the halogens will best be stabilized by forming Br^- ions rather than Cl^- or I^-, while the Cl/I system will form Cl^- ions. Furthermore, the heats of promotion to the two states of carbon described by the equation

$$\mathrm{C}(\mathrm{tr\,tr\,tr}\,\pi) \rightarrow \mathrm{C}^+(\mathrm{tr\,tr\,tr}) + \mathrm{e}^- \tag{9.3}$$

(in which, using the *Mulliken* notation, tr describes an electron in trigonal hybridization) are, respectively, 28.3 and 75.0 kJ, as established by *Hinze* and *Jaffe* [9.26]. Assuming now that the entropy of a graphene layer is the same as that of the charged layer, the only entropy change in the reaction is due to the formation of the free electron. The free energy of the reaction then works out to $\Delta G = 46.7$ kJ. By comparing this value to the free energy of formation of the charged ion, it is obvious that, within the rather gross approximations of this simple computation, there should be about 46.7/237, or about 1/5 of an ionized bromine (or chlorine) atom, a value in satisfactory agreement with the excess found experimentally.

9.2 Ternary Compounds of Metals with a Nitrogen-Containing Base

9.2.1 Ternary Metal–Ammonia–Graphite Intercalation Compounds

Metal–ammonia–graphite intercalation compounds were probably the first ternary GICs to be prepared and recognized as such. By dissolving an alkali metal (AM) (Li, Na, K, Rb or Cs) in liquid NH_3 (LNH_3), *Rüdorff* et al. [9.27] obtained the blue metal–ammonia solution, which produced, in a fast reaction with powdered graphite, a ternary $M(\mathrm{NH_3})_x\mathrm{C}_y$. Another method used sucessfully with the first-stage binary compounds consisted in letting them react with pure LNH_3; the ternary GIC thus formed contained less metal than the starting binary since some of the AM was always leached out. However, this did not happen with the higher-stage binaries which retained all the metal originally present. Table 9.1 gives the composition and I_c values of the AM derivatives obtained. All the stage-1 compounds are dark blue, while the known stage-2 derivatives are grayish blue. The K, Rb and Cs ternaries were reported to be quite stable under nitrogen, in contrast to the Li and Na compounds which were reported to decompose spontaneously after a few days, releasing hydrogen and forming the metal amide.

Table 9.1. Ternary graphite intercalation compounds of alkali metals with ammonia and ammonia derivatives

Compound[a]	Stage	I_c [Å]	d_{CC} [Å]	References and remarks
$Li(NH_3)_{1.6}C_{10.6}$	1	6.62	1.429	[9.28]
$Li(NH_3)_{1.7}C_{28.8}$	2	9.94	1.433	[9.28]
$Na(NH_3)_{2.2}C_{16.5}$	1	6.64	1.4302	[9.28]
$Na(NH_3)_{2.3}C_{26.9}$	2	9.95	1.4264	[9.28]
$K(NH_3)_{2.1}C_{12.5}$	1	6.56	1.4272	[9.28]
$K(NH_3)_{2.3}C_{23.4}$	2	9.95		[9.27] Small amounts of stage 1
$Rb(NH_3)_{2.1}C_{11.9}$	1	6.58		[9.27] Average composition, stage 1
$Cs(NH_3)_{2.2}C_{12.8}$	1	6.58		[9.27]
$K(NH_3)_{4.33}C_{23.6}$	1	6.62		[9.29]
$K(NH_3)_xC_y$	2	9.94		
$K(NH_3)_{x'}C_{y'}$	2	9.65		[9.30] Composition depends
$K(NH_3)_{x''}C_{y''}$	3	13.3		on method of preparation
$Rb(NH_3)_{4.1}C_{23.3}$	1	6.64		[9.29]
$Cs(NH_3)_{3.65}C_{23.5}$	1	6.67		[9.29]
$Li(MeNH_2)_{1.8}C_{12.9}$	1	6.66	1.437	[9.27]
	2	2 × 9.95	1.426	[9.27]
$Li(NH_2CH_2CH_2NH_2)C_{28}$	2	11.85		[9.31]
$Li(HMPT)C_{32}$	1	4 × 7.48		[9.32]
$Na(HMPT)C_{27}$	1	4 × 7.63		[9.32]
$K(K222)C_{36}$	1	15.50		[9.33, 34]
$K[N\,Et_3]_{0.43}C_{36}$	1	12.40	1.4247	[9.34]

[a] Me = CH_3; HMPT = $(NMe_2)_3PO$; K222 = $N(CH_2CH_2OCH_2CH_2OCH_2CH_2)_3N$; Et = C_2H_5

The values of I_c for the stage-1 and stage-2 ternaries given in Table 9.1 do not seem to depend on the nature of the AM, except perhaps for a slight difference in the values for the stage-1 Li and Na compounds, with $I_c \simeq 6.63$ Å, and those shown by the other three alkali metals, with $I_c \simeq 6.57$ Å; in fact, the measurements led *Rüdorff* et al. to the conclusion [9.27, 31] that I_c was determined by the bulk of NH_3. Yet, recent accurate measurements [9.30, 35] on compounds obtained by allowing gaseous NH_3 at 0.85 MPa to react with stage-1 MC_{24} GICs show a systematic variation in I_c and in the composition as the AM changes [9.36], as shown in Table 9.1. The value of I_c is also highly sensitive to P_{NH_3}, the pressure of gaseous NH_3 above the ternary GIC, and it changes from 6.515 Å to 6.633 Å as P_{NH_3} is modified from about 4×10^{-1} to 10^6 Pa, with the stage-2 compounds showing similar variations [9.30, 35]. Note that there are important differences in the methods of preparation used by *Rüdorff* et al. [9.27] and by *Solin* and co-workers [9.36]. In the earlier preparations, the final composition (corresponding to $x \simeq 2.1$, C/K $\simeq 12.5$ and $I_c = 6.56$ Å) was obtained by letting graphite react with an excess of K/NH_3 solution and allowing the washed solid phase to reach its final composition by equilibration against gaseous NH_3 at 1 atm, whereas, in the new preparations, C/M is fixed by the composition of the binary reacting with the gaseous ammonia. In fact, the weight vs P_{NH_3} measurements clearly indicate a phase transition from stage-2 to stage-1 with increasing pressure for $P_{NH_3} > 0.1$ MPa, with a corresponding change from $x \simeq 2$ to $x > 4$ [9.35]. The stage-3 and stage-4 binaries behave in much the same way [9.30] and a tentative Gibbs phase diagram has been suggested to represent the domains of existence of the mixtures met under different conditions [9.37].

The deammoniation of the ternary compounds is a complex process which varies with the metal component: *Akuzawa* and co-workers [9.38] found that:

- Li and the alkaline earths (Sect. 9.2.2) have a strong tendency to form the amide;
- Na and K release the greater part of the ammonia, but also form higher-stage and/or residue compounds, depending on the temperature and the type of graphite used in the preparation of the binary;
- Rb and Cs release all the ammonia in the ternary compound.

On the basis of these observations, they prepared a new series of ternary M–NH_3–GICs (with M = Na or K) by the following procedure [9.38, 39].

(1) Using weighed quantities of metal and graphite, stage-1 or stage-2 compounds $M(NH_3)_2C_{12}$ were prepared by condensing gaseous NH_3 on the two components.
(2) The initial ternary GIC thus formed was vacuum-deammoniated at different temperatures, under 0.1 mPa.
(3) The deammoniated (and exfoliated) residue compound was submitted at 350°C to the vapor of the metal from a source kept at 300°C in the usual two-bulb apparatus for 48 h, upon which a new metal-rich compound was formed.

The new reddish yellow compounds were formulated $K(NH_3)_xC_{5.3}$ (with $x \simeq$ 0.03–0.08, $I_c \simeq 5.4$ Å and a minor component with $I_c \simeq 8.3$ Å, and $Na_4(NH_3)C_{28}$ ($I_c \simeq 10.7$ Å). With sodium, compounds with different compositions were formed when HOPG or low-HTT carbons were used as starting materials. Two interesting features of these compounds are, on the one hand, their remarkable stability in air, water or even some acid solutions and, on the other hand, the high metal content: C/M is only 7 for the Na compound, and even as low as 5.2 for the K GIC. In the latter case, this is all the more remarkable in view of the fact that the interlamellar distance is given as being practically identical to that of KC_8. There are, however, important differences between the intensities of the reflections in the published X-ray diffractograms of these NH_3-AM-GICs [9.38, 39] and that of KC_8 [9.40], and future research may reveal less simple structures for these ternary GICs.

9.2.2 Other Ternary Compounds of Metals with a Nitrogen Base

Table 9.2 gives the characteristics of a number of ternary GICs in which the metal co-intercalated with ammonia is not an AM. Most of them were prepared

Table 9.2. Ternary graphite intercalation compounds of metals (other than alkali metals) with ammonia

Compound	Stage	I_c [Å]	d_{CC} [Å]	References and remarks
$Be(NH_3)_{0.77}C_{6.1}$	3	12.24	1.4271	[9.31, 41]
$Mg(NH_3)_{2-3}C_{32}$	4	15.95		[9.31]
$Mg(NH_3)_{5.2}C_{40\pm5}$	~ 6.0	~ 16.0		[9.42]
$Ca(NH_3)_{2.2}C_{12.1}$	1	6.62	1.428	[9.31, 43]
$Ca(NH_3)_{4.1}C_{26.5}$	2	9.95		
$Sr(NH_3)_{2.4}C_{11.3}$	1	6.36		[9.31, 43]
$Sr(NH_3)_{3.4}C_{11.3}$	2	9.87		[9.31, 43]
$Ba(NH_3)_{2.5}C_{10.9}$	1	6.36		[9.31, 43]
$Ba(NH_3)_{3.9}C_{28.3}$	2	9.79		[9.31, 43]
$Ba(NH_3)_xC_y$	1	6.316		[9.44]
$Ba(NH_3)_{x'}C_{y'}$	2	9.78		[9.44]
$Eu(NH_3)_{3.31}C_{17.5}$	2	9.80		[9.41]
$Eu(NH_3)_{3.9}H_{0.9}C_{32}$	1?	5.54		[9.45]
$Al(NH_3)_{1.3-1.6}C_{8-10}$	4	16.0	1.4263	[9.41]
$Yb(NH_3)_{1.9}H_{0.3}C_{23.6}$	2?			[9.41][a]

[a] Compounds of Gd, Tb, Dy, Ho, Er, and Tm that apparently contain about 0.5 mole of NH_4I have been prepared, but these give unexploitable X-ray data.

by *Rüdorff* and co-workers either by the action on graphite of a solution of the metal in LNH_3 (Ca, Sr, Ba, Eu or Yb), or by the electrolysis of a suspension of graphite in a solution of the metal iodide in LNH_3 (Be, Al, Mg, Gd, Tb, Dy, Ho, or Er) [9.41]; the action of the naphthalene-Mg complex in LNH_3 was also used later [9.42]. The ternary Al and Eu GICs were found to be different from the others in that no hydrogen was evolved when they were treated with water.

Other nitrogen bases can be intercalated with metal atoms, as shown in Table 9.1; in most cases, data on the physical or chemical properties of these compounds is rather sparse.

9.2.3 Physical Properties of the Ternary Metal–Base Compounds

Some of the dimensional effects associated with the presence of variable quantities of NH_3 in the ternary GICs have already been mentioned.

The in-plane X-ray diffraction pattern of the K-NH_3 layer in the stage-1 $K(NH_3)_{4.33}C_{24}$ was measured [9.46, 47] and found to be in semiquantitative agreement with the scattering calculated for a random distribution of coplanar hard disks corresponding to K^+ ions placed at the center of squares formed by four coordinating NH_3 molecules, with about 2.9 Å between the center of K^+ and each NH_3 disk, and with the balance of NH_3 uncoordinated. The intercalated layer is the 2d analog of the 3d structures met in the blue K/NH_3 solutions in which the K^+ ions are sixfold coordinated with solvating ammonia molecules. The intercalated layer behaves essentially like a simple 2d liquid at room temperature, with no correlation between the intercalated layers. Further work [9.35] has emphasized the role of elastic energy in the relation linking interlayer distance with ammonia concentration, while ^{13}C NMR measurements at 15 MHz [9.48] and optical reflectance studies on $K(NH_3)_{4.33}C_{24}$ [9.49] confirmed that part of the charge is back-transferred to the intercalated layer.

Equally significant is the finding that, in the same ternary GIC, there is a marked drop in a-axis resistivity at about $P_{NH_3} = 0.5$ MPa [9.50], which is the 2d analog of the metal–insulator transition in bulk 3d metal–ammonia solutions.

9.3 Metal–Organic-Molecule–Graphite Intercalation Compounds

The group of ternary graphite derivatives containing an AM and an organic molecule (other than ammonia or one of its derivatives) has been intensively investigated. Four methods of preparation were used to obtain the compounds in this series, but the following point should be stressed: in two of the methods to be described, the solid ternary GIC is filtered from the liquid suspension in which it was formed and dried under specific conditions of pressure and temperature. This may lead to compounds different from the ones obtained by other methods, owing to the fact that the dynamic drying process eliminates an unknown proportion of the volatile third component. Many of the apparent discrepancies in results reported by different authors can be traced to this fact, since this elimination does not occur in other preparations.

(1) A solution of an AM complex in the appropriate solvent is allowed to react on graphite. At the end of the reaction, both the AM and solvent are found to be intercalated, sometimes accompanied by traces or, in some cases, substantial quantities of the "activating molecule" used to form the complex. The solid phase is filtered off, washed with some of the pure solvent, and dried. The soluble complexes used are the AM-naphthalene ion radical combination in tetrahydrofuran (THF) [9.51, 52] or dimethoxyethane (DME) [9.52], or the AM-benzophenone ketyl in THF [9.53], or cobalt (0) complexes of the type $[L_3L'\ \mathrm{Co}]$ in which L is a phosphane, such as $P(CH_3)_3$, and L' is an olefin [9.54]. In nonpolar solvents (e.g. pentane), the soluble complex of this molecule with alkali metals yields binary GICs, but solutions in benzene or toluene lead to ternary GICs. With some activating molecules, the GIC formed is not a ternary but rather a quaternary compound: this happens, for instance, with benzonitrile in THF when both molecules are simultaneously intercalated with Li [9.55].

(2) A variation on the first method consists in allowing a solution of the activating molecule to react on a binary GIC. Some of the intercalated metal is leached out and replaced by molecules of the solvent [9.56], after which the solid phase is isolated. The amount of metal removed from the binary depends on the reduction potential [9.57] and the electron affinity [9.58] of the hydrocarbon used as the activated molecule; a wide variety of compounds

can be prepared by the judicious choice of the starting binary and of the hydrocarbon.

(3) Electrolysis in the presence of a graphite cathode or merely of suspended graphite of a salt soluble in propylene carbonate (PC), nitromethane (NM), dimethylsulfoxide (DMSO) or DME yields ternary GICs [9.59, 60]. The stepwise changes of the potential at the graphite cathode indicate the successive formation of stages of the GIC, and it is possible to stop the electrolysis at any point. The overall stoichiometry of the organic component (which is not related to the total amount of current passed) is obtained by weighing the solid phase after centrifugation to eliminate the excess of adsorbed solvent [9.61].

(4) The simplest method of preparation is the direct action of the third component on a binary AM GIC. There is as yet no hard and fast rule to predict whether a given organic molecule will be intercalated into a specific binary: for instance, both benzene (Bz) [9.62] and THF [9.63] are absorbed into KC_{24} but not into KC_8. Exploratory studies of the possibility of a reaction between a number of organic molecules (cyclic aromatic or olefinic hydrocarbons, heterocycles, epoxides, etc.) and KC_8 or KC_{24} revealed that a reaction occurred in most cases, but very seldom with the first binary [9.64], and that the organic molecule-GICs are probably all metastable complexes whose equilibrium is disrupted by deintercalation, high temperatures, or mere aging. (The problems associated with chemical reactions involving co-intercalated species will be discussed later.) Note, in this respect, that the formation of a complex between the two nongraphitic components, with the inevitable and simultaneous redistribution of electric charges and with no possibility of modifying the global C/M ratio, may lead to a mixture of phases or be the very cause of the instability often observed, since the number of graphitic C atoms may not be the one required to provide or support the counter-charge needed for proper thermodynamic equilibrium.

The kinetics of intercalation of the ligand from the gaseous phase can be obtained by a number of methods. One of the simplest involves the use of a cathetometer to monitor the height of the meniscus in a calibrated capillary of constant cross section containing the liquid ligand at a well-controlled temperature. Knowledge of the amount of binary involved in the reaction and of the density of the liquid ligand gives direct access to the global composition during the course of the intercalation [9.65]. The use of neutron diffraction is more sophisticated but pays greater dividends: full diffraction sweeps can be obtained in less than two minutes [9.66].

9.3.1 Ternary Donor Compounds Containing Aromatic Hydrocarbons

(a) Kinetics of Formation and Structure

The mechanism of intercalation of benzene into KC_{24} is now well known: under a pressure of about 300 torr from liquid benzene at 16°C, the powdered stage-2 binary at 50°C will form $K(Bz)_{1.2}C_{24}$ in less than an hour, with I_c changing from 8.8 to 12.6 Å, corresponding to a stage-2 ternary compound. When the temperature of the binary is only slightly above that of the benzene, the reaction goes much further and the stage-1 $K(Bz)_{2.2}C_{24}$ is obtained in an equally short time. Even then, it has been shown by neutron diffraction studies [9.68] that the stage-2 compound is always formed before the stage-1 one, which has I_c = 9.24 Å. This value changes to 9.31 Å over a period of 100 days [9.69] but the cause of this change is likely to be the irreversible chemical reactions occurring among the co-intercalated species, rather than a rearrangement of the intercalated layer, as was believed originally.

Early work on this ternary GIC suggested two possibilities for the position of the benzene molecules in the interlamellar space: in the first, the principal plane of the hydrocarbon was taken to be parallel to the layers of graphene, with a metal ion sandwiched between a molecule of benzene and a layer of graphene [9.70]; in the second, the Bz molecules, placed perpendicularly to the graphene, acted as spacers, leaving the metal ion in the cavities bounded by two planes of graphene and the perpendicular walls of benzene [9.71]. The fact that the ternary GICs prepared from K, Rb or Cs all gave extremely similar diffractograms and nearly identical values of I_c (namely 9.24–9.29 Å) ruled out the first possibility, in which I_c should change with the diameter of the ion. Later, detailed geometrical calculations and comparison of the computed and experimentally observed intensities of the (00ℓ) series of reflections indicated that the Bz molecules are canted at an angle of 60° with respect to the graphene and that the K^+ ion is in a cavity created by two benzene rings tilted towards each other, at the intersection of their hexad axes, and in maximum interaction with the two rings and a plane of graphene [9.69]. This structure was confirmed by real-time neutron diffraction studies on the intercalation of deuterobenzene into RbC_{24} [9.72].

A characteristic of the benzene ternary GICs is the in situ formation of biphenyl which occurs on standing: for a short while after $K(Bz)_{\sim 1}C_{24}$ or $K(Bz)_{\sim 2}C_{24}$ are formed from the binary, the intercalated hydrocarbon can readily be "washed away" unchanged by merely submitting the ternary compound to the repeated action of THF vapor [9.67]. After aging, THF will no longer replace all the benzene and prolonged cryopumping leaves a "residue" GIC $K(Bz)_{2/3}C_{24}$. Hydrolysis of the solid phase shows that biphenyl and a number of its hydrogenated derivatives were present in the ternary compound, as confirmed by Raman spectroscopy of the ternary prior to hydrolysis [9.73]. However, since some of the benzene could be deintercalated by cryopumping under vacuum, another structure has been suggested for the K-Bz-GICs [9.74], which distinguishes between "strongly" and "weakly" bonded benzene, the latter being that removed by pumping. In this mode, the structure of the ternary is stage 1 with respect to Bz molecules and stage 2 with respect to K. The "weakly bonded" benzene is that present in the interlamellar space devoid of potassium, whereas the "strongly bonded" hydrocarbon is directly in contact with the intercalated

K. The occurrence of two proton spin–lattice relaxation times (T_1) was given as further proof of the presence of two types of Bz molecules [9.75].

The main argument presented in favor of this model is that "the existence of two types of benzene molecules in (the ternary K-Bz-GIC) is difficult to explain in terms of the (previous) models, since ... the benzene molecules are coordinated with potassium in approximately the same way". In spite of this, there is now a vast body of knowledge that speaks against the new structure. Thus, deintercalation of a large part of the benzene from a ternary, shown by X-ray diffraction to be stage 2, is possible [9.67]. According to the model under discussion, this corresponds to Bz molecules in the metal-free spaces, i.e., to a stage 1 compound whose presence was apparently not detectable in the diffractogram. Another argument is provided by the intensities of the (00ℓ) reflections in the diffractograms [9.76], which point unambiguously to the model presented previously, in which the intercalated species are uniformly distributed in all the available space. A third argument is found in the results of the ternarization of the stage-4 KC_{48}: according to the new model, "the motive force enabling the intercalation of benzene into KC_{24}, and its stabilization in the interlayer of graphite, is the ion–dipole interaction between the negatively charged layers of graphite and the C-H bonds of the benzene molecules". One can then expect the ternarization of a stage-4 compound to proceed according to the scheme

$$\ldots G\,G^*\,K\,G\,G\,G\,G^*\,K\,G\,G\ldots$$

$$\xrightarrow{Bz}\ldots G\,Bz\,G^*\,(Bz+K)\,G\,Bz\,G\,G\,Bz\,G\,(Bz+K)\,G^*\,Bz\,G\ldots$$

in which the Bz molecules invade all the interlamellar spaces adjacent to negatively charged graphene layers G, and in which I_c (the distance between the G* layers) changes from $5.33 + 3(3.35) = 15.43$ to $3(9.24) + 3.35 = 31.05$ Å (where 9.24 Å is the contribution of the GBz or G(Bz+K) layers). The ternarization of KC_{48}, performed under conditions similar to the ones used to obtain the ternary from KC_{24}, led to a compound with $I_c = 12.68$ Å, i.e., a value corresponding to a stage-2 compound with the structural sequence

$$\ldots GG^*(Bz+K)GG^*(Bz+K)G\ldots$$

with $d_1 = 9.33$ Å, not very different from the range of values found in the stage-1 compound [9.76]. In conclusion, the new model does not seem to give a correct description of the Bz ternaries. The presence of two minima in the T_1 vs $1/T$ plot could be due to any one of the three following causes, none of which requires the structure suggested: the presence of Bz in close interaction with the K^+ ions and of "free" Bz, all in the same interlamellar space; or the presence of Bz and biphenyl; or the effect of the bounded two-dimensional motion of the Bz molecules, which has been shown experimentally and theoretically [9.77] to give rise to such multi-minima curves.

Other benzenoid hydrocarbons have been found to react with KC_{24}, as shown in Table 9.3, which summarizes some of the structural characteristics of the

Table 9.3. Ternary graphite intercalation compounds containing aromatic hydrocarbons

Compound[a]	Stage	I_c [Å]	d_{CC} [Å]	References and remarks
$K(Bz)_{2.8-3}C_{24-26}$	1	9.27–9.31		[9.62, 71, 75]
$K(Bz)_{\sim 2.2}C_{24}$	1	9.27–9.30	1.42[b]	[9.64, 67–69]
$K(Bz)_yC_{48}$	2	12.6		[9.52, 62]
$K(Bz)_{1.2}C_{24}$	2	12.2–12.6		[9.68, 75, 77]
$K(Bz')_{0.66-1}C_{24}$	2	12.2		[9.67, 75]
$Rb(DBz)_yC_{24}$ [c]	1	9.29	1.424	[9.71]
$Rb(Bz)_yC_{24}$ [c]	2	12.6		[9.71]
$Cs(Bz)_yC_{24}$	1	9.29		[9.71]
$K(Tol)_3C_{24}$	1	9.08		[9.54]
$K(Tol)_{1.2}C_{24}$	1	9.14		[9.78]
$K(Tol)_yC_n$	2	12.37		[9.54]
$K(Tol)_{0.7}C_n$	2	12.3		[9.78]
$K(Tol)_yC_n$	3	15.72		[9.54]
$K(Tol)_{0.7}C_{24}$	4	19.01		[9.54]
$K(O\text{-}Xyl)_{2.0}C_{24}$	1	10.60		[9.78]

[a] Bz = C_6H_6; Bz′ denotes compounds derived from benzene; DBz = C_6D_6; Tol = $C_6H_5CH_3$; O-Xyl is O-$C_6H_4(CH_3)_2$

[b] In [9.68], the a and b parameters of RbC_{24} give d_{CC} = 1.415 Å, but d_{CC} is given as 1.421 Å; d_{CC} is reported to increase by 0.004 Å when Bz reacts with RbC_{24}.

[c] In [9.71], all three alkali metal derivatives have $2.75 < y < 3.31$.

ternary GICs thus formed. Note that the toluene derivative $K(C_6H_5CH_3)_{1.2}C_{24}$ also forms a residue compound $K(C_6H_5CH_3)_{0.7}C_{24}$ [9.78], probably through a process similar to the one described above, and that the presence of the bulky methyl group does not affect the overall formula of the compound.

(b) Chemical Properties

The chemical properties of the K-Bz-GICs are dominated by the chemical reactions spontaneously occurring among the co-intercalated species. Attempts at removing the benzene from an aged stage-1 $K(Bz)_{2.2}C_{24}$ always result in the formation of the "residue compound" $K(Bz')_{2/3}C_{24}$ [9.67], in which the notation Bz′ indicates that the organic moiety is no longer the original hydrocarbon. Heating the residue compound releases large quantities of hydrogen, while hydrolysis yields a mixture which contains over 70 % of biphenyl, with the balance composed of a number of hydrogenated biphenyl derivatives, all of them having one fully aromatic ring. (A similar mixture is formed upon hydrolysis of the compound $K(THF)_{0.4}(Bz')_{0.37}C_8$ which will be discussed later.) A detailed study of the energetics of the transformation of benzene into the set of derivatives found was conducted by *Setton* [9.79], who showed that the hydrogen formed by the self-condensation of Bz into biphenyl must be fixed, partly on some of the available aromatic double bonds and partly as "intercalated potassium hydride $KH_{2/3}$". In fact, the global reactions occurring in the interlamellar space must be

$$2C_6H_6 \rightarrow C_6H_5 - C_6H_7\ , \quad \Delta H \simeq 121\,\mathrm{kJ}\ , \tag{9.4}$$

which needs the driving force provided by the reaction

$$2C_6H_6 + 3K \rightarrow C_6H_5 - C_6H_5 + 3KH_{2/3}\,, \quad \Delta H \simeq -127\,\text{kJ}\,. \tag{9.5}$$

Note that, at this point, there are three fully aromatic rings available; if each is coordinated with a K^+ ion, the ratio of the total number of Bz molecules to total K atoms will be 4/6, as found experimentally. Recent work with KC_{24} and RbC_{24} [9.80] showed that the coupling of benzene can proceed as far as the formation of terphenyl and quaterphenyl, though the latter is formed in only very small quantities.

Using a solution in pentane of the complex formed by the cobalt (0) compound with potassium, *Besenhard* was able to enrich a stage-1 K-Bz-GIC (or its toluene homologue) in potassium, thus transforming the dark blue ternary into a dark red compound, with hardly any change in the interlayer spacing. However, the same treatment applied to a stage-2 K-Tol-GIC yielded a new compound with $I_c = 14.47$ Å, which probably corresponds to the structure ...GKG(Tol)GKG... (Tol denotes toluene) [9.54].

(c) Physical Properties

Since biphenyl is spontaneously and gradually formed by mere aging of the ternary Bz-GIC, it should be borne in mind that many of the results reported refer to mixtures containing benzene and biphenyl derivatives in unknown proportions.

As determined by differential scanning calorimetry (DSC), the formation of $K(Bz)_{2.5}C_{24}$ from *gaseous* Bz and KC_{24} is exothermic, so that $\Delta H = -38$ kJ [9.81]. The same technique also reveals a second-order transition at 280 K, not far from the melting point of free benzene.

At room temperature, the ^{1}H NMR derivative absorption of the compound richest in Bz, $K(Bz)_{\sim 2}C_{24}$, shows a single line of peak-to-peak linewidth $\Delta H_{pp} =$ 2 G, independent of temperature down to 77 K [9.69]. The main difference with bulk benzene lies in the absence of isotropic motion of the Bz molecules: the K^+ ions seem to pin the aromatic rings so that only reorientation around the hexad axis is allowed. The fact that $K(Bz')_{2/3}C_{24}$, which no longer contains Bz molecules but rather biphenyl, shows a much broader signal with $\Delta H_{pp} = 6$ G is particularly significant [9.76].

High-field ^{13}C NMR spectra have also been obtained for some of these derivatives [9.76]. The δ_{ab} component of the absorption in the spectrum of $K(Bz)_{2.5}C_{24}$ is found at -28 ppm (with respect to a liquid benzene standard), indicating a 16 % loss of charge on the graphene C atoms as compared to KC_{24}, where it occurs at 33 ppm. Unfortunately, because of the ^{13}C/H interaction, the contribution of the organic moiety to the absorption is not visible. The situation changes with $K(C_6D_6)_{\sim 1}C_{24}$: firstly the δ_c component, which was at +65 ppm in the previous spectrum, is now at +50 ppm, indicating a partial loss of the charge on the graphene; furthermore, a new component at -40 ppm is found, probably associated with the presence of charge on the benzene rings, now visible thanks to the replacement of the protons by deuterium. With $K(Bz')_{2/3}C_{24}$, there is a very slight upshift of δ_c to +53 ppm, but since the quantity of Bz′ is only 2/3 that in $K(C_6C_6)_{\sim 1}C_{24}$, there must be a correspondingly larger amount of charge on the aromatic rings.

The EPR spectra of the nonresidue compounds are Dysonian, with $\Delta H_{1/2}$ values always smaller than those in KC_{24} [9.69]. This shows an important decrease of the spin-orbit coupling, probably due to the shielding of the K^+ ion provided by the organic ligand.

9.3.2 Ternary Metal Compounds of Oxygen Derivatives

The ligands involved in these compounds can be aliphatic (tetrahydrofuran, dimethoxyethane, etc.) or unsaturated (furan). Their preparation, which presents no particular difficulty, follows the methods outlined in Sect. 9.3.

(a) Structural Characteristics and Kinetics of Formation

Among all the compounds in this series (Table 9.4), those containing THF are probably the most studied and best known. They share, with those containing DME, a particular feature in that it is often possible to obtain, for a given stage, two different nonisomeric phases whose characteristics are, for the so-called "lean" or "ligand-poor" phase, THF/K $\simeq$ 1.2 and $d_1 \simeq 7.2$ Å, and for the "ligand-rich" phase, THF/K $\simeq$ 2.5 and $d_1 \simeq 8.9$ Å. In the lean phase, the mean plane of each THF is parallel to the graphene layer, whereas it is perpendicular to them in the rich phase [9.93 – 95]. Because of the closer packing achieved in the second position and the gain in volume due to the larger interlayer spacing, the THF/K ratio is higher; similarly, the larger dimension of the THF molecule compared to its thickness as it stands on its oxygen atom accounts very nicely for the increase in d_1. In both cases, the M^+ ion is roughly at the same distance (2.7–2.8 Å) from the oxygen atom of THF, at least for M = Rb [9.86, 96].

The kinetics of formation of the ternary compounds by the action of THF vapor on KC_{24} has been followed by X-ray diffraction on a HOPG platelet [9.97], and on a powder sample (containing large quantities of KC_8) by neutron diffraction [9.66]. The initial intake of THF results in the rapid disruption of the structure of the binary, as judged by the disappearance of the (003) reflection, strongest in its spectrum. A transient new phase appears, formulated $K(THF)_{\sim 1}C_{24}$ (n = 2, ||), i.e. stage-2 with the THF molecules parallel to the graphene and I_c = 10.73 Å, which gradually disappears to the benefit of another transient phase $K(THF)_{\sim 2}C_{24}$ ($n = 2, \perp$), with the THF molecules perpendicular to the graphene and I_c = 12.12–12.2 Å. This second transient phase, which forms extremely slowly, is not observed in the neutron diffraction spectra from powders since the reaction is about 35 times faster than with HOPG-based binaries. On the other hand, small amounts of $K(THF)_{\sim 1}C_{24}$ (n = 1, ||), with I_c = 7.15 Å, are visible using the latter technique; they soon disappear, leaving only the rich phase $K(THF)_{\sim 2}C_{24}$ ($n = 1, \perp$), I_c = 8.84–8.89 Å.

Table 9.4. Ternary and quaternary graphite intercalation compounds of alkali metals with oxygen-containing molecules

Compound[a]	Stage	I_c [Å]	d_{CC} [Å]	References and remarks
$Li(THF)_{2.27}C_{207}$	1	7.24		[9.51]
		12.45		[9.51]
$Li(THF)_{1.4}C_6$	1	12.53		[9.82]
$Li(THF)_{2.3}C_{12}$	1	12.44		[9.82]
$Li(THF)_{3.4}C_{18}$	1	12.45		[9.82]
$Na(THF)_{3.57}C_{32}$	1	7.24		[9.51]
$Na(THF)_1C_{64}$	2	14.68		[9.51]
$K(THF)_{1.1-1.6}C_{24-37}$	1	7.16–7.18	1.4213	[9.51, 56, 83]
$K(THF)_{\sim1}C_{24}$	1	7.15	1.428	d_{CC} = 1.421 in KC_{24} [9.66]
$K(THF)_{1.1-1.7}C_{48-49.5}$	2	10.56	1.4216	[9.56, 83]
$K(THF)_{1-1.9}C_{72-75}$	3	13.86–13.92	1.4218	[9.53[b], 56, 83]
$K(THF)_{2.2}C_{32}$				[9.63, 84]
$K(THF)_{2.2-2.5}C_{24}$	1	8.84–8.89	1.4216	[9.56, 65]
$K(THF)_{\sim2}C_{24}$	1	8.84	1.427	d_{CC} = 1.421 in KC_{24} [9.66]
$K(THF)_{2.5}C_{48}$	2	12.27	1.4221	[9.83]
$K(THF)_{\sim2}C_{24}$	2	12.12		[9.66]
$K(THF)_{2.6}C_{72}$	3	15.64		[9.83]
$Rb(THF)_{\sim1}C_{24}$	1	7.16	1.429	d_{CC} = 1.425 in RbC_{24} [9.66]
$Rb(THF)_{\sim2}C_{24}$	2	10.43		[9.66]
$Rb(THF)_{\sim2}C_{24}$	1	8.89	1.429	[9.66]
$Rb(THF)_{2-3}C_{24}$	2	8.90–8.95		[9.85, 86]
$Rb(THF)_{\sim2}C_{24}$	2	12.17		[9.66]
$Cs(THF)_3C_{24}$	1	8.95		[9.85]
$Cs(THF)_{1.75}C_{24}$	1	7.046		[9.87, 88]
$Cs(THF)_xC_{24}$	2	10.192		[9.88]
$K(DOX)_{0.91}C_{36}$	1	7.30		[9.63]
$K(DOX)_yC_{36}$	2	10.65		[9.63]
$K(DOX)_{y'}C_{36}$	3	14.0		[9.63]
$K(DOX)_zC_{37}$	1	8.9		[9.89]
$Li(DME)_{2.8}C_{33}$	1	7.32		[9.90] Both phases present
	1	11.62		simultaneously
$Na(DME)_{2.8}C_{32}$	1	7.26		[9.90] Both phases present
		11.77		simultaneously
$Na(DME)C_{32}$	1	7.25		[9.91]
$K(DME)_{2.1}C_{31}$	1	7.28		[9.90] Both phases present
	1	11.94		simultaneously
$K(DME)C_{32}$	1	7.24		[9.91]
$Li_2(BzN)(THF)_{1.5}C_{54}$	1	7.44		[9.91]
$K(THF)_{0.4}(Bz)_{0.37}C_8$	1	9.00	1.4275	[9.93, 109]
$K(THF)_x(Bz)_yC_{24}$	1	9.00	1.4281	$x \simeq 1.2$, $y \simeq 1$ [9.111]

[a] THF: tetrahydrofuran; DOX: dioxan; DME: dimethoxyethane, BzN: benzonitrile
[b] Recalculated

During this intercalation process, small quantities of KC_8 were found to be formed (in excess of any of this binary initially present) [9.66] which did not disappear at the end. Partial deintercalation of THF from the rich phase to form the lean phase increased the amount of KC_8. Similar occurrences were observed during equivalent experiments on Rb homologues. This was interpreted as proof that the value of the M/C ratio must be greater in the rich phase than it is in the lean phase, and that deintercalation to form the latter must occur according to

$$\begin{aligned}\mathrm{K(THF)}_{x_1}\mathrm{C}_{24} &\rightarrow (1-y)[\mathrm{K(THF)}_{x_2/(1-y)}\mathrm{C}_{(24-8y)/(1-y)}] \\ &+ y\mathrm{KC}_8 + (x_1 - x_2)\mathrm{THF}\,.\end{aligned} \tag{9.6}$$

In fact, it was argued that the limiting and "ideal" formulas of the two phases would be $M(\mathrm{THF})_2\mathrm{C}_{30}\,(n = 1, \|)$ and $M(\mathrm{THF})_2\mathrm{C}_{20}\,(n = 1, \perp)$. This interesting hypothesis warrants further verification, especially since many of the stage-1 ternary K-THF-GICs prepared from solutions of the metal tend to show values of the C/K ratio often higher than 24 (Table 9.4). Nevertheless, if these formulas are correct, it is hard to understand why deintercalation of the rich phase $\mathrm{K(THF)}_{2.2}\mathrm{C}_{24}$ does not end by forming the mixture $[(3/11)\mathrm{K(THF)}_2\mathrm{C}_{30}$ $(n = 1, \|) + (8/11)\mathrm{KC}_8]$, rather than by going to $\mathrm{K(THF)}_{\sim 1.2}\mathrm{C}_{24}\,(n = 1, \|)$.

The formation of $\mathrm{K(THF)}_2\mathrm{C}_{20}$ from liquid (rather than gaseous) THF has also been followed by neutron diffraction [9.98], with the metal ions supplied by the electrolysis of potassium cyanotriethylborate $\mathrm{K[B(CN)(C_2H_5)_3]}$. The results are somewhat different from those described above: in all the intermediate stages, the molecules of THF are perpendicular to the layers of graphene and the "lean" phases are not formed. Furthermore, a new phase, corresponding to the fractional stage 4/3, was found to appear transiently between the stage-2 and stage-1 compounds. Note also that the value of the K/C ratio is not 1/24 but rather 1/20, as determined by the fit of the intensities. This would be in agreement with the contention mentioned above that this ratio might have a higher value in the rich compound than in the lean compound. Furthermore, ternary K-THF-GICs obtained in liquid THF using the soluble K-naphthalene complex [9.63, 84] have been found to have K:C = 1:32; this ratio, which was free to reach its equilibrium value, corresponds to that of a lean compound, the structure established by X-ray diffraction.

Among the alkali metals, Li stands by itself in its ability to form a series of stage-1 compounds with values of I_c much larger than those found in the derivatives of other alkali metals. This has been attributed to the formation of tetrahedral structures of four THF molecules surrounding a central Li^+ ion [9.94, 99]. The high coordination number is undoubtedly connected with the extreme polarizing power of Li. (A counter example is provided by the cesium compound, which only exhibits the structure in which THF lies parallel to the graphene [9.88].) Also noteworthy in this series is the fact that the ternarization of $\mathrm{LiC_6}$, $\mathrm{LiC_{12}}$ and $\mathrm{LiC_{18}}$ gives three ternary compounds in which, though they are all stage 1 and have nearly identical I_c values, the THF/Li ratio is quite different (Table 9.4). If the tetrahedral structures are the same in all three compounds, the amount of complexed lithium in each must be different.

The reaction between $\mathrm{KC_8}$ and the chlorides of cobalt or iron dissolved in DME or THF is said to give ternary compounds of the transition metal [9.100]: thus, stage-1 and stage-2 compounds of Co-THF with I_c = 8.90 and 12.6 Å respectively could be formed, which were found to be strongly ferromagnetic. However, these results conflict with those found by *Boehm* et al. [9.101] who could not confirm intercalation of Co (among other metals) by reduction of the hydrated chloride with KC_8.

Dimethoxyethane, 2,5-dimethyltetrahydrofuran and dioxan readily form ternary GICs with KC_{24} and the higher-stage binaries [9.63] but only mixtures of uncertain compositions seem to have been isolated. Nevertheless, it proved possible to obtain I_c values from the X-ray diffractograms, from which steric models of the metal/DME complex could be surmised [9.90], leading to a lean phase containing one mole of DME roughly parallel to the graphene for the ternary compounds with the smaller value of I_c, and to a rich phase with DME/M $\simeq 2$ for M = Li or Na, but DME/K $\simeq 3$ for the rich phase.

Table 9.5. Other ternary donor intercalation compounds of alkali metals or quaternary ammonium with oxygen-bearing molecules

Compound[a]	Stage	I_c [Å]	d_{CC} [Å]	References and remarks
$Na(DMSO)_yC_{24}$	1	15.81		[9.60] $y \simeq 6$
$K(DMSO)_yC_{24}$	1	14.93		[9.60, 103]
$K(DMSO)_yC_z$	2	18.24	1.4223 [b]	$y = 5.1?$, $z = 40$ [9.103]; $z = 48$ [9.60]
	2			$I_c = 18.4$, $y = 3$ [9.61]
$K(DMSO)_yC_z$	3	21.61	1.4212 [b]	$y = 6.4$, $z = 75$ [9.103]; $z = 72$ [9.60]
$K(DMSO)_yC_z$	4	25.00	1.4212 [b]	[9.103]; $z = 96$ [9.60]; $y = 3$, $z = 48$ [9.61]
$Rb(DMSO)_yC_{24}$	1	14.98		$y \simeq 6$ [9.60]
$Cs(DMSO)_yC_{24}$	1	15.82		$y \simeq 6$ [9.60]
$N\,Me_4\,(DMSO)_yC_{24}$	1	15.82		$y = 6$; $I_c = 15.82 + (s-1)3.346$ [9.60]

[a] DMSO: dimethylsulfoxide
[b] For graphite, d_{CC} is given as 1.4217 Å [9.103]

Another oxygen compound found to form ternary derivatives fairly readily is dimethylsulfoxide, [$(CH_3)_2SO$, DMSO] (Table 9.5). The GICs are obtained by electrolysis of a salt of one of the alkali metals (except Li) or of a quaternary ammonium salt in DMSO [9.60, 61, 102, 103] and are characterized by a large value of I_c, i.e. 11.5–12.5 Å more than the 3.35 Å of the interlayer spacing in graphite. There are, however, many discrepancies in the data presented by different authors: for example, is the stage-2 compound blue, with the composition $K(DMSO)_3C_{24}$ [9.61] or black, with the composition $K(DMSO)_{6.05}C_{48}$ [9.60]? Furthermore, only in one case [9.61] was the DMSO content obtained gravimetrically; in other cases, the number of intercalated moles was estimated by portioning out the space in the expanded interlamellar layer, assuming that each molecule of sulfoxide needed as much space when intercalated as it did in the standard (bulk) state. As will be seen later, there are reasons to believe that this may lead to erroneous results.

In contrast to the difficulties encountered with DME, many pure phases were obtained with furan [9.65,92], an unsaturated heterocycle which is one of the few compounds to form ternary GICs with the stage-1 KC_8 and RbC_8, as well as with the higher-stage derivatives. There is a fundamental difference between the ternarization processes by the vapor of THF and of furan (Fu). With the former, even-stage binaries yield ternaries whose stage is half that of the parent binary, while odd-stage binaries yield mixtures of lesser stages. With furan, a binary of stage n yields a ternary GIC of stage $n-1$ $(2 < n < 6)$, the only exception (besides $n = 1$) being $n = 2$, since the stage-1 ternary formed initially decomposes spontaneously to a stage-2 GIC. Only one type of structure is known for the Fu derivatives: much as with the Bz compounds, the molecules stand at an angle of 15° to the c-axis (Table 9.6) [9.92].

Table 9.6. Ternary graphite compounds of alkali metals with furan[a]

Compound[b]	Stage	I_c [Å]	d_{CC} [Å]	V_A [c] [Å^3]
$KFu_{0.76}C_8$	1	8.78	1.4279	138
$KFu_{1.44}C_{24}$	2	12.12	1.4250	112
$KFu_{2.17}C_{36}$	2	12.16	1.4235	115
$KFu_{2.34}C_{48}$	3	15.48	1.4234	93
$KFu_{2.28}C_{60}$	4	18.85	1.4239	90
$KFu_{2.35}C_{72}$	5	22.30	1.4233	85
$RbFu_{0.90}C_8$	1	9.00		

[a] All values from [9.92]
[b] Fu: furan, C_4H_4O
[c] V_A: apparent volume available to each furan molecule

(b) Chemical Properties

The chemical properties of the ternary GICs containing tetrahydrofuran, furan or dimethylsulfoxide are dominated by the metastability of the co-intercalated combination. Although it was originally thought that the uptake of THF by KC_{24} was reversible [9.85], later attempts to run through repeated cycles of intercalation and deintercalation showed that irreversible reactions had occurred, as proved by the increasing difficulty of distilling the THF to and from the ternary compound, owing to the presence of gases given off by the solid, and by modifications found in the diffractogram of the solid phase. Mass-spectrometric analysis of the gases established the presence of large quantities of hydrogen, ethane and ethylene among other components [9.104], and hydrolysis of the solid released THF, n-butanol and ethyl acetate. The reactions, which occur exclusively between the co-intercalated species, can be explained by fairly simple processes [9.105] and are not quite identical to those observed when THF is maintained over KC_8 or KC_{24} [9.106]. Although this decomposition process is accelerated when the GIC is heated to expel the THF, it even occurs (but at a much lesser rate) during deintercalation by cryopumping. Similar results were observed with the Fu ternary GICs [9.110], the main product of the reaction being crotonaldehyde.

(c) Physical Properties

The enthalpy of formation of $K(THF)_{2.5}C_{24}$ from gaseous THF and solid KC_{24} was determined by differential scanning calorimetry to be -119.5 kJ/mol at 331 K [9.107], equivalent to -47.8 kJ/THF mol. This is quite comparable, for instance, to the heat evolved during the formation of $CuSO_4 \cdot NH_3$ from gaseous NH_3 and solid $CuSO_4$ (-53 kJ).

The low-temperature specific heat of the lean K/THF ternary is characteristic of metallic behavior [9.95]. Both THF ternaries are paramagnetic [9.95, 99], with Dysonian EPR signals narrower than the signal from KC_{24}. The ^1H NMR signal is strongly dependent on the THF/K ratio, but not on the stage of the compound: during deintercalation, ΔH_{pp} changes from 7 to 11 G as the ratio changes from 2.6 to 1.8 [9.83], demonstrating the gradual loss of freedom of the intercalated molecule. Similarly, the transition occurring at 130–150 K and visible in the NMR spectra of the rich compound shows that part of the molecular motion of the intercalated THF gets "frozen out" in that temperature range [9.108]. The lean Cs compound $Cs(THF)_{1.6}C_{24}$ was found, by neutron diffraction [9.107], to have a liquid-like intercalated layer at room-temperature; cooling to 250 K resulted in a first-order transition, with gradual separation of two phases and partial loss of the freedom of motion of THF.

The ^1H NMR spectra of the three stage-1 ternaries [9.82] $Li(THF)_{1.4}C_6$, $Li(THF)_{2.3}C_{12}$ and $Li(THF)_{3.4}C_{18}$ are fairly similar: they are composed of a narrow ($\Delta H_{1/2} < 1$ G) line superposed on a broader line whose width abruptly increases below 180 K. Extensive analysis proves that the THF molecules have a fairly high degree of freedom at room temperature. The Li NMR of the same complexes [9.82, 99] shows the presence of two populations of Li atoms, one of which is due to the Li ions in the tetrahedral $[Li(THF)_4]^+$ complex while the other could be attributable to Li^+ involved in nonsymmetrical complexes or even to noncomplexed Li^+. Unfortunately, the quantitative interpretation of these spectra does not yet tally with the conclusions drawn from the chemical formula and the structural data of the ternary compounds, and the problem of the actual status of all the Li atoms is still partly unsolved.

By working with single carbon fibers, *Besenhard* and co-workers [9.60] were able to obtain direct comparisons of the influence of various intercalated combinations on the ϱ_a (in-plane) resistance. Expressed as R_0/R, in which R_0 is the resistance of the pristine fiber and R the value after intercalation, the measurements showed a twenty- to thirtyfold decrease of resistance in all the stage-1 DMSO compounds, irrespective of the cation (Na^+, K^+, Rb^+, Cs^+, NMe_4^+): practically identical values were shown by Li^+-DME or K^+-Bz combinations [9.102], but the improvement in conductivity was much poorer in other cases, being about 17 for Na^+-DME and about 10 for K^+-DEM or Li^+-THF combinations.

9.3.3 Synergetic Intercalation

Neither benzene nor THF are able to produce, by themselves, an intercalation compound with KC_8, yet their mixture reacts quite readily by a still unknown mechanism, yielding a bright blue solid with $I_c = 9.0$ Å [9.93, 109], formulated $K(THF)_{0.4}(Bz)_{0.37}C_8$. This "synergetic intercalation", noticed for the first time with a Bz-THF combination, also occurs with the combination thiophene-THF [9.110]. In the former case, hydrolysis of the solid phase yields biphenyl and some of its hydrogenated derivatives, whose formation follows the same mechanism as the one already given for the K-Bz-GICs [9.79]. It is interesting to note that a quaternary $K(THF)_x(Bz)_yC_{24}$, with an apparently identical X-ray diffractogram, has recently been obtained by the action of Bz on the lean THF ternary GIC. In contrast with the behavior shown by the quaternary prepared from KC_8, this newer quaternary is reported to release the intercalated benzene easily, with no formation of biphenyl [9.111].

9.3.4 Problems Associated with the Stoichiometry of the Ternary Compounds

Much of the early work on ternary GICs gave the results of preparations in terms of "ideal formulas" with integral stoichiometries deemed to be more correct than the nonintegral values obtained experimentally. Yet, weight gain vs time measurements or X-ray diffractograms taken as the THF/K ratio straddled a whole number (e.g. at the values 1.9, 2.0 and 2.1) did not reveal any particular phenomenon as the integral value was crossed [9.65, 67, 83]. A *prima facie* analysis tended to show that the invasion of the interlamellar space by the third component went on until all the available space was occupied [9.65]. Strong justification for this was found in the following argument: once I_c is known (From 00ℓ X-ray data) and if the number n of carbon atoms of the ternary MS_mC_n (where S is the third component) and its stage s are known, it is easy to calculate the volume V (in Å^3) of the prism bounded by the mid-planes of the graphenes on either side of the intercalated layer, from

$$V = 2.62(n/s)d_1 , \tag{9.7}$$

where 2.62 is the surface area in Å^2, of a single C atom in the graphene and d_1 is the distance in Å between the two mid-planes. If the metal is assumed to be ionized, the volume V_S available to the third component is

$$\begin{aligned} V_S &= V - V_M^+ - V_G \\ &= 2.62(n/s)(d_1 - 3.35) - V_M^+ , \end{aligned} \tag{9.8}$$

where V_G is the volume occupied by the graphene. A further assumption is then made, namely that the molecular volume of S is the same when the molecule is intercalated and when in the bulk, nonintercalated state. Dividing the available volume given above by the molecular volume gives the maximum number of molecules that can be co-intercalated with the metal:

$$m = [2.62(n/s)(d_1 - 3.35) - V_M^+]/V_S \,, \tag{9.9}$$

where V_S is the bulk molecular volume of S.

This simple approach often gave surprisingly good agreement with the experimentally determined values, which implied that the stoichiometry of the third component was dictated by its bulk rather than by the coordination relationship with the metal atom or ion [9.112]. In order to give (9.9) a thorough test, the stoichiometry and the value of I_c of the ternary GICs obtained by the reaction of furan on KC_8 to KC_{72} (the first six stages of the potassium binaries) were carefully determined [9.92]. The results, shown in Table 9.6, are expressed as values of V_A, the volume apparently available to each molecule of furan, obtained from (9.9) by interchanging m, in the left hand side of the equation with V_S, in the right hand side, so as to obtain a result free of any assumption as to the actual density of furan when intercalated. Keeping in mind that the room-temperature bulk molecular volume of furan is 120 Å^3, the results in Table 9.6 show that V_A is, at first, 15 % larger, but finally 30 % less than the standard molecular volume. A plot of V_A vs the stage s showed a hyperbolic variation with a lower asymptote at 74.4 Å^3 close to the *van der Waals* volume of furan, namely 63 Å^3 [9.113]. This was explained as due to a varying amount of electric charge carried by the molecules of furan, which are competing with the atoms in the graphene layers for the electron lost by the atom of potassium. Each C atom in the ternary compound is a potential well for the electron; as the total number of C atoms increases, i.e. as the stage increases, the probability of finding an electron on a molecule of furan decreases, as does the Coulomb repulsion among the charged molecules. The fact that the molecules of furan are, on the one hand, somehow bound to the potassium ions and, on the other, restricted in their motion by the graphene layers must result in a considerable loss of freedom of motion, as shown by the limiting value of the molecular volume close to the van der Waals volume, i.e. to a molecule completely motionless. Early data on the THF and the Bz ternaries led to similar conclusions [9.120].

An indirect proof of the presence of charge on the molecules of furan was provided by monitoring the C-C in-plane bond length in the graphene layers. Using the semi-empirical approach of *Pietronero* et al. linking bond length to charge borne by the C atoms [9.114] and assuming that the charge which is not on the graphene is on the furan, it was found that V_A increased with the charge on the molecule, as expected [9.64].

A more direct proof was provded by the *Raman* spectrum of the stage-2 KC_{24}-furan ternary, in which the intralayer vibration mode of the graphene layers was found at 1606 cm^{-1} [9.115], at a frequency identical to the one found with KC_{24} but with a much decreased width at half-height, namely 23 cm^{-1} against 29 cm^{-1} for the pure binary. This was interpreted as due to a decrease in the charge borne by the layers of graphene. Similar results were obtained with a stage-1 $K(THF)_2C_{24}$ compound [9.115] and with a ternary containing benzene [9.80], indicating electron back-transfer to the organic molecule. In the ternary system formed by graphite with potassium and ammonia, measurement of the optical reflectance [9.116] or of the ^{1}H NMR absorption [9.117], as well as detailed analysis of the effect of composition on charge exchange and lattice expansion [9.118, 119], led to the conclusion that 1/5 to 1/4 of the total charge available was solvated by molecules of ammonia in $K(NH_3)_{3.34}C_{24}$.

9.4 Ternary Acceptor Intercalation Compounds

It is convenient to classify the acceptor intercalation compounds of graphite into derivatives of the metal halides and salt-like derivatives of strong acids. This is a purely arbitrary division since current evidence points to the formation of stable anionic species whenever a metal halide is intercalated, with or without the help of an oxidizing agent.

The ternary acceptor GICs have been obtained by the following methods:

- direct action of the third component on a metal halide GIC;
- direct and simultaneous intercalation of two metal halides;
- co-intercalation of a metal halide (or of an ionized acid molecule) and of the solvent during the action of a solution on graphite.

9.4.1 Ternary Metal-Halide–Ammonia–Graphite Intercalation Compounds

Although gaseous ammonia did not react with the stage-1 C_9FeCl_3, liquid ammonia was found by *Rüdorff* et al. to do so quite readily, giving a hexammine complex whose presence between the layers of graphene modified I_c from 6.37 to 16.7 Å [9.121]. The same method was later used by *Touzain* and co-workers [9.122, 123] to obtain hexammine complexes of C_7FeCl_3, C_7CoCl_2, C_6MnCl_2 and $C_{12}NiCl_2$, all of which were found to have interesting possibilities for heat-energy storage.

9.4.2 Ternary Intercalation Compounds Containing a Solid Solution of Two Metal Halides

Attempting to separate $FeCl_3$ from $AlCl_3$ by intercalation, *Croft* obtained a compound formulated $C_9(FeCl_3)_{0.62}(AlCl_3)_{0.38}$ [9.124, 125]. This was confirmed by *Dzurus* and *Hennig* [9.126], who gave, however, a ratio $AlCl_3/FeCl_3 = 3$.

Later, *Stumpp* used the fact that the volatility of vapor halides with low vapor pressure is often considerably increased in the presence of $AlCl_3$ to obtain the easy intercalation of $CoCl_2$ [9.127]; this probably occurs via the formation of a gaseous complex by the reaction

$$MCl_2(s) + nAlCl_3(g) \rightarrow MAl_nCl_{2+3n}(g) ,$$

which facilitates the transport of the chloride MCl_2 [9.128]. According to the results presented, $AlCl_3$ was not co-intercalated with cobalt chloride. However, many metal halide–$AlCl_3$ combinations which have now been tried lead to the simultaneous intercalation of both halides.

The $FeCl_3$-$AlCl_3$-graphite system has been further investigated [9.129] and compounds showing a large range of values for the $FeCl_3/AlCl_3$ ratio were obtained (Table 9.7). Mössbauer spectroscopy results on these compounds were interpreted as favoring a transfer of π electrons to the intercalated Fe^{3+}, thus forming Fe^{2+} and freeing a chlorine atom to give $C_n^+Cl^-FeCl_2(AlCl_3)$, a structure which had been suggested in early results on this system [9.125]. Similar conclusions were drawn, also from Mössbauer spectroscopy, as to the presence of Fe^{2+} atoms in $FeCl_3$-$MoCl_5$-graphite intercalation compounds [9.130], but the presence of Fe^{2+} ions, which were said to bear no relation to the presence of graphite, could well be due to the formation of $(MoCl_6)^-$ ions since their amount increased when the Mo:Fe ratio also increased.

Table 9.7. Intercalated solid solutions of transition metal chlorides

Compound	Stage	I_c [Å]	References
$C_9(FeCl_3)_{0.62}(AlCl_3)_{0.38}$	1 ?		[9.124, 125]
$C_n(FeO_3)(AlCl_3)_3$	1		[9.126]
$C_{8.9}FeCl_3(AlCl_3)_{0.25}$	1 ?		[9.129]
$C_{73}(FeCl_3)_{0.3}AlCl_3$	?		[9.129]
$C_{5.95}MnCl_2(AlCl_3)_{0.21}$	1	9.51	[9.131]
$C_{12.68}MnCl_2(AlCl_{2.4})_{0.70}$	1		[9.132]
$C_{15}MnCl_2(AlCl_{3.8})_{0.3}$	2	9.52	[9.132]
$C_{11.7}MnCl_2(AlCl_{3.2})_{0.33}$	2		[9.132]
$C_{40}CuAl_2Cl_8$	1	9.42	[9.135]
$C_{17.4}CuAlCl_5$	1		[9.135]
$C_{25}Cu_2AlCl_7$	1		[9.135]
$C_{33}NiCl_2(AlCl_3)_{0.5}$	3		[9.135]
$C_{21}NiCl_2(AlCl_3)_{0.25}$	2		[9.135]
$C_{17}PdAl_2Cl_8$	2	9.47	[9.135]
$C_{21}AgCl(AlCl_3)_{1.12}$	2	9.74	[9.135]
$C_{29}AgCl(FeCl_3)_{1.6}$	2		[9.135]

The kinetics of formation and characteristics of the compounds formed in the system $MnCl_2$-$AlCl_3$-graphite have also been studied. At 325°C, $AlCl_3$ is rapidly taken up until a stage-2 GIC is formed: this is followed by a period during which the Mn halide is intercalated, with no loss of $AlCl_3$; the last step in the formation of the final $C_{7.15}MnCl_2(AlCl_{2.5})_{0.3}$ is a concomitant decrease of the $AlCl_3$ content and increase of intercalated $MnCl_2$ [9.131]. The compound is stage 1, with I_c = 9.51 Å and the presence of an *ABC* hexagonal stacking (as in solid $MnCl_2$) requires that c be $3 \times I_c$. The formula of the compound can vary within wide limits without apparent modification of the stage (Table 9.7) [9.132, 133].

The intercalation of cupric chloride $CuCl_2$ has also been found to be accelerated by the presence of $AlCl_3$ [9.134]. As with $MnCl_2$, the kinetics of the reaction implies the gradual replacement of $AlCl_3$ by $CuCl_2$ in the aluminium-rich mixture which is first intercalated. Latest results show that (a) the intercalation reaction is even faster when graphite is in contact with the solid mixture $MCl_n/AlCl_3$; (b) the composition of the intercalated species is often close to that of the complexes formed by the two halides in the bulk or in the gas phase [9.135]. This is particularly true for the systems $CuCl_2/AlCl_3$ and $PdCl_2/AlCl_3$.

A different technique was used by *Inagaki* and co-workers [9.136]: by reaction of a molten $NiCl_2$-$FeCl_3$ mixture on graphite, stage-1 and stage-2 compounds were obtained at 400°C in 3 days. The stage of the compounds could be controlled by changing the initial value of the ratios graphite/total chloride and $NiCl_2/FeCl_3$. The intercalation compounds obtained were said to be particularly resistant to the action of air or water.

9.4.3 Ternary Systems Involving the Fluorides of Transition Metals

The direct intercalation of chromyl fluoride leads to the stage-3 $C_{22-24}CrO_2F_2$ [9.137] with I_c = 14.8 Å, equivalent to d_1 = 8.1 Å. Many other fluorides or oxyfluorides, which react only sluggishly or under extreme conditions, have been found to react quite readily when dissolved in chromyl fluoride, thus yielding stage-1 compounds in which the ratio M/Cr is close to 2 (M =V, Nb, Ta) or 1 (Mo, W) [9.138]. A striking feature of all the compounds in Table 9.8 is their I_c value, which is always smaller than that of the corresponding binary GIC devoid of chromyl fluoride. Also noteworthy is the absence of infrared absorption attributable to Cr-O bonds.

Table 9.8. Intercalation of metal fluorides from their solution in chromyl fluoride [9.138]

Compound	Stage	I_c [Å]	Remarks
$C_{7.2}VOF_3(CrOF_2)_{0.5}$	1	8.15	Presence of CF and CF_2 bands in IR spectrum
$C_{6.5}NbF_5(CrO_2)_{0.6}$	1	8.36	
$C_{8.4}TaF_{4.5}(CrO_2)_{0.5}$	1	8.36	No Cr-O bands in IR spectrum. Presence of bands due to TaF_6^-
$C_{15}MoF_6(CrO_2F)$	1	8.25	
$C_{19}MoOF_4(CrO_2F_2)_{0.95}$	1	8.25	Only weak Cr-O bands in IR spectrum
$C_{22}WOF_4(CrO_2F_2)$	1	8.25	Composition uncertain

A completely different type of behavior is shown by other fluorides: intercalation of MF_m in the presence of F_2 leads to C_nMF_{m+1} [9.139, 140]. When treated with another fluoride, $M'F_p$, the GIC can react

- either by taking up $M'F_p$ reversibly, as happens when $C_{12n}GeF_5$ forms $C_{12n}GeF_5(SiF_4)_{0.5}$ or $C_{12n}GeF_5(SF_6)_{0.16}$ [9.140], or when $C_{12}AsF_6$ gives $C_{12}AsF_6(AsF_5)_{0.4}$ or $C_{12}AsF_6(Br_2)_{0.4}$ [9.141].

- or by taking up $M'F_p$ irreversibly, while simultaneously expelling MF_m, as happens when $C_{24}SiF_5$ reacts with BF_3 to yield C_7BF_4 or with PF_5 to form $C_{24}PF_6{\cdot}PF_5$ [9.139].

9.4.4 Intercalation of Metal Halides from Solutions

Many metal halides are known to dissolve in nonaqueous solvents, and the resulting solutions have been used to obtain intercalation into graphite. The possibility of forming such a solution is not a condition necessary or sufficient for intercalation; in fact, it has been shown [9.142] that high solubility of the halide in the solvent is often detrimental to the formation of lower-stage GICs, since the partitioning of the halide between the solid and the liquid phase is then heavily biased in favor of the latter. Furthermore, the chemical or physical properties of the halide may be modified by the formation of a complex with the solvent to such an extent that intercalation may no longer be possible.

When the dissolved halide is sufficiently oxidizing, its solution will readily react with graphite, and $FeCl_3$ in nitromethane (CH_3NO_2, denoted by NM) gave a product which, after vacuum-drying, was a stage-3 $C_{26.3}FeCl_3$ [9.143] but a careful check of the preparation conditions led *Hooley* [9.144] to conclude that the reaction had probably yielded $C_{24\pm2}FeCl_3(NM)_{0.73\pm0.08}$. In the same study, it was shown that at least 11 other halides can be intercalated into graphite from their solution in NM, and since many of them are known to form complexes with NM [9.145], the chances are high that the solvent was co-intercalated. This reaction was further investigated by *Stumpp* and co-workers [9.146] who confirmed the detrimental effect of high solubility by comparing the results obtained with solutions of $TlCl_3$ in NM and in thionyl chloride ($SOCl_2$) in which the halide is much less soluble: only in the latter case could the first-stage GIC be prepared; moreover, in the former case, the GICs often contained large amounts of NM.

Thionyl chloride has been used to obtain the intercalation of many halides [9.127], but co-intercalation of the solvent is not always the case. Thus, depending on the type of graphite used, solutions of trithiazyl trichloride $S_3N_3Cl_3$ and $FeCl_3$ in $SOCl_2$ yield $C_{11}S_3N_3Cl_3{\cdot}(FeCl_2)_{2.4}$ (mainly stage 1, $I_c = 9.17 \pm 0.05$ Å) or $C_{9.5}(S_3N_3Cl_3)_{0.2}(SOCl_2{\cdot}FeCl_2)_{0.8}$ (stage 2, $I_c = 12.77 \pm 0.03$ Å) [9.147]. Similarly, $AlCl_3$ yields $C_{24}(S_3N_3Cl_3 \cdot 2AlCl_2)_{0.4} - (SOCl_2 \cdot 2AlCl_2)_{0.6}$ (a mixture of stage 2, $I_c = 12.89 \pm 0.03$ Å, and stage 3, $I_c = 16.23 \pm 0.03$ Å).

Closely connected with the above preparations is the formation of $C_{24}S_4N_4 \cdot 2SbCl_5$ obtained by allowing the compound $S_4N_4SbCl_5$ dissolved in CCl_4 to react with graphite. The resulting GIC is brown, extremely stable and inert towards air, with an electrical conductivity about an order of magnitude greater than that of graphite [9.147].

Other halocarbon solvents have been used. Thus, UF_6 in *sym*-difluorotetrachloroethane leads to a ternary GIC with $I_c = 8.48$–8.75 Å [9.148]. These solvents are, furthermore, a valuable source of free chlorine (or of other free radicals) under irradiation by ultraviolet light. This has led to the elaboration of a new technique, namely that of *photochemical intercalation* [9.149]: the suspension of graphite in a solution of the halide in CCl_4, $C_2H_2Cl_4$, or some other predried and purified chlorinated solvent is stirred and submitted to near-ultraviolet radiation (310–320 nm) for the appropriate time, as checked by X-ray diffractograms on samples of the dried solid phase. The preparation – which can be carried out on a large scale – can take from a few hours to some days. Proof of the intermediate formation of free radicals is found in the presence of numerous substituted derivatives of the solvent molecule detected by gas-phase chromatography, and in the C-Cl bonds on the graphene layers detected by XPS. The solvent is always co-intercalated with the metal halide but the main disadvantage lies in the difficulty of obtaining a single pure phase. Moreover, the presence of co-intercalated solvent affects both the amount of intercalated halide and, as could be expected, the I_c value. Thus, for intercalated $FeCl_3$, $I_c = 9.88$ Å (from a solution in $C_2H_2Cl_4$), or $I_c = 10.88$ Å (from CCl_4) [9.149]. Other halides successfully intercalated using this method are $SbCl_5$ [9.150], $SnCl_4$ [9.151] and $Sn(CH_3)_3Cl$ [9.152, 153].

It is interesting to note that, as a variation on this procedure, the successful intercalation into graphite of the iron carbonyls $Fe(CO)_5$, $Fe_2(CO)_9$ and $Fe_3(CO)_{12}$ was obtained by irradiation of solutions in hydrocarbons or, better, in THF [9.156]. In this case, the active species created by photolysis were probably of types similar to $Fe(CO)_4$, which would be very powerful electron acceptors because of their incomplete electron shell and would be intercalated as $Fe(CO)_4^{2-}$.

The intercalation of metal halides from their solution in a halogen leads to derivatives which are at the border between the binary and the ternary compounds since many halogen or interhalogen combinations can be intercalated by themselves. By allowing graphite to react on solutions of $AlBr_3$ in Br_2, *Sasa* and co-workers obtained $C_9AlBr_3{\cdot}Br_2$ (stage 1, $I_c = 10.24$ Å) [9.155], as well as $C_xAlBr_3(Br_2)_{2.5}$ (stage 2, $x = 25.5$–31, $I_c = 13.72$–13.68 Å) and $C_{24}AlBr_3{\cdot}Br_{0.3}$ (stage 2, $I_c = 13.35$ Å). In the first two compounds, the overall packing density of the intercalated species is about 4.4 g/cm^3, i.e. much higher than that of either species by itself. Later work [9.156, 157] tended to confirm the I_c values and suggested that the ligand entering the graphite varied between $AlBr_4^-$ and $Al_2Br_6AlBr_4^-$, but the high concentrations of bromine were not achieved. Similar gallium derivatives, formulated $C_xGaBr_3{\cdot}Br_y$ (stage 2, $x = 13$ and $y = 2.5$, with $I_c = 13.55$ Å, or $x = 16.5$ and $y = 2.2$, with $I_c = 13.3$ Å) have also been prepared [9.158] as well as gold bromide derivatives recently shown to be $C_{40}AuBr_{8.2}$ (stage 1, $I_c = 7.07$ Å), which could be thermally degraded to $C_{40}AuBr_{4.9}$ (stage 2, $I_c = 10.36$ Å) [9.158]. The system $FeBr_3/Br_2$/graphite has also been thoroughly investigated [9.160], but the behavior of solutions of TlBr in Br_2 towards graphite is particularly interesting. Although $TlBr_3$ does not exist in the free state, it seems to be sufficiently stabilized by intercalation in compounds such as $C_{18.6}TlBr_{3.4}$ (stage 2, $I_c = 13.4$ Å), in which it probably forms sheet-like triple layers of Tl_2Br_6 dimers, to resist temperatures of 150°C or even to be involved in ternaries like $C_{22}TlBr_{3.3}Br_2$ [9.161].

Attempts were also made to obtain intercalation of metal halides from their solution in chlorosulfuric acid HSO_3Cl [9.162] or fluorosulfuric acid HSO_3F

[9.163]. Both acids can form binary GICs when a suitable oxidizing agent is present [9.164] and intercalation of the acid did indeed occur with all the metal halide solutions, probably owing to the possibility of having the equilibrium

$$HSO_3X \leftrightarrows SO_3 + HX$$

(known to be catalyzed by the presence of certain metallic salts) to provide SO_3 as the necessary oxidizing agent. However, the formation of ternary GICs was only obtained in a very limited number of cases: with BCl_3/HSO_3Cl, yielding a GIC with I_c = 15.06 Å (stage 3, composition unknown); with $SbCl_5/HSO_3Cl$, when each component of the solution was intercalated separately to form a two-phase mixture of stage-2 GICs; with $AuCl_3/HSO_3Cl$, which also yielded a mixture of the two stage-2 GICs as one of the steps in an involved process ending with the formation of the pure stage-2 metal halide GIC; and finally, with SbF_5/HSO_3F, when true ternary compounds were obtained whose stage (1, 2 or 3) depended on the temperature of preparation and whose composition and I_c value depended on the composition of the solution. For the stage-1 GIC, $C_7(HSO_3F)_{0.35}(SbF_5)_{0.65}$ seemed to be particularly stable, with I_c = 8.18–8.32 Å.

9.4.5 Intercalation Compounds Containing Solvated Complex Anions

The electrochemical oxidation of graphite immersed in solutions of complex anion salts in dry propylene carbonate (PC) led to GICs formulated $C_{24}^{+}X^{-} \cdot (PC)_y$, with $X = BF_4^-$, PF_6^- or SbF_6^- [9.165]. The bulky molecules of solvent and/or their number impose large values of I_c: for the compounds $C_nPF_6^- \cdot (PC)_y$, $I_c(s) = 14.08 + 3.34(s-1)$ Å, where s is the stage, $n = 24s$ and y is probably 4 [9.166]. Structure factor calculations have been performed for the (00ℓ) series in the X-ray diffractograms, which lead, for the intercalated complex $[X(PC)_x]^-$ with $X = PF_6$, AsF_6 or SbF_6, to X at the center of four PC molecules placed tetrahedrally around it [9.167]. Similar results were obtained with the series of GICs formulated $C_{24s}TaF_6 \cdot (PC)_4$, for which $I_c(s) = 13.9 + 3.35(s-1)$ Å [9.168], which show a relatively poor stability of the stage-1 compound similar to the one observed with the other series. Electrochemical intercalation of PF_6^- and AsF_6^- dissolved in NM yielded, for instance, $C_{48}^{+}PF_6^- \cdot (NM)_{2-4}$ (stage 2, I_c = 11.1 Å) [9.169] and $C_{24}^{+}AsF_6^- \cdot (NM)_2$ (stage 1, I_c = 7.99 Å) [9.170].

The formation of salts of graphite by chemical – rather than electrochemical – oxidation in the presence of strongly acidic anions was, of course, the first method used for their preparation. It has been extended to nonaqueous solvents by using salts of the strongly oxidizing nitronium (NO_2^+) [9.171] or nitrosonium (NO^+) [9.172] cations which are soluble in nitromethane or tetramethylene sulfone. In the presence of graphite, the cations cede their positive charge to the layers and are thereby transformed into gaseous nitrogen oxides, which are evolved, while the anions and some solvent molecules are co-intercalated. Although some of the reactions required several weeks for completion, this mode of preparation enabled *Billaud* and co-workers [9.173] to obtain many stages of a number of salt-like GICs with the following characteristics:

$C_n^+BF_4^- \cdot (NM)_y$ $:s = 2-7;\ I_c(s)[\text{in Å}] = 7.90 + (s-1)3.35;$

for $s = 2,\ n \simeq 50, y \simeq 2.5$ [9.173, 174]

$C_n^+PF_6^- \cdot (NM)_y$ $:s = 2-8;\ I_c(s)[\text{in Å}] = 7.75 + (s-1)3.35;$

for $s = 2,\ n = 48,\ y = 2.4$ [9.169], or 2.1 [9.175],

or 1.64 – 1.14 [9.176]

$C_n^+SbF_6^- \cdot (NM)_y$ $:s = 1-8;\ I_c(s)[\text{in Å}] = 8.05 + (s-1)3.35;$

for $s = 1,\ n = 23,\ y = 1.7$ [9.176, 177]

Nitrosylsulfate ($NOHSO_4$) suspended in NM has been found to give the ternary compounds $C_{23s}HSO_4 \cdot (NM)_y$ (stages 4–10, $I_c(s) = 7.92 + 3.36s$) which had $y = 1.8$ for $s = 5$ [9.178]. Equally interesting is the action of $[NO][Fe(NO_3)_4]$ dissolved in nitromethane, to give the stage-3 compound $C_x[Fe(NO_3)_{3.5}(NM)]$, (x = 34–39, I_c = 18.0 Å) [9.179], in which one of the NO groups of the $[Fe(NO_3)_4]$ ion is reported to have been replaced by a molecule of nitromethane through the reaction

$$xC + [NO][Fe(NO_3)_4] + NM \rightarrow C_x^+[Fe(NO_3)_3(NM)]^- + N_2O_4\ .$$

Ternary compounds are also formed when dilute acids are intercalated [9.180]. In the presence of water, the interlayer distance of the HNO_3/H_2O/graphite ternaries decreases from 7.8 to 7.72 Å, and there is a gradual inversion of the intensity of the (001) and (004) lines in the X-ray diffractograms. Similar changes are observed for the sulfuric acid derivatives.

9.4.6 Physical Properties

Measurements of the electrical resistivity of graphite fibers containing $CuCl_2$ and $AlCl_3$ show a good parallelism between conductivity and halide content, and a much increased stability towards air or moisture [9.134], but a truly neutral third component, not involved in solvation or electronic exchanges – such as AsF_3 intercalated in $C_{8s}AsF_5$ – can be removed or added back without significant change in the conductivity [9.126]. The improvement of conductivity due to the formation of the ternary compound is quite often about tenfold at room temperature: this is the case, for instance, for the thiazyl/$SbCl_5$ derivatives [9.147], the $AlBr_3/Br_2$ compounds [9.155, 182], and most of the complex anion GICs containing nitromethane or propylene carbonate [9.166, 185]. There is, of course, a stage dependence of the conductivity, which is often maximum for the stage-4 or stage-5 compound [9.169, 173], but this is relatively small. Quenching phenomena, quite general in acceptor compounds and associated with order–disorder transitions, have also been found in $C_{24}^+SbF_6^-(NM)_x$ (stage 1, $x \simeq 2$) [9.126].

The Mössbauer results on some mixed halide GICs containing Fe species [9.128, 129] have already been mentioned, Similarly, ^{121}Sb Mössbauer spectroscopy on $C_{23s}SbF_6(NM)_{1.7}$ (s = 1, 2) led to the conclusion that the metal in the GIC was present as $[SbF_6]^-$ ions [9.184].

Nuclear magnetic resonance of ^{19}F has been used fairly extensively to establish the status of the various intercalated fluorine-bearing species. Thus, $C_{20n}AsF_6 \cdot (NM)_x$ ($n = 1, 2, \ldots, x \simeq 2$) was shown [9.185] to contain $[AsF_6]^-$ ions exclusively, with no chemical exchange with the F atoms of AsF_3 when this was subsequently intercalated into the GIC, in contrast to earlier reports [9.186]. The same technique led to the assignment of a positive charge on the graphite lattice equal to the concentration of $[PF_6]^-$ ions in $C_{48}PF_6^- \cdot (NM)_2$ [9.187]. Similarly, ^1H NMR was used [9.176] to establish the position of the molecular symmetry axis of the co-intercalated NM as parallel to the graphitic planes in the GICs containing $[MF_6]^-$ ions (M = P, As, Sb) [9.176]. Low-temperature magnetic properties of the $MnCl_2/AlCl_3$ GICs have been determined [9.132]: the effective magnetic moment depends on the Mn/Al ratio, which might indicate that the volatile gaseous complexes also exist in the intercalated layers.

9.5 Multi-Intercalation Compounds

Ideal multi-intercalation compounds contain different, separate and distinct intercalated layers within a single crystal structure, with each layer in its own interlamellar space (in the *Rüdorff* representation). When there are two (three, or four...) types of intercalated layers, the structure is a bi- (tri-, quadri-...) intercalation compound. The multi-intercalation compounds differ from the heterostructures discussed in Sect. 8.4 in that the latter are composed of two or more phases which can each be characterized by its own set of crystallographic parameters.

When the intercalated layers of a multi-intercalation compound are of different types, i.e. if there are both donor and acceptor layers present simultaneously, the compound can be said to be "mixed". Otherwise, if all the layers are of the same type, the multi-intercalation compound will be a "donor" or an "acceptor" multi-intercalation compound.

9.5.1 Donor Bi-Intercalation Compounds

The first donor bi-intercalation compounds were prepared by the reaction of an alkali metal M with a donor GIC (*Method I*), as in

$$2KH_{2/3}C_8(\text{stage 2}) + M \rightarrow MK_2H_{4/3}C_{16}(\text{stage 1}) \; (M = \text{K, Rb or Cs}) ,$$

in which the result is better described by $K_2H_{4/3}C_8 \cdot MC_8$ to indicate the composition of each intercalated layer and its concentration with respect to the underlying graphene [9.188]. The color of the compounds obtained with the three alkali metals is copper-red, quite different from that of the three binaries, and the three I_c values (Table 9.9) correspond exactly to those calculated from $I_c = 11.88 - 3.35 + I_c(MC_8)$ [Å]. As could be expected, the behavior of the three bi-intercalation compounds towards hydrogen is not similar to that of the corresponding binaries: they all form a mixture of $KH_{2/3}C_8$ and the hydride MH.

Using a solution in pentane of the complex formed by potassium and $[P(CH_3)_3]_3C_2H_4Co$ (*Method II*), *Besenhard* and co-workers [9.189] obtained the intercalation of K into the free interlayer space of the stage-2 ternary compound $K(Tol)_yC_n$ (Tol = $C_6H_5CH_3$), whose I_c value changed from 12.37 to 14.47 Å owing to the formation of $K(Tol)_yC_{n/2}{\cdot}K_xC_{n/2}$.

9.5.2 Acceptor Bi-Intercalation Compounds

The action of gaseous reagents on metal halide GICs (*Method III*) is a simple method often leading to the formation of bi-intercalation derivatives. The reaction of N_2O_5 with $C_{31}FeCl_3$ (presumed to be stage 4) gave $C_{31}FeCl_3{\cdot}(N_2O_5)_{1.7}$ [9.190]. From the fact that the Mössbauer spectra of the two derivatives were identical and that the environment of the iron had therefore not been changed by intercalation of the oxide, it was concluded that the latter had entered previously empty layer spaces. A possible sequence for the bi-intercalation compound is

$$\ldots(N)G^*(Fe)G(N)G(N)G(N)G^*(Fe)G(N)\ldots ,$$

in which (Fe) and (N) represent the layers of $FeCl_3$ and N_2O_5 respectively. This would correspond to I_c = 28.6 Å between the starred layers (rather than 31.9 Å as stated in [9.190]), but no X-ray measurements were presented. Similarly, the reaction of gaseous bromine on $C_{10s}Cu(NO_3)_2$ (idealized formula, $s = 4$–6, $I_c = 10.25 + (s-1)3.35$ Å) gave ...(Cu)GG(Br)GG(Br)GG(Cu)... and ... (Cu)GG(Br)GG(Cu)... from the stage-6 and stage-4 compounds respectively [9.191], while $C_{28}Zn(NO_3)_2$ (stage 4, I_c = 21.05 Å) yielded ...(Zn)GG(Br)GG (Zn)... (I_c = 24.74 Å) [9.192].

It is possible to obtain sequential intercalation by the reaction of a dissolved halide on a metal halide GIC (*Method IV*): when the second-stage $C_{22.3}TlBr_{3.3}$ is treated with a saturated solution of $TlCl_3$ in $SOCl_2$, a stage-1 $C_{12.5}TlCl_{3.6}C_{12.5}{\cdot}TlBr_{3.2}$ is formed [9.194], with I_c = (2 × 9.83 Å), about 1 % smaller than the value calculated by summing the interplanar distances of the binary chloride and bromide GICs. More commonly, sequential intercalation is performed by the direct intercalation of the second halide into the free interlayer spaces of a stage-2 (or higher) metal halide GIC (*Method V*), often in the presence of an atmosphere of chlorine when both halides are chlorides. The temperature T_G of the solid phase must be higher than that of the source of gaseous halide, but sufficiently low that the intercalated species will not be expelled from the GIC and replaced by the incoming halide.

In connection with this last point, it should be pointed out that intercalation of a halide followed by partial deintercalation (*Method VI*) has been used to prepare higher-stage multi-intercalation compounds, when there is enough difference in activity of the intercalated species [9.195], as in the preparation of $C_xBiCl_3{\cdot}(AuBr_y)_r$ (stage 3, d_I = 23.55 Å) from the stage-4 C_xBiCl_3 (I_c = 19.55 Å).

Electrochemical techniques (*Method VII*) can be used to obtain sequential intercalation: starting from a stage-2 $InCl_3$ GIC, a stage-1 blue bi-intercalation derivative was produced by electrochemical oxidation in sulfuric acid [9.195]. Using the same technique, a more complex structure was obtained, starting from a stage-3 $BiCl_3$ intercalation compound, namely the GIC with the sequence

$$\ldots G(BiCl_3)G(H_2SO_4)G(H_2SO_4)G(BiCl_3)G\ldots ,$$

which is stage 3 in $BiCl_3$ and stage 3/2 in H_2SO_4 [9.195].

A general feature of most of the bi-intercalation compounds in Table 9.9 must be underlined. If the parameters of the intercalated sheet of halide in the host second stage compound C_mMCl_n are known, the index m can be calculated [9.202]. For the graphite-$CdCl_2$ compound, the theoretical value of m is 9.7, whereas the experimental values lie between 11.3 and 14. This implies that the 2d packing of the halide molecules is not maximum and that, with reference to the *Rüdorff* model of structure, 1/3–1/6 of the available surface in the occupied interlamellar space is free. For other halides, the proportion of "free" surface is about 1/5 (with $CoCl_2$ or $InCl_3$) to 1/20 (with $FeCl_3$). Yet, when the second halide – say $AlCl_3$ – is intercalated, the C/$AlCl_3$ ratio is practically the same in the bi-intercalation compound as it is in the stage-2 $AlCl_3$ binary. In other words, and again with reference to the *Rüdorff* model, the second halide only goes into the unoccupied interlamellar spaces.

Another important point is the reciprocal destabilization of both halides in the bi-intercalation compound. It is reasonable to suppose that the intercalation of the second halide must modify the charge transfer, and the compromise which is reached is not necessarily satisfactory to both halides. This is illustrated by the fact that heating a G-$FeCl_3$-$AlCl_3$ bi-intercalation compound in a sealed tube at 200°C under a chlorine atmosphere leads to a pure stage-2 binary $AlCl_3$ GIC [9.202], so that the bi-intercalation compound must have been a metastable structure. On the other hand, this modification of the charge transfer can permit the formation of compounds otherwise unobtainable: whereas Br_2 does not intercalate into a stage-2 $NiCl_2$ GIC, it can be co-intercalated with $AuBr_3$ in $C_{11}NiCl_2$ (stage 2, I_c = 16.34 Å) [9.193], as shown in Table 9.9.

9.5.3 Mixed Bi-Intercalation Compounds

Two groups simultaneously and independently prepared the first mixed bi-intercalation compounds by the reaction of an alkali metal on a stage-2 halide GIC (*Method VIII*). The species prepared by *Furdin* and co-workers [9.201] and by *Suzuki* et al. [9.200] correspond to

$$\ldots G(M)G(M'Cl_n)G(M)G(M'Cl_n)G\ldots .$$

The second group investigated the reaction of K on $CoCl_2$ while the first group

Table 9.9. Bi-intercalation compounds

Compound[a]	Method and stage[b]	I_c [Å]	Remarks and references
$K_2H_{4/3}C_8 \cdot KC_8$	I, $2 \to 1$	13.80	[9.188]
$K_2H_{4/3}C_8 \cdot RbC_8$	I, $2 \to 1$	14.15	[9.188]
$K_2H_{4/3}C_8 \cdot CsC_8$	I, $2 \to 1$	14.45	[9.188]
$K(Tol)_y C_{n/2} \cdot K_x C_{n/2}$	II, $2 \to 1$	14.47	[9.189]
$C_{31}FeCl_3 \cdot (N_2O_5)_{1.7}$	III, $4 \to 1?$	28.6?	The compound is probably stage 4 in $FeCl_3$ and stage 4/3 in N_2O_5 [9.190]
$C_{40}Cu(NO_3)_2 \cdot (Br_2)_r$	III, $4 \to 2$		Stage 4 in $Cu(NO_3)_2$ and stage 4 in Br_2 [9.191]
$C_{60}Cu(NO_3)_2 \cdot (Br_2)_r$	III, $6 \to 2$		Stage 6 in $Cu(NO_3)_2$ and stage 3 in Br_2 [9.191]
$C_{28}Zn(NO_3)_2 \cdot (Br_2)_r$	III, $4 \to 2$	24.74	Stage 4 in $Zn(NO_3)_2$ and stage 4 in Br_2 [9.192]
$C_{11}NiCl_2 \cdot (AuBr_{7.5})_{0.14}$	IV, $2 \to 1$	16.34	$AuBr_3$ dissolved in liquid Br_2 [9.193]
$C_{12.5}TlCl_{3.6} \cdot C_{12.5}TlBr_{3.2}$	IV, $2 \to 1$	19.66	[9.194]
$C_x BiCl_3 \cdot (AuBr_x)_r$	IV, VI, $4 \to 2$	23.35	Stage 4 in $BiCl_3$ and stage 4 in $AuBr_x$ [9.195]
$C_x BiCl_3 \cdot (AuBr_x)_r \cdot (TlCl_3)_s$	IV, $2 \to 4/3$	29.93	Same as above, and stage 4 in $TlCl_3$ [9.195]
$C_{12.8}FeCl_3 \cdot (AlCl_{3.3})_{0.75}$	V, $2 \to 1$	18.8	Average composition [9.196]
$C_{13.2}FeCl_3 \cdot (GaCl_{3.3})_{0.81}$	V, $2 \to 1$	18.93	Average composition [9.196]
$C_{10.6}CoCl_{2.17} \cdot (AlCl_{3.6})_{0.65}$	V, $2 \to 1$	18.94	[9.197]
$C_x CoCl_2 \cdot (FeCl_3)_r$	V, $2 \to 1$	18.60	[9.198]
$C_{10.6}CoCl_{2.1} \cdot (GaCl_{3.6})_{0.56}$	V, $2 \to 1$	18.95	[9.197]
$C_x YCl_3 \cdot (FeCl_3)_r$	V, $2 \to 1$		[9.199]
$C_{13.9}CdCl_{2.1} \cdot (AlCl_{3.6})_{0.78}$	V, $2 \to 1$	19.11	Composition uncertain [9.197]
$C_{13.9}CdCl_{2.1} \cdot (GaCl_{3.6})_{0.70}$	V, $2 \to 1$	19.2	Average composition [9.197]
$C_{14}In_{0.8}Cl_3 \cdot (FeCl_3)_{1.1}$	V, $2 \to 1$	18.9	[9.196]
$C_{16}InCl_{3.1} \cdot (H_2SO_4)_r$	VII, $2 \to 1$	17.23	[9.195]
$C_{23}BiCl_{3.2} \cdot (H_2SO_4)_r$	VII, $3 \to 1$	25.38	[9.195]
$C_x BiCl_3 \cdot (AuBr_x)_r \cdot (TlCl_3)_s \cdot (H_2SO_4)_r$	VII, $4/3 \to 1$	33.34	Very unstable [9.195]
$C_{10.6}CoCl_2 \cdot (K)$	VIII, $2 \to 1$	14.71	[9.197]
$C_x CoCl_2 \cdot KC_{12}$	VIII, $2 \to 1$	14.65	In admixture with initial stage 2 compound [9.200]
$C_{13.9}CdCl_2 \cdot (Na)$	VIII, $2 \to 1$	13.80	[9.197,201]
$C_{13.9}CdCl_2 \cdot (K)$	VIII, $2 \to 1$	14.90	[9.197,201]
$C_{13.9}CdCl_2 \cdot (Rb)$	VIII, $2 \to 1$	15.25	[9.197,201]
$C_{13.9}CdCl_2 \cdot (Cs)$	VIII, $2 \to 1$	15.10	[9.197,201]

[a] In the formulation of the compounds, the host structure is always given first, to the left of the separator point. When the concentration of each layer with respect to the underlying graphene layer is known (or has been assumed), it is indicated.
[b] The change in stage number due to the bi-intercalation process is shown after the number(s) in Roman numerals identifying the method of preparation (see text).

took M = Na, K, Rb or Cs and $M'Cl_n$ = $CdCl_2$ or $CoCl_2$. The four repeat distances of the compounds obtained with the $CdCl_2$ GIC (which had I_c = 12.85 Å) were in good agreement with the calculated values except for the compound obtained with Cs, which, against all expectations, had I_c smaller than that obtained with Rb. With the $CoCl_2$ GIC, the I_c distance was 14.65 Å [9.200] or 14.71 Å [9.197]. The course of the reaction was followed by X-ray diffraction [9.196, 200] and was found to go through the following stages.

– Initially, the intercalation of the alkali metal results in the appearance of a large number of new reflections due to the formation of a high-stage derivative co-existing with the stage-2 halide compound.

– Most of the new reflections disappear and are replaced by a set of lines due to the new bi-intercalation compound. The reflections due to the host structure begin to weaken.

– The intensity of the reflections from the new compound increases, while that of the original reflections decreases. Reduction of the intercalated halide begins and traces of the metal M' (i.e. Cd or Co) appear.

– The reflections due to the original host structure have completely disappeared, with only a slight increase in the amount of metal formed by the reduction. At that point, the bi-intercalation reaction is over, since beyond it extensive reduction of the halide occurs.

9.5.4 Physical Properties

Magnetic studies have been performed on the $CoCl_2$-$AlCl_3$ [9.196] and $CoCl_2$-$FeCl_3$ [9.198] bi-intercalation compounds. Above 6 K, the first compound is paramagnetic, but below 5 K there seems to be three-dimensional in-plane ferromagnetic order. Between 5 and 6 K, there is an intermediate phase, with two-dimensional order, which could be characteristic of a *Kosterlitz-Thouless* transition in a 2d XY phase. In the second bi-intercalation compound, competition may occur between $CoCl_2$ and $FeCl_3$ for the charge donated by the graphite. There are two magnetic phase transitions at 8 and 8.8 K. The intermediate phase may be regarded as the equivalent of the 2d XY system seen above. Below 8 K, the 2d sheets stacked along the c-axis form an antiferromagnetic phase.

9.6 Ternary Systems Involving Fluorinated Graphite

In a number of reactions involving fluorine and/or some of its derivatives, C-F bonds are formed, which modify, at least partly, the sp^3 hybridization of the graphene layers. The "fluorinated graphite" thus formed is often capable of harboring a third component present at the time of the fluorination.

9.6.1 Compounds of Fluorinated Graphite with Metal Fluorides

By heating a mixture of graphite and a metal fluoride $MF_m(M = \text{Al, Mg, Li or Cu})$ under 1 atm of fluorine to about 350°C, a ternary $C_xF(MF_m)_y$ is formed. Considering that the fluorides all have extremely low vapor pressure, it is believed that intercalation occurs via the formation of a more volatile $(MF_m)\cdot(F_2)$ complex [9.203]. The following are typical values of the stoichiometries obtained.

$C_6F(AlF_3)_{0.15}$: stage 1, $I_c = 9.4$ Å [9.204 – 207]

$C_9F(MgF_2)_{0.1}$: stage 1, $I_c = 9.3$ Å;stage 2, $I_c = 12.82$ Å [9.205 – 209]

$C_7F(CuF_2)_{0.02}$:·stage 1, $I_c = 9.4$ Å [9.209 – 211]

$C_9F(LiF)_{0.1}$: stage 1, $I_c = 9.4$ Å; stage 2, $I_c = 12.7$ Å [9.212] .

The Cu and Mg derivatives show an intriguing behavior: as the intercalation reaction progresses, the I_c value changes from 9.3–9.4 Å to 10.7–10.8 Å, indicating a change of structure [9.209]. From the ESCA spectra [9.203–205, 208, 209, 212], it was found that the bond between carbon and fluorine is nearly covalent in the surface layers, but nearly ionic in the bulk of the crystal. This conclusion was bolstered by the ^{19}F NMR spectra [9.208, 209, 212], which indicated highly mobile fluorine species. All the compounds show good in-plane electrical conductivity, about 20 times higher than that of the original graphite and not much affected by short immersion in water [9.207, 210, 212].

A most important chemical property associated with these ternary compounds is the fact that heating them at higher temperatures (300°–400°C) in the presence of excess fluorine expels the metal fluoride and transforms them into C_2F, a technologically important graphite fluoride used as the cathode material in the manufacture of high electromotive force lithium batteries [9.213, 214].

9.6.2 Compounds of Fluorinated Graphite with Nonmetal Fluorides

By bubbling ClF_3 through a suspension of graphite in liquid HF, a green compound is formed, formulated $C_{14}F_n\cdot 3HF_n\cdot ClF_3$ [9.215], which can also be obtained by allowing graphite to react with a solution of ClF_3 in NOF·3HF [9.216]. Infrared spectroscopy results point to the structure $C_{14}^+(H_2F_3)^-ClF_2^+(HF_2)^-$, but the presence of C-F bonds cannot be excluded [9.215]. Later work [9.217, 218] gave $C_{14.4}ClF_{6.3}(HF)_{1.5}$ for the compound resulting from the reaction in anhydrous HF, in which the main difference in the overall composition lies in the amount of hydrogen, which is halved. It is interesting that all the HF molecules can be replaced by strong acids such as HNO_3, CF_3COOH, etc. [9.219].

One can infer, from the nature of the reagents used above, that the ion F^- must be involved – at least catalytically – in the fluorination of graphite by ClF_3. This is confirmed by the reaction of solutions of cesium fluoride in ClF_3 on graphite, when $C_4F\cdot ClF_3$ is formed [9.220], or rather $C_4F\cdot(ClF_3)_{0.46}$ [9.221], which can be thermally degraded to $C_4F_{2.23}$. The decomposition was shown to involve the formation of an intermediate compound $C_4F_{1.92}\cdot(ClF)_{0.46}$ [9.222, 223] in which ClF could be replaced by ClF_3. Both compounds could also be made to react with N_2O_4 [9.224] to yield $C_4F_{1.88}Cl_{0.19}\cdot(N_2O_4)_{0.2}$, a new derivative ($I_c$ = 7.6 Å) whose thermal decomposition [9.225, 226] led to a solid phase (I_c = 6.3 Å) containing C-F and C-Cl bonds. Both $C_4F\cdot(ClF_3)_{0.46}$ and $C_4F_{2.23}$ were found to react with NH_3 to give $C_xF_y(NH_2)_z$ [9.227] which could take in large quantities of water, with I_c changing from 5.1 Å for the dry material to 27 Å for a solid phase containing up to 60 % H_2O, in which the NH_2 groups are bound rigidly to the C lattice [9.228].

The mobility of various molecules (BrF_3, ClF_3, HF) intercalated in C_4F or in some of the species described above has been studied by ^1H and ^{19}F NMR spectroscopy [9.229, 230].

References

9.1 S.A. Solin, H. Zabel: Adv. Phys. **37**, 87 (1988)
9.2 D. Billaud, G. Furdin, A. Hérold, J.F. Marêché, E. McRae: In Extended Abstracts, 14th Biennial Conf. on Carbon, American Carbon Society (1979) p. 268
9.3 A. Hérold, D. Billaud, D. Guérard, P. Lagrange: Mater. Sci. Eng. **31**, 25 (1977)
9.4 D. Billaud, A. Hérold, F.L. Vogel: Mater. Sci. Eng. **45**, 55 (1980)
9.5 G. Furdin, B. Carton, D. Billaud, C. Zeller, A. Hérold: C.R. Acad. Sci. B **278**, 1025 (1974)
9.6 G. Furdin: Thesis, Nancy (1974)
9.7 G. Furdin, D. Billaud, A. Hérold, C. Zeller, A. Marchand: J. Solid State Chem. **34**, 361 (1980)
9.8 H.P. Böhm, R. Setton, E. Stumpp: Synth. Met. **11**, 363 (1985)
9.9 S.A. Solin, P. Chow, H. Zabel: Phys. Rev. Lett. **53**, 1927 (1984)
9.10 D. Neumann, H. Zabel, J.J. Rush, N. Berk: Phys. Rev. Lett. **53**, 56 (1984)
9.11 P.C. Chow, H. Zabel: Synth. Met. **7**, 243 (1983)
9.12 P.C. Chow, D.A. Neumann, D.G. Wiesler, H. Zabel, S.B. DiCenzo: Mater. Res. Soc. Ext. Abstr. **2**, 240 (1984)
9.13 M.E. Misenheimer, P. Chow, H. Zabel: Mater. Res. Soc. Symp. Proc. **20**, 283 (1983)
9.14 R.C. Tatar, S. Rabii: Mater. Res. Soc. Ext. Abstr. **2**, 77 (1984)
9.15 R.C. Tatar, S. Rabii: Synth. Met. **12**, 245 (1985)
9.16 H. Akera, H. Kamimura: Synth. Met. **12**, 275 (1985)
9.17 D. Medjahed, R. Merlin, R. Clarke: Phys. Rev. B **36**, 9345 (1987)
9.18 H. Selig, L.B. Ebert: "Graphite Intercalation Compounds", in *Advances in Inorganic Chemistry and Radiochemistry*, Vol. 23, ed. by H.J. Emeleus, A.G. Sharpe (Academic, Orlando 1980) p. 281
9.19 G. Furdin, B. Bach, A. Hérold: C.R. Acad. Sci. C **271**, 683 (1970)
9.20 G. Furdin, A. Hérold: Bull. Soc. Chim. Fr. 1768 (1972)
9.21 G. Furdin: Thesis, Nancy (1974)
9.22 W. Rüdorff, U. Sils, R. Zeller: Z. Anorg. Allg. Chem. **46**, 282 (1957)
9.23 B. Bach: Thesis, Nancy (1968)
9.24 B. Bach, A. Hérold: Bull. Soc. Chim. Fr. 1978 (1968)
9.25 G. Colin, A. Hérold: Bull. Soc. Chim. Fr. 3345 (1972)
9.26 J. Hinze, H.H. Jaffe: J. Am. Chem. Soc. **84**, 540 (1962)
9.27 W. Rüdorff, E. Schulze, O. Rubisch: Z. Anorg. Allg. Chem. **282**, 232 (1955)
9.28 E. Schulze: "Untersuchungen über Alkaligraphitverbindungen"; Thesis, Tübingen (1954)
9.29 S.A. Solin, Y.B. Fan, X.W. Qian: Synth. Met. **12**, 181 (1985)
9.30 S.K. Hark, B.R. York, S.A. Solin: Synth. Met. **7**, 257 (1983)
9.31 W. Rüdorff: Chimia **19**, 489 (1965)
9.32 D. Ginderow: Ann. Chim. **6**, 5 (1971)

9.33 R. Setton, F. Beguin, L. Facchini, M.F. Quinton, A.P. Legrand, B. Ruisinger, H.P. Böhm: J. Chem. Soc., Chem. Commun. 36 (1983)
9.34 G. Merle, J.M. Letoffe, I.B. Rashkov, P. Claudy: J. Therm. Anal. **13**, 293 (1978)
9.35 B.R. York, S.A. Solin: Phys. Rev. **31**, 8206 (1985)
9.36 S.A. Solin, Y.B. Fan, X.W. Qian: Synth. Met. **12**, 239 (1985)
9.37 Y.B. Fan, S.A. Solin: In Extended Abstracts, 17th Biennial Conf. on Carbon, American Carbon Society (1985) p. 130
9.38 N. Akuzawa, T. Amemiya, Y. Takahashi: Nippon Kagaku Kaishi 287 (1983)
9.39 N. Akuzawa, M. Ikeda, T. Amemiya, Y. Takahashi: Synth. Met. **7**, 65 (1983)
9.40 T. Trewern, R.K. Thomas, G. Naylor, J.W. White: J. Chem. Soc., Faraday Trans. 1, **78**, 2369 (1982)
9.41 W. Ostertag: Diplomarbeit, Tübingen (1963); see also W.E. Craven, W. Ostertag: Carbon **4**, 223 (1966)
9.42 J.P. Pascault, M. Prost, C. Stein, J. Gole: Bull. Soc. Chim. Fr. 1167 (1971)
9.43 W. Rüdorff: "Graphite Intercalation Compounds", in *Advances in Inorganic Chemistry and Radiochemistry*, Vol. 1, ed. by H.J. Emeleus, A.G. Sharpe (Academic, New York 1959) p. 223
9.44 N. Akuzawa, Y. Takahashi: Tanso **102**, 101 (1980)
9.45 H. Schäffer-Stahl, G. von Eynatten: Synth. Met. **7**, 73 (1983)
9.46 S.A. Solin, X.B. Fan, B.R. York: Mater. Res. Soc. Ext. Abstr. **2**, 234 (1984)
9.47 X.W. Qian, D.R. Stump, B.R. York, S.A. Solin: Phys. Rev. Lett. **54**, 127 (1985)
9.48 H.A. Resing, B.R. York, S.A. Solin, R.M. Fronko: In Extended Abstracts, 17th Biennial Conf. on Carbon, American Carbon Society (1985) p. 192
9.49 D.M. Hoffman, A.M. Rao, G.L. Doll, P.C. Eklund, B.R. York, S.A. Solin: Mater. Res. Soc. Ext. Abstr. **2**, 237 (1984)
9.50 Y.Y. Huang, Y.B. Fan, S.A. Solin, J. Heremans, G.G. Tibbetts, J.M. Zhang, P.C. Eklund: Mater. Res. Soc. Ext. Abstr. **8**, 146 (1986)
9.51 M. Nomine, L. Bonnetain: C.R. Acad. Sci. C **264**, 2084 (1967)
9.52 Co-Minh-Duc, M. Rose, J.P. Pascault: C.R. Acad. Sci. C **270**, 569 (1970)
9.53 D. Ginderow, R. Setton: C.R. Acad. Sci. C **266**, 1515 (1968)
9.54 J.O. Besenhard, I. Kain, H.F. Klein, H. Möhwald, H. Witty: Mater. Res. Soc. Symp. Proc. **20**, 221 (1983)
9.55 D. Ginderow, R Setton: C.R. Acad. Sci. C **270**, 135 (1970)
9.56 F. Beguin, R. Setton: Carbon **13**, 293 (1975)
9.57 I. Rashkov, V. Shishkova, I. Panayotov, G. Merle, J.M. Letoffe: In Proc. 3rd Int. Carbon Conf., Baden-Baden (1980) p. 96
9.58 M. Inagaki, K. Uchida, M. Sakai, K. Itoh: Nippon Kagaku Kaishi 312 (1983)
9.59 J.O. Besenhard: Carbon **14**, 111 (1976)
9.60 J.O. Besenhard, H. Möhwald, J.J. Nickl: Carbon **18**, 399 (1980)
9.61 N. Okuyama, T. Takahashi, S. Kanayama, H. Yasunaga: Physica B+C **105**, 298 (1981)
9.62 A. Hamwi, P. Touzain, L. Bonnetain: In Proc. 2nd Int. Carbon Conf., Baden-Baden (1976) p. 143
9.63 G. Merle, J.M. Letoffe, I.B. Rashkov, P. Claudy: J. Therm. Anal. **13**, 293 (1978)
9.64 J. Jegoudez, C. Mazieres, R. Setton: Synth. Met. **7**, 85 (1983)
9.65 R. Setton, F. Beguin, J. Jegoudez, C. Mazieres: Rev. Chim. Miner. **19**, 360 (1982)
9.66 B. Marcus, P. Touzain, H. Hamwi: Carbon **24**, 403 (1986)
9.67 G. Merle, F. Beguin: Carbon **18**, 371 (1980)
9.68 A. Hamwi, P. Touzain, C. Riekel: Synth. Met. **2**, 153 (1980)
9.69 F. Beguin, R. Setton, L. Facchini, A.P. Legrand, G. Merle, C. Mai: Synth. Met. **2**, 161 (1980)
9.70 G. Merle, I. Rashkov, C. Mai, J. Gole: Mater. Sci. Eng. **31**, 39 (1977)
9.71 L. Bonnetain, P. Touzain, A. Hamwi: Mater. Sci. Eng. **31**, 45 (1977)
9.72 C. Riekel, A. Hamwi, P. Touzain: Synth. Met. **6**, 265 (1983)
9.73 S. Matsuzaki, H. Sano: Chem. Phys. Lett. **115**, 424 (1985)
9.74 Y.V. Isaev, Y.N. Novikov, M.E. Vol'pin: Synth. Met. **5**, 23 (1982)
9.75 A.L. Blumenfeld, E.I. Fedin, Y.V. Isaev, Y.N. Novikov: Synth. Met. **6**, 15 (1983)
9.76 F. Beguin, R. Setton, L. Facchini: Synth. Met. **7**, 263 (1983)
9.77 J. Tabony, J.P. Korb: Mol. Phys. **56**, 1281 (1985)
9.78 Y.V. Isaev, Y.N. Novikov, M.E. Vol'pin, I. Rashkov, I. Panayotov: Synth. Met. **6**, 9 (1983)
9.79 R. Setton: J. Mol. Catal. **27**, 263 (1984)

9.80 S. Matsuzaki, M. Tanagushi, M. Sano: Synth. Met. **16**, 343 (1986)
9.81 F. Beguin, P. Vast: Synth. Met. **23**, 427 (1988)
9.82 F. Beguin, B. Gonzalez, J. Conard, H. Estrade-Szwarckopf, D. Guérard: Synth. Met. **12**, 187 (1985)
9.83 M.F. Quinton, A.P. Legrand, F. Beguin: Synth. Met. **14**, 179 (1986)
9.84 Co-Minh-Duc: "Contribution à l'étude des composés d'insertion ternaires Graphite-Métal Alcalin-Solvant"; Thèse, Lyon (1971)
9.85 F. Beguin, R. Setton, A. Hamwi, P. Touzain: Mater. Sci. Eng. **40**, 167 (1970)
9.86 J. Bouat, D. Bonnin, L. Facchini, F. Beguin: Synth. Met. **7**, 233 (1983)
9.87 F. Beguin, C. Laroche, B. Gonzalez, M. Goldmann, M.F. Quinton: Synth. Met. **23**, 155 (1988)
9.88 M. Goldmann, J. Pannetier, F. Beguin, B. Gonzalez: Synth. Met. **23**, 133 (1988)
9.89 G. Merle, J. Gole: Bull. Soc. Chim. Fr. 1082 (1976)
9.90 Co-Minh-Duc, C. Mai, R. Riviere, J. Gole: J. Chim. Phys. 991 (1972)
9.91 D. Ginderow: Ann. Chim. **6**, 5 (1971)
9.92 J. Jegoudez, C. Mazieres, R. Setton: Carbon **24**, 747 (1986)
9.93 F. Beguin: "Ion Complexes 'Potassium-Molecule Organique' en insertion dans le Graphite"; Thesis, Orléans (1980)
9.94 M. Nomine, L. Bonnetain: J. Chim. Phys. 173 (1969)
9.95 J. Amiell, P. Delhaes, F. Beguin, R. Setton: Mater. Sci. Eng. **31**, 243 (1977)
9.96 A.P. Legrand, L. Facchini, D. Bonnin, J. Bouat, M.F. Quinton, F. Beguin: Synth. Met. **12**, 175 (1985)
9.97 F. Beguin, L. Gatineau, R. Setton: In Proc. 6th London Int. Carbon and Graphite Conf. (1982) p. 97
9.98 B. Marcus, P. Touzain, J.L. Soubeyroux: Synth. Met. **23**, 13 (1988)
9.99 F. Beguin, H. Estrade-Szwarckopf, J. Conard, P. Lauginie, P. Marceau, D. Guérard, L. Facchini: Synth. Met. **7**, 77 (1983)
9.100 M. Inagaki, Y. Shiwachi, Y. Maeda: J. Chim. Phys. **81**, 847 (1984)
9.101 H.P. Böhm, Y. Ko, B. Ruisinger, R. Schlögl: In *Chemical Reactions in Organic and Inorganic Constrained Systems*, ed. by R. Setton (Reidel, Dordrecht 1986) p. 429
9.102 J.O. Besenhard, H.P. Fritz, H. Möhwald, J.J. Nickl: In Proc. 3rd Int. Carbon Conf., Baden-Baden (1980) p. 147
9.103 B. Marcus, P. Touzain: In Proc. 4th Int. Carbon Conf., Baden-Baden (1986) p. 430
9.104 F. Beguin, J. Jegoudez, C. Mazieres, R. Setton: C.R. Acad. Sci. **293 II**, 969 (1981)
9.105 R. Setton: Tanso 180 (1983)
9.106 R. Schögl, H.P. Böhm: Carbon **22**, 341 (1984)
9.107 F. Beguin, P. Vast: In Proc. 4th Int. Carbon Conf., Baden-Baden (1986) p. 424
9.108 L. Facchini, M.F. Quinton, A.P. Legrand, F. Beguin, R. Setton: Physica **99B**, 525 (1980)
9.109 F. Beguin, R. Setton: J. Chem. Soc., Chem. Commun. 611 (1976)
9.110 J. Jegoudez, C. Mazieres, R. Setton: Unpublished results
9.111 A. Hamwi, P. Touzain, L. Bonnetain: C.R. Acad. Sci. **299 II**, 1385 (1984)
9.112 S.A. Solin, Y.B. F[Aan, X.W. Qian: Synth. Met. **12**, 181 (1985)
9.113 A. Bondi: J. Chem. Phys. **68**, 441 (1964)
9.114 L. Pietronero, S. Strässler: Phys. Rev. Lett. **47**, 593 (1981)
9.115 P. Vora, B.R. York, S.A. Solin: Synth. Met. **7**, 355 (1983)
9.116 D.M. Hoffman, A.M. Rao, L.L. Dole, P.C. Eklund, B.R. York, S.A. Solin: Mater. Res. Soc. Ext. Abstr. **2**, 237 (1984)
9.117 H.A. Resing,œ[A B.R. York, S.A. Solin, R.M. Fronko: In Extended Abstracts, 17th Biennial Conf. on Carbon, American Carbon Society (1985) p. 192
9.118 B.R. York, S.A. Solin: Phys. Rev. B **31**, 8206 (1985)
9.119 T. Tsang, R.M. Fronko, H.A. Resing, X.W. Qian, S.A. Solin: Solid State Commun. [A**62**, 117 (1987)
9.120 R. Setton: In *Chemical Reactions in Organic and Inorganic Constrained Systems*, ed. by R. Setton (Reidel, Dordrecht 1986) p. 381
9.121 W. Rüdorff, H. Schulz: Z. Anorg. Allg. Chem. **245**, 121 (1940)
9.122 P. Touzain, J. Michel, P. Blum: Synth. Met. **8**, 313 (1983)
9.123 A. El Atifi, P. Touzain, L. Bonnetain: Proc. 4th Int. Carbon Conf., Baden-Baden (1986) p. 468
9.124 R.C. Croft: Nature **172**, 725 (1953)
9.125 R.C. Croft: J. Appl. Chem. **2**, 557 (1952)

9.126 M. Dzurus, G.R. Hennig: J. Am. Chem. Soc. **79**, 1051 (1957)
9.127 E. Stumpp: Mater. Sci. Eng. **31**, 53 (1977)
9.128 H. Schäfer: Angew. Chemie **88**, 775 (1976)
9.129 T. Tominaga, T. Sakai, T. Kimura: Bull. Chem. Soc. Jpn. **49**, 2755 (1976)
9.130 A.V. Nefed'ev, N.D. Lapkina, R.A. Stukan, Y.T. Stuchkov, O.L. Lependina, Y.N. Novikov, M.E. Vol'pin: Zh. Strukt. Khim. **20**, 835 (1979)
9.131 T. Dziemianowicz, R. Vangelisti, A. Hérold, W. Forsman: In Extended Abstracts, 15th Biennial Conf. on Carbon, American Carbon Society (1981) p. 379
9.132 R. Vangelisti, J.F. Marêché, G. Furdin, T. Dziemianowicz: C.R. Acad. Sci. **296 II**, 691 (1983)
9.133 T. Dziemianowicz, W. Forsman, R. Vangelisti, A. Hérold: Carbon **22**, 53 (1984)
9.134 D.E. Wessbecher, W.C. Forsman: Mater. Res. Soc. Ext. Abstr. **8**, 203 (1986)
9.135 E. Stumpp, R. Kebschull, G. Nietfeld: In Extended Abstracts, 18th Biennial Conf. on Carbon, American Carbon Society (1987)
9.136 M. Inagaki, Z.D. Wang, Y. Okamoto, M. Ohira: Synth. Met. **20**, 9 (1987)
9.137 E. Buscarlet: Thesis, Grenoble (1976)
9.138 A. Hamwi, P. Touzain, L. Bonnetain: Rev. Chim. Miner. **19**, 651 (1982)
9.139 G. Rosenthal, T.F. Mallouk, N. Bartlett: Synth. Met. **9**, 433 (1984)
9.140 E.M. McCarron, Y.J. Grannec, N. Bartlett: J. Chem. Soc., Chem. Commun. 890 (1980)
9.141 J.G. Hooley: In Extended Abstracts, 16th Biennial Conf. on Carbon, American Carbon Society (1983) p. 240
9.142 J.M. Lalancette, J.M. Roy, L. Lafontaine: Can. J. Chem. **50**, 3058 (1972)
9.143 D. Ginderow, R. Setton: C.R. Acad. Sci. **257**, 687 (1963)
9.144 J.G. Hooley: Carbon **10**, 155 (1977)
9.145 R.C. Paul, R. Kanshal, S.S. Pahil: J. Indian Chem. Soc. **42**, 483 (1965)
9.146 E. Stumpp, R. Niess: Carbon **16**, 259 (1978)
9.147 T. Nakajima, M. Kawaguchi, N. Watanabe: Bull. Chem. Soc. Jpn. **54**, 2295 (1981)
9.148 K. Klouda, V. Rak, J. Vachuska: Collect. Czech. Chem. Commun. **50**, 947 (1985)
9.149 R. Schlögl, H.P. Böhm: Synth. Met. **23**, 407 (1988)
9.150 R. Schlögl, W. Jones: J. Chim. Phys. **81**, 877 (1984)
9.151 R. Schlögl, H.P. Böhm: Z. Naturforsch. **39B**, 112 (1984)
9.152 R. Schlögl, H.P. Böhm: Z. Naturforsch. **39B**, 788 (1984)
9.153 P. Bowen, W. Jones, J.M. Thomas, R. Schlögl: J. Chem. Soc., Chem. Commun. 677 (1981)
9.154 E. Schlögl, H.P. Böhm: Proc. 6th London Int. Conf. on Carbon and Graphite (1982) p. 73
9.155 T. Sasa, Y. Takahashi, T. Mukaibo: Bull. Chem. Soc. Jpn. **45**, 2250 (1972)
9.156 V.V. Avdeev, V.A. Mukhanov, K.N. Semenenko: Izv. Akad. Nauk SSSR, Neorg. Mater. **19**, 174 (1983)
9.157 K.N. Semenenko, V.A. Mukhonov, V.V. Avdeev: Izv. Akad. Nauk SSSR, Neorg. Mater. **19**, 609 (1983)
9.158 C. Balestreri, R. Vangelisti, J. Melin, A. Hérold: C.R. Acad. Sci. C **279**, 279 (1974)
9.159 E. Stumpp, H.J. Hummel, C. Ehrhardt: Synth. Met. **23**, 441 (1988)
9.160 J. Stahl: Z. Anorg. Allg. Chem. **428**, 269 (1977)
9.161 R. Niess, E. Stumpp: Carbon **16**, 265 (1978)
9.162 J. Melin, G. Furdin, H. Fuzellier, R. Vasse, A. Hérold: Mater. Sci. Eng. **31**, 61 (1977)
9.163 R. Vasse, G. Furdin, J. Melin, E. McRae, A. Hérold: Synth. Met. **2**, 185 (1980)
9.164 A.R. Ubbelohde: Carbon **6**, 177 (1968)
9.165 J.O. Besenhard, H.P. Fritz: Naturforsch. **27B**, 1294 (1972)
9.166 J.O. Besenhard, H.P. Fritz, H. Möhwald, J.J. Nickl: In Proc. 3rd Int. Carbon Conf., Baden-Baden (1980) p. 147
9.167 A. Jobert, P. Touzain, L. Bonnetain: Carbon **19**, 193 (1982)
9.168 P. Touzain, A. Jobert: Bull. Soc. Chim. Fr. 110 (1983)
9.169 D. Billaud, A. Metrot, P. Willmann, A. Hérold: In Proc. 3rd Int. Carbon Conf., Baden-Baden (1980) p. 140
9.170 D. Billaud, A. Chemti, A. Metrot: Carbon **20**, 493 (1982)
9.171 W.C. Forsman, H.E. Mertwoy: Synth. Met. **2**, 171 (1980)
9.172 W.C. Forsman, H.E. Mertwoy: Carbon **20**, 205 (1982)
9.173 D. Billaud, A. Pron, F.L. Vogel: Synth. Met. **2**, 177 (1980)
9.174 R. Setton: Unpublished results
9.175 D. Billaud, P.J. Flanders, A. Pron, J.E. Fischer: Mater. Sci. Eng. **54**, 31 (1982)

9.176 G.R. Miller, H.A. Resing, M.J. Moran, L. Banks, F.L. Vogel, A. Pron, D. Billaud: Synth. Met. **8**, 77 (1983)
9.177 H. Zaleski, P.K. Unmat, W.R. Datars: Synth. Met. **11**, 183 (1985)
9.178 D. Billaud, A. Pron: In Extended Abstracts, 15th Biennial Conf. on Carbon, American Carbon Society (1981) p. 395
9.179 E. Stumpp, S. Cieschinger, G. Nietfeld: Proc. 4th Int. Carbon Conf. Baden-Baden (1986) p. 451
9.180 H. Fuzellier, A. Metrot, B. Bouayad, A. Hérold: In Extended Abstracts, 15th Biennial Conf. on Carbon, American Carbon Society (1981) p. 391
9.181 T.F. Thomson, E.M. McCarron, N. Bartlett: Synth. Met. **3**, 255 (1981)
9.182 T. Sasa, Y. Takahashi, T. Mukaibo: Bull. Chem. Soc. Jpn. **45**, 2657 (1972)
9.183 P. Flanders, M.J. Moran, J.E. Fischer, D. Billaud, G. Furdin, H. Fuzellier, A. Metrot: In Extended Abstracts, 15th Biennial Conf. on Carbon (1981) p. 98
9.184 P. Boolchand, W. Bresser, D. McDaniel, P.C. Eklund, D. Billaud, J.E. Fischer: Mater. Res. Soc. Symp. Proc. **20**, 389 (1983)
9.185 M.J. Moran, G.R. Miller, R.A. Demarco, H.A. Resing: J. Phys. Chem. **88**, 1580 (1984)
9.186 G.R. Miller, H.A. Resing, P. Brant, M.J. Moran, F.L. Vogel, T.L. Wu, D. Billaud, A. Pron: Synth. Met. **2**, 237 (1980)
9.187 G.R. Miller, H.A. Resing, F.L. Vogel, A. Pron, T.C. Wu, D. Billaud: J. Phys. Chem. **84**, 3333 (1980)
9.188 P. Lagrange, A. Metrot, A. Hérold: C.R. Acad. Sci. C **270**, 701 (1974)
9.189 J.O. Besenhard, I. Kain, H.F. Klein, H. Möhwald, A. Witty: Mater. Res. Soc. Symp. Proc. **20**, 221 (1983)
9.190 A.G. Freeman: J. Chem. Soc., Chem. Commun. 746 (1974)
9.191 E. Stumpp, K.-D. Wageringel: In Extended Abstracts, 14th Biennial Conf. on Carbon (1979) p. 288
9.192 K.-D. Wageringel, E. Stumpp: In Proc. 3rd Int. Carbon Conf., Baden-Baden (1980) p. 132
9.193 E. Stumpp, H.J. Hummel, C. Ehrhardt: Synth. Met. **23**, 441 (1988)
9.194 R. Niess, E. Stumpp: Carbon **16**, 265 (1978)
9.195 P. Scharff, E. Stumpp, C. Ehrhardt: Synth. Met. **23**, 415 (1988)
9.196 N. Nadi: "Insertion de trichlorures dans le graphite: composés binaires, composés ternaires à couches alternées ou composés de bi-insertion"; Thesis, Nancy (1985)
9.197 L. Hachim: "Recherches sur les composés du graphite à couches insérées alternées"; Thesis, Nancy (1984)
9.198 M. Suzuki, I. Oguro, Y. Jinzaki: J. Phys. Chem. **17**, L575 (1984)
9.199 E. Stumpp: Physica **105B**, 9 (1981)
9.200 M. Suzuki, P.C. Chow, H. Zabel: Phys. Rev. B **32**, 6800 (1985)
9.201 G. Furdin, L. Hachim, D. Guérard, A. Hérold: C.R. Acad. Sci. **301 II**, 579 (1985)
9.202 A. Hérold, G. Furdin, D. Guérard, L. Hachim, M. Lelaurain, N.E. Nadi, R. Vangelisti: Synth. Met. **12**, 11 (1985); the error in Table 1 of this reference is corrected in Table 9.9
9.203 T. Nakajima, M. Kawaguchi, N. Watanabe: Z. Naturforsch. **36B**, 1419 (1981)
9.204 T. Nakajima, M. Kawaguchi, N. Watanabe: Chem. Lett. 1045 (1981)
9.205 N. Watanabe: Electrochim. Acta **27**, 1535 (1982)
9.206 T. Nakajima, M. Kawaguchi, N. Watanabe: In Proc. Int. Symp. Carbon, Toyohashi, Japan (1982) p. 481
9.207 N. Watanabe, T. Nakajima, M. Kawaguchi: German Patent DE 3, 235, 596 (1983)
9.208 T. Nakajima, M. Kawaguchi, N. Watanabe: Carbon **20**, 287 (1982)
9.209 T. Nakajima, M. Kawaguchi, N. Watanabe: Solid State Ionics **11**, 65 (1983)
9.210 N. Watanabe: In Extended Abstracts, 16th Biennial Conf. Carbon, American Carbon Society (1983) p. 250
9.211 N. Watanabe, T. Nakajima, M. Kawaguchi: German Patents DE 3, 330, 695 (1984) and DE 3, 319, 111 (1984)
9.212 T. Nakajima, M. Kawaguchi, N. Watanabe: Synth. Met. **7**, 117 (1983)
9.213 N. Watanabe, T. Nakajima, M. Kawaguchi: German Patent DE 3, 201, 116 (1983)
9.214 R. Setton: Synth. Met. **23**, 467 (1988)
9.215 A.A. Opalcvskii, A.S. Nazarov, A.A. Uminskii: Zh. Neorg. Khim. **17**, 2608 (1972)
9.216 A.S. Nazarov, U.G. Mokotchenko, I.I. Yakovlev: Zh. Neorg. Khim. **23**, 621 (1978)
9.217 Y.I. Nikonorov, L.L. Gornotsaev: Izv. Sib. Otd. Akad. Nauk SSSR, Ser. Khim Nauk 48 (1980)

9.218 Y.I. Nikonorov, L.L. Gornotsaev: USSR Patent SU 707, 889 (1980)
9.219 A.S. Nazarov, N.F. Yudanov, Y.V. Chicagov: Russ. J. Inorg. Chem. (Engl. Transl.) **21**, 1248 (1976)
9.220 A.S. Nazarov, V.G. Mokotchenko, I.I. Yakovlev: Zh. Neorg. Khim. **23**, 1680 (1978)
9.221 A.S. Nazarov, I.I. Yakovlev, A.F. Antimonov, V.M. Grankin, V.B. Anrasov, P.P. Semyanikov: Zh. Neorg. Khim. **25**, 1506 (1980)
9.222 A.S. Nazarov, A.F. Antimonov: Zh. Neorg. Khim. **26**, 1563 (1981)
9.223 V.V. Nedelko, M.Y. Verkhovskaya, B.L. Korsunskii, A.S. Nazarov: Izv. Akad. Nauk SSSR, Ser. Khim. 1902 (1981)
9.224 A.S. Nazarov, A.F. Antimonov, I.I. Yakovlev: Zh. Neorg. Khim. **26**, 3269 (1981)
9.225 V.A. Nadolinnyi, A.F. Antimonov, A.S. Nazarov, V.M. Moralev, I.I. Yakovlev: Zh. Neorg. Khim. **27**, 1981 (1982)
9.226 A.F. Antimonov, V.M. Grankin, P.P. Semyanikov, A.S. Nazarov, I.I. Yakovlev: Zh. Neorg. Khim. **31**, 365 (1986)
9.227 A.S. Nazarov, A.F. Antimonov: Zh. Neorg. Khim. **25**, 2123 (1980)
9.228 A.F. Antimonov, A.S. Nazarov: Izv. Sib. Otd. Akad. Nauk SSSR, Ser. Khim. Nauk **2**, 78 (1982)
9.229 Y.I. Nikorov, B.I. Obmoin: Zh. Neorg. Khim. **25**, 1511 (1980)
9.230 A.M. Panich, A.M. Danilenko, S.P. Gabuda: Dokl. Akad. Nauk SSSR **281**, 389 (1985)

Subject Index

FSC
www.fsc.org
MIX
Papier aus verantwortungsvollen Quellen
Paper from responsible sources
FSC® C105338

If you have any concerns about our products,
you can contact us on
ProductSafety@springernature.com

In case Publisher is established outside the EU,
the EU authorized representative is:
Springer Nature Customer Service Center GmbH
Europaplatz 3, 69115 Heidelberg, Germany

Printed by Libri Plureos GmbH
in Hamburg, Germany